Communications and Control Engineering

Series Editors
A. Isidori • J.H. van Schuppen • E.D. Sontag • M. Thoma • M. Krstić

Published titles include:

Daizhan Cheng · Hongsheng Qi · Zhiqiang Li

Analysis and Control of Boolean Networks

A Semi-tensor Product Approach

Springer

Dr. Daizhan Cheng
Academy of Mathematics and Systems
Science (AMSS), Institute of Systems
Science
Chinese Academy of Sciences
100190 Beijing
China, People's Republic
dcheng@iss.ac.cn

Hongsheng Qi
Academy of Mathematics and Systems
Science (AMSS), Institute of Systems
Science
Chinese Academy of Sciences
100190 Beijing
China, People's Republic

Zhiqiang Li
Academy of Mathematics and Systems
Science (AMSS), Institute of Systems
Science
Chinese Academy of Sciences
100190 Beijing
China, People's Republic

ISSN 0178-5354
ISBN 978-1-4471-2611-9   ISBN 978-0-85729-097-7 (eBook)
DOI 10.1007/978-0-85729-097-7
Springer London Dordrecht Heidelberg New York

British Library Cataloguing in Publication Data
A catalogue record for this book is available from the British Library

Cover design: eStudio Calamar S.L.

Printed on acid-free paper

Springer is part of Springer Science+Business Media (www.springer.com)

Preface

Motivated by the Human Genome Project, a new view of biology, called systems biology, is emerging [5]. Systems biology does not investigate individual genes, proteins or cells in isolation. Rather, it studies the behavior and relationships of all of the cells, proteins, DNA and RNA in a biological system called a cellular network. The most active networks may be those associated with genetic regulation, which regulate the growth, replication, and death of cells in response to changes in the environment.

How do these genetic regulatory networks function? In the early 1960s Jacob and Monod showed that any cell contains a number of "regulatory" genes that act as switches and which can turn each another on and off. This shows that a genetic network is of "on–off" type [7].

Boolean networks, first introduced by Kauffman, have become powerful tools for describing, analyzing, and simulating cellular networks [2, 3]. Hence, they have received much attention, not only from the biology community, but also from researchers with backgrounds in physics, systems science, etc.

The purpose of this book is to present a new approach to the investigation of Boolean (control) networks. In this new approach, a logical relation is expressed as an algebraic equation, and a logical dynamical system, such as a Boolean network, is converted into a standard discrete-time linear system. Similarly, a Boolean control network is converted into a discrete-time bilinear system. In this way, various tools for solving conventional algebraic equations and dealing with difference or differential equations can be used to solve logic-based problems. Under this framework, the topological structures of Boolean networks are revealed via the structures of their network transition matrices. The state space, subspaces, etc., are then defined as sets of logical functions. This framework makes the state-space approach to dynamical (control) systems applicable to Boolean (control) networks. Using this new technique, we investigate the properties and control design of Boolean networks. Many basic problems in control theory are studied, such as controllability, observability, realization, stabilization, disturbance decoupling and optimal control.

The fundamental tool in this approach is a new matrix product, called the semitensor product (STP). The STP of matrices is a generalization of the conventional

v

matrix product to the case where the dimension-matching condition is not satisfied. That is, we extend the matrix product AB to the case where the column number of A and the row number of B are different. This generalization preserves all the major properties of the conventional matrix product.

Using the STP, a logical function can be converted into a multilinear mapping, called the matrix expression of logical relations. Under this construction, the dynamics of a Boolean network can be expressed as a conventional discrete-time linear system. In the light of this linear expression, certain major features of the topology of a Boolean network, such as fixed points, cycles, transient time, and basins of attractors, can be easily revealed via a set of formulas.

When the control of a Boolean network is considered, the bilinear system representation of a Boolean control network makes it possible to apply most techniques developed in modern control theory to the analysis and synthesis of a Boolean control network.

The main contents of this book are as follows.

Chapter 1 consists of a brief introduction to propositional logic. This is very elementary and involves only the propositional logic required in this book. A reader who is familiar with mathematical logic can skip it.

In Chap. 2 we introduce some basic concepts and properties of the STP, which is the principal tool used in this book. The STP is a generalization of the conventional matrix product in cases where the dimension-matching requirement for the factor matrices fails. This generalization preserves the major properties of the conventional matrix product.

In Chap. 3 we consider the matrix expression of logical relations. Identifying T (true) and F (false) with vectors $[1, 0]^T$ and $[0, 1]^T$, respectively, a logical variable becomes a 2-dimensional vector variable. Using the STP, a logical function can be expressed as a multilinear mapping with respect to its logical arguments so that each logical function is uniquely determined by a matrix, called its structure matrix.

Chapter 4 is devoted to solving logical equations. Using the matrix expression of logic a system of logical equations can be converted into a linear algebraic equation. Ignoring the complexity of computation, the solution of systems of logical equations becomes theoretically equivalent to the solution of algebraic equations, which can be achieved with straightforward computation.

Chapter 5 considers the linear expression of Boolean networks. Using the technique developed in previous chapters, the dynamics of a Boolean network is converted into a conventional discrete-time linear system. In the light of this linear expression, the topological structures of Boolean networks are investigated via their transition matrices. Formulas are obtained to calculate the fixed points, cycles of different lengths, transient period, and the basin of each attractor.

The input-state structures of Boolean control networks are studied in Chap. 6. The compounded structure of cycles in input-state space is obtained. This approach is applied to the analysis of Boolean networks with cascading structure. The "rolling gear" structure of cycles is revealed, which explains the phenomenon that tiny attractors can determine the vast order of the network [4].

Chapter 7 presents a technique to build the dynamic model of a Boolean network via observed data. Instead of building the logical dynamics of a Boolean network, we first identify its algebraic form, so the conversion of the algebraic form of a Boolean network back to its logical form is first investigated. After a general model construction technique is introduced, several special cases are studied, including the known network graph case, the least in-degree model, the uniform model, etc. The problem of dealing with data containing errors is also discussed.

In Chap. 8 a systematic state-space description is developed. The state space (and its subspaces) of a Boolean (control) network are defined in a dual way, i.e., they are defined as sets of logical functions. It is shown that this description is very convenient in revealing the properties of Boolean networks and in the control design of Boolean control networks.

Chapter 9 is devoted to Boolean control networks. Using linear expressions, it is shown that Boolean control networks can be converted into linear control systems. Some basic control problems such as controllability and observability of Boolean control networks are then investigated via their equivalent forms for linear control systems.

Chapter 10 considers the realization problem of Boolean control networks. First, coordinate transformations are considered, and then the Kalman decomposition of Boolean input–output networks is proposed. Using the Kalman decomposition, the minimum realization of a Boolean input–output mapping is obtained.

The stability and stabilization problem is discussed in Chap. 11. The applicable set from metric-based convergence analysis [6] is enlarged by the use of coordinate transformations. Based on the analysis of the network transition matrix, necessary and sufficient conditions are then obtained for stability and stabilization by either open-loop control or closed-loop control. Several examples are included.

Chapter 12 considers the disturbance decoupling problem. First, the output-friendly subspace is introduced. Formulas and algorithms are provided to construct a minimum regular subspace, which is called the "friend" of output y. The design technique for constructing the feedback and solving the disturbance decoupling problem is presented. To construct a constant stabilizing control, the canalizing mapping, which is a generalization of the canalizing function, is proposed and its main properties are revealed.

In Chap. 13 we consider the coordinate-independent geometric structure of Boolean (control) networks. Based on this structure, the feedback decomposition of Boolean control networks is studied. The input-state decomposition, including cascading and parallel decompositions, and input–output decomposition of Boolean control networks are investigated, and necessary and sufficient conditions are presented.

Chapter 14 deals with the multivalued logic which could provide a more precise description for real networks such as gene regulation networks, etc. The structure of k-valued logical networks is first investigated. Controllability and observability of k-valued logical networks are then considered. In fact, almost all the arguments and results about Boolean networks can be extended to the k-valued logic setting.

Chapter 15 considers the optimal control of Boolean control networks. To deal with Boolean (or k-valued) games with s-memory, higher-order Boolean (control) networks are introduced, and their algebraic forms are also presented. The one-to-one correspondence between the cycles of the original network and the cycles of its algebraic form is established. The optimal control problem is then investigated and the optimal control is designed.

Chapter 16 introduces a useful tool, called the input-state incidence matrix, which is an algebraic description of the input-state transfer graph. Controllability and observability of Boolean control networks are revisited and some further results are presented. The topological structures of Boolean control networks with free controls are also investigated. Finally, the results are extended to mix-valued logical dynamical systems.

Chapter 17 investigates the identification of Boolean control networks. First, a new observability condition is obtained which provides a way to construct the initial state of a trajectory from its input–output data. A necessary and sufficient condition for identifiability is then presented. A numerical algorithm is proposed for practical application.

Chapter 18 considers an application to game theory. We consider a game with finitely many players and where each player has finitely many possible actions. When the game is infinitely repeated, a strategy using finite memory becomes a logical dynamical system. Hence, the results obtained for Boolean or logical networks are applicable to finding Nash or sub-Nash solutions for the infinitely repeated games.

The primary objects of this book are deterministic Boolean networks, but in Chap. 19 we provide a brief introduction to random Boolean networks. Basic concepts are presented and then the steady-state distribution of a random Boolean network is investigated. Finally, the stabilization of a random Boolean network is studied. Recently, random Boolean networks have been the subject of much research, and so a detailed discussion is beyond the scope of this work.

Appendix A explains relevant numerical calculations. A software toolbox for the algorithms is available at http://lsc.amss.ac.cn/~dcheng/.

Appendix B contains proofs of some key properties of the semi-tensor product, which are translated from [1], with the permission of Science Press.

This book is self-contained. The prerequisites for its use are linear algebra and some basic knowledge of the control theory of linear systems. The manuscript was originally prepared when the first author was visiting Kyoto University. The first author would like to express his hearty thanks to Professor Yutaka Takahashi for his proof-reading and useful suggestions for parts of the manuscript. The manuscript has been used as lecture notes in a series of seminars organized jointly by the Academy of Mathematics and Systems Science, Tsinghua University, and Peking University. Many colleagues and students attending these seminars have contributed to this book via useful discussions, suggestions, and corrections. Particularly, Dr. Yin Zhao helped in the preparation of Chaps. 15–17. Dr. Yifen Mu, Dr. Zhenning Zhang, Dr. Yin Zhao, Dr. Xiangru Xu, and Dr. Jiangbo Zhang helped with the final galley proof of the manuscript. The authors are also indebted to Mr. Oliver Jackson for his warmhearted support.

The research presented in this book was partly supported by the Chinese National Natural Science Foundation under grant number G60736022.

Beijing   Daizhan Cheng
                                                                       Hongsheng Qi
                                                                       Zhiqiang Li

References

1. Cheng, D., Qi, H.: Semi-tensor Product of Matrices—Theory and Applications. Science Press, Beijing (2007) (in Chinese)
2. Kauffman, S.: Metabolic stability and epigenesis in randomly constructed genetic nets. J. Theor. Biol. **22**(3), 437 (1969)
3. Kauffman, S.: The Origins of Order: Self-organization and Selection in Evolution. Oxford University Press, London (1993)
4. Kauffman, S.: At Home in the Universe. Oxford University Press, London (1995)
5. Kitano, H.: Systems biology: a brief overview. Science **259**, 1662–1664 (2002)
6. Robert, F.: Discrete Iterations: A Metric Study. Springer, Berlin (1986). Translated by J. Rolne
7. Waldrop, M.: Complexity: The Emerging Science at the Edge of Order and Chaos. Touchstone, New York (1992)

Contents

Notation

$\mathbb{C}$	set of complex numbers
$\mathbb{R}$	set of real numbers
$\mathbb{Q}$	set of rational numbers
$\mathbb{Z}$	set of integers
$\mathbb{N}$	set of natural numbers
$\mathbb{Z}_n$	finite group $\{1, \ldots, n\}$ equipped with $+(\bmod\ n)$
$:=$	"is defined as"
$\mathcal{M}_{m \times n}$	set of $m \times n$ real matrices
$\mathcal{M}_n$	set of $n \times n$ real matrices
$|S|$	cardinal number of set S
$Id(i_1, \ldots, i_k; n_1, \ldots, n_k)$	ordered multi-index
$A \succ_t B$	column number of A is t times the row number of B
$A \prec_t B$	row number of B is t times the column number of A
$\neg$	negation
$\vee$	disjunction
$\wedge$	conjunction
$\rightarrow$	conditional
$\leftrightarrow$	biconditional
$\bar{\vee}$	exclusive or (EOR)
$\uparrow$	not and (NAND)
$\downarrow$	not or (NOR)
$\oslash_k$	rotator in k-valued logic
$\nabla_{i,k}$	i-confirmor in k-valued logic
$[a]$	largest integer less than or equal to a
$\mathrm{lcm}(p, q)$	least common multiple of p and q
$\Rightarrow$	implication
$\Leftrightarrow$	equivalence
δ_n^k	kth column of I_n
$\mathcal{D}$	set $\{T, F\}$ or $\{1, 0\}$
$\mathcal{D}_k$	set $\{0, \frac{1}{k-1}, \ldots, \frac{k-2}{k-1}, 1\}$
$\mathcal{D}_f$	set $\{r \in \mathbb{R} \mid 0 \leq r \leq 1\}$
Δ	set $\{\delta_2^1, \delta_2^2\}$

Δ_k	set $\{\delta_k^i \mid 1 \leq i \leq k\}$
$R(x_0)$	reachable set from x_0
$R_s(x_0)$	reachable set from x_0 at the sth step
$\otimes$	tensor (or Kronecker) product
$\ltimes$	left semi-tensor product
$\rtimes$	right semi-tensor product
$L(U, V)$	set of linear mappings from U to V
$\mathscr{T}_t^s$	set of tensors with covariant order s and contravariant order t
$\mathscr{L}_{m \times n}$	set of $m \times n$ logical matrices
$\delta_k[i_1 \cdots i_s]$	logical matrix with $\delta_k^{i_j}$ as its jth column
$\delta_k\{i_1, \ldots, i_s\}$	$\{\delta_k^{i_1}, \ldots, \delta_k^{i_s}\} \subset \Delta_k$
$\mathrm{Col}(A)$	set of columns of matrix A
$\mathrm{Col}_i(A)$	ith column of matrix A
$\mathrm{Row}(A)$	set of rows of matrix A
$\mathrm{Row}_i(A)$	ith row of matrix A
$\mathrm{Blk}_i(A)$	ith block of matrix A
$\mathrm{diag}(A_1, \ldots, A_k)$	block diagonal matrix whose diagonal blocks are A_i, $i = 1, \ldots, k$
$V_c(A)$	column-stacking form of matrix A
$V_r(A)$	row-stacking form of matrix A
$\det(t)$	determinant of A
$\mathrm{tr}(A)$	trace of A
$\mathscr{P}(k)$	set of proper factors of k
$\mathbf{S}_k$	permutation group of k elements
$W_{[m,n]}$	swap matrix with index (m, n)
T_t	transient period
$\mathbf{1}_k$	$[\underbrace{1, 1, \ldots, 1}_k]^\mathrm{T}$
Ω	limit set
$\mathscr{B}_{m \times n}$	set of $m \times n$ Boolean matrices
$+_{\mathscr{B}}$	Boolean addition for Boolean matrices
$\sum_{\mathscr{B}}$	Boolean sum for Boolean matrices
$\ltimes_{\mathscr{B}}$	Boolean product for Boolean matrices
$A^{(k)}$	Boolean power of Boolean matrix A
$D_v(A, B)$	vector distance of $A, B \in \mathscr{B}_{m \times n}$
$\mathscr{X}$	state space
$\mathscr{F}_\ell\{\cdots\}$	subspace generated by $\cdots$
$\mathscr{I}(\Sigma)$	incidence matrix of Σ
$wt(\cdot)$	weight function of a Boolean matrix
$wb(\cdot)$	Boolean weight function of a Boolean matrix

Chapter 1
Propositional Logic

1.1 Statements

Mathematical logic uses mathematical methods to perform logical deduction and logical reasoning. It is now the subject of a fundamental course for students of pure mathematics, computer science, etc., and there are many standard textbooks on the topic. We will use [2] as one of our main references. The use of mathematical methods means that concepts are expressed through mathematical symbols, and that reasoning and deduction are performed by means of mathematical calculations. The objects studied in propositional logic are statements. A simple statement is a simple sentence which could be either "true" or "false". Such statements are also called propositions. We give some examples.

Example 1.1

1. The Earth is round.
2. The Earth is square.
3. If $n > 2$, then $x^n + y^n = z^n$ has no integer solutions (x, y, z).
4. There are beings in outer space.
5. Bridge, stream, village.

It is easy to see that statement 1 is "true" and statement 2 is "false". Statement 3 is Fermat's Last Theorem, which was proven by Andrew Wiles in 1995, so we now know that it is "true". For statement 4, the answer could be "true" or "false", although we still do not know which. Hence statements 1–4 are all propositions. Statement 5 is not a proposition because neither "true" nor "false" can be meaningfully applied to it.

We now consider some other examples.

Example 1.2

1. Mr. Martin is an old man.
2. Today is hot.

D. Cheng et al., *Analysis and Control of Boolean Networks*,
Communications and Control Engineering,
DOI 10.1007/978-0-85729-097-7_1, © Springer-Verlag London Limited 2011

First, we would like to emphasize that these two statements are well-defined propositions because "Is the statement true?" is a logically meaningful question in both cases. However, it is also worth noting that this does not mean the answer is obvious. Consider statement 1. If this man is in his eighties or nineties, the answer is obviously "true". If he is a teenager, the statement is "false". But what if he is in his forties or fifties? The answer is not clear. An analogous argument shows that the second statement has a similar status. Hence, we may need a value between 0 ("false") and 1 ("true") to describe such propositions. This is a topic discussed in the study of multivalued logic or fuzzy logic.

In classical logic we make the basic assumption that a proposition must be either "true" or "false". For compactness we use "T" or "1" for "true", and "F" or "0" for "false". We use capital letters $A, B, C, \ldots$ to represent simple statements. In the following example all the statements are propositions.

Example 1.3 Consider the following statements: A. Beijing is a city in China; B. Beijing is a city in Europe; C. Beijing is a city in Asia; D. Beijing is a city outside China; E. Either Beijing or Moscow is in Europe.

In Example 1.3 the propositions seem to be related. For instance, if A is "T" then B is "F" and vice versa. Similarly, A is "T" if and only if D is "F". We now introduce some symbols, called connectives, to express relationships between propositions.

The following are five commonly used connectives:

- *Negation.* The negation of proposition A is denoted by $\neg A$ and is its opposite. A is true if and only if $\neg A$ is false and vice versa.
- *Conjunction.* The conjunction of A and B, denoted by $A \wedge B$, is a proposition which is true only if both A and B are true.
- *Disjunction.* The disjunction of A and B, denoted by $A \vee B$, is a proposition which is true if either A or B or both A and B are true.
- *Conditional.* The conditional of A to B, denoted by $A \rightarrow B$, means that A implies B (equivalently, if A then B).
- *Biconditional.* The biconditional of A and B, denoted by $A \leftrightarrow B$, means that A is true if and only if B is true.

A connective is also called a logical operator.

Example 1.4 Recall Example 1.3. One sees easily that the following relations are true:

1.
$$A \rightarrow (\neg B), \qquad B \rightarrow (\neg A).$$

2.
$$A \rightarrow C, \qquad (\neg C) \rightarrow D.$$

3.
$$A \leftrightarrow (\neg D), \qquad B \rightarrow E.$$

Simple statements can be compounded by connectives to form compound statements. To investigate general relationships between propositions with connectives, we may use logical variables to replace particular statements. This is the same as in simple algebra where we use letters $x, y, \ldots$ or $a, b, \ldots$ to replace particular numbers. Logical variables are also called statement variables. Statement variables are denoted by $p, q, r, \ldots$ or $x_1, x_2, \ldots$. A valid logical relation (logical identity) for some logical variables is true when the variables are replaced by any particular propositions. This is the same as in simple algebra: for example, if we have $x^2 - y^2 = (x + y)(x - y)$, then no matter what values x and y are replaced with, the equality is always true. It is easy to check that

$$p \vee (\neg p) = T, \tag{1.1}$$

$$p \wedge (\neg p) = F. \tag{1.2}$$

That is, no matter what logical value p takes, logical equations (1.1) and (1.2) always hold.

Consider the set

$$\mathscr{D} = \{T, F\} \quad (\text{equivalently, } \mathscr{D} = \{1, 0\}). \tag{1.3}$$

Definition 1.1

1. A logical variable is a variable which can take values from $\mathscr{D}$.
2. A set of logical variables $x_1, \ldots, x_n$ are independent if, for any fixed values x_j, $j \neq i$, the logical variable x_i can still take value either 1 or 0.
3. A logical function of logical variables, $x_1, \ldots, x_n$ is a logical expression involving $x_1, \ldots, x_n$ and some possible statements (called constants), joined by connectives. Hence a logical function is a mapping $f : \mathscr{D}^n \to \mathscr{D}$. It is also called an n-ary operator [1].

Example 1.5

$$y = (p \wedge q) \leftrightarrow (T \wedge r) \tag{1.4}$$

is a logical function of p, q, r. Using conventional notation we have $y = f(p, q, r)$, $p, q, r \in \mathscr{D}$. The only difference is that here, f is a logical function.

In general, a known constant can be removed from the function. For example, (1.4) is equivalent to

$$y = (p \wedge q) \leftrightarrow r. \tag{1.5}$$

Remark 1.1 Let $x_1, \ldots, x_n$ be logical variables.

1. y is said to be independent of $x_1, \ldots, x_n$ if y can take either F or T as its value, regardless of what values the $x_1, \ldots, x_n$ take.
2. It is said that y depends on $x_1, \ldots, x_n$ (completely) if, as long as the values of $x_1, \ldots, x_n$ are fixed, y can take only a unique value. In this case, y is a logical function of $x_1, \ldots, x_n$. Alternatively, $y : \mathscr{D}^n \to \mathscr{D}$ is a logical mapping.

Table 1.1 Truth table for negation

p	$\neg p$
1	0
0	1

Table 1.2 Truth table for $\wedge$, $\vee$, $\rightarrow$, $\leftrightarrow$, $\bar{\vee}$, $\uparrow$, $\downarrow$

p	q	$p \wedge q$	$p \vee q$	$p \rightarrow q$	$p \leftrightarrow q$	$p \bar{\vee} q$	$p \uparrow q$	$p \downarrow q$
1	1	1	1	1	1	0	0	0
1	0	0	1	0	0	1	1	0
0	1	0	1	1	0	1	1	0
0	0	0	0	1	1	0	1	1

3. y can be neither independent of $x_1,\ldots,x_n$ nor dependent on $x_1,\ldots,x_n$. For example, $y = x_1 \wedge x_2$ is neither independent of x_1 nor dependent on x_1 because when $x_1 = F$, $y = F$, but when $x_1 = T$, we can say nothing about y.

Note that an operator is also a logical function. Recall the basic connectives defined earlier. It is clear that negation, $\neg$, is a 1-ary operator and that conjunction, $\wedge$, disjunction, $\vee$, conditional, $\rightarrow$, and biconditional, $\leftrightarrow$, are all 2-ary operators.

Remark 1.2 Operating priority is defined such that 1-ary operators take priority over 2-ary operators. So (1.1) and (1.2) can be expressed respectively as

$$p \vee \neg p = T \tag{1.6}$$

and

$$p \wedge \neg p = F. \tag{1.7}$$

A connective or a logical operator can easily be expressed by a table, called a truth table. For instance, for negation, we have Table 1.1.

Similarly, we can give truth tables for conjunction, disjunction, conditional, biconditional, and three others, as in Table 1.2.

The truth value of a logical function can easily be obtained from the truth tables of basic connectives. We use an example to illustrate this point.

Example 1.6

1. Let $x = p \wedge (\neg q)$. The truth table of x is shown in Table 1.3.
2. Let $y = (\neg p) \rightarrow (q \vee r)$. The truth table of y is shown in Table 1.4.

Table 1.3 Truth table for x

p	q	$\neg q$	$x = p \wedge (\neg q)$
1	1	0	0
1	0	1	1
0	1	0	0
0	0	1	0

Table 1.4 Truth table for y

p	q	r	$\neg p$	$q \vee r$	$y = (\neg p) \rightarrow (q \vee r)$
1	1	1	0	1	1
1	1	0	0	1	1
1	0	1	0	1	1
1	0	0	0	0	1
0	1	1	1	1	1
0	1	0	1	1	1
0	0	1	1	1	1
0	0	0	1	0	0

1.2 Implication and Equivalence

Definition 1.2

1. A logical function involving certain logical variables is said to be a tautology if it is always true no matter what values the logical variables take.
2. A logical function involving certain logical variables is said to be a contradiction if it is always false no matter what values the logical variables take.

From (1.1) we know $p \vee \neg p$ is a tautology, and from (1.2) we know $p \wedge \neg p$ is a contradiction. According to the definition, it is clear that if x is a tautology, then $\neg x$ is a contradiction. Conversely, if x is a contradiction, then $\neg x$ is a tautology. Both tautology and contradiction are extreme cases. A logical expression which is neither tautology nor contradiction is called a possibly true form.

In the following example we give some useful tautologies and contradictions.

Example 1.7

1. (Law of excluded middle) $p \vee \neg p$ is a tautology.
2. (Law of contradiction) $p \wedge \neg p$ is a contradiction.
3. (Law of negation of negation) $p \leftrightarrow \neg(\neg p)$ is a tautology.
4. $(p \rightarrow (q \rightarrow r)) \rightarrow ((p \rightarrow q) \rightarrow (p \rightarrow r))$ is a tautology.

To prove a tautology or a contradiction, we simply use a truth table. For example, if we set $x = (p \rightarrow (q \rightarrow r)) \rightarrow ((p \rightarrow q) \rightarrow (p \rightarrow r))$, then the truth table for x is shown in Table 1.5.

Table 1.5 Truth table for x

p	q	r	$q \to r$	$p \to (q \to r)$	$p \to q$	$p \to r$	$(p \to q) \to (p \to r)$	x
1	1	1	1	1	1	1	1	1
1	1	0	0	0	1	0	0	1
1	0	1	1	1	0	1	1	1
1	0	0	1	1	0	0	1	1
0	1	1	1	1	1	1	1	1
0	1	0	0	1	1	1	1	1
0	0	1	1	1	1	1	1	1
0	0	0	1	1	1	1	1	1

Definition 1.3 Let x, y be two logical variables.

- x is said to logically imply y if $x \to y$ is a tautology. Logical implication is denoted by $\Rightarrow$, as in $x \Rightarrow y$.
- x and y are logically equivalent if $x \leftrightarrow y$ is a tautology. Logical equivalence is denoted by $\Leftrightarrow$, as in $x \Leftrightarrow y$ (or $x = y$).

In the following example we give some useful laws involving logical implication.

Example 1.8

1.
$$((p \to q) \wedge \neg q) \Rightarrow \neg p. \tag{1.8}$$

2.
$$((p \vee q) \wedge \neg p) \Rightarrow q, \tag{1.9}$$
$$((p \vee q) \wedge \neg q) \Rightarrow p. \tag{1.10}$$

3.
$$(p \wedge q) \Rightarrow p, \tag{1.11}$$
$$(p \wedge q) \Rightarrow q. \tag{1.12}$$

4.
$$((p \to q) \wedge (q \to r)) \Rightarrow (p \to r). \tag{1.13}$$

5.
$$(p \to (r \wedge \neg r)) \Rightarrow \neg p. \tag{1.14}$$

6.
$$p \Rightarrow (p \vee q). \tag{1.15}$$

We leave the proofs to the reader.

In the following example some useful laws involving logical equivalence are presented.

Example 1.9

1. (De Morgan's law)

$$\neg\left(\bigwedge_{i=1}^{n} p_i\right) \Leftrightarrow \bigvee_{i=1}^{n} (\neg p_i), \tag{1.16}$$

$$\neg\left(\bigvee_{i=1}^{n} p_i\right) \Leftrightarrow \bigwedge_{i=1}^{n} (\neg p_i). \tag{1.17}$$

2. (Commutativity)

$$(p \wedge q) \Leftrightarrow (q \wedge p), \tag{1.18}$$

$$(p \vee q) \Leftrightarrow (q \vee p). \tag{1.19}$$

3. (Distributive law)

$$\big(p \wedge (q \vee r)\big) \Leftrightarrow \big((p \wedge q) \vee (p \wedge r)\big), \tag{1.20}$$

$$\big(p \vee (q \wedge r)\big) \Leftrightarrow \big((p \vee q) \wedge (p \vee r)\big). \tag{1.21}$$

4.

$$(p \to q) \Leftrightarrow (\neg p \vee q). \tag{1.22}$$

5.

$$p \Leftrightarrow \big(p \wedge (q \vee \neg q)\big), \tag{1.23}$$

$$p \Leftrightarrow \big(p \vee (q \wedge \neg q)\big). \tag{1.24}$$

6.

$$(p \leftrightarrow q) \Leftrightarrow \big((p \to q) \wedge (q \to p)\big), \tag{1.25}$$

$$(p \leftrightarrow q) \Leftrightarrow \big((p \wedge q) \vee (\neg p \wedge \neg q)\big). \tag{1.26}$$

7.

$$(p \vee p) \Leftrightarrow p, \tag{1.27}$$

$$(p \wedge p) \Leftrightarrow p. \tag{1.28}$$

We give one more example, which is useful in normal form deduction.

Example 1.10

$$(a \wedge p) \vee (\neg a \wedge q) = (a \vee q) \wedge (\neg a \vee p). \tag{1.29}$$

Table 1.6 Truth table for (1.29)

a	p	q	$\neg a$	$a \wedge p$	$\neg a \wedge q$	LHS	$a \vee q$	$\neg a \vee p$	RHS
1	1	1	0	1	0	1	1	1	1
1	1	0	0	1	0	1	1	1	1
1	0	1	0	0	0	0	1	0	0
1	0	0	0	0	0	0	1	0	0
0	1	1	1	0	1	1	1	1	1
0	1	0	1	0	0	0	0	1	0
0	0	1	1	0	1	1	1	1	1
0	0	0	1	0	0	0	0	1	0

Table 1.7 Truth table for binary operators

p	q	σ_1	σ_2	σ_3	σ_4	σ_5	σ_6	σ_7	σ_8	σ_9	σ_{10}	σ_{11}	σ_{12}	σ_{13}	σ_{14}	σ_{15}	σ_{16}
1	1	1	1	1	1	1	1	1	1	0	0	0	0	0	0	0	0
1	0	1	1	1	1	0	0	0	0	1	1	1	1	0	0	0	0
0	1	1	1	0	0	1	1	0	0	1	1	0	0	1	1	0	0
0	0	1	0	1	0	1	0	1	0	1	0	1	0	1	0	1	0

Note that we have here used equality ("="), which is an alternative expression of logical equivalence. That is, two logical expressions are equal if and only if they are logically equivalent.

We use a truth table to prove (1.29). Denote the left- (resp., right-) hand side of (1.29) by *LHS* (resp., *RHS*).

From Table 1.6 it is clear that $LHS = RHS$.

1.3 Adequate Sets of Connectives

In the previous section a 1-ary (unary) connective, $\neg$, and some 2-ary (binary) connectives, $\wedge$, $\vee$, $\rightarrow$, $\leftrightarrow$, etc. were discussed. Note that for an n-ary connective there are n logical variables and each variable can take two possible values, so an n-ary operator is a mapping from a set (domain) of 2^n different elements to a set (region) of two elements. Hence there are 2^{2^n} different connectives. When $n = 2$, we know that there are $2^{2^2} = 16$ different binary connectives. We list them all in Table 1.7, where they are denoted by $\sigma_1, \sigma_2, \ldots, \sigma_{16}$.

Remark 1.3

1. σ_8 is $\wedge$, σ_2 is $\vee$, σ_5 is $\rightarrow$, and σ_7 is $\leftrightarrow$.
2. $\sigma_1(p, q) = T$ and $\sigma_{16}(p, q) = F$. They are 0-ary connectives (constant operators).

3. $\sigma_{13}(p, q) = \neg p$ and $\sigma_{11}(p, q) = \neg q$. They are 1-ary connectives.
4. In general, a k-ary $(k < s)$ connective can be formally expressed as an s-ary connective. The above two operators, σ_{13} and σ_{11}, are such examples.
5. σ_{10} is called the "exclusive or" (EOR), denoted by "$\bar{\vee}$" [4]:

$$\sigma_{10}(p, q) = p \, \bar{\vee} \, q = \neg(p \leftrightarrow q).$$

6. σ_9 is called the "not and" (NAND), denoted by "$\uparrow$":

$$\sigma_9(p, q) = p \uparrow q = \neg(p \wedge q).$$

7. σ_{15} is called the "not or" (NOR), denoted by "$\downarrow$":

$$\sigma_{15}(p, q) = p \downarrow q = \neg(p \vee q).$$

The following proposition provides two important tautologies. They may be used as alternative definitions of the conditional and biconditional, respectively.

Proposition 1.1

1.

$$p \rightarrow q \Leftrightarrow (\neg p) \vee q. \tag{1.30}$$

2.

$$p \leftrightarrow q \Leftrightarrow (p \rightarrow q) \wedge (q \rightarrow p). \tag{1.31}$$

Definition 1.4 A set of connectives is called an adequate set if any connective can be expressed in terms of its elements.

Proposition 1.2 *The following four sets are all adequate sets*: (i) $\{\neg, \wedge\}$, (ii) $\{\neg, \vee\}$, (iii) $\{\neg, \wedge, \vee\}$, (iv) $\{\neg, \wedge, \vee, \rightarrow, \leftrightarrow\}$.

Proof Since (iii) is a subset of (iv), if (iii) is adequate then so is (iv).
    According to De Morgan's law, we have

$$x \vee y \Leftrightarrow \neg\big((\neg x) \wedge (\neg y)\big),$$
$$x \wedge y \Leftrightarrow \neg\big((\neg x) \vee (\neg y)\big).$$

Hence if (iii) is adequate, so are (i) and (ii). Therefore, it is enough to prove that (iii) is adequate.
    Note that it is easy to check that Table 1.7 is "antisymmetric", meaning that $\sigma_1 \Leftrightarrow \neg\sigma_{16}$, $\sigma_2 \Leftrightarrow \neg\sigma_{15}$, etc. In general,

$$\sigma_i \Leftrightarrow \neg\sigma_{17-i}, \quad i = 1, \ldots, 8.$$

Hence it is enough to prove that σ_i, $i = 1, \ldots, 8$, can be expressed in terms of $\{\neg, \wedge, \vee\}$. Since $\sigma_2(p, q) = p \vee q$ and $\sigma_8(p, q) = p \wedge q$, these do not need proof.

Table 1.8 Truth value to logical form

$\sigma(p,q)$	σ	$\sigma(p,q)$	σ
1111	T	0000	F
1110	$p \vee q$	0001	$\neg(p \vee q)$ or $\neg p \wedge \neg q$ or $p \downarrow q$
1101	$q \to p$	0010	$\neg(q \to p)$ or $\neg p \wedge q$
1100	p	0011	$\neg p$
1011	$p \to q$	0100	$\neg(p \to q)$ or $p \wedge \neg q$
1010	q	0101	$\neg q$
1001	$p \leftrightarrow q$	0110	$\neg(p \leftrightarrow q)$ or $(p \wedge \neg q) \vee (q \wedge \neg p)$ or $p \barvee q$
1000	$p \wedge q$	0111	$\neg(p \wedge q)$ or $\neg p \vee \neg q$ or $p \uparrow q$

We also have $\sigma_5(p, q) = p \to q$ and so, using Proposition 1.1, σ_5 can be expressed in terms of them. Furthermore, since $\sigma_7(p, q) \Leftrightarrow (p \leftrightarrow q)$, the second identity in Proposition 1.1 ensures that σ_7 can be expressed in terms of them as

$$(p \leftrightarrow q) \Leftrightarrow \big((p \to q) \wedge (q \to p)\big).$$

We still need to prove $\sigma_1, \sigma_3, \sigma_4, \sigma_6$. In fact we have

$$\sigma_1(p, q) \Leftrightarrow (p \wedge q) \vee (p \wedge \neg q) \vee (\neg p \wedge q) \vee (\neg p \wedge \neg q),$$

$$\sigma_3(p, q) \Leftrightarrow (p \vee \neg q) \quad \text{or} \quad \sigma_3(p, q) \Leftrightarrow q \to p,$$

$$\sigma_4(p, q) \Leftrightarrow p,$$

$$\sigma_6(p, q) \Leftrightarrow q. \qquad \qquad \qquad \square$$

In the sequel it will be very useful to find a logical operator from its truth values in a truth table. We use four $\{0, 1\}$ numbers to denote the truth values of a binary operator. For instance, referring to Table 1.8, $p \wedge q$ takes four values: $(1, 0, 0, 0)^T$. (We use superscript T for transpose.) We then use "1000" to denote its truth values. The following table shows the mapping from the four numerical truth values to their corresponding logical operators, which may have several equivalent forms.

A single connective can form an adequate set, as we will see in the following example.

Example 1.11

1. If we define

$$\sigma_9(p, q) := p|q,$$

   then $\{|\}$ is an adequate set. First, we have

$$(\neg p) \Leftrightarrow (p|p);$$

second, we have

$$(p \vee q) \Leftrightarrow ((p|p)|(q|q)).$$

According to Proposition 1.2, the conclusion follows.

2. If we define

$$\sigma_{15}(p,q) := p \downarrow q,$$

then $\{\downarrow\}$ is an adequate set. Note that

$$(\neg p) \Leftrightarrow (p \downarrow p).$$

We also have

$$(p \wedge q) \Leftrightarrow ((p \downarrow p) \downarrow (q \downarrow q)).$$

The conclusion then follows.

Remark 1.4 In the study of Boolean networks, mod 2 addition "+(mod 2)" and mod 2 multiplication "×(mod 2)" are commonly used as logical operators. It is obvious that "×(mod 2)" is the same as conjunction, "∧", and that "+(mod 2)" is the same as EOR, "$\bar{\vee}$". They form an adequate set, so they are sufficient to describe all logical expressions.

1.4 Normal Form

Definition 1.5 Let $\{p_1, p_2, \ldots, p_n\}$ be a set of logical variables. Define a set of logical variables by also including their negations, as follows:

$$P := \{p_1, \neg p_1, p_2, \neg p_2, \ldots, p_n, \neg p_n\}.$$

1. If

$$c := \bigwedge_{i=1}^{s} a_i, \quad a_i \in P,$$

then c is called a basic conjunctive form.

2. If

$$d := \bigvee_{i=1}^{s} a_i, \quad a_i \in P,$$

then d is called a basic disjunctive form.

3. If

$$\ell := \bigvee_{i=1}^{s} c_i,$$

where c_i are basic conjunctive forms, then ℓ is called a disjunctive normal form.

4. If

$$\ell := \bigwedge_{i=1}^{s} d_i,$$

where d_i are basic disjunctive forms, then ℓ is called a conjunctive normal form.

We give some examples.

Example 1.12
Let p, q, r be three logical variables. Then:

1. p, $\neg p$, and $p \wedge (\neg q) \wedge (\neg r)$ are basic conjunctive forms.
2. $(\neg p) \vee r$, $p \vee (\neg p)$, and $(\neg q) \vee p \vee (\neg r)$ are basic disjunctive forms.
3. $p \wedge q$, $p \vee q$, and $(\neg p) \vee (p \wedge q) \vee ((\neg q) \wedge p \wedge (\neg r))$ are disjunctive normal forms.
4. $p \wedge q$, $p \vee q$, and $(\neg p) \wedge (p \vee q) \wedge ((\neg q) \vee (\neg q) \vee (\neg r))$ are conjunctive normal forms.

Proposition 1.3 *Any logical expression can be expressed in disjunctive normal form as well as conjunctive normal form.*

Proof Let ℓ be a logical expression with $p_1, p_2, \ldots, p_n$ as its logical variables. We first prove that it can be expressed as a disjunctive normal form. If, for any i and any value of p_i, it is always F, then it is a contradiction. Hence it can be expressed as

$$\ell = p_1 \wedge (\neg p_1) \wedge p_2 \wedge \cdots \wedge p_n.$$

Assume that when $(p_1, \ldots, p_n) = \alpha := (\alpha_1, \ldots, \alpha_n)$ (i.e., $p_i = \alpha_i$, $i = 1, \ldots, n$), ℓ is T. We construct a basic conjunctive form as

$$b_\alpha := c_1 \wedge c_2 \wedge \cdots \wedge c_n,$$

where

$$c_i := \begin{cases} p_i, & \alpha_i = T, \\ \neg p_i, & \alpha_i = F. \end{cases}$$

Now assume the set of values of logical variables for which ℓ is T to be $\alpha^i = (\alpha_1^i, \alpha_2^i, \ldots, \alpha_n^i)$, $i = 1, 2, \ldots, s$. Using the above method, we can construct for each α^i a corresponding b_{α_i}. It is obvious that

$$\ell = \bigvee_{i=1}^{s} b_{\alpha_i}. \tag{1.32}$$

This is a disjunctive normal form.

Table 1.9 Truth table of ℓ

p	q	r	$(p \vee q) \to \neg r$	$(r \to p) \wedge (r \vee q)$	ℓ
1	1	1	0	1	1
1	1	0	1	1	1
1	0	1	0	1	1
1	0	0	1	0	0
0	1	1	0	0	1
0	1	0	1	1	1
0	0	1	1	0	0
0	0	0	1	0	0

Next, we construct a conjunctive normal form. Because of the existence of a disjunctive normal form of $\neg\ell$, we have

$$\neg\ell = \bigvee_{i=1}^{s} b_i,$$

where $b_i = c_1^i \wedge \cdots \wedge c_{n_i}^i$ are basic conjunctive forms. Using De Morgan's law,

$$\ell = (\neg b_1) \wedge (\neg b_2) \wedge \cdots \wedge (\neg b_k). \tag{1.33}$$

Note that $\neg b_i = \neg c_1^i \vee \cdots \vee \neg c_{n_i}^i$ is a basic disjunctive form. It follows that (1.33) is a conjunctive normal form. □

The proof of the above proposition is constructive, so we can use it to construct normal forms. We show this by means of the following example.

Example 1.13 Consider

$$\ell := \big((p \vee q) \to \neg r\big) \to \big((r \to p) \wedge (r \vee q)\big). \tag{1.34}$$

We will convert this into a disjunctive normal form and a conjunctive normal form. We give the truth table of ℓ in Table 1.9.

When p, q, and r take values from rows $1, 2, 3, 5$, and 6, ℓ is true. According to the values of the variables in each row, we can construct a basic conjunctive form. Then, the disjunction of all such terms yields the disjunctive normal form, as follows:

$$\ell = (p \wedge q \wedge r) \vee (p \wedge q \wedge \neg r) \vee (p \wedge \neg q \wedge r) \vee (\neg p \wedge q \wedge r) \vee (\neg p \wedge q \wedge \neg r).$$

When p, q, and r take values from rows $4, 7$, and 8, $\neg\ell$ is true. As before, the disjunctive form of $\neg\ell$ can be constructed as

$$\neg\ell = (p \wedge \neg q \wedge \neg r) \vee (\neg p \wedge \neg q \wedge r) \vee (\neg p \wedge \neg q \wedge \neg r).$$

Using De Morgan's law, the conjunctive normal form of ℓ is obtained as

$$\ell = (\neg p \vee q \vee r) \wedge (p \vee q \vee \neg r) \wedge (p \vee q \vee r).$$

According to Definition 1.5, neither the disjunctive normal form nor the conjunctive normal form is unique. To get unique expressions, we give the following definition.

Definition 1.6

1. A disjunctive normal form is said to be optimized if it satisfies the following conditions:

   - If a variable appears in the normal form, then it appears in all basic conjunctive forms.
   - There is no basic conjunctive form that is a contradiction.
   - There are no identical variables in each basic conjunctive form.
   - There are no identical basic conjunctive forms.
   - In the normal form, the variables, their negations, and basic conjunctive forms are all arranged in alphabetical order.

2. A conjunctive normal form is said to be optimized if it satisfies the following conditions:

   - If a variable appears in the normal form, then it appears in all basic disjunctive forms.
   - There is no basic disjunctive form that is a tautology.
   - There are no identical variables in each basic disjunctive form.
   - There are no identical basic disjunctive forms.
   - In the normal form, the variables, their negations, and basic disjunctive forms are all arranged in alphabetical order.

Theorem 1.1 *For each logical expression there exist a unique optimized disjunctive normal form and a unique optimized conjunctive normal form.*

Proof In fact the constructive proof of Proposition 1.3 provides a way to construct the optimal normal forms. For instance, in Example 1.13, the disjunctive normal form and conjunctive normal form obtained there are optimal, as long as we reorder the variables in alphabetical order, i.e.,

$$(p \wedge q)\sigma_1(p \wedge \neg q)\sigma_2(\neg p \wedge q)\sigma_3(\neg p \wedge \neg q),$$

where σ_i, $i = 1, 2, 3$, are connectives. From the construction it is clear that the optimized normal forms are unique.   □

1.5 Multivalued Logic

Hereafter, two-valued logic will be called Boolean logic. It was mentioned earlier that in the real world, "true" and "false" may not be sufficient to describe a statement. We give some additional simple examples.

Table 1.10 Logical values with respect to age

a	≥ 70	$[60, 70)$	$[40, 60)$	$[30, 40)$	$[20, 30)$	< 20
A	1	0.8	0.6	0.4	0.2	0

Example 1.14 Consider the following statements:

1. The temperature in the stove is high.
2. The air pollution is severe.
3. Smith's family is rich.
4. She is an old lady.

All the statements are propositions, but "true" or "false" may not be enough to characterize them. For instance, in a chemical factory the stove temperature may be classified as "very high", "high", "average", "low", or "very low". Here, we could use "true" or "1" for the first case, and "false" or "0" for the fifth case. But what of the intermediate cases? It would be natural to define some logical values between 1 and 0 to describe them, e.g., "0.75" for "high", "0.5" for "average", and "0.25" for "low".

In Beijing, the television broadcast uses "clean", "mildly polluted", and "severely polluted" to describe the air quality. The broadcasters thus provide three values (which we could label "0", "0.5", and "1") to classify statement 2.

For wealth, "below the poverty line", "low income", "middle class", "high income", etc. may be used to describe statement 3.

Finally, we consider the last statement. According to the person's age, we may assign a logical value to it. We refer to Table 1.10 for this.

In this way, we need six different values to describe a statement. This yields multivalued logic. If we allow the values to be anything between 1 and 0, we have fuzzy logical values.

Next, we define multivalued logic and fuzzy logic rigorously.
Define

$$\mathscr{D}_k = \left\{ T = 1, \frac{k-2}{k-1}, \frac{k-3}{k-1}, \dots, F = 0 \right\}$$

and

$$\mathscr{D}_f = \{ r \mid 0 \leq r \leq 1 \}.$$

Note that $\mathscr{D}_2 = \mathscr{D}$ is what we defined before for Boolean logic.

Definition 1.7

1. A logical system is called a k-valued logic if its logical variables may take any values from $\mathscr{D}_k$.
2. A logical system is called a fuzzy logic if its logical variables may take any values from $\mathscr{D}_f$.

Table 1.11 k-valued unary operators

p	$\neg p$	$\oslash_k(p)$	$\nabla_{i,k}(p)$
1	0	$(k-2)/(k-1)$	0
$(k-2)/(k-1)$	$1/(k-1)$	$(k-3)/(k-1)$	0
$\vdots$	$\vdots$	$\vdots$	$\vdots$
$(i-1)/(k-1)$	$(k-i)/(k-1)$	$(i-2)/(k-1)$	1
$\vdots$	$\vdots$	$\vdots$	$\vdots$
$2/(k-1)$	$(k-3)/(k-1)$	$1/(k-1)$	0
$1/(k-1)$	$(k-2)/(k-1)$	0	0
0	1	1	0

3. A logical operator $\sigma : \mathscr{D}_k^s \to \mathscr{D}_k$ is an s-ary k-valued logical operator; a logical operator $\sigma : \mathscr{D}_f^s \to \mathscr{D}_f$ is an s-ary fuzzy logical operator.

For the remainder of this section we mainly consider k-valued logic.

First, we define some unary operators. It is easy to see that there are k^k unary operators. We define some which will be useful in the sequel: (1) "negation", $\neg$, (2) "rotator", $\oslash_k$, (3) "i-confirmer", $\nabla_{i,k}$, $i = 1, 2, \ldots, k$.

Definition 1.8

1. Let $p = \frac{i}{k-1}$. Then

$$\neg p = \frac{(k-1)-i}{k-1}. \tag{1.35}$$

2. Let $p = \frac{i}{k-1}$. Then

$$\oslash_k p = \begin{cases} \frac{i-1}{k-1}, & i > 0, \\ 1, & i = 0. \end{cases} \tag{1.36}$$

3.

$$\nabla_{i,k} p = \begin{cases} p, & p = \frac{k-i}{k-1}, \\ 0, & p \neq \frac{k-i}{k-1}. \end{cases} \tag{1.37}$$

Table 1.11 shows the truth values of these unary operators.
Next, we define some binary operators.

Definition 1.9 Let p and q be two k-valued logical variables. Define their disjunction as

$$p \vee q = \max(p, q) \tag{1.38}$$

Table 1.12 3-valued extended logic

p	q	$\neg p$	$p \to q$	$\neg q$	$q \to p$	$p \leftrightarrow q$
1	1	0	1	0	1	1
1	0.5	0	0.5	0.5	1	0.5
1	0	0	0	1	1	0
0.5	1	0.5	1	0	0.5	0.5
0.5	0.5	0.5	0.5	0.5	0.5	0.5
0.5	0	0.5	0.5	1	1	0.5
0	1	1	1	0	0	0
0	0.5	1	1	0.5	0.5	0.5
0	0	1	1	1	1	1

Table 1.13 Some other 3-valued logics

p	q	KD		L		B	
		$\to$	$\leftrightarrow$	$\to$	$\leftrightarrow$	$\to$	$\leftrightarrow$
1	1	1	1	1	1	1	1
1	0.5	0.5	0.5	0.5	0.5	0.5	0.5
1	0	0	0	0	0	0	0
0.5	1	1	0.5	1	0.5	0.5	0.5
0.5	0.5	0.5	0.5	1	1	0.5	0.5
0.5	0	0.5	0.5	0.5	0.5	0.5	0.5
0	1	1	0	1	0	1	0
0	0.5	1	0.5	1	0.5	0.5	0.5
0	0	1	1	1	1	1	1

and their conjunction as

$$p \wedge q = \min(p, q). \tag{1.39}$$

Definition 1.9 is a natural generalization of Boolean logic. When $k = 2$, it is obvious that these definitions of disjunction and conjunction coincide with those in Boolean logic. Definition 1.9 is widely accepted, but others exist. For implication, there are many different definitions.

A natural way to define the "conditional" and "biconditional" is by using equations (1.30) and (1.31) of Proposition 1.1, respectively. We call this the extended logic. Using (1.30), (1.31), (1.38), and (1.39), the truth table for the conditional and biconditional in the 3-valued extended logic can be easily calculated, as in Table 1.12.

There are several other types of 3-valued logic. They may have different "conditionals", but the "biconditional" is usually defined by (1.31). In the following table we give three different 3-valued logics: (1) Kleene–Dienes type (KD), (2) Łukasiewicz type (L), (3) Bochvar type (B), as in Table 1.13 of [3].

From Tables 1.12 and 1.13, one easily sees that the Kleene–Dienes logic is the same as the extended logic. Throughout this book, our default multivalued logic is the extended logic, unless otherwise stated.

References

1. Barnes, D., Mac, J.: An Algebraic Introduction to Mathematical Logic. Springer, New York (1975)
2. Hamilton, A.: Logic for Mathematicians. Cambridge University Press, Cambridge (1988). Revised edn.
3. Liu, Z., Liu, Y.: Fuzzy Logic and Neural Network. BUAA Press, Beijing (1996) (in Chinese)
4. Rade, L., Westergren, B.: Mathematics Handbook. Studentlitteratur, Lund (1989)

Chapter 2
Semi-tensor Product of Matrices

2.1 Multiple-Dimensional Data

Roughly speaking, linear algebra mainly concerns two kinds of objects: vectors and matrices. An n-dimensional vector is expressed as $X = (x_1, x_2, \ldots, x_n)$. Its elements are labeled by one index, i, where x_i is the ith element of X. For an $m \times n$ matrix

$$A = \begin{bmatrix} a_{11} & a_{12} & \cdots & a_{1n} \\ a_{21} & a_{22} & \cdots & a_{2n} \\ \vdots & & & \\ a_{m1} & a_{m2} & \cdots & a_{mn} \end{bmatrix},$$

elements are labeled by two indices, i and j, where $a_{i,j}$ is the element of A located in the ith row and jth column. In this way, it is easy to connect the dimension of a set of data with the number of indices. We define the dimension of a set of data as follows.

Definition 2.1 A set of data, labeled by k indices, is called a set of k-dimensional data. Precisely,

$$X = \{x_{i_1, i_2, \ldots, i_k} \mid 1 \leq i_j \leq n_j, j = 1, 2, \ldots, k\} \tag{2.1}$$

is a set of k-dimensional data. The cardinal number of X, denoted by $|X|$, is $|X| = n_1 n_2 \cdots n_k$.

In the following example we give an example of 3-dimensional data.

Example 2.1 Consider $\mathbb{R}^3$, with its canonical basis $\{e_1, e_2, e_3\}$. Any vector $X \in \mathbb{R}^3$ may then be expressed as $X = x_1 e_1 + x_2 e_2 + x_3 e_3$. When the basis is fixed, we simply use $X = (x_1, x_2, x_3)^T$ to represent it. From simple vector algebra we know that in $\mathbb{R}^3$ there is a cross product, $\times$, such that for any two vectors $X, Y \in \mathbb{R}^3$ we

D. Cheng et al., *Analysis and Control of Boolean Networks*,
Communications and Control Engineering,
DOI 10.1007/978-0-85729-097-7_2, © Springer-Verlag London Limited 2011

have $X \times Y \in \mathbb{R}^3$, defined as follows:

$$X \times Y = \det\left(\begin{bmatrix} e_1 & e_2 & e_3 \\ x_1 & x_2 & x_3 \\ y_1 & y_2 & y_3 \end{bmatrix}\right). \tag{2.2}$$

Since the cross product is linear with respect to X as well as Y, it is a bilinear mapping. The value of the cross product is thus uniquely determined by its value on the basis. Write

$$e_i \times e_j = c_{ij}^1 e_1 + c_{ij}^2 e_2 + c_{ij}^3 e_3, \quad i, j = 1, 2, 3.$$

The coefficients form a set of 3-dimensional data,

$$\{c_{ij}^k \mid i, j, k = 1, 2, 3\},$$

which are called the structure constants. Structure constants are easily computable. For instance,

$$e_1 \times e_2 = \det\left(\begin{bmatrix} e_1 & e_2 & e_3 \\ 1 & 0 & 0 \\ 0 & 1 & 0 \end{bmatrix}\right) = e_3,$$

which means that $c_{12}^1 = c_{12}^2 = 0$, $c_{12}^3 = 1$. Similarly, we can determine all the structure constants:

$$
\begin{array}{llllll}
c_{11}^1 = 0, & c_{11}^2 = 0, & c_{11}^3 = 0, & c_{12}^1 = 0, & c_{12}^2 = 0, & c_{12}^3 = 1, \\
c_{13}^1 = 0, & c_{13}^2 = -1, & c_{13}^3 = 0, & c_{21}^1 = 0, & c_{21}^2 = 0, & c_{21}^3 = -1, \\
c_{22}^1 = 0, & c_{22}^2 = 0, & c_{22}^3 = 0, & c_{23}^1 = 1, & c_{23}^2 = 0, & c_{23}^3 = 0, \\
c_{31}^1 = 0, & c_{31}^2 = 1, & c_{31}^3 = 0, & c_{32}^1 = -1, & c_{32}^2 = 0, & c_{32}^3 = 0, \\
c_{33}^1 = 0, & c_{33}^2 = 0, & c_{33}^3 = 0.
\end{array}
$$

Since the cross product is linear with respect to the coefficients of each vector, the structure constants uniquely determine the cross product. For instance, let $X = 3e_1 - e_3$ and $Y = 2e_2 + 3e_3$. Then

$$X \times Y = 6e_1 \times e_2 + 9e_1 \times e_3 - 2e_3 \times e_2 - 3e_3 \times e_3$$
$$= 6\left(c_{12}^1 e_1 + c_{12}^2 e_2 + c_{12}^3 e_3\right) + 9\left(c_{13}^1 e_1 + c_{13}^2 e_2 + c_{13}^3 e_3\right)$$
$$- 2\left(c_{32}^1 e_1 + c_{32}^2 e_2 + c_{32}^3 e_3\right) - 3\left(e_{33}^1 e_1 + c_{33}^2 e_2 + c_{33}^3 e_3\right)$$
$$= 2e_1 - 9e_2 + 6e_3.$$

It is obvious that using structure constants to calculate the cross product in this way is very inconvenient, but the example shows that the cross product is uniquely determined by structure constants. So, in general, to define a multilinear mapping it is enough to give its structure constants.

Using structure constants to describe an algebraic structure is a powerful method.

Definition 2.2 [6] Let V be an n-dimensional vector space with coefficients in $\mathbb{R}$. If there is a mapping $* : V \times V \to V$, called the product of two vectors, satisfying

$$\begin{cases} (\alpha X + \beta Y) * Z = \alpha(X * Z) + \beta(Y * Z), \\ X * (\alpha Y + \beta Z) = \alpha(X * Y) + \beta(X * Z) \end{cases} \tag{2.3}$$

(where $\alpha, \beta \in \mathbb{R}$, $X, Y, Z \in V$), then $(V, *)$ is called an algebra.

Let $(V, *)$ be an algebra. If the product satisfies associative law, i.e.,

$$(X * Y) * Z = X * (Y * Z), \quad X, Y, Z \in V, \tag{2.4}$$

then it is called an associative algebra.

$\mathbb{R}^3$ with the cross product is obviously an algebra. It is also easy to check that it is not an associative algebra.

Let V be an n-dimensional vector space and $(V, *)$ an algebra. Choosing a basis $\{e_1, e_2, \ldots, e_n\}$, the structure constants can be obtained as

$$e_i * e_j = \sum_{k=1}^{n} c_{ij}^k e_k, \quad i, j = 1, 2, \ldots, n.$$

Although the structure constants $\{c_{ij}^k \mid i, j, k = 1, 2, \ldots, n\}$ depend on the choice of basis, they uniquely determine the structure of the algebra. It is also easy to convert a set of structure constants, which correspond to a basis, to another set of structure constants, which correspond to another basis. For an algebra, the structure constants are always a set of 3-dimensional data.

Next, we consider an s-linear mapping on an n-dimensional vector space. Let V be an n-dimensional vector space and let $\phi : \underbrace{V \times V \times \cdots \times V}_{s} \to \mathbb{R}$, satisfying (for any $1 \leq i \leq s$, $\alpha, \beta \in \mathbb{R}$)

$$\phi(X_1, X_2, \ldots, \alpha X_i + \beta Y_i, \ldots, X_{s-1}, X_s)$$
$$= \alpha\phi(X_1, X_2, \ldots, X_i, \ldots, X_{s-1}, X_s) + \beta\phi(X_1, X_2, \ldots, Y_i, \ldots, X_{s-1}, X_s). \tag{2.5}$$

Equation (2.5) shows the linearity of ϕ with respect to each vector argument. Choosing a basis of V, $\{e_1, e_2, \ldots, e_n\}$, the structure constants of ϕ are defined as

$$\phi(e_{i_1}, e_{i_2}, \ldots, e_{i_s}) = c_{i_1, i_2, \ldots, i_s}, \quad i_j = 1, 2, \ldots, n, j = 1, 2, \ldots, s.$$

Similarly, the structure constants, $\{c_{i_1, i_2, \ldots, i_s} \mid i_1, \ldots, i_s = 1, 2, \ldots, n\}$, uniquely determine ϕ. Conventionally, ϕ is called a tensor, where s is called its covariant degree.

It is clear that for a tensor with covariant degree s, its structure constants form a set of s-dimensional data.

Example 2.2

1. In $\mathbb{R}^3$ we define a three linear mapping as

$$\phi(X, Y, Z) = \langle X \times Y, Z \rangle, \quad X, Y, Z \in \mathbb{R}^3,$$

where $\langle \cdot, \cdot \rangle$ denotes the inner product. Its geometric interpretation is the volume of the parallelogram with X, Y, Z as three adjacent edges [when (X, Y, Z) form a right-hand system, the volume is positive, otherwise, the volume is negative]. It is obvious that ϕ is a tensor with covariant degree 3.

2. In $\mathbb{R}^3$ we can define a four linear mapping as

$$\psi(X, Y, Z, W) = \langle X \times Y, Z \times W \rangle, \quad X, Y, Z, W \in \mathbb{R}^3.$$

Obviously, ψ is a tensor of covariant degree 4.

Next, we consider a more general case. Let $\mu : V \to \mathbb{R}$ be a linear mapping on V,

$$\mu(e_i) = c_i, \quad i = 1, \ldots, n.$$

Then, μ can be expressed as

$$\mu = c_1 e_1^* + c_2 e_2^* + \cdots + c_n e_n^*,$$

where $e_i^* : V \to \mathbb{R}$ satisfies

$$e_i^*(e_j) = \delta_{i,j} = \begin{cases} 1, & i = j, \\ 0, & i \neq j. \end{cases}$$

It can be seen easily that the set of linear mappings on V forms a vector space, called the dual space of V and denoted by V^*.

Let $X = x_1 e_1 + x_2 e_2 + \cdots + x_n e_n \in V$ and $\mu = \mu_1 e_1^* + \mu_2 e_2^* + \cdots + \mu_n e_n^* \in V^*$. When the basis and the dual basis are fixed, $X \in V$ can be expressed as a column vector and $\mu \in V^*$ can be expressed as a row vector, i.e.,

$$X = (a_1, a_2, \ldots, a_n)^\mathrm{T}, \qquad \mu = (c_1, c_2, \ldots, c_n).$$

Using these vector forms, the action of μ on X can be expressed as their matrix product:

$$\mu(X) = \mu X = \sum_{i=1}^{n} a_i c_i, \quad \mu \in V^*, X \in V.$$

Let $\phi : \underbrace{V^* \times \cdots \times V^*}_{t} \times \underbrace{V \times \cdots \times V}_{s} \to \mathbb{R}$ be an $(s + t)$-fold multilinear mapping. Then, ϕ is said to be a tensor on V with covariant degree s and contravariant degree t. Denote by $\mathcal{T}_t^s$ the set of tensors on V with covariant degree s and contravariant degree t.

If we define

$$c_{j_1, j_2, \ldots, j_t}^{i_1, i_2, \ldots, i_s} := \phi\left(e_{i_1}, e_{i_2}, \ldots, e_{i_s}, e_{j_1}^*, e_{j_2}^*, \ldots, e_{j_t}^*\right),$$

then

$$\left\{ c_{j_1, j_2, \ldots, j_t}^{i_1, i_2, \ldots, i_s} \mid 1 \leq i_1, \ldots, i_s, j_1, \ldots, j_t \leq n \right\}$$

is the set of structure constants of ϕ. Structure constants of $\phi \in \mathcal{T}_t^s$ form a set of $(s + t)$-dimensional data.

Next, we consider how to arrange higher-dimensional data. In linear algebra one-dimensional data are arranged as a column or a row, called a vector, while two-dimensional data are arranged as a rectangle, called a matrix. In these forms matrix computation becomes a very convenient and powerful tool for dealing with one- or two-dimensional data. A question which then naturally arises is how to arrange three-dimensional data. A cubic matrix approach has been proposed for this purpose [1, 2] and has been used in some statistics problems [8–10], but, in general, has not been very successful. The problem is: (1) cubic matrices cannot be clearly expressed in a plane (i.e., on paper), (2) the conventional matrix product does not apply, hence some new product rules have to be produced, (3) it is very difficult to generalize this approach to even higher-dimensional cases.

The basic idea concerning the semi-tensor product of matrices is that no matter what the dimension of the data, they are arranged in one- or two-dimensional form. By then properly defining the product, the hierarchy structure of the data can be automatically determined. Hence the data arrangement is important for the semi-tensor product of data.

Definition 2.3 Suppose we are given a set of data S with $\prod_{i=1}^k n_i$ elements and, as in (2.1), the elements of x are labeled by k indices. Moreover, suppose the elements of x are arranged in a row (or a column). It is said that the data are labeled by indices $i_1, \ldots, i_k$ according to an ordered multi-index, denoted by Id or, more precisely,

$$Id(i_1, \ldots, i_k; n_1, \ldots, n_k),$$

if the elements are labeled by $i_1, \ldots, i_k$ and arranged as follows: Let i_t, $t = 1, \ldots, k$, run from 1 to n_t with the order that $t = k$ first, then $t = k - 1$, and so on, until $t = 1$. Hence, $x_{\alpha_1, \ldots, \alpha_k}$ is ahead of $x_{\beta_1, \ldots, \beta_k}$ if and only if there exists $1 \leq j \leq k$ such that

$$\alpha_i = \beta_i, \quad i = 1, \ldots, j - 1, \qquad \alpha_j < \beta_j.$$

If the numbers $n_1, \ldots, n_k$ of $i_1, \ldots, i_k$ are equal, we may use

$$Id(i_1, \ldots, i_k; n) := Id(i_1, \ldots, i_k; n, \ldots, n).$$

If n_i are obviously known, the expression of Id can be simplified as

$$Id(i_1, \ldots, i_k) := Id(i_1, \ldots, i_k; n_1, \ldots, n_k).$$

Example 2.3

1. Assume $x = \{x_{ijk} \mid i = 1, 2, 3; j = 1, 2; k = 1, 2\}$. If we arrange the data according to the ordered multi-index $Id(i, j, k)$, they are

$$x_{111}, x_{112}, x_{121}, x_{122}, x_{211}, x_{212}, x_{221}, x_{222}, x_{311}, x_{312}, x_{321}, x_{322}.$$

If they are arranged by $Id(j, k, i)$, they become

$$x_{111}, x_{211}, x_{311}, x_{112}, x_{212}, x_{312}, x_{121}, x_{221}, x_{321}, x_{122}, x_{222}, x_{322}.$$

2. Let $x = \{x_1, x_2, \ldots, x_{24}\}$. If we use $\lambda_1, \lambda_2, \lambda_3$ to express the data in the form $a_i = a_{\lambda_1, \lambda_2, \lambda_3}$, then under different Id's they have different arrangements:
   (a) Using the ordered multi-index $Id(\lambda_1, \lambda_2, \lambda_3; 2, 3, 4)$, the elements are arranged as

$$
\begin{array}{cccc}
x_{111} & x_{112} & x_{113} & x_{114} \\
x_{121} & x_{122} & x_{123} & x_{124} \\
x_{131} & x_{132} & x_{133} & x_{134} \\
& & \vdots & \\
x_{231} & x_{232} & x_{233} & x_{234}.
\end{array}
$$

   (b) Using the ordered multi-index $Id(\lambda_1, \lambda_2, \lambda_3; 3, 2, 4)$, the elements are arranged as

$$
\begin{array}{cccc}
x_{111} & x_{112} & x_{113} & x_{114} \\
x_{121} & x_{122} & x_{123} & x_{124} \\
x_{211} & x_{212} & x_{213} & x_{214} \\
& & \vdots & \\
x_{321} & x_{322} & x_{323} & x_{324}.
\end{array}
$$

   (c) Using the ordered multi-index $Id(\lambda_1, \lambda_2, \lambda_3; 4, 2, 3)$, the elements are arranged as

$$
\begin{array}{ccc}
x_{111} & x_{112} & x_{113} \\
x_{121} & x_{122} & x_{123} \\
x_{211} & x_{212} & x_{213} \\
& \vdots & \\
x_{421} & x_{422} & x_{423}.
\end{array}
$$

Note that in the above arrangements the data are divided into several rows, but this is simply because of spatial restrictions. Also, in this arrangement the hierarchy structure of the data is clear. In fact, the data should be arranged into one row.

Different Id's, corresponding to certain index permutations, cause certain permutations of the data. For convenience, we now present a brief introduction to the permutation group. Denote by S_k the permutations of k elements, which form a group called the kth order permutation group. We use $1, \ldots, k$ to denote the k elements. If we suppose that $k = 5$, then $\mathbf{S}_5$ consists of all possible permutations of five elements: $\{1, 2, 3, 4, 5\}$. An element $\sigma \in \mathbf{S}_5$ can be expressed as

$$
\sigma = \begin{bmatrix} 1 & 2 & 3 & 4 & 5 \\ \downarrow & \downarrow & \downarrow & \downarrow & \downarrow \\ 2 & 3 & 1 & 5 & 4 \end{bmatrix} \in \mathbf{S}_5.
$$

That is, σ changes 1 to 2, 2 to 3, 3 to 1, 4 to 5, and 5 to 4. σ can also be simply expressed in a rotational form as

$$
\sigma = (1, 2, 3)(4, 5).
$$

Let $\mu \in \mathbf{S}_5$ and

$$
\mu = \begin{bmatrix} 1 & 2 & 3 & 4 & 5 \\ \downarrow & \downarrow & \downarrow & \downarrow & \downarrow \\ 4 & 3 & 2 & 1 & 5 \end{bmatrix}.
$$

The product (group operation) on $\mathbf{S}_5$ is then defined as

$$
\mu\sigma = \begin{bmatrix} 1 & 2 & 3 & 4 & 5 \\ \downarrow & \downarrow & \downarrow & \downarrow & \downarrow \\ 2 & 3 & 1 & 5 & 4 \\ \downarrow & \downarrow & \downarrow & \downarrow & \downarrow \\ 3 & 2 & 4 & 5 & 1 \end{bmatrix},
$$

that is, $\mu\sigma = (1, 3, 4, 5)$.

If, in (2.1), the data x are arranged according to the ordered multi-index $Id(i_1, \ldots, i_k)$, it is said that the data are arranged in a natural order. Of course, they may be arranged in the order of $(i_{\sigma(1)}, \ldots, i_{\sigma(k)})$, that is, letting index $i_{\sigma(k)}$ run from 1 to $n_{\sigma(k)}$ first, then letting $i_{\sigma(k-1)}$ run from 1 to $n_{\sigma(k-1)}$, and so on. It is obvious that a different Id corresponds to a different data arrangement.

Definition 2.4 Let $\sigma \in \mathbf{S}_k$ and x be a set of data with $\prod_{i=1}^{k} n_i$ elements. Arrange x in a row or a column. It is said that x is arranged by the ordered multi-index $Id(i_{\sigma(1)}, \ldots, i_{\sigma(k)}; n_{\sigma(1)}, \ldots, n_{\sigma(k)})$ if the indices $i_1, \ldots, i_k$ in the sequence are running in the following order: first, $i_{\sigma(k)}$ runs from 1 to $n_{\sigma(k)}$, then $i_{\sigma(k-1)}$ runs from 1 to $n_{\sigma(k-1)}$, and so on, until, finally, $i_{\sigma(1)}$ runs from 1 to $n_{\sigma(1)}$.

We now introduce some notation. Let $a \in \mathbb{Z}$ and $b \in \mathbb{Z}_+$. As in the programming language C, we use $a\%b$ to denote the remainder of a/b, which is always nonnegative, and $[t]$ for the largest integer that is less than or equal to t. For instance,

$$
100\%3 = 1, \qquad 100\%7 = 2, \qquad (-7)\%3 = 2,
$$

$$
\left[\frac{7}{3}\right] = 2, \qquad [-1.25] = -2.
$$

It is easy to see that

$$a = \left[\frac{a}{b}\right] b + a\%b. \tag{2.6}$$

Next, we consider the index-conversion problem. That is, we sometimes need to convert a single index into a multi-index, or vice versa. Particularly, when we need to deform a matrix into a designed form using computer, index conversion is necessary. The following conversion formulas can easily be proven by mathematical induction.

Proposition 2.1 *Let S be a set of data with $n = \prod_{i=1}^{k} n_i$ elements. The data are labeled by single index as $\{x_i\}$ and by k-fold index, by the ordered multi-index $Id(\lambda_1, \ldots, \lambda_k; n_1, \ldots, n_k)$, as*

$$S = \{s_p \mid p = 1, \ldots, n\} = \{s_{\lambda_1, \ldots, \lambda_k} \mid 1 \leq \lambda_i \leq n_i; i = 1, \ldots, k\}.$$

We then have the following conversion formulas:

1. *Single index to multi-index. Defining $p_k := p - 1$, the single index p can be converted into the order of the ordered multi-index $Id(i_1, \ldots, i_k; n_1, \ldots, n_k)$ as $(\lambda_1, \ldots, \lambda_k)$, where λ_i can be calculated recursively as*

$$\begin{cases} \lambda_k = p_k \% n_k + 1, \\ p_j = [\frac{p_{j+1}}{n_{j+1}}], \quad \lambda_j = p_j \% n_j + 1, \quad j = k-1, \ldots, 1. \end{cases} \tag{2.7}$$

2. *Multi-index to single index. From multi-index $(\lambda_1, \ldots, \lambda_k)$ in the order of $Id(i_1, \ldots, i_k; n_1, \ldots, n_k)$ back to the single index, we have*

$$p = \sum_{j=1}^{k-1} (\lambda_j - 1) n_{j+1} n_{j+2} \cdots n_k + \lambda_k. \tag{2.8}$$

The following example illustrates the conversion between different types of indices.

Example 2.4 Recalling the second part of Example 2.3, we may use different types of indices to label the elements.

1. Consider an element which is x_{11} in single-index form. Converting it into the order of $Id(\lambda_1, \lambda_2, \lambda_3; 2, 3, 4)$ by using (2.7), we have

$$p_3 = p - 1 = 10,$$

$$\lambda_3 = p_3 \% n_3 + 1 = 10\%4 + 1 = 2 + 1 = 3,$$

$$p_2 = \left[\frac{p_3}{n_3}\right] = \left[\frac{10}{2}\right] = 2,$$

$$\lambda_2 = p_2 \% n_2 + 1 = 2\%3 + 1 = 2 + 1 = 3,$$

$$p_1 = \left[\frac{p_2}{n_2}\right] = \left[\frac{2}{4}\right] = 0,$$

$$\lambda_1 = p_1 \% n_1 + 1 = 0\%2 + 1 = 1.$$

Hence $x_{11} = x_{133}$.

2. Consider the element x_{214} in the order of $Id(\lambda_1, \lambda_2, \lambda_3; 2, 3, 4)$. Using (2.8), we have

$$p = (\lambda_1 - 1)n_2 n_3 + (\lambda_2 - 1)n_3 + \lambda_3 = 1 \cdot 3 \cdot 4 + 0 + 4 = 16.$$

Hence $x_{214} = x_{16}$.

3. In the order of $Id(\lambda_2, \lambda_3, \lambda_1; 3, 4, 2)$, the data are arranged as

$$
\begin{array}{cccc}
x_{111} & x_{211} & x_{112} & x_{212} \\
x_{113} & x_{213} & x_{114} & x_{214} \\
& \vdots & & \\
x_{131} & x_{231} & x_{132} & x_{232} \\
x_{133} & x_{233} & x_{134} & x_{234}.
\end{array}
$$

For this index, if we want to use the formulas for conversion between natural multi-index and single index, we can construct an auxiliary natural multi-index $y_{\Lambda_1, \Lambda_2, \Lambda_3}$, where $\Lambda_1 = \lambda_2$, $\Lambda_2 = \lambda_3$, $\Lambda_3 = \lambda_1$ and $N_1 = n_2 = 3$, $N_2 = n_3 = 4$, $N_3 = n_1 = 2$. Then, $b_{i,j,k}$ is indexed by $(\Lambda_1, \Lambda_2, \Lambda_3)$ in the order of $Id(\Lambda_1, \Lambda_2, \Lambda_3; N_1, N_2, N_3)$. In this way, we can use (2.7) and (2.8) to convert the indices.

For instance, consider x_{124}. Let $x_{124} = y_{241}$. For y_{241}, using (2.7), we have

$$p = (\Lambda_1 - 1)N_2 N_3 + (\Lambda_2 - 1)N_3 + \Lambda_3$$

$$= (2 - 1) \times 4 \times 2 + (4 - 1) \times 2 + 1 = 8 + 6 + 1 = 15.$$

Hence

$$x_{124} = y_{241} = y_{15} = x_{15}.$$

Consider x_{17} again. Since $x_{17} = y_{17}$, using (2.6), we have

$$p_3 = p - 1 = 16, \qquad \Lambda_3 = p_3 \% N_3 + 1 = 1,$$

$$p_2 = [p_3/N_3] = 8, \qquad \Lambda_2 = p_2 \% N_2 + 1 = 1,$$

$$p_1 = [p_2/N_2] = 2, \qquad \Lambda_1 = p_1 \% N_1 + 1 = 3.$$

Hence $x_{17} = y_{17} = y_{311} = x_{131}$.

From the above argument one sees that a set of higher-dimensional data, labeled by a multi-index, can be converted into a set of 1-dimensional data, labeled by

Table 2.1 The prisoner's dilemma

$P_1 \backslash P_2$	A_1	A_2
A_1	$-1, -1$	$-9, 0$
A_2	$0, -9$	$-6, -6$

single-index. A matrix, as a set of 2-dimensional data, can certainly be converted into a set of 1-dimensional data. Consider a matrix

$$A = \begin{bmatrix} a_{11} & a_{12} & \cdots & a_{1n} \\ \vdots & \vdots & & \vdots \\ a_{m1} & a_{m2} & \cdots & a_{mn} \end{bmatrix}.$$

The row-stacking form of A, denoted by $V_r(A)$, is a row-by-row arranged nm-vector, i.e.,

$$V_r(A) = (a_{11}, a_{12}, \ldots, a_{1n}, \ldots, a_{m1}, a_{m2}, \ldots, a_{mn})^T. \qquad (2.9)$$

The column-stacking form of A, denoted by $V_c(A)$, is the following nm-vector:

$$V_c(A) = (a_{11}, a_{21}, \ldots, a_{m1}, \ldots, a_{1n}, a_{2n}, \ldots, a_{mn})^T. \qquad (2.10)$$

From the definition it is clear that we have the following result.

Proposition 2.2

$$V_c(A) = V_r(A^T), \qquad V_r(A) = V_c(A^T). \qquad (2.11)$$

Finally, we give an example for multidimensional data labeled by an ordered multi-index.

Example 2.5

1. Consider the so-called prisoner's dilemma [5]. Two suspects are arrested and charged with a crime and each prisoner has two possible strategies:

A_1: not confess (or be mum); A_2: confess (or fink).

The payoffs are described by a payoff bi-matrix, given in Table 2.1.
  For instance, if prisoner P_1 chooses "mum" (A_1) and P_2 chooses "fink" (A_2), P_1 will be sentenced to jail for nine months and P_2 will be released. Now, if we denote by

$$r^i_{j,k}, \quad i = 1, 2, j = 1, 2, k = 1, 2,$$

the payoff of P_i as P_1 takes strategy j and P_2 takes strategy k, then $\{r^i_{j,k}\}$ is a set of 3-dimensional data. We may arrange it into a payoff matrix as

$$M_p = \begin{bmatrix} r^1_{11} & r^1_{12} & r^1_{21} & r^1_{22} \\ r^2_{11} & r^2_{12} & r^2_{21} & r^2_{22} \end{bmatrix} = \begin{bmatrix} -1 & -9 & 0 & -6 \\ -1 & 0 & -9 & -6 \end{bmatrix}. \qquad (2.12)$$

2. Consider a game with n players. Player P_i has k_i strategies and the payoff of P_i as P_j takes strategy s_j, $j = 1, \ldots, n$, is

$$r^i_{s_1,\ldots,s_n}, \quad i = 1,\ldots,n; \ s_j = 1,\ldots,k_j, \ j = 1,\ldots,n.$$

Then, $\{r^i_{s_1,\ldots,s_n}\}$ is a set of $(n+1)$-dimensional data. Arranging it with i as the row index and its column by the ordered multi-index $Id(s_1, \ldots, s_n; k_1, \ldots, k_n)$, we have

$$
M_g =
\begin{bmatrix}
r^1_{11\cdots1} & \cdots & r^1_{11\cdots k_n} & \cdots & r^1_{1k_2\cdots k_n} & \cdots & r^1_{k_1 k_2\cdots k_n} \\
\vdots & & & & & & \\
r^n_{11\cdots1} & \cdots & r^n_{11\cdots k_n} & \cdots & r^n_{1k_2\cdots k_n} & \cdots & r^n_{k_1 k_2\cdots k_n}
\end{bmatrix}.
\tag{2.13}
$$

M_g is called the payoff matrix of game g.

2.2 Semi-tensor Product of Matrices

We consider the conventional matrix product first.

Example 2.6 Let U and V be m- and n-dimensional vector spaces, respectively. Assume $F \in L(U \times V, \mathbb{R})$, that is, F is a bilinear mapping from $U \times V$ to $\mathbb{R}$. Denote by $\{u_1, \ldots, u_m\}$ and $\{v_1, \ldots, v_n\}$ the bases of U and V, respectively. We call $S = (s_{ij})$ the structure matrix of F, where

$$s_{ij} = F(u_i, v_j), \quad i = 1,\ldots,m, \ j = 1,\ldots,n.$$

If we let $X = \sum_{i=1}^m x_i u_i \in U$, otherwise written as $X = (x_1, \ldots, x_m)^T \in U$, and $Y = \sum_{i=1}^n y_i v_i \in V$, otherwise written as $Y = (y_1, \ldots, y_n)^T \in V$, then

$$F(X, Y) = X^T S Y. \tag{2.14}$$

Denoting the rows of S by $S^1, \ldots, S^m$, we can alternatively calculate F in two steps.

Step 1: Calculate $x_1 S^1, x_2 S^2, \ldots, x_m S^m$ and take their sum.
Step 2: Multiply $\sum_{i=1}^m x_i S^i$ by Y (which is a standard inner product).

It is easy to check that this algorithm produces the same result. Now, in the first step it seems that we have $(S^1 \cdots S^n) \times X$. This calculation motivates a new algorithm, which is defined as follows.

Definition 2.5 Let T be an np-dimensional row vector and X a p-dimensional column vector. Split T into p equal blocks, named $T^1, \ldots, T^p$, which are $1 \times n$ matrices. Define a left semi-tensor product, denoted by $\ltimes$, as

$$T \ltimes X = \sum_{i=1}^p T^i x_i \in \mathbb{R}^n. \tag{2.15}$$

Using this new product, we reconsider Example 2.6 and propose another algorithm.

Example 2.7 (Example 2.6 continued) We rearrange the structure constants of F into a row as

$$T : V_r(S) = (s_{11}, \ldots, s_{1n}, \ldots, s_{m1}, \ldots, s_{mn}),$$

called the structure matrix of F. This is a row vector of dimension mn, labeled by the ordered multi-index $Id(i, j; m, n)$. The following algorithm provides the same result as (2.14):

$$F(X, Y) = T \ltimes X \ltimes Y. \tag{2.16}$$

It is easy to check the correctness of (2.16), but what is its advantage? Note that (2.16) realized the product of 2-dimensional data (a matrix) with 1-dimensional data by using the product of two sets of 1-dimensional data. If, in this product, 2-dimensional data can be converted into 1-dimensional data, we would expect that the same thing can be done for higher-dimensional data. If this is true, then (2.16) is superior to (2.14) because it allows the product of higher-dimensional data to be taken. Let us see one more example.

Example 2.8 Let U, V, and W be m-, n-, and t-dimensional vector spaces, respectively, and let $F \in L(U \times V \times W, \mathbb{R})$. Assume $\{u_1, \ldots, u_m\}$, $\{v_1, \ldots, v_n\}$, and $\{w_1, \ldots, w_t\}$ are the bases of U, V, and W, respectively. We define the structure constants as

$$s_{ijk} = F(u_i, v_j, w_k), \quad i = 1, \ldots, m, \ j = 1, \ldots, n, \ k = 1, \ldots, t.$$

The structure matrix S of F can be constructed as follows. Its data are labeled by the ordered multi-index $Id(i, j, k; m, n, t)$ to form an mnt-dimensional row vector as

$$S = (s_{111}, \ldots, s_{11t}, \ldots, s_{1n1}, \ldots, s_{1nt}, \ldots, s_{mn1}, \ldots, s_{mnt}).$$

Then, for $X \in U$, $Y \in V$, $Z \in W$, it is easy to verify that

$$F(X, Y, Z) = S \ltimes X \ltimes Y \ltimes Z.$$

Observe that in a semi-tensor product, $\ltimes$ can automatically find the "pointer" of different hierarchies and then perform the required computation.

It is obvious that the structure and algorithm developed in Example 2.8 can be used for any multilinear mapping. Unlike the conventional matrix product, which can generally treat only one- or two-dimensional data, the semi-tensor product of matrices can be used to deal with any finite-dimensional data.

Next, we give a general definition of semi-tensor product.

Definition 2.6

(1) Let $X = (x_1, \ldots, x_s)$ be a row vector, $Y = (y_1, \ldots, y_t)^T$ a column vector.

Case 1: If t is a factor of s, say, $s = t \times n$, then the n-dimensional row vector defined as

$$X \ltimes Y := \sum_{k=1}^{t} X^k y_k \in \mathbb{R}^n \qquad (2.17)$$

is called the left semi-tensor inner product of X and Y, where

$$X = \left(X^1, \ldots, X^t\right), \quad X^i \in \mathbb{R}^n, i = 1, \ldots, t.$$

Case 2: If s is a factor of t, say, $t = s \times n$, then the n-dimensional column vector defined as

$$X \ltimes Y := \sum_{k=1}^{t} x_k Y^k \in \mathbb{R}^n \qquad (2.18)$$

is called the left semi-tensor inner product of X and Y, where

$$Y = \left(\left(Y^1\right)^T, \ldots, \left(Y^t\right)^T\right)^T, \quad Y^i \in \mathbb{R}^n, i = 1, \ldots, t.$$

(2) Let $M \in \mathcal{M}_{m \times n}$ and $N \in \mathcal{M}_{p \times q}$. If n is a factor of p or p is a factor of n, then $C = M \ltimes N$ is called the left semi-tensor product of M and N, where C consists of $m \times q$ blocks as $C = (C^{ij})$, and

$$C^{ij} = M^i \ltimes N_j, \quad i = 1, \ldots, m, \ j = 1, \ldots, q,$$

where $M^i = \mathrm{Row}_i(M)$ and $N_j = \mathrm{Col}_j(N)$.

Remark 2.1

1. In the first item of Definition 2.6, if $t = s$, the left semi-tensor inner product becomes the conventional inner product. Hence, in the second item of Definition 2.6, if $n = p$, the left semi-tensor product becomes the conventional matrix product. Therefore, the left semi-tensor product is a generalization of the conventional matrix product. Equivalently, the conventional matrix product is a special case of the left semi-tensor product.
2. Throughout this book, the default matrix product is the left semi-tensor product, so we simply call it the "semi-tensor product" (or just "product").
3. Let $A \in \mathcal{M}_{m \times n}$ and $B \in \mathcal{M}_{p \times q}$. For convenience, when $n = p$, A and B are said to satisfy the "equal dimension" condition, and when $n = tp$ or $p = tn$, A and B are said to satisfy the "multiple dimension" condition.
4. When $n = tp$, we write $A \succ_t B$; when $p = tn$, we write $A \prec_t B$.
5. So far, the semi-tensor product is a generalization of the matrix product from the equal dimension case to the multiple dimension case.

Example 2.9

1. Let $X = [2 \; {-1} \; 1 \; 2]$, $Y = [-2 \; 1]^T$. Then

$$X \ltimes Y = \begin{bmatrix} 2 & -1 \end{bmatrix} \times (-2) + \begin{bmatrix} 1 & 2 \end{bmatrix} \times 1 = \begin{bmatrix} -3 & 4 \end{bmatrix}.$$

2. Let

$$X = \begin{bmatrix} 2 & 1 & -1 & 3 \\ 0 & 1 & 2 & -1 \\ 2 & -1 & 1 & 1 \end{bmatrix}, \qquad Y = \begin{bmatrix} -1 & 2 \\ 3 & 2 \end{bmatrix}.$$

Then

$$X \ltimes Y = \begin{bmatrix} (21) \times (-1) + (-13) \times 3 & (21) \times 2 + (-13) \times 2 \\ (01) \times (-1) + (2-1) \times 3 & (01) \times 2 + (2-1) \times 2 \\ (2-1) \times (-1) + (11) \times 3 & (2-1) \times 2 + (11) \times 2 \end{bmatrix}$$

$$= \begin{bmatrix} -5 & 8 & 2 & 8 \\ 6 & -4 & 4 & 0 \\ 1 & 4 & 6 & 0 \end{bmatrix}.$$

Remark 2.2

1. The dimension of the semi-tensor product of two matrices can be determined by deleting the largest common factor of the dimensions of the two factor matrices. For instance,

$$A_{p \times qr} \ltimes B_{r \times s} \ltimes C_{qst \times l} = (A \ltimes B)_{p \times qs} \ltimes C_{qst \times l} = (A \ltimes B \ltimes C)_{pt \times l}.$$

   In the first product, r is deleted, and in the second product, qs is deleted. This is a generalization of the conventional matrix product: for the conventional matrix product, $A_{p \times s} B_{s \times q} = (AB)_{p \times q}$, where s is deleted.

2. Unlike the conventional matrix product, for the semi-tensor product even $A \ltimes B$ and $B \ltimes C$ are well defined, but $A \ltimes B \ltimes C = (A \ltimes B) \ltimes C$ may not be well defined. For instance, $A \in \mathcal{M}_{3 \times 4}$, $B \in \mathcal{M}_{2 \times 3}$, $C \in \mathcal{M}_{9 \times 1}$.

In the conventional matrix product the equal dimension condition has certain physical interpretation. For instance, inner product, linear mapping, or differential of compound multiple variable function, etc. Similarly, the multiple dimension condition has its physical interpretation, e.g., the product of different-dimensional data, tensor product, etc.

We give one more example.

Example 2.10 Denote by Δ_k the set of columns of the identity matrix I_k, i.e.,

$$\Delta_k = \text{Col}\{I_k\} = \{\delta_k^i \mid i = 1, 2, \dots, k\}.$$

Define

$$\mathcal{L} = \{B \in \mathcal{M}_{2^m \times 2^n} \mid m \geq 1, n \geq 0, \text{Col}(B) \subset \Delta_{2^m}\}. \tag{2.19}$$

The elements of $\mathscr{L}$ are called logical matrices. It is easy to verify that the semi-tensor product $\ltimes : \mathscr{L} \times \mathscr{L} \to \mathscr{L}$ is always well defined. So, when we are considering matrices in $\mathscr{L}$, we have full freedom to use the semi-tensor product. (The formal definition of a logical matrix is given in the next chapter.)

Comparing the conventional matrix product, the tensor product, and the semi-tensor product of matrices, it is easily seen that there are significant differences between them. For the conventional matrix product, the product is element-to-element, for the tensor product, it is a product of one element to a whole matrix, while for the semi-tensor product, it is one element times a block of the other matrix. This is one reason why we call this new product the "semi-tensor product".

The following example shows that in the conventional matrix product, an illegal term may appear after some legal computations. This introduces some confusion into the otherwise seemingly perfect matrix theory. However, if we extend the conventional matrix product to the semi-tensor product, it becomes consistent again. This may give some support to the necessity of introducing the semi-tensor product.

Example 2.11 Let $X, Y, Z, W \in \mathbb{R}^n$ be column vectors. Since $Y^T Z$ is a scalar, we have

$$(XY^T)(ZW^T) = X(Y^T Z)W^T = (Y^T Z)(XW^T) \in \mathscr{M}_n. \tag{2.20}$$

Again using the associative law, we have

$$(Y^T Z)(XW^T) = Y^T(ZX)W^T. \tag{2.21}$$

A problem now arises: What is ZX? It seems that the conventional matrix product is flawed.

If we consider the conventional matrix product as a particular case of the semi-tensor product, then we have

$$(XY^T)(ZW^T) = Y^T \ltimes (Z \ltimes X) \ltimes W^T. \tag{2.22}$$

It is easy to prove that (2.22) holds. Hence, when the conventional matrix product is extended to the semi-tensor product, the previous inconsistency disappears.

The following two examples show how to use the semi-tensor product to perform multilinear computations.

Example 2.12

1. Let $(V, *)$ be an algebra (refer to Definition 2.2) and $\{e_1, e_2, \ldots, e_n\}$ a basis of V. For any two elements in this basis we calculate the product as

$$e_i * e_j = \sum_{k=1}^{n} c_{ij}^k e_k, \quad i, j, k = 1, 2, \ldots, n. \tag{2.23}$$

We then have the structure constants $\{c_{ij}^k\}$. We arrange the constants into a matrix as follows:

$$M = \begin{bmatrix} c_{11}^1 & c_{12}^1 & \cdots & c_{1n}^1 & \cdots & c_{nn}^1 \\ c_{11}^2 & c_{12}^2 & \cdots & c_{1n}^2 & \cdots & c_{nn}^2 \\ \vdots & & & & & \\ c_{11}^n & c_{12}^n & \cdots & c_{1n}^n & \cdots & c_{nn}^n \end{bmatrix}. \tag{2.24}$$

M is called the structure matrix of the algebra.

Let $X, Y \in V$ be given as

$$X = \sum_{i=1}^n a_i e_i, \qquad Y = \sum_{i=1}^n b_i e_i.$$

If we fix the basis, then X, Y can be expressed in vector form as

$$X = (a_1, a_2, \ldots, a_n)^{\mathrm{T}}, \qquad Y = (b_1, b_2, \ldots, b_n)^{\mathrm{T}}.$$

In vector form, the vector product of X and Y can be simply calculated as

$$X * Y = M \ltimes X \ltimes Y. \tag{2.25}$$

2. Consider the cross product on $\mathbb{R}^3$. Its structure constants were obtained in Example 2.1. We can arrange them into a matrix as

$$M_c = \begin{bmatrix} 0 & 0 & 0 & 0 & 0 & 1 & 0 & -1 & 0 \\ 0 & 0 & -1 & 0 & 0 & 0 & 1 & 0 & 0 \\ 0 & 1 & 0 & -1 & 0 & 0 & 0 & 0 & 0 \end{bmatrix}. \tag{2.26}$$

Now, if

$$X = \frac{1}{\sqrt{3}} \begin{bmatrix} 1 \\ -1 \\ 1 \end{bmatrix}, \qquad Y = \frac{1}{\sqrt{2}} \begin{bmatrix} 1 \\ 0 \\ -1 \end{bmatrix},$$

then we have

$$X \times Y = M_c XY = \begin{bmatrix} 0.4082 \\ 0.8165 \\ 0.4082 \end{bmatrix}.$$

When a multifold cross product is considered, this form becomes very convenient. For instance,

$$X \times \underbrace{Y \times \cdots \times Y}_{100} = M_c^{100} XY^{100} = \begin{bmatrix} 0.5774 \\ -0.5774 \\ 0.5774 \end{bmatrix}.$$

Example 2.13 Let $\phi \in \mathscr{T}_t^s(V)$. That is, ϕ is a tensor on V with covariant order s and contra-variant order t. Suppose that its structure constants are $\{c_{j_1,j_2,\ldots,j_t}^{i_1,i_2,\ldots,i_s}\}$. Arrange it into a matrix by using the ordered multi-index $Id(i_1, i_2, \ldots, i_s; n)$ for columns and the ordered multi-index $Id(j_1, j_2, \ldots, j_t; n)$ for rows. The matrix turns out to be

$$M_\phi = \begin{bmatrix} c_{11\cdots1}^{11\cdots1} & \cdots & c_{11\cdots1}^{11\cdots n} & \cdots & c_{11\cdots1}^{nn\cdots n} \\ c_{11\cdots2}^{11\cdots1} & \cdots & c_{11\cdots2}^{11\cdots n} & \cdots & c_{11\cdots2}^{nn\cdots n} \\ \vdots & & & & \\ c_{nn\cdots n}^{11\cdots1} & \cdots & c_{nn\cdots n}^{11\cdots n} & \cdots & c_{nn\cdots n}^{nn\cdots n} \end{bmatrix}. \tag{2.27}$$

It is the structure matrix of the tensor ϕ. Now, assume $\omega_i \in V^*$, $i = 1, 2, \ldots, t$, and $X_j \in V$, $j = 1, 2, \ldots, s$, where ω_i are expressed as rows, and X_j are expressed as columns. Then

$$\phi(\omega_1, \ldots, \omega_t, X_1, \ldots, X_s) = \omega_t \omega_{t-1} \cdots \omega_1 M_\phi X_1 X_2 \cdots X_s, \tag{2.28}$$

where the product symbol $\ltimes$ is omitted.

Next, we define the power of a matrix. The definition is natural and was used in the previous example.

Definition 2.7 Given a matrix $A \in \mathscr{M}_{p \times q}$ such that $p\%q = 0$ or $q\%p = 0$, we define A^n, $n > 0$, inductively as

$$\begin{cases} A^1 = A, \\ A^{k+1} = A^k \ltimes A, \quad k = 1, 2, \ldots. \end{cases}$$

Remark 2.3 It is easy to verify that the above A^n is well defined. Moreover, if $p = sq$, where $s \in \mathbb{N}$, then the dimension of A^k is $s^k q \times q$; if $q = sp$, then the dimension of A^k is $p \times s^k p$.

Example 2.14

1. If X is a row or a column, then according to Definition 2.7, X^n is always well defined. Particularly, when X, Y are columns, we have

$$X \ltimes Y = X \otimes Y. \tag{2.29}$$

When X, Y are rows,

$$X \ltimes Y = Y \otimes X. \tag{2.30}$$

In both cases,

$$X^k = \underbrace{X \otimes \cdots \otimes X}_{k}. \tag{2.31}$$

2. Let $X \in \mathbb{R}^n$, $Y \in \mathbb{R}^q$ be column vectors and $A \in \mathcal{M}_{m \times n}$, $B \in \mathcal{M}_{p \times q}$. Then,

$$(AX) \ltimes (BY) = (A \otimes B)(X \ltimes Y). \tag{2.32}$$

Particularly,

$$(AX)^k = (\underbrace{A \otimes \cdots \otimes A}_{k})X^k. \tag{2.33}$$

3. Let $X \in \mathbb{R}^m$, $Y \in \mathbb{R}^p$ be row vectors and A, B be matrices (as in 2. above). Then

$$(XA) \ltimes (YB) = (X \ltimes Y)(B \otimes A). \tag{2.34}$$

Hence,

$$(XA)^k = X^k(\underbrace{A \otimes \cdots \otimes A}_{k}). \tag{2.35}$$

4. Consider the set of real kth order homogeneous polynomials of $x \in \mathbb{R}^n$ and denote it by B_n^k. Under conventional addition and real number multiplication, B_n^k is a vector space. It is obvious that x^k contains a basis (x^k itself is not a basis because it contains redundant elements). Hence, every $p(x) \in B_n^k$ can be expressed as $p(x) = Cx^k$, where the coefficients $C \in \mathbb{R}^{n^k}$ are not unique. Note that here $x = (x_1, x_2, \ldots, x_n)^T$ is a column vector.

In the rest of this section we describe some basic properties of the semi-tensor product.

Theorem 2.1 *As long as $\ltimes$ is well defined, i.e., the factor matrices have proper dimensions, then $\ltimes$ satisfies the following laws:*

1. *Distributive law:*

$$\begin{cases} F \ltimes (aG \pm bH) = aF \ltimes G \pm bF \ltimes H, \\ (aF \pm bG) \ltimes H = aF \ltimes H \pm bG \ltimes H, \quad a, b \in \mathbb{R}. \end{cases} \tag{2.36}$$

2. *Associative law:*

$$(F \ltimes G) \ltimes H = F \ltimes (G \ltimes H). \tag{2.37}$$

(We refer to Appendix B for the proof.)
The block multiplication law also holds for the semi-tensor product.

Proposition 2.3 *Assume $A \succ_t B$ (or $A \prec_t B$). Split A and B into blockwise forms as*

$$A = \begin{bmatrix} A^{11} & \cdots & A^{1s} \\ \vdots & & \vdots \\ A^{r1} & \cdots & A^{rs} \end{bmatrix}, \quad B = \begin{bmatrix} B^{11} & \cdots & B^{1t} \\ \vdots & & \vdots \\ B^{s1} & \cdots & B^{st} \end{bmatrix}.$$

If we assume $A^{ik} \succ_t B^{kj}, \forall i, j, k$ (correspondingly, $A^{ik} \prec_t B^{kj}, \forall i, j, k$), then

$$A \ltimes B = \begin{bmatrix} C^{11} & \cdots & C^{1t} \\ \vdots & & \vdots \\ C^{r1} & \cdots & C^{rt} \end{bmatrix}, \tag{2.38}$$

where

$$C^{ij} = \sum_{k=1}^{s} A^{ik} \ltimes B^{kj}.$$

Remark 2.4 We have mentioned that the semi-tensor product of matrices is a generalization of the conventional matrix product. That is, if we assume $A \in \mathcal{M}_{m \times n}$, $B \in \mathcal{M}_{p \times q}$, and $n = p$, then

$$A \ltimes B = AB.$$

Hence, in the following discussion the symbol $\ltimes$ will be omitted, unless we want to emphasize it. Throughout this book, unless otherwise stated, the matrix product will be the semi-tensor product, and the conventional matrix product is its particular case.

As a simple application of the semi-tensor product, we recall an earlier example.

Example 2.15 Recall Example 2.5. To use a matrix expression, we introduce the following notation. Let δ_n^i be the ith column of the identity matrix I_n. Denote by P the variable of players, where $P = \delta_n^i$ means $P = P_i$, i.e., the player under consideration is P_i. Similarly, denote by x_i the strategy chosen by the ith player, where $x_i = \delta_{k_i}^j$ means that the jth strategy of player i is chosen.

1. Consider the prisoner's dilemma. The payoff function can then be expressed as

$$r_p(P, x_1, x_2) = P^T \ltimes M_p \ltimes x_1 \ltimes x_2, \tag{2.39}$$

where M_p is the payoff matrix, as defined in (2.12).
2. Consider the general case. The payoff function is then

$$r_g(P, x_1, x_2, \ldots, x_m) = P^T \ltimes M_g \ltimes_{i=1}^{n} x_i, \tag{2.40}$$

where M_g is defined in (2.13).

2.3 Swap Matrix

One of the major differences between the matrix product and the scalar product is that the scalar (number) product is commutative but the matrix product is not. That is, in general,

$$AB \neq BA.$$

Since the semi-tensor product is a generalization of the conventional matrix product, it would be absurd to expect it to be commutative. Fortunately, with some auxiliary tools, the semi-tensor product has some "commutative" properties, called pseudo-commutative properties. In the sequel, we will see that the pseudo-commutative properties play important roles, such as separating coefficients from the variables, which makes it possible for the calculation of polynomials of multiple variables to be treated in a similar way as the calculation of polynomials of a single variable. The swap matrix is the key tool for pseudo-commutativity of the semi-tensor product.

Definition 2.8 A swap matrix $W_{[m,n]}$ is an $mn \times mn$ matrix, defined as follows. Its rows and columns are labeled by double index (i, j), the columns are arranged by the ordered multi-index $Id(i, j; m, n)$, and the rows are arranged by the ordered multi-index $Id(j, i; n, m)$. The element at position $[(I, J), (i, j)]$ is then

$$w_{(IJ),(ij)} = \delta_{i,j}^{I,J} = \begin{cases} 1, & I = i \text{ and } J = j, \\ 0, & \text{otherwise.} \end{cases} \tag{2.41}$$

Example 2.16

1. Letting $m = 2$, $n = 3$, the swap matrix $W_{[m,n]}$ can be constructed as follows. Using double index (i, j) to label its columns and rows, the columns of W are labeled by $Id(i, j; 2, 3)$, that is, $(11, 12, 13, 21, 22, 23)$, and the rows of W are labeled by $Id(j, i; 3, 2)$, that is, $(11, 21, 12, 22, 13, 23)$. According to (2.41), we have

$$W_{[2,3]} = \begin{array}{c} \\ \\ \\ \\ \\ \end{array} \begin{bmatrix} 1 & 0 & 0 & 0 & 0 & 0 \\ 0 & 0 & 0 & 1 & 0 & 0 \\ 0 & 1 & 0 & 0 & 0 & 0 \\ 0 & 0 & 0 & 0 & 1 & 0 \\ 0 & 0 & 1 & 0 & 0 & 0 \\ 0 & 0 & 0 & 0 & 0 & 1 \end{bmatrix} \begin{array}{l} (11) \\ (21) \\ (12) \\ (22) \\ (13) \\ (23) \end{array}.$$

with column labels $(11)\ (12)\ (13)\ (21)\ (22)\ (23)$

2. Consider $W_{[3,2]}$. Its columns are labeled by $Id(i, j; 3, 2)$, and its rows are labeled by $Id(j, i; 2, 3)$. We then have

$$W_{[3,2]} = \begin{bmatrix} 1 & 0 & 0 & 0 & 0 & 0 \\ 0 & 0 & 1 & 0 & 0 & 0 \\ 0 & 0 & 0 & 0 & 1 & 0 \\ 0 & 1 & 0 & 0 & 0 & 0 \\ 0 & 0 & 0 & 1 & 0 & 0 \\ 0 & 0 & 0 & 0 & 0 & 1 \end{bmatrix} \begin{array}{l} (11) \\ (21) \\ (31) \\ (12) \\ (22) \\ (32) \end{array}.$$

with column labels $(11)\ (12)\ (21)\ (22)\ (31)\ (32)$

The swap matrix is a special orthogonal matrix. A straightforward computation shows the following properties.

Proposition 2.4

1. *The inverse and the transpose of a swap matrix are also swap matrices. That is,*

$$W_{[m,n]}^{\mathrm{T}} = W_{[m,n]}^{-1} = W_{[n,m]}. \tag{2.42}$$

2. *When $m = n$, (2.42) becomes*

$$W_{[n,n]} = W_{[n,n]}^{\mathrm{T}} = W_{[n,n]}^{-1}. \tag{2.43}$$

3.

$$W_{[1,n]} = W_{[n,1]} = I_n. \tag{2.44}$$

Since the case of $m = n$ is particularly important, for compactness, we denote it as

$$W_{[n]} := W_{[n,n]}.$$

From (2.42) it is clear that $W_{[m,n]}$ is an orthogonal matrix. This is because, when used as a linear mapping from $\mathbb{R}^{mn}$ to $\mathbb{R}^{mn}$, it changes only the positions of the elements but not the values.

A swap matrix can be used to convert the matrix stacking forms, as described in the following result.

Proposition 2.5 *Let $A \in \mathcal{M}_{m \times n}$. Then*

$$\begin{cases} W_{[m,n]}V_{\mathrm{r}}(A) = V_{\mathrm{c}}(A), \\ W_{[n,m]}V_{\mathrm{c}}(A) = V_{\mathrm{r}}(A). \end{cases} \tag{2.45}$$

For double-index-labeled data $\{a_{ij}\}$, if it is arranged by $Id(i, j; m, n)$, then the swap matrix $W_{[m,n]}$ can convert its arrangement to the order of $Id(j, i; n, m)$ and vice versa. This is what the "swap" refers to. This property can also be extended to the multiple index-case. We give a rigorous statement for this.

Corollary 2.1 *Let the data $\{a_{ij} \mid 1 \leq i \leq m, 1 \leq j \leq n\}$ be arranged by the ordered multi-index $Id(i, j; m, n)$ as a column X. Then*

$$Y = W_{[m,n]}X$$

is the same data $\{a_{ij}\}$ arranged in the order of $Id(j, i; n, m)$.

Example 2.17

1. Let $X = (x_{11}, x_{12}, x_{13}, x_{21}, x_{22}, x_{23})$. That is, $\{x_{ij}\}$ is arranged by the ordered multi-index $Id(i, j; 2, 3)$. A straightforward computation shows

$$Y = W_{[23]}X = \begin{bmatrix} 1 & 0 & 0 & 0 & 0 & 0 \\ 0 & 0 & 0 & 1 & 0 & 0 \\ 0 & 1 & 0 & 0 & 0 & 0 \\ 0 & 0 & 0 & 0 & 1 & 0 \\ 0 & 0 & 1 & 0 & 0 & 0 \\ 0 & 0 & 0 & 0 & 0 & 1 \end{bmatrix} \begin{bmatrix} x_{11} \\ x_{12} \\ x_{13} \\ x_{21} \\ x_{22} \\ x_{23} \end{bmatrix} = \begin{bmatrix} x_{11} \\ x_{21} \\ x_{12} \\ x_{22} \\ x_{13} \\ x_{23} \end{bmatrix}.$$

That is, Y is the rearrangement of the elements x_{ij} in the order of $Id(j, i; 3, 2)$.

2. Let $X = (x_1, x_2, \ldots, x_m)^T \in \mathbb{R}^m$, $Y = (y_1, y_2, \ldots, y_n)^T \in \mathbb{R}^n$. We then have

$$X \otimes Y = (x_1y_1, x_1y_2, \ldots, x_1y_n, \ldots, x_my_1, x_my_2, \ldots, x_my_n)^T,$$

$$Y \otimes X = (y_1x_1, y_1x_2, \ldots, y_1x_m, \ldots, y_nx_1, y_nx_2, \ldots, y_nx_m)^T$$
$$= (x_1y_1, x_2y_1, \ldots, x_my_1, \ldots, x_1y_n, x_2y_n, \ldots, x_my_n)^T.$$

They both consist of $\{x_iy_j\}$. However, in $X \otimes Y$ the elements are arranged in the order of $Id(i, j; m, n)$, while in $Y \otimes X$ the elements are arranged in the order of $Id(j, i; n, m)$. According to Corollary 2.1 we have

$$Y \otimes X = W_{[m,n]}(X \otimes Y). \tag{2.46}$$

It is easy to check that $XY = X \otimes Y$, so we have

$$YX = W_{[m,n]}XY. \tag{2.47}$$

The following proposition comes from the definition.

Proposition 2.6

1. *Let $X = (x_{ij})$ be a set of data arranged as a column vector by the ordered multi-index $Id(i, j; m, n)$. Then $W_{[m,n]}X$ is a column with the same data, arranged by the ordered multi-index $Id(j, i; n, m)$.*
2. *Let $\omega = (\omega_{ij})$ be a set of data arranged by the ordered multi-index $Id(i, j; m, n)$. Then $\omega W_{[n,m]}$ is a row with the same set of data, arranged by the ordered multi-index $Id(j, i; n, m)$.*

A swap matrix can be used for multiple-index-labeled data and can swap two special indices. This allows a very useful generalization of the previous proposition, which we now state as a theorem.

Theorem 2.2

1. Let $X = (x_{i_1,\ldots,i_k})$ be a column vector with its elements arranged by the ordered multi-index $Id(i_1,\ldots,i_k; n_1,\ldots,n_k)$. Then

$$[I_{n_1+\cdots+n_{t-1}} \otimes W_{[n_t,n_{t+1}]} \otimes I_{n_{t+2}+\cdots+n_k}]X$$

   is a column vector consisting of the same elements, arranged by the ordered multi-index $Id(i_1,\ldots,i_{t+1},i_t,\ldots,i_k; n_1,\ldots,n_{t+1},n_t,\ldots,n_k)$.

2. Let $\omega = (\omega_{i_1,\ldots,i_k})$ be a row vector with its elements arranged by the ordered multi-index $Id(i_1,\ldots,i_k; n_1,\ldots,n_k)$. Then

$$\omega[I_{n_1+\cdots+n_{t-1}} \otimes W_{[n_{t+1},n_t]} \otimes I_{n_{t+2}+\cdots+n_k}]$$

   is a row vector consisting of the same elements, arranged by the ordered multi-index $Id(i_1,\ldots,i_{t+1},i_t,\ldots,i_k; n_1,\ldots,n_{t+1},n_t,\ldots,n_k)$.

$W_{[m,n]}$ can be constructed in an alternative way which is convenient in some applications. Denoting by δ_n^i the ith column of the identity matrix I_n, we have the following.

Proposition 2.7

$$W_{[m,n]} = \begin{bmatrix} \delta_n^1 \ltimes \delta_m^1 & \cdots & \delta_n^n \ltimes \delta_m^1 & \cdots & \delta_n^1 \ltimes \delta_m^m & \cdots & \delta_n^n \ltimes \delta_m^m \end{bmatrix}. \quad (2.48)$$

For convenience, we provide two more forms of swap matrix:

$$W_{[m,n]} = \begin{bmatrix} I_m \otimes \delta_n^{1\mathrm{T}} \\ \vdots \\ I_m \otimes \delta_n^{n\mathrm{T}} \end{bmatrix} \quad (2.49)$$

and, similarly,

$$W_{[m,n]} = \begin{bmatrix} I_n \otimes \delta_m^1, \ldots, I_n \otimes \delta_m^m \end{bmatrix}. \quad (2.50)$$

The following factorization properties reflect the blockwise permutation property of the swap matrix.

Proposition 2.8 *The swap matrix has the following factorization properties*:

$$W_{[p,qr]} = (I_q \otimes W_{[p,r]})(W_{[p,q]} \otimes I_r) = (I_r \otimes W_{[p,q]})(W_{[p,r]} \otimes I_q), \quad (2.51)$$

$$W_{[pq,r]} = (W_{[p,r]} \otimes I_q)(I_p \otimes W_{[q,r]}) = (W_{[q,r]} \otimes I_p)(I_q \otimes W_{[p,r]}). \quad (2.52)$$

2.4 Properties of the Semi-tensor Product

In this section some fundamental properties of the semi-tensor product of matrices are introduced. Throughout, it is easily seen that when the conventional matrix product is extended to the semi-tensor product, almost all its properties continue to hold. This is a significant advantage of the semi-tensor product.

Proposition 2.9 *Assuming that A and B have proper dimensions such that $\ltimes$ is well defined, then*

$$(A \ltimes B)^{\mathrm{T}} = B^{\mathrm{T}} \ltimes A^{\mathrm{T}}. \tag{2.53}$$

The following property shows that the semi-tensor product can be expressed by the conventional matrix product plus the Kronecker product.

Proposition 2.10

1. *If $A \in \mathcal{M}_{m \times np}$, $B \in \mathcal{M}_{p \times q}$, then*

$$A \ltimes B = A(B \otimes I_n). \tag{2.54}$$

2. *If $A \in \mathcal{M}_{m \times n}$, $B \in \mathcal{M}_{np \times q}$, then*

$$A \ltimes B = (A \otimes I_p)B. \tag{2.55}$$

(We refer to Appendix B for the proof.)

Proposition 2.10 is a fundamental result. Many properties of the semi-tensor product can be obtained through it. We may consider equations (2.54) and (2.55) as providing an alternative definition of the semi-tensor product. In fact, the name "semi-tensor product" comes from this proposition. Recall that for $A \in \mathcal{M}_{m \times n}$ and $B \in \mathcal{M}_{p \times q}$, their tensor product satisfies

$$A \otimes B = (A \otimes I_p)(I_n \otimes B). \tag{2.56}$$

Intuitively, it seems that the semi-tensor product takes the "left half" of the product in the right-hand side of (2.56) to form the product.

The following property may be considered as a direct corollary of Proposition 2.10.

Proposition 2.11 *Let A and B be matrices with proper dimensions such that $A \ltimes B$ is well defined. Then:*

1. *$A \ltimes B$ and $B \ltimes A$ have the same characteristic functions.*
2. *$\mathrm{tr}(A \ltimes B) = \mathrm{tr}(B \ltimes A)$.*
3. *If A and B are invertible, then $A \ltimes B \sim B \ltimes A$, where "$\sim$" stands for matrix similarity.*
4. *If both A and B are upper triangular (resp., lower triangular, diagonal, orthogonal) matrices, then $A \ltimes B$ is also an upper triangular (resp., lower triangular, diagonal, orthogonal) matrix.*
5. *If both A and B are invertible, then $A \ltimes B$ is also invertible. Moreover,*

$$(A \ltimes B)^{-1} = B^{-1} \ltimes A^{-1}. \tag{2.57}$$

6. *If $A \prec_t B$, then*

$$\det(A \ltimes B) = [\det(A)]^t \det(B). \qquad (2.58)$$

If $A \succ_t B$, then

$$\det(A \ltimes B) = \det(A)[\det(B)]^t. \qquad (2.59)$$

The following proposition shows that the swap matrix can also perform the swap of blocks in a matrix.

Proposition 2.12

1. *Assume*

$$A = (A_{11}, \ldots, A_{1n}, \ldots, A_{m1}, \ldots, A_{mn}),$$

   where each block has the same dimension and the blocks are labeled by double index $\{i, j\}$ and arranged by the ordered multi-index $Id(i, j; m, n)$. Then

$$A W_{[n,m]} = (A_{11}, \ldots, A_{m1}, \ldots, A_{1n}, \ldots, A_{mn})$$

   consists of the same set of blocks, which are arranged by the ordered multi-index $Id(j, i; n, m)$.
2. *Let*

$$B = \left(B_{11}^{\mathrm{T}}, \ldots, B_{1n}^{\mathrm{T}}, \ldots, B_{m1}^{\mathrm{T}}, \ldots, B_{mn}^{\mathrm{T}} \right)^{\mathrm{T}},$$

   where each block has the same dimension and the blocks are labeled by double index $\{i, j\}$ and arranged by the ordered multi-index $Id(i, j; m, n)$. Then

$$W_{[m,n]} B = \left(B_{11}^{\mathrm{T}}, \ldots, B_{m1}^{\mathrm{T}}, \ldots, B_{1n}^{\mathrm{T}}, \ldots, M_{mn}^{\mathrm{T}} \right)^{\mathrm{T}}$$

   consists of the same set of blocks, which are arranged by the ordered multi-index $Id(j, i; n, m)$.

The product of a matrix with an identity matrix I has some special properties.

Proposition 2.13

1. *Let $M \in \mathcal{M}_{m \times pn}$. Then*

$$M \ltimes I_n = M. \qquad (2.60)$$

2. *Let $M \in \mathcal{M}_{m \times n}$. Then*

$$M \ltimes I_{pn} = M \otimes I_p. \qquad (2.61)$$

3. *Let $M \in \mathcal{M}_{pm \times n}$. Then*

$$I_p \ltimes M = M. \qquad (2.62)$$

4. *Let $M \in \mathcal{M}_{m \times n}$. Then*

$$I_{pm} \ltimes M = M \otimes I_p.$$
(2.63)

In the following, some linear mappings of matrices are expressed in their stacking form via the semi-tensor product.

Proposition 2.14 *Let $A \in \mathcal{M}_{m \times n}$, $X \in \mathcal{M}_{n \times q}$, $Y \in \mathcal{M}_{p \times m}$. Then*

$$V_r(AX) = A \ltimes V_r(X),$$
(2.64)

$$V_c(YA) = A^T \ltimes V_c(Y).$$
(2.65)

Note that (2.64) is similar to a linear mapping over a linear space (e.g., $\mathbb{R}^n$). In fact, as X is a vector, (2.64) becomes a standard linear mapping.

Using (2.64) and (2.65), the stacking expression of a matrix polynomial may also be obtained.

Corollary 2.2 *Let X be a square matrix and $p(x)$ be a polynomial, expressible as $p(x) = q(x)x + p_0$. Then*

$$V_r\big(p(X)\big) = q(X)V_r(X) + p_0 V_r(I).$$
(2.66)

Using linear mappings on matrices, some other useful formulas may be obtained [4].

Proposition 2.15 *Let $A \in \mathcal{M}_{m \times n}$ and $B \in \mathcal{M}_{p \times q}$. Then*

$$(I_p \otimes A)W_{[n,p]} = W_{[m,p]}(A \otimes I_p),$$
(2.67)

$$W_{[m,p]}(A \otimes B)W_{[q,n]} = (B \otimes A).$$
(2.68)

In fact, (2.67) can be obtained from (2.68).

Proposition 2.16 *Let $X \in \mathcal{M}_{m \times n}$ and $A \in \mathcal{M}_{n \times s}$. Then*

$$XA = \big(I_m \otimes V_r^T(I_s)\big)W_{[s,m]}A^T V_c(X).$$
(2.69)

Roughly speaking, a swap matrix can swap a matrix with a vector. This is sometimes useful.

Proposition 2.17

1. *Let Z be a t-dimensional row vector and $A \in \mathcal{M}_{m \times n}$. Then*

$$ZW_{[m,t]}A = AZW_{[n,t]} = A \otimes Z.$$
(2.70)

2. *Let Y be a t-dimensional column vector and $A \in \mathcal{M}_{m \times n}$. Then*

$$AW_{[t,n]}Y = W_{[t,m]}YA = A \otimes Y.$$
(2.71)

The following lemma is useful for simplifying some expressions.

Lemma 2.1 *Let $A \in \mathcal{M}_{m \times n}$. Then*

$$W_{[m,q]} \ltimes A \ltimes W_{[q,n]} = I_q \otimes A. \tag{2.72}$$

The semi-tensor product has some pseudo-commutative properties. The following are some useful pseudo-commutative properties. Their usefulness will become apparent later.

Proposition 2.18 *Suppose we are given a matrix $A \in \mathcal{M}_{m \times n}$.*

1. *Let $Z \in \mathbb{R}^t$ be a column vector. Then*

$$A Z^{\mathrm{T}} = Z^{\mathrm{T}} W_{[m,t]} A W_{[t,n]} = Z^{\mathrm{T}} (I_t \otimes A). \tag{2.73}$$

2. *Let $Z \in \mathbb{R}^t$ be a column vector. Then*

$$ZA = W_{[m,t]} A W_{[t,n]} Z = (I_t \otimes A) Z. \tag{2.74}$$

3. *Let $X \in \mathbb{R}^m$ be a row vector. Then*

$$X^{\mathrm{T}} A = \left[V_{\mathrm{r}}(A) \right]^{\mathrm{T}} X. \tag{2.75}$$

4. *Let $Y \in \mathbb{R}^n$ be a row vector. Then*

$$AY = Y^{\mathrm{T}} V_{\mathrm{c}}(A). \tag{2.76}$$

5. *Let $X \in \mathbb{R}^m$ be a column vector and $Y \in \mathbb{R}^n$ a row vector. Then*

$$XY = Y W_{[m,n]} X. \tag{2.77}$$

Proposition 2.19 *Let $A \in \mathcal{M}_{m \times n}$ and $B \in \mathcal{M}_{s \times t}$. Then*

$$A \otimes B = W_{[s,m]} \ltimes B \ltimes W_{[m,t]} \ltimes A = (I_m \otimes B) \ltimes A. \tag{2.78}$$

Example 2.18 Assume

$$A = \begin{bmatrix} a_{11} & a_{12} \\ a_{21} & a_{22} \end{bmatrix}, \qquad B = \begin{bmatrix} b_{11} & b_{12} \\ b_{21} & b_{22} \\ b_{31} & b_{32} \end{bmatrix},$$

where $m = n = 2$, $s = 3$ and $t = 2$. Then

$$W_{[3,2]} = \begin{bmatrix} 1 & 0 & 0 & 0 & 0 & 0 \\ 0 & 0 & 1 & 0 & 0 & 0 \\ 0 & 0 & 0 & 0 & 1 & 0 \\ 0 & 1 & 0 & 0 & 0 & 0 \\ 0 & 0 & 0 & 1 & 0 & 0 \\ 0 & 0 & 0 & 0 & 0 & 1 \end{bmatrix},$$

$$W_{[2,2]} = \begin{bmatrix} 1 & 0 & 0 & 0 \\ 0 & 0 & 1 & 0 \\ 0 & 1 & 0 & 0 \\ 0 & 0 & 0 & 1 \end{bmatrix},$$

$$W_{[3,2]} \ltimes B \ltimes W_{[2,2]} \ltimes A = \begin{bmatrix} b_{11} & b_{12} & 0 & 0 \\ b_{21} & b_{22} & 0 & 0 \\ b_{31} & b_{32} & 0 & 0 \\ 0 & 0 & b_{11} & b_{12} \\ 0 & 0 & b_{21} & b_{22} \\ 0 & 0 & b_{31} & b_{32} \end{bmatrix} \ltimes \begin{bmatrix} a_{11} & a_{12} \\ a_{21} & a_{22} \end{bmatrix}$$

$$= \begin{bmatrix} a_{11}b_{11} & a_{11}b_{12} & a_{12}b_{11} & a_{12}b_{12} \\ a_{11}b_{21} & a_{11}b_{22} & a_{12}b_{21} & a_{12}b_{22} \\ a_{11}b_{31} & a_{11}b_{32} & a_{12}b_{31} & a_{12}b_{32} \\ a_{21}b_{11} & a_{21}b_{12} & a_{22}b_{11} & a_{22}b_{12} \\ a_{21}b_{21} & a_{21}b_{22} & a_{22}b_{21} & a_{22}b_{22} \\ a_{21}b_{31} & a_{21}b_{32} & a_{22}b_{31} & a_{22}b_{32} \end{bmatrix} = A \otimes B.$$

As a corollary of the previous proposition, we have the following.

Corollary 2.3 *Let $C \in \mathcal{M}_{s \times t}$. Then, for any integer $m > 0$, we have*

$$W_{[s,m]} \ltimes C \ltimes W_{[m,t]} = I_m \otimes C. \tag{2.79}$$

Finally, we consider how to express a matrix in stacking form and vice versa, via the semi-tensor product.

Proposition 2.20 *Let $A \in \mathcal{M}_{m \times n}$. Then*

$$V_r(A) = A \ltimes V_r(I_n), \tag{2.80}$$

$$V_c(A) = W_{[m,n]} \ltimes A \ltimes V_c(I_n). \tag{2.81}$$

Conversely, we can retrieve A from its row- or column-stacking form.

Proposition 2.21 *Let $A \in \mathcal{M}_{m \times n}$. Then*

$$A = \left[I_m \otimes V_r^{\mathrm{T}}(I_n) \right] \ltimes V_r(A) = \left[I_m \otimes V_r^{\mathrm{T}}(I_n) \right] \ltimes W_{[n,m]} \ltimes V_c(A). \tag{2.82}$$

As an elementary application of semi-tensor product, we consider the following example.

Example 2.19 In mechanics, it is well known that the angular momentum of a rigid body about its mass center is

$$H = \int r \times (\omega \times r) \, \mathrm{d}m, \tag{2.83}$$

Fig. 2.1 Rotation

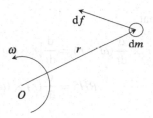

where $r = (x, y, z)$ is the position vector, starting from the mass center, and $\omega = (\omega_x, \omega_y, \omega_z)^{\mathrm{T}}$ is the angular speed. We want to prove the following equation for angular momentum (2.84), which often appears in the literature:

$$
\begin{bmatrix} H_x \\ H_y \\ H_z \end{bmatrix} = \begin{bmatrix} I_x & -I_{xy} & -I_{zx} \\ -I_{xy} & I_y & -I_{yz} \\ I_{zx} & -I_{yz} & I_z \end{bmatrix} \begin{bmatrix} \omega_x \\ \omega_y \\ \omega_z \end{bmatrix},
\tag{2.84}
$$

where

$$
I_x = \int \left(y^2 + z^2 \right) dm, \qquad I_y = \int \left(z^2 + x^2 \right) dm, \qquad I_z = \int \left(x^2 + y^2 \right) dm,
$$

$$
I_{xy} = \int xy \, dm, \qquad I_{yz} = \int yz \, dm, \qquad I_{zx} = \int zx \, dm.
$$

Let M be the moment of the force acting on the rigid body. We first prove that the dynamic equation of a rotating solid body is

$$
\frac{dH}{dt} = M.
\tag{2.85}
$$

Consider a mass dm, with O as its rotating center, r as the position vector (from O to dm) and df the force acting on it (see Fig. 2.1). From Newton's second law,

$$
df = a \, dm = \frac{dv}{dt} \, dm
$$

$$
= \frac{d}{dt} (\omega \times r) \, dm.
$$

Now, consider the moment of force on it, which is

$$
dM = r \times df = r \times \frac{d}{dt} (\omega \times r) \, dm.
$$

Integrating this over the solid body, we have

$$
M = \int r \times \frac{d}{dt} (\omega \times r) \, dm.
\tag{2.86}
$$

We claim that

$$r \times \frac{d}{dt}(\omega \times r) = \frac{d}{dt}[r \times (\omega \times r)],$$

$$RHS = \frac{d}{dt}(r) \times (\omega \times r) + r \times \frac{d}{dt}(\omega \times r)$$

$$= (\omega \times r) \times (\omega \times r) + r \times \frac{d}{dt}(\omega \times r) \qquad (2.87)$$

$$= 0 + r \times \frac{d}{dt}(\omega \times r) = LHS.$$

Applying this to (2.86), we have

$$M = \int \frac{d}{dt}[r \times (\omega \times r)]\,dm$$

$$= \frac{d}{dt}\int r \times (\omega \times r)\,dm$$

$$= \frac{d}{dt}H.$$

Next, we prove the angular momentum equation (2.84). Recall that the structure matrix of the cross product (2.26) is

$$M_c = \begin{bmatrix} 0 & 0 & 0 & 0 & 0 & 1 & 0 & -1 & 0 \\ 0 & 0 & -1 & 0 & 0 & 0 & 1 & 0 & 0 \\ 0 & 1 & 0 & -1 & 0 & 0 & 0 & 0 & 0 \end{bmatrix},$$

and for any two vectors $X, Y \in \mathbb{R}^3$, their cross product is

$$X \times Y = M_c XY. \qquad (2.88)$$

Using this, we have

$$H = \int r \times (\omega \times r)\,dm$$

$$= \int M_c r M_c \omega r\,dm$$

$$= \int M_c (I_3 \otimes M_c) r\omega r\,dm$$

$$= \int M_c (I_3 \otimes M_c) W_{[3,9]} r^2 \omega\,dm$$

$$= \int M_c (I_3 \otimes M_c) W_{[3,9]} r^2\,dm\omega$$

$$:= \int \Psi r^2\,dm\omega,$$

where

$$\Psi = M_c(I_3 \otimes M_c)W_{[3,9]}$$

$$= \begin{bmatrix} 0 & 0 & 0 & 0 & -1 & 0 & 0 & 0 & -1 & 0 & 0 & 0 & 1 & 0 & 0 & 0 & 0 & 0 & 0 & 0 & 0 & 0 & 0 & 0 & 1 & 0 & 0 \\ 0 & 1 & 0 & 0 & 0 & 0 & 0 & 0 & 0 & -1 & 0 & 0 & 0 & 0 & 0 & 0 & 0 & -1 & 0 & 0 & 0 & 0 & 0 & 0 & 0 & 1 & 0 \\ 0 & 0 & 1 & 0 & 0 & 0 & 0 & 0 & 0 & 0 & 0 & 0 & 0 & 0 & 1 & 0 & 0 & 0 & -1 & 0 & 0 & 0 & -1 & 0 & 0 & 0 & 0 \end{bmatrix}.$$

We then have

$$\Psi r^2 = \begin{bmatrix} y^2 + z^2 & xy & -xz \\ -xy & x^2 + z^2 & -yz \\ -xz & -yz & x^2 + y^2 \end{bmatrix}.$$

Equation (2.84) follows immediately.

2.5 General Semi-tensor Product

In previous sections of this chapter the semi-tensor product considered was the left semi-tensor product of matrices. Throughout this book the default semi-tensor product is the left semi-tensor product, unless otherwise stated. In this section we will discuss some other kinds of semi-tensor products.

According to Proposition 2.10, an alternative definition of the left semi-tensor product is

$$A \ltimes B = \begin{cases} (A \otimes I_t)B, & A \prec_t B, \\ A(B \otimes I_t), & A \succ_t B. \end{cases} \tag{2.89}$$

This proceeds as follows. For a smaller-dimensional factor matrix, we match it on the right with an identity matrix of proper dimension such that the conventional matrix product is possible. The following definition then becomes natural.

Definition 2.9 Suppose we are given matrices A and B. Assuming $A \prec_t B$ or $A \succ_t B$, we define the right semi-tensor product of A and B as

$$A \rtimes B = \begin{cases} (I_t \otimes A)B, & A \prec_t B, \\ A(I_t \otimes B), & A \succ_t B. \end{cases} \tag{2.90}$$

Most properties of the left semi-tensor product hold for the right semi-tensor product. In Proposition 2.22 we assume the matrices have proper dimensions such the product $\rtimes$ is defined. In addition, for items 5–10 A and B are assumed to be two square matrices.

Proposition 2.22

1. *Associative law:*

$$(mA \rtimes B) \rtimes C = A \rtimes (B \rtimes C). \tag{2.91}$$

Distributive law:

$$(A + B) \ltimes C = A \ltimes C + B \ltimes C, \qquad C \ltimes (A + B) = C \ltimes A + C \ltimes B. \quad (2.92)$$

2. *Let X and Y be column vectors. Then,*

$$X \ltimes Y = Y \otimes X. \qquad (2.93)$$

Let X and Y be row vectors. Then,

$$X \ltimes Y = X \otimes Y. \qquad (2.94)$$

3.

$$(A \ltimes B)^{\mathrm{T}} = B^{\mathrm{T}} \ltimes A^{\mathrm{T}}. \qquad (2.95)$$

4. *Let $M \in \mathcal{M}_{m \times pn}$. Then,*

$$M \ltimes I_n = M. \qquad (2.96)$$

Let $M \in \mathcal{M}_{m \times n}$. Then,

$$M \ltimes I_{pn} = I_p \otimes M. \qquad (2.97)$$

Let $M \in \mathcal{M}_{pm \times n}$. Then,

$$I_p \ltimes M = M. \qquad (2.98)$$

Let $M \in \mathcal{M}_{m \times n}$. Then,

$$I_{pm} \ltimes M = I_p \otimes M. \qquad (2.99)$$

5. *$A \ltimes B$ and $B \ltimes A$ have the same characteristic function.*
6.

$$\mathrm{tr}(A \ltimes B) = \mathrm{tr}(B \ltimes A). \qquad (2.100)$$

7. *If A and B are orthogonal (upper triangular, lower triangular) matrices, then so is $A \ltimes B$.*
8. *If A and B are invertible, then $A \ltimes B \sim B \ltimes A$.*
9. *If A and B are invertible, then*

$$(A \ltimes B)^{-1} = B^{-1} \ltimes A^{-1}. \qquad (2.101)$$

10. *If $A \prec_t B$, then*

$$\det(A \ltimes B) = \left[\det(A)\right]^t \det(B). \qquad (2.102)$$

If $A \succ_t B$, then

$$\det(A \ltimes B) = \det(A)\left[\det(B)\right]^t. \qquad (2.103)$$

A question which naturally arises is whether we can define the right semi-tensor product in a similar way as in Definition 2.6, i.e., in a "row-times-column" way. The answer is that we cannot. In fact, a basic difference between the right and the left semi-tensor products is that the right semi-tensor product does not satisfy the block product law. The row-times-column rule is ensured by the block product law. This difference makes the left semi-tensor product more useful. However, it is sometimes convenient to use the right semi-tensor product.

We now consider some relationships between left and right semi-tensor products.

Proposition 2.23 *Let X be a row vector of dimension np, Y a column vector of dimension p. Then,*

$$X \rtimes Y = X W_{[p,n]} \ltimes Y. \tag{2.104}$$

Conversely, we also have

$$X \ltimes Y = X W_{[n,p]} \rtimes Y. \tag{2.105}$$

If $\dim(X) = p$ and $\dim(Y) = pn$, then

$$X \rtimes Y = X \ltimes W_{[n,p]} Y. \tag{2.106}$$

Conversely, we also have

$$X \ltimes Y = X \rtimes W_{[p,n]} Y. \tag{2.107}$$

In the following, we introduce the left and right semi-tensor products of matrices of arbitrary dimensions. This will not be discussed beyond this section since we have not found any meaningful use for semi-tensor products of arbitrary dimensions.

Definition 2.10 Let $A \in \mathcal{M}_{m \times n}$, $B \in \mathcal{M}_{p \times q}$, and $\alpha = \mathrm{lcm}(n, p)$ be the least common multiple of n and p. The left semi-tensor product of A and B is defined as

$$A \ltimes B = (A \otimes I_{\frac{\alpha}{n}})(B \otimes I_{\frac{\alpha}{p}}). \tag{2.108}$$

The right semi-tensor product of A and B is defined as

$$A \rtimes B = (I_{\frac{\alpha}{n}} \otimes A)(I_{\frac{\alpha}{p}} \otimes B). \tag{2.109}$$

Note that if $n = p$, then both the left and right semi-tensor products of arbitrary matrices become the conventional matrix product. When the dimensions of the two factor matrices satisfy the multiple dimension condition, they become the multiple dimension semi-tensor products, as defined earlier.

Proposition 2.24 *The semi-tensor products of arbitrary matrices satisfy the following laws:*

1. *Distributive law*:

$$(A + B) \ltimes C = (A \ltimes C) + (B \ltimes C), \tag{2.110}$$

$$(A + B) \rtimes C = (A \rtimes C) + (B \rtimes C), \tag{2.111}$$

$$C \ltimes (A + B) = (C \ltimes A) + (C \ltimes B), \tag{2.112}$$

$$C \rtimes (A + B) = (C \rtimes A) + (C \rtimes B). \tag{2.113}$$

2. *Associative law*:

$$(A \ltimes B) \ltimes C = A \ltimes (B \ltimes C), \tag{2.114}$$

$$(A \rtimes B) \rtimes C = A \rtimes (B \rtimes C). \tag{2.115}$$

Almost all of the properties of the conventional matrix product hold for the left or right semi-tensor product of arbitrary matrices. For instance, we have the following.

Proposition 2.25

1.

$$\begin{cases} (A \ltimes B)^{\mathrm{T}} = B^{\mathrm{T}} \ltimes A^{\mathrm{T}}, \\ (A \rtimes B)^{\mathrm{T}} = B^{\mathrm{T}} \rtimes A^{\mathrm{T}}. \end{cases} \tag{2.116}$$

2. *If $M \in \mathcal{M}_{m \times pn}$, then*

$$\begin{cases} M \ltimes I_n = M, \\ M \rtimes I_n = M. \end{cases} \tag{2.117}$$

If $M \in \mathcal{M}_{pm \times n}$, then

$$\begin{cases} I_m \ltimes M = M, \\ I_m \rtimes M = M. \end{cases} \tag{2.118}$$

In the following, A and B are square matrices.

3. *$A \ltimes B$ and $B \ltimes A$ ($A \rtimes B$ and $B \rtimes A$) have the same characteristic function.*

4.

$$\begin{cases} \mathrm{tr}(A \ltimes B) = \mathrm{tr}(B \ltimes A), \\ \mathrm{tr}(A \rtimes B) = \mathrm{tr}(B \rtimes A). \end{cases} \tag{2.119}$$

5. *If both A and B are orthogonal (resp., upper triangular, lower triangular, diagonal) matrices, then $A \ltimes B$ ($A \rtimes B$) is orthogonal (resp., upper triangular, lower triangular, diagonal).*

6. *If both A and B are invertible, then $A \ltimes B \sim B \ltimes A$ ($A \rtimes B \sim B \rtimes A$).*

7. *If both A and B are invertible, then*

$$\begin{cases} (A \ltimes B)^{-1} = B^{-1} \ltimes A^{-1}, \\ (A \rtimes B)^{-1} = B^{-1} \rtimes A^{-1}. \end{cases} \tag{2.120}$$

8. *The determinant of the product satisfies*

$$\begin{cases} \det(A \ltimes B) = [\det(A)]^{\frac{\alpha}{n}} [\det(B)]^{\frac{\alpha}{p}}, \\ \det(A \rtimes B) = [\det(A)]^{\frac{\alpha}{n}} [\det(B)]^{\frac{\alpha}{p}}. \end{cases} \tag{2.121}$$

Corollary 2.4 *Let* $A \in \mathcal{M}_{m \times n}$, $B \in \mathcal{M}_{p \times q}$. *Then*

$$C = A \ltimes B = \left(C^{ij}\right), \quad i = 1, \dots, m, j = 1, \dots, q, \tag{2.122}$$

where

$$C^{ij} = A^i \ltimes B_j,$$

$A^i = \mathrm{Row}_i(A)$, *and* $B_j = \mathrm{Col}_j(B)$.

References

1. Bates, D., Watts, D.: Relative curvature measures of nonlinearity. J. R. Stat. Soc. Ser. B (Methodol.) **42**, 1–25 (1980)
2. Bates, D., Watts, D.: Parameter transformations for improved approximate confidence regions in nonlinear least squares. Ann. Stat. **9**, 1152–1167 (1981)
3. Cheng, D.: Sime-tensor product of matrices and its applications: a survey. In: Proc. 4th International Congress of Chinese Mathematicians, pp. 641–668. Higher Edu. Press, Int. Press, Hangzhou (2007)
4. Cheng, D., Qi, H.: Semi-tensor Product of Matrices—Theory and Applications. Science Press, Beijing (2007) (in Chinese)
5. Gibbons, R.: A Primer in Game Theory. Prentice Hall, New York (1992)
6. Lang, S.: Algebra, 3rd edn. Springer, New York (2002)
7. Mei, S., Liu, F., Xue, A.: Semi-tensor Product Approach to Transient Analysis of Power Systems. Tsinghua Univ. Press, Beijing (2010) (in Chinese)
8. Tsai, C.: Contributions to the design and analysis of nonlinear models. Ph.D. thesis, Univ. of Minnesota (1983)
9. Wang, X.: Parameter Estimate of Nonlinear Models—Theory and Applications. Wuhan Univ. Press, Wuhan (2002) (in Chinese)
10. Wei, B.: Second moments of LS estimate of nonlinear regressive model. J. Univ. Appl. Math. **1**, 279–285 (1986) (in Chinese)

Chapter 3
Matrix Expression of Logic

3.1 Structure Matrix of a Logical Operator

Recall that a logical variable takes value from $\mathscr{D} = \{T, F\}$ or, equivalently, $\mathscr{D} = \{1, 0\}$. To obtain a matrix expression we identify "T" and "F", respectively, with the vectors

$$T := 1 \sim \begin{bmatrix} 1 \\ 0 \end{bmatrix}, \qquad F := 0 \sim \begin{bmatrix} 0 \\ 1 \end{bmatrix}. \tag{3.1}$$

To describe the vector form of logic we first recall some notation:

- δ_k^i is the ith column of the identity matrix I_k,
- $\Delta_k := \{\delta_k^i \mid i = 1, 2, \ldots, k\}$.

For notational ease, let $\Delta := \Delta_2$. Then,

$$\Delta = \{\delta_2^1, \delta_2^2\} = \left\{ \begin{bmatrix} 1 \\ 0 \end{bmatrix}, \begin{bmatrix} 0 \\ 1 \end{bmatrix} \right\},$$

and an r-ary logical operator is a mapping $\sigma : \Delta^r \to \Delta$.

Definition 3.1 A matrix $L \in \mathscr{M}_{n \times m}$ is called a logical matrix if $\mathrm{Col}(L) \subset \Delta_n$. The set of $n \times m$ logical matrices is denoted by $\mathscr{L}_{n \times m}$.

If $L \in \mathscr{L}_{n \times m}$, then it has the form

$$L = \begin{bmatrix} \delta_n^{i_1} & \delta_n^{i_2} & \cdots & \delta_n^{i_m} \end{bmatrix}.$$

For notational compactness we write this as

$$L = \delta_n[i_1 \, i_2 \, \cdots \, i_m].$$

Definition 3.2 A 2×2^r matrix M_σ is said to be the structure matrix of the r-ary logical operator σ if

$$\sigma(p_1, \ldots, p_r) = M_\sigma \ltimes p_1 \ltimes \cdots \ltimes p_r := M_\sigma \ltimes_{i=1}^r p_i. \tag{3.2}$$

D. Cheng et al., *Analysis and Control of Boolean Networks*,
Communications and Control Engineering,
DOI 10.1007/978-0-85729-097-7_3, © Springer-Verlag London Limited 2011

Note that throughout this book we assume the matrix product is the (left) semi-tensor product, and hereafter the symbol $\ltimes$ will be omitted in most cases. However, we use

$$\ltimes_{i=1}^{r} p_i := p_1 \ltimes p_2 \ltimes \cdots \ltimes p_r.$$

We start by constructing the structure matrices for some fundamental logical operators. We define the structure matrix for negation, $\neg$, denoted by M_n, as

$$M_n = \begin{bmatrix} 0 & 1 \\ 1 & 0 \end{bmatrix} = \delta_2[2\ 1]. \tag{3.3}$$

It is then easy to check that when a logical variable p is expressed in vector form, we have

$$\neg p = M_n p. \tag{3.4}$$

To see this, when $p = T$,

$$p = T \sim \delta_2^1 \quad \Longrightarrow \quad M_n p = \delta_2^2 \sim F,$$

and when $p = F$,

$$p = F \sim \delta_2^2 \quad \Longrightarrow \quad M_n p = \delta_2^1 \sim T.$$

Similarly, for conjunction, $\wedge$, disjunction, $\vee$, conditional, $\rightarrow$, and biconditional, $\leftrightarrow$, we define their corresponding structure matrices, denoted by M_c, M_d, M_i, and M_e, respectively, as follows:

$$M_c = \delta_2[1\ 2\ 2\ 2], \tag{3.5}$$

$$M_d = \delta_2[1\ 1\ 1\ 2], \tag{3.6}$$

$$M_i = \delta_2[1\ 2\ 1\ 1], \tag{3.7}$$

$$M_e = \delta_2[1\ 2\ 2\ 1]. \tag{3.8}$$

A straightforward computation then shows that for any two logical variables p and q, we have

$$p \wedge q = M_c p q, \tag{3.9}$$

$$p \vee q = M_d p q, \tag{3.10}$$

$$p \rightarrow q = M_i p q, \tag{3.11}$$

$$p \leftrightarrow q = M_e p q. \tag{3.12}$$

In the following we will show that for any logical function, f, there exists a unique structure matrix M_f of f such that (3.2) holds.

We need an auxiliary tool.

Define a matrix, M_r, called the power-reducing matrix, as

$$M_r = \delta_4[1\ 4]. \tag{3.13}$$

Proposition 3.1 *Let $p \in \Delta$. Then*

$$p^2 = M_r p. \tag{3.14}$$

Proof Let $p = [t, 1 - t]^T$. Then,

$$p^2 = \left[t^2, t(1 - t), (1 - t)t, (1 - t)^2\right]^T.$$

Since $t \in \{0, 1\}$ it is clear that $t^2 = t$, $(1 - t)^2 = 1 - t$, and $t(1 - t) = 0$. Then,

$$p^2 = [t, 0, 0, 1 - t]^T = M_r p. \qquad \square$$

Lemma 3.1 *Let $f(p_1, p_2, \ldots, p_r)$ be a logical function with logical variables (arguments) $p_1, p_2, \ldots, p_r$. Then, f can be expressed as*

$$f(p_1, p_2, \ldots, p_r) = \ltimes_i \xi_i, \tag{3.15}$$

where

$$\xi_i \in \{M_n, M_d, M_c, p_1, p_2, \ldots, p_r\}.$$

Proof Using the disjunctive (or conjunctive) normal form, $f(p_1, p_2, \ldots, p_r)$ can be written as a logical expression involving only $\wedge$, $\vee$, $\neg$, and p_i, $i = 1, 2, \ldots, r$. Using the corresponding structure matrices, the semi-tensor product form (3.15) can be obtained. $\qquad \square$

We now give an example to illustrate this.

Example 3.1 Consider

$$f(p, q, r) = (p \wedge \neg q) \vee (r \wedge p).$$

This can be expressed as

$$\begin{aligned}
f(p, q, r) &= (p \wedge \neg q) \vee (r \wedge p) \\
&= M_d(p \wedge \neg q)(r \wedge p) \\
&= M_d\left(M_c p (M_n q)\right)(M_c r p) \\
&= M_d M_c p M_n q M_c r p.
\end{aligned}$$

We are now ready to present a general result.

Theorem 3.1 *Given a logical function $f(p_1, p_2, \ldots, p_r)$ with logical variables $p_1, p_2, \ldots, p_r$, there exists a unique 2×2^r matrix M_f, called the structure matrix of f, such that*

$$f(p_1, p_2, \ldots, p_r) = M_f p_1 p_2 \cdots p_r. \tag{3.16}$$

Moreover, $M_f \in \mathscr{L}_{2 \times 2^r}$.

Proof We first prove the existence of M_f. Using Lemma 3.1 we only have to prove that $\ltimes_i \xi_i$ can be expressed as the right-hand side of (3.16). This can be done in three steps.

- Step 1. Using the fact that

$$pM = (I_2 \otimes M)p \qquad (3.17)$$

we can move all factors of structure matrices, such as M_j or $I_2 \otimes M_j$, to the front and move all variables, p_i, to the rear of the product:

$$\ltimes_i \xi_i = \ltimes_j N_j \ltimes_k p_{i_k},$$

where

$$N_j \in \{ I_{2^s} \otimes M_n, I_{2^s} \otimes M_d, I_{2^s} \otimes M_c \mid s = 0, 1, 2, \ldots \}, \quad i_k \in \{1, 2, \ldots, r\}.$$

- Step 2. Using a swap matrix we can change the order of two logical variables:

$$W_{[2]} p_i p_j = p_j p_i.$$

Using (3.17) it is easy to obtain the following form:

$$\ltimes_k p_{i_k} = M p_1^{k_1} p_2^{k_2} \cdots p_r^{k_r}.$$

- Step 3. Using a power-reducing matrix, the powers of the p_i's can all be reduced to 1. Again using (3.17), the coefficient matrices, generated by reducing orders, can be moved to the front part.

Following this procedure, a structure matrix will be produced.

To prove uniqueness, assume there are two structure matrices, $M_f \neq M'_f$. These must differ in at least one column, say the ith column, $c_i \neq c'_i$. Choose $p_1, \ldots, p_r$ such that $\ltimes_{i=1}^r p_i = \delta_{2^r}^i$. We then have

$$f(p_1, \ldots, p_r) = M_f \ltimes_{i=1}^r p_i = c_i.$$

Meanwhile,

$$f(p_1, \ldots, p_r) = M'_f \ltimes_{i=1}^r p_i = c'_i.$$

This then leads to a contradiction.

As for $M_f \in \mathscr{L}_{2 \times 2^r}$, this follows from the properties of logical matrices (we refer to Sect. 3.3 for details). $\square$

We now reconsider Example 3.1.

Example 3.2 In Example 3.1 we already have

$$f(p, q, r) = M_d M_c p M_n q M_c r p.$$

We continue by converting this into canonical form:

$$
\begin{aligned}
f(p,q,r) &= M_d M_c p M_n q M_c r p \\
&= M_d M_c (I_2 \otimes M_n) pq M_c r p \\
&= M_d M_c (I_2 \otimes M_n)(I_4 \otimes M_c) pqrp \\
&= M_d M_c (I_2 \otimes M_n)(I_4 \otimes M_c) p W_{[2,4]} pqr \\
&= M_d M_c (I_2 \otimes M_n)(I_4 \otimes M_c)(I_2 \otimes W_{[2,4]}) p^2 qr \\
&= M_d M_c (I_2 \otimes M_n)(I_4 \otimes M_c)(I_2 \otimes W_{[2,4]}) M_r pqr \\
&:= M_f pqr.
\end{aligned}
$$

Then,

$$
\begin{aligned}
M_f &= M_d M_c (I_2 \otimes M_n)(I_4 \otimes M_c)(I_2 \otimes W_{[2,4]}) M_r \\
&= \delta_2[1\,2\,1\,1\,2\,2\,2\,2].
\end{aligned}
$$

Note that, for convenience, we will hereafter identify T with 1 or δ_2^1, and F with 0 or δ_2^2. Hence, we will also identify $\mathscr{D}$ with Δ. It should be clear from the text which form we are using.

3.2 Structure Matrix for k-valued Logic

In this section we consider the matrix expression of k-valued logic [3]. Observe that a k-valued logical variable takes values from

$$
\mathscr{D}_k = \left\{ T = 1, \frac{k-2}{k-1}, \frac{k-3}{k-1}, \ldots, F = 0 \right\}.
$$

To use a matrix expression, we identify each k-valued logical value with a vector as follows:

$$
\frac{i}{k-1} \sim \delta_k^{k-i}, \quad i = 1, 2, \ldots, k-1. \tag{3.18}
$$

For instance, $T = 1 \sim \delta_k^1$, $F = 0 \sim \delta_k^k$, etc. Similarly to the Boolean case, we identify $\mathscr{D}_k$ with Δ_k. An r-ary k-valued logical operator is then a mapping $\sigma : \mathscr{D}_k^r \to \mathscr{D}_k$, and in vector form it is a mapping $\sigma : \Delta_k^r \to \Delta_k$.

Definition 3.3 A $k \times k^r$ matrix M_σ is said to be the structure matrix of the r-ary logical operator σ if

$$
\sigma(p_1, \ldots, p_r) = M_\sigma p_1 \cdots p_r. \tag{3.19}
$$

Similarly to the Boolean case, we can construct the structure matrices of some fundamental logical operators. For negation, $\neg$, we define its structure matrix, denoted by $M_{n,k}$, as

$$M_{n,k} = \delta_k[k\ k-1\ \cdots\ 1]. \tag{3.20}$$

It is then easy to check that when a logical variable p is expressed in k-dimensional vector form, we have

$$\neg p = M_{n,k}p. \tag{3.21}$$

The structure matrix of the rotator $\oslash_k$, denoted by $M_{o,k}$, can be easily shown to be

$$M_{o,k} = \delta_k[2\ 3\ \cdots\ k\ 1]. \tag{3.22}$$

For instance, when $k = 3$ we have

$$M_{o,3} = \delta_3[2\ 3\ 1].$$

When $k = 4$,

$$M_{o,4} = \delta_4[2\ 3\ 4\ 1].$$

Consider the i-confirmer, $\nabla_{i,k}$. A straightforward verification shows that its structure matrix is

$$M_{\nabla_{i,k}} = \delta_k[\underbrace{k \cdots k}_{i-1}\ 1\ \underbrace{k \cdots k}_{k-i}], \quad i = 1, 2, \ldots, k. \tag{3.23}$$

For instance, let $k = 3$ or 4 and $i = 2$. We then have

$$M_{\nabla_{2,3}} = \delta_3[3\ 1\ 3], \qquad M_{\nabla_{2,4}} = \delta_4[4\ 1\ 4\ 4].$$

For conjunction, $\wedge$, and disjunction, $\vee$, as defined in Chap. 1 for the k-valued case, we define their respective structure matrices, $M_{c,k}$ and $M_{d,k}$, as

$$M_{c,k} = \delta_k[\underbrace{1\ 2\ 3 \cdots k}_{k}\ \underbrace{2\ 2\ 3\ \cdots\ k}_{k}\ \underbrace{3\ 3\ 3\ \cdots\ k}_{k}\ \cdots\ \underbrace{k\ k\ k\ \cdots k}_{k}], \tag{3.24}$$

$$M_{d,k} = \delta_k[\underbrace{1\ 1\ 1 \cdots 1}_{k}\ \underbrace{1\ 2\ 2 \cdots 2}_{k}\ \underbrace{1\ 2\ 3 \cdots 3}_{k}\ \cdots\ \underbrace{1\ 2\ 3 \cdots k}_{k}]. \tag{3.25}$$

For instance, when $k = 3$ we have

$$M_{c,3} = \delta_3[1\ 2\ 3\ 2\ 2\ 3\ 3\ 3\ 3], \tag{3.26}$$

$$M_{d,3} = \delta_3[1\ 1\ 1\ 1\ 2\ 2\ 1\ 2\ 3]. \tag{3.27}$$

Next, assume that we use equations (1.30) and (1.31) in Proposition 1.1 as the respective definitions of conditional and biconditional. That is, define $p \to q = (\neg p) \vee q$ and $p \leftrightarrow q = (p \to q) \wedge (q \to p)$. Since

$$p \to q = M_{i,k}pq = M_{d,k}(M_{n,k}p)q$$

we have

$$M_{i,k} = M_{d,k}M_{n,k}. \tag{3.28}$$

When $k = 3$ we have

$$M_{i,3} = \delta_3[1\ 2\ 3\ 1\ 2\ 2\ 1\ 1\ 1]. \tag{3.29}$$

When $k = 4$ we have

$$M_{i,4} = \delta_4[1\ 2\ 3\ 4\ 1\ 2\ 3\ 3\ 1\ 2\ 2\ 2\ 1\ 1\ 1\ 1]. \tag{3.30}$$

Next, we consider M_e. To do this we need the k-valued power-reducing matrix. Define the k-valued power-reducing matrix, $M_{r,k}$, as

$$M_{r,k} = \begin{bmatrix} \delta_k^1 & 0_k & \cdots & 0_k \\ 0_k & \delta_k^2 & \cdots & 0_k \\ \vdots & & & \\ 0_k & 0_k & \cdots & \delta_k^k \end{bmatrix}, \tag{3.31}$$

where $0_k \in \mathbb{R}^k$ is a zero vector. When $k = 3$,

$$M_{r,3} = \delta_9[1\ 5\ 9]. \tag{3.32}$$

When $k = 4$,

$$M_{r,4} = \delta_{16}[1\ 6\ 11\ 16]. \tag{3.33}$$

Similarly to the Boolean case, it is easy to prove the following.

Proposition 3.2 *If $p \in \Delta_k$, then*

$$p^2 = M_{r,k}p. \tag{3.34}$$

We are now ready to calculate $M_{e,k}$. Using (3.31) we have

$$M_{e,k}pq = M_{c,k}M_{i,k}pqM_{i,k}qp$$

$$= M_{c,k}M_{i,k}(I_{k^2} \otimes M_{i,k})pqqp$$

$$= M_{c,k}M_{i,k}(I_{k^2} \otimes M_{i,k})pW_{[k,k^2]}pq^2$$

$$= M_{c,k}M_{i,k}(I_{k^2} \otimes M_{i,k})(I_k \otimes W_{[k,k^2]})p^2q^2$$

$$= M_{c,k}M_{i,k}(I_{k^2} \otimes M_{i,k})(I_k \otimes W_{[k,k^2]})M_{r,k}pM_{r,k}q$$

$$= M_{c,k}M_{i,k}(I_{k^2} \otimes M_{i,k})(I_k \otimes W_{[k,k^2]})M_{r,k}(I_k \otimes M_{r,k})pq. \tag{3.35}$$

Hence,

$$M_{e,k} = M_{c,k} M_{i,k} (I_{k^2} \otimes M_{i,k})(I_k \otimes W_{[k,k^2]}) M_{r,k} (I_k \otimes M_{r,k}). \qquad (3.36)$$

When $k = 3$ we have

$$M_{e,3} = \delta_3 [1\,2\,3\,2\,2\,2\,3\,2\,1]. \qquad (3.37)$$

When $k = 4$ we have

$$M_{e,4} = \delta_4 [1\,2\,3\,4\,2\,2\,3\,3\,3\,3\,2\,2\,4\,3\,2\,1]. \qquad (3.38)$$

Similarly to the Boolean case, it is easy to prove the following theorem, which is the counterpart of Theorem 3.1 for the k-valued logic case.

Theorem 3.2 *Given a k-valued logical function $f(p_1, p_2, \ldots, p_r)$ with k-valued logical variables $p_1, p_2, \ldots, p_r$, there exists a unique logical matrix $M_f \in \mathscr{L}_{k \times k^r}$, called the structure matrix of f, such that*

$$f(p_1, p_2, \ldots, p_r) = M_f p_1 p_2 \cdots p_r. \qquad (3.39)$$

We now give an example to illustrate this.

Example 3.3 Assume that

$$f(p, q, r) = (p \vee q) \wedge (q \vee r) \wedge (r \vee p),$$

where $p, q, r \in \Delta_k$. Then,

$$
\begin{aligned}
f(p,q,r) &\\
&= (M_{c,k})^2 (M_{d,k} p q)(M_{d,k} q r)(M_{d,k} r p) \\
&= (M_{c,k})^2 M_{d,k} (I_{k^2} \otimes M_{d,k}) p q^2 r (M_{d,k} r p) \\
&= (M_{c,k})^2 M_{d,k} (I_{k^2} \otimes M_{d,k})(I_{k^4} \otimes M_{d,k}) p q^2 r^2 p \\
&= (M_{c,k})^2 M_{d,k} (I_{k^2} \otimes M_{d,k})(I_{k^4} \otimes M_{d,k}) p W_{[k,k^4]} p q^2 r^2 \\
&= (M_{c,k})^2 M_{d,k} (I_{k^2} \otimes M_{d,k})(I_{k^4} \otimes M_{d,k})(I_k \otimes W_{[k,k^4]}) p^2 q^2 r^2 \\
&= (M_{c,k})^2 M_{d,k} (I_{k^2} \otimes M_{d,k})(I_{k^4} \otimes M_{d,k})(I_k \otimes W_{[k,k^4]}) M_{r,k} p M_{r,k} q M_{r,k} r \\
&= (M_{c,k})^2 M_{d,k} (I_{k^2} \otimes M_{d,k})(I_{k^4} \otimes M_{d,k}) \\
&\qquad (I_k \otimes W_{[k,k^4]}) M_{r,k} (I_k \otimes M_{r,k})(I_{k^2} \otimes M_{r,k}) p q r \\
&:= M_f p q r.
\end{aligned}
$$

It follows that

$$
\begin{aligned}
M_f &= (M_{c,k})^2 M_{d,k} (I_{k^2} \otimes M_{d,k})(I_{k^4} \otimes M_{d,k}) \\
&\qquad (I_k \otimes W_{[k,k^4]}) M_{r,k} (I_k \otimes M_{r,k})(I_{k^2} \otimes M_{r,k}).
\end{aligned}
$$

When $k = 3$, M_f is a 3×3^3 matrix, which is

$$M_f = \delta_3[1\ 1\ 1\ 1\ 2\ 2\ 1\ 2\ 3\ 1\ 2\ 2\ 2\ 2\ 2\ 2\ 2\ 3\ 1\ 2\ 3\ 2\ 2\ 3\ 3\ 3\ 3].$$

When $k = 4$, M_f is a 4×4^4 matrix, which is

$$M_f = \delta_4[1\ 1\ 1\ 1\ 1\ 2\ 2\ 2\ 1\ 2\ 3\ 3\ 1\ 2\ 3\ 4\ 1\ 2\ 2\ 2\ 2\ 2\ 2\ 2\ 2\ 3\ 3\ 2\ 2\ 3\ 4$$
$$1\ 2\ 3\ 3\ 2\ 2\ 3\ 3\ 3\ 3\ 3\ 3\ 3\ 3\ 4\ 1\ 2\ 3\ 4\ 2\ 2\ 3\ 4\ 3\ 3\ 3\ 4\ 4\ 4\ 4\ 4].$$

3.3 Logical Matrices

Recall that in Definition 3.1, a logical matrix was defined as a matrix whose columns are in Δ_m. In this section we will show that most of the matrices encountered in the algebraic expression of logic are logical matrices.

Proposition 3.3

1. *A swap matrix is a logical matrix:*

$$W_{[m,n]} \in \mathscr{L}_{mn \times mn}.$$

2. *The identity matrix is a logical matrix:*

$$I_m \in \mathscr{L}_{m \times m}.$$

3. *The (k-valued) power-reducing matrix is a logical matrix:*

$$M_{r,k} \in \mathscr{L}_{k^2 \times k}.$$

4. *The structure matrices of rotator, i-confirmer, and negation are logical matrices:*

$$M_{o,k} \in \mathscr{L}_{k \times k}, \qquad M_{\nabla_{i,k}} \in \mathscr{L}_{k \times k}, \qquad M_{n,k} \in \mathscr{L}_{k \times k}.$$

5. *The structure matrices of conjunction, disjunction, conditional, and biconditional are logical matrices:*

$$M_{c,k} \in \mathscr{L}_{k \times k^2}, \qquad M_{d,k} \in \mathscr{L}_{k \times k^2}, \qquad M_{i,k} \in \mathscr{L}_{k \times k^2}, \qquad M_{e,k} \in \mathscr{L}_{k \times k^2}.$$

Next, we investigate some fundamental properties of logical matrices. First, we will show that the product of two logical matrices is itself a logical matrix. Later, we will see that this property is extremely important because a certain set of logical matrices is closed under the semi-tensor product.

Denote the set of all logical matrices by $\mathscr{L}$. That is,

$$\mathscr{L} = \bigcup_{i,j=1,2,\dots} \mathscr{L}_{i \times j}. \tag{3.40}$$

We are also interested in certain subsets, each of which is related to a k-valued logic:

$$\mathscr{L}_k = \bigcup_{i,j=0,1,2,\dots} \mathscr{L}_{k^i \times k^j}. \tag{3.41}$$

Proposition 3.4

1. *Let $L \in \mathscr{L}$ and I be an identity matrix. Then,*

$$L \otimes I \in \mathscr{L}, \qquad I \otimes L \in \mathscr{L}. \tag{3.42}$$

2. *Let $L \in \mathscr{L}_k$ and $I = I_{k^s}$. Then,*

$$L \otimes I \in \mathscr{L}_k, \qquad I \otimes L \in \mathscr{L}_k. \tag{3.43}$$

Proof We prove only (3.43). Note that both $I \otimes L$ and $L \otimes I$ are of dimension $k^{p+s} \times k^{q+s}$, satisfying the multiple dimension requirement. A straightforward computation shows that

$$L_j \otimes I = \delta_{k^{p+s}}\big[(i_j - 1)k^s + 1, (i_j - 1)k^s + 2, \dots, i_j k^s\big]. \tag{3.44}$$

Then,

$$L \otimes I = [L_1 \otimes I, L_2 \otimes I, \dots, L_{k^q} \otimes I], \tag{3.45}$$

which is obviously an element of $\mathscr{L}_k$.

Similarly, a straightforward computation shows that

$$\delta_{k^s}^j \otimes L = \delta_{k^{p+s}}\big[(j - 1)k^p + 1, (j - 1)k^p + 2, \dots, jk^p\big]. \tag{3.46}$$

Then,

$$I \otimes L = \big[\delta_{k^s}^1 \otimes L, \delta_{k^s}^2 \otimes L, \dots, \delta_{k^s}^{k^s} \otimes L\big], \tag{3.47}$$

which is also an element of $\mathscr{L}_k$. $\qquad\square$

Note that formulas (3.44)–(3.47) are useful in computer-based calculations.

Proposition 3.5 $\mathscr{L}_k$ *is closed under the semi-tensor product $\ltimes$. That is, if $A, B \in \mathscr{L}_k$, then $AB := A \ltimes B$ is always well defined, and $AB \in \mathscr{L}_k$.*

Proof Using Propositions 2.10 and 3.4, we only have to prove this for the conventional product case. So, we assume that $A \in M_{k^p \times k^q}$ and $B \in M_{k^q \times k^r}$, and write

$$A = \delta_{k^p}[i_1 \, i_2 \, \cdots \, i_{k^q}], \qquad B = \delta_{k^q}[j_1 \, j_2 \, \cdots \, j_{k^r}].$$

A straightforward computation then shows that

$$AB = \delta_{k^p}[i_{j_1} \, i_{j_2} \, \cdots \, i_{j_{k^r}}] \in \mathscr{L}_k. \tag{3.48}$$

$\qquad\square$

Note that equation (3.48) is itself a useful formula.

Combining the above arguments and taking the constructive proof of Theorem 3.1 into consideration, the following result clearly follows.

Theorem 3.3

1. *The structure matrix of a k-valued (n-ary for any n) logical operator M_f is a logical matrix, i.e., $M_f \in \mathscr{L}_k$, which is called a k-valued logical matrix.*
2. *A (semi-tensor) product of several structure matrices of k-valued (n-ary for any n) logical operators is also a k-valued logical matrix.*

Remark 3.1

1. The first statement of Theorem 3.3 is based on the following fact: any operation performed in obtaining the structure matrix of a logical function is $\mathscr{L}_k$ closed. The second statement of Theorem 3.3 implies that further multiplication of structure matrices is legal and that $\mathscr{L}_k$ is closed with respect to the semi-tensor product. This fact will be used in the sequel.
2. In the previous two sections, all matrix products were taken without checking the "multiple dimension" requirement. Theorem 3.3 ensures the legality of this.
3. We can also say that the set of logical matrices, $\mathscr{L}$, is closed under the semi-tensor product. When two matrices satisfy the multiple dimension requirement, this is obvious. However, when this requirement is not satisfied, the general definition of the semi-tensor product must be used.

References

1. Cheng, D.: Sime-tensor product of matrices and its applications: a survey. In: Proc. 4th International Congress of Chinese Mathematicians, pp. 641–668. Higher Edu. Press, Int. Press, Hangzhou (2007)
2. Cheng, D., Qi, H.: Semi-tensor Product of Matrices—Theory and Applications. Science Press, Beijing (2007) (in Chinese)
3. Li, Z., Cheng, D.: Algebraic approach to dynamics of multi-valued networks. Int. J. Bifurc. Chaos **20**(3), 561–582 (2010)

Chapter 4
Logical Equations

4.1 Solution of a Logical Equation

A logical variable p is called an logical argument or logical unknown if it can take a value $p \in \mathscr{D} = \{T, F\}$ to satisfy certain logical requirements. A logical constant c is a fixed value $c \in \mathscr{D}$.

Definition 4.1 A standard system of logical equations is expressed as

$$\begin{cases} f_1(p_1, p_2, \ldots, p_n) = c_1, \\ f_2(p_1, p_2, \ldots, p_n) = c_2, \\ \vdots \\ f_m(p_1, p_2, \ldots, p_n) = c_m, \end{cases} \qquad (4.1)$$

where f_i, $i = 1, \ldots, m$, are logical functions, p_i, $i = 1, \ldots, n$, are logical arguments (unknowns), and c_i, $i = 1, \ldots, m$, are logical constants. A set of logical constants d_i, $i = 1, \ldots, n$, such that

$$p_i = d_i, \quad i = 1, \ldots, n, \qquad (4.2)$$

satisfy (4.1) is said to be a solution of (4.1).

We now give an illustrative example.

Example 4.1 Consider the following system:

$$\begin{cases} p \wedge q = c_1, \\ q \vee r = c_2, \\ r \leftrightarrow (\neg p) = c_3. \end{cases} \qquad (4.3)$$

D. Cheng et al., *Analysis and Control of Boolean Networks*,
Communications and Control Engineering,
DOI 10.1007/978-0-85729-097-7_4, © Springer-Verlag London Limited 2011

1. Assume the logical constants are

$$c_1 = 1, \qquad c_2 = 1, \qquad c_3 = 1.$$

A straightforward verification shows that

$$\begin{cases} p = 1, \\ q = 1, \\ r = 0 \end{cases}$$

is the only solution.

2. Assume the logical constants are

$$c_1 = 1, \qquad c_2 = 0, \qquad c_3 = 1.$$

It can then be checked that there is no solution.

3. Assume the logical constants are

$$c_1 = 0, \qquad c_2 = 1, \qquad c_3 = 0.$$

There are then two solutions:

$$\begin{cases} p_1 = 1, \\ q_1 = 0, \\ r_1 = 1 \end{cases}$$

and

$$\begin{cases} p_2 = 0, \\ q_2 = 1, \\ r_2 = 0. \end{cases}$$

Example 4.1 is heuristic. It shows that the solutions of systems of logical equations are quite different from those of linear algebraic equations where the type of solution depends only on the coefficients of the system.

4.2 Equivalent Algebraic Equations

This section considers how to solve the system (4.1). The basic idea is first to convert (4.1) into an equivalent linear algebraic equation and then to solve this algebraic equation, thereby providing the solution(s) to the system of logical equations. To do this, we first need some preparatory results.

Lemma 4.1 *Let p_i, $i = 1, 2, \ldots, n$, be logical variables in vector form, i.e., $p_i \in \Delta$. We define*

$$x = \ltimes_{i=1}^{n} p_i.$$

Then, p_i, $i = 1, 2, \ldots, n$, are uniquely determined by x.

Proof We prove this by giving a formula to calculate p_i. First, since $p_i \in \Delta$, it follows that $x \in \Delta_{2^n}$. We can now assume that $x = \delta_{2^n}^i$. Split x into two equal-sized segments as

$$x = [x_1^T, x_2^T]^T,$$

where either $0 \neq x_1 \in \Delta_{2^{n-1}}$ and $x_2 = 0$, or $x_1 = 0$ and $0 \neq x_2 \in \Delta_{2^{n-1}}$. According to the definition of the semi-tensor product, if $x_2 = 0$, then $p_1 = 1$, and if $x_2 = 1$, then $p_1 = 0$. We can then split a nonzero segment, say $x_1 \neq 0$, into two equal-sized parts as $x_1 = [x_{11}^T, x_{12}^T]^T$, then apply the same judgment to p_2, and so on. The result follows. □

Based on the argument in the proof of the last lemma, we give the following algorithm.

Algorithm 4.1 Let $\ltimes_{j=1}^{n} p_j = \delta_{2^n}^i$, where $p_j \in \Delta$ are in vector form. Then:

1. The scalar form of $\{p_j\}$ can be calculated from i inductively as follows:

   - Step 1. Set $q_0 := 2^n - i$.
   - Step 2. Calculate p_j and q_j, $j = 1, 2, \ldots, n$, recursively by

   $$\begin{cases} p_j = [\frac{q_{j-1}}{2^{n-j}}], \\ q_j = q_{j-1} - p_j * 2^{n-j}, \quad j = 1, 2, \ldots, n, \end{cases} \tag{4.4}$$

   where, in the first equation, $[a]$ denotes the largest integer less than or equal to a.

2. i can be calculated from the scalar form of $\{p_j\}$ by

$$i = \sum_{j=1}^{n} (1 - p_j) 2^{n-j} + 1. \tag{4.5}$$

We now give an example to demonstrate the formulas.

Example 4.2 Assume $x = p_1 p_2 p_3 p_4 p_5$.

1. The value of x is known to be $x = \delta_{32}^7$. We then try to obtain the values of p_i, $i = 1, \ldots, 5$. Using the first part of Algorithm 4.1, we have

$$q_0 = 2^5 - 7 = 32 - 7 = 25.$$

It follows that

$$p_1 = [q_0/16] = 1, \qquad q_1 = q_0 - p_1 * (16) = 9,$$
$$p_2 = [q_1/8] = 1, \qquad q_2 = q_1 - p_2 * 8 = 1,$$
$$p_3 = [q_2/4] = 0, \qquad q_3 = q_2 - p_3 * 4 = 1,$$
$$p_4 = [q_3/2] = 0, \qquad q_4 = q_3 - p_4 * 2 = 1,$$
$$p_5 = [q_4/1] = 1.$$

We conclude that $p_1 = 1 \sim \delta_2^1$, $p_2 = 1 \sim \delta_2^1$, $p_3 = 0 \sim \delta_2^2$, $p_4 = 0 \sim \delta_2^2$, and $p_5 = 1 \sim \delta_2^1$.

2. Assume $p_1 = 0$, $p_2 = 1$, $p_3 = 0$, $p_4 = 1$, and $p_5 = 1$. Using (4.5), we have

$$i = 2^4 + 2^2 + 1 = 21.$$

Therefore, $x = \delta_{32}^{21}$.

Next, we construct a matrix, which may be called the group power-reducing matrix, as follows. For $j \geq 1$, define

$$\Phi_j = \prod_{i=1}^{j} I_{2^{i-1}} \otimes \left[(I_2 \otimes W_{[2,2^{j-i}]}) M_r \right]. \tag{4.6}$$

We then have the following result.

Lemma 4.2 *If* $z_j = p_1 p_2 \cdots p_j$, *where* $p_i \in \Delta$, $i = 1, 2, \ldots, j$, *then*

$$z_j^2 = \Phi_j z_j. \tag{4.7}$$

Proof We prove this by mathematical induction. When $j = 1$, using Proposition 3.1, we have

$$z_1^2 = p_1^2 = M_r p_1.$$

In the above formula

$$\Phi_1 = (I_2 \otimes W_{[2,1]}) M_r.$$

Note that $W_{[2,1]} = I_2$, so it follows that $\Phi_1 = M_r$. Hence, (4.7) is true for $j = 1$. If we assume (4.7) is true for $j = s$, then for $j = s + 1$ we have

$$P_{s+1}^2 = p_1 p_2 \cdots p_{s+1} p_1 p_2 \cdots p_{s+1}$$
$$= p_1 W_{[2,2^s]} p_1 [p_2 \cdots p_{s+1}]^2$$
$$= (I_2 \otimes W_{[2,2^s]}) p_1^2 [p_2 \cdots p_{s+1}]^2$$
$$= \left[(I_2 \otimes W_{[2,2^s]}) M_r \right] p_1 [p_2 \cdots p_{s+1}]^2.$$

Applying the induction assumption to the last equality, we have

$$z_{s+1}^2 = (I_2 \otimes W_{[2,2^s]}) M_r p_1$$

$$\times \left(\prod_{i=1}^{s} I_{2^{i-1}} \otimes \left[(I_2 \otimes W_{[2,2^{s-i}]}) M_r \right] \right) p_2 p_3 \cdots p_{s+1}$$

$$= \left[(I_2 \otimes W_{[2,2^s]}) M_r \right]$$

$$\times \left(\prod_{i=1}^{s} I_{2^i} \otimes \left[(I_2 \otimes W_{[2,2^{s-i}]}) M_r \right] \right) p_1 p_2 \cdots p_{s+1}.$$

The conclusion then follows. □

Before presenting the next lemma we require another concept, called a dummy operator, σ_d, defined by

$$\sigma_d(p, q) = q, \quad \forall p, q \in \mathscr{D}. \tag{4.8}$$

It is easy to show that the structure matrix of the dummy operator σ_d is

$$E_d := \begin{bmatrix} 1 & 0 & 1 & 0 \\ 0 & 1 & 0 & 1 \end{bmatrix}. \tag{4.9}$$

It follows from the definition that for any two logical variables X, Y,

$$E_d XY = Y \quad \text{or} \quad E_d W_{[2]} XY = X. \tag{4.10}$$

A logical variable which only formally appears in a logical function, but does not affect the value of that function, is called a fabricated variable.

Lemma 4.3 *Let*

$$x = \ltimes_{i=1}^{n} p_i.$$

Using vector form, each logical equation

$$f_i(p_1, p_2, \ldots, p_n) = c_i, \quad i = 1, 2, \ldots, m,$$

in the system (4.1) *can be expressed as*

$$M_i x = c_i, \quad i = 1, 2, \ldots, m, \tag{4.11}$$

where $M_i \in \mathscr{L}_{2 \times 2^n}$.

Proof Assume f_i is a logical equation of $p_1, \ldots, p_n$. Let M_i be the structure matrix of f_i. Then, (4.11) immediately follows. Assume some p_j's do not appear in f_i. Using the dummy operator technique we can still obtain (4.11) by introducing fabricated variables. □

We are now ready to present the main result, which converts the system of logical equations (4.1) into an algebraic equation.

Theorem 4.1 *Let* $x = \ltimes_{i=1}^{n} p_i$, $b = \ltimes_{i=1}^{m} c_i$. *The system of logical equations* (4.1) *can then be converted into a linear algebraic equation as*

$$Lx = b, \tag{4.12}$$

where

$$L = M_1 \ltimes_{j=2}^{n} \left[(I_{2^n} \otimes M_j) \Phi_n \right], \tag{4.13}$$

M_i *being defined as in* (4.11).

Proof Note that from Lemma 4.2 we have

$$x^2 = \Phi_n x.$$

Multiplying (4.11) together yields

$$
\begin{aligned}
b &= M_1 x M_2 x \cdots M_n x \\
&= M_1 (I_{2^n} \otimes M_2) x^2 M_3 x \cdots M_n x \\
&= M_1 (I_{2^n} \otimes M_2) \Phi_n x M_3 x \cdots M_n x \\
&= \cdots \\
&= M_1 (I_{2^n} \otimes M_2) \Phi_n (I_{2^n} \otimes M_3) \Phi_n \cdots (I_{2^n} \otimes M_n) \Phi_n x.
\end{aligned}
$$

(4.13) then follows.   □

Remark 4.1

1. To obtain the algebraic form for a particular logical equation, we may not need to use formula (4.13). In most cases, L can be obtained by a direct computation.
2. Using Lemma 4.1 and Algorithm 4.1, as long as algebraic equation (4.12) is solved for x, the logical unknowns p_i, $i = 1, 2, \ldots, n$, can be easily calculated.
3. As discussed in Chap. 3, in equation (4.12), the coefficient matrix $L \in \mathscr{L}_{2^m \times 2^n}$, and the constant vector $b \in \Delta_{2^m}$.

Denote by $\mathrm{Col}(L)$ the set of columns of matrix L. Since $L \in \mathscr{L}_{2^m \times 2^n}$ and $b \in \Delta_{2^m}$, it is clear that algebraic equation (4.12) has solution $x \in \Delta_{2^n}$ if and only if

$$b \in \mathrm{Col}(L).$$

Express L in condensed form as

$$L = \delta_{2^m}[i_1, i_2, \ldots, i_{2^n}].$$

We define a set

$$\Lambda = \left\{ \lambda \,\middle|\, \delta_{2^m}^{i\lambda} = b, 1 \le \lambda \le 2^n \right\}.$$

The following result is then obvious.

Theorem 4.2 *Using the above notation, the solution of (4.12) is*

$$x = \delta_{2^n}^{\lambda}, \quad \lambda \in \Lambda. \tag{4.14}$$

As an application, we reconsider Example 4.1.

Example 4.3 Consider system (4.3) again. We have its algebraic form as

$$\begin{cases} M_c pq = c_1, \\ M_d qr = c_2, \\ M_e r(M_n p) = c_3. \end{cases} \tag{4.15}$$

Multiplying these three equations together yields

$$M_c pq M_d qr M_e r M_n p = c_1 c_2 c_3 := b. \tag{4.16}$$

Next, set $x = p \ltimes q \ltimes r$. We simplify the left-hand side of (4.16) as follows:

$$
\begin{aligned}
&M_c pq M_d qr M_e r M_n p \\
&= M_c(I_4 \otimes M_d) pq^2 r M_e r M_n p \\
&= M_c(I_4 \otimes M_d)(I_{16} \otimes M_e) pq^2 r^2 M_n p \\
&= M_c(I_4 \otimes M_d)(I_{16} \otimes M_e)(I_{32} \otimes M_n) pq^2 r^2 p \\
&= M_c(I_4 \otimes M_d)(I_{16} \otimes M_e)(I_{32} \otimes M_n) p W_{[2,16]} pq^2 r^2 \\
&= M_c(I_4 \otimes M_d)(I_{16} \otimes M_e)(I_{32} \otimes M_n)(I_2 \otimes W_{[2,16]}) p^2 q^2 r^2 \\
&= M_c(I_4 \otimes M_d)(I_{16} \otimes M_e)(I_{32} \otimes M_n)(I_2 \otimes W_{[2,16]}) M_r p M_r q M_r r \\
&= M_c(I_4 \otimes M_d)(I_{16} \otimes M_e)(I_{32} \otimes M_n)(I_2 \otimes W_{[2,16]}) M_r (I_2 \otimes M_r)(I_4 \otimes M_r) pqr \\
&:= Lx.
\end{aligned}
$$

It is easy to calculate that

$$
\begin{aligned}
L &= M_c(I_4 \otimes M_d)(I_{16} \otimes M_e)(I_{32} \otimes M_n)(I_2 \otimes W_{[2,16]}) M_r (I_2 \otimes M_r)(I_4 \otimes M_r) \\
&= \delta_{2^3}[2, 1, 6, 7, 5, 6, 5, 8].
\end{aligned}
$$

Now, if $b = \delta_8^1$, then $\Lambda = \{2\}$. That is, the second column of L equals b. We have the solution $x = \delta_8^2$. Returning to Boolean form we have

$$
\begin{aligned}
b = \delta_8^1 &\iff c_1 = 1, c_2 = 1, c_3 = 1, \\
x = \delta_8^2 &\iff p_1 = 1, p_2 = 1, p_3 = 0.
\end{aligned}
$$

We list all possible constants and their corresponding solutions in Table 4.1.

Table 4.1 Solutions of (4.3)

b	(c_1, c_2, c_3)	Λ	x	(p_1, p_2, p_3)
δ_8^1	$(1, 1, 1)$	$\{2\}$	δ_8^2	$(1, 1, 0)$
δ_8^2	$(1, 1, 0)$	$\{1\}$	δ_8^1	$(1, 1, 1)$
δ_8^3	$(1, 0, 1)$	$\emptyset$		
δ_8^4	$(1, 0, 0)$	$\emptyset$		
δ_8^5	$(0, 1, 1)$	$\{5, 7\}$	δ_8^5, δ_8^7	$(0, 1, 0), (0, 0, 1)$
δ_8^6	$(0, 1, 0)$	$\{3, 6\}$	δ_8^3, δ_8^6	$(1, 0, 1), (0, 1, 0)$
δ_8^7	$(0, 0, 1)$	$\{4\}$	δ_8^4	$(1, 0, 0)$
δ_8^8	$(0, 0, 0)$	$\{8\}$	δ_8^8	$(0, 0, 0)$

When the number of unknowns is not very small, calculating the coefficient matrix by hand will be very difficult, but a simple routine can do this easily. We now give another example.

Example 4.4 Consider the following system of logical equations:

$$\begin{cases} p_1 \wedge p_2 = c_1, \\ p_2 \vee (p_3 \leftrightarrow p_2) = c_2, \\ p_5 \rightarrow (p_4 \vee p_3) = c_3, \\ \neg p_3 = c_4, \\ p_4 \vee (p_5 \wedge p_2) = c_5, \\ (p_6 \vee p_2) \wedge p_6 = c_6, \\ (\neg p_9) \rightarrow p_7 = c_7, \\ p_5 \wedge p_6 \wedge p_7 = c_8, \\ (p_6 \vee p_8) \leftrightarrow p_3 = c_9. \end{cases} \qquad (4.17)$$

Its algebraic form is

$$\begin{cases} M_c p_1 p_2 = c_1, \\ M_d p_2 M_e p_3 p_2 = c_2, \\ M_i p_5 M_d p_4 p_3 = c_3, \\ M_n p_3 = c_4, \\ M_d p_4 M_c p_5 p_2 = c_5, \\ M_c M_d p_6 p_2 p_6 = c_6, \\ M_i M_n p_9 p_7 = c_7, \\ M_c^2 p_5 p_6 p_7 = c_8, \\ M_e M_d p_6 p_8 p_3 = c_9. \end{cases} \qquad (4.18)$$

Of course, we can convert (4.18) into an algebraic equation as

$$L \ltimes_{i=1}^{9} p_i = \ltimes_{i=1}^{9} c_i \quad \text{or} \quad Lx = b.$$

Then, $L \in \mathscr{L}_{512 \times 1024}$. To save space, we give the first and last few columns (in condensed form) as follows:

$$
\begin{aligned}
\delta_{2^9} \ [& 33\ 33\ 33\ 33\ 35\ 39\ 35\ 39\ 43\ 43\ 44\ 44\ 43\ 47\ 44\ 48 \\
& 35\ 35\ 35\ 35\ 35\ 39\ 35\ 39\ 43\ 43\ 44\ 44\ 43\ 47\ 44\ 48 \\
& 33\ 33\ 33\ 33\ 35\ 39\ 35\ 39\ 43\ 43\ 44\ 44\ 43\ 47\ 44\ 48 \\
& 51\ 51\ 51\ 51\ 51\ 55\ 51\ 55\ 59\ 59\ 60\ 60\ 59\ 63\ 60\ 64 \\
& \ \ 2\ \ \ 2\ \ \ 2\ \ \ 2\ \ \ 4\ \ \ 8\ \ \ 4\ \ \ 8\ 12\ 12\ 11\ 11\ 12\ 16\ 11\ 15 \\
& \qquad\qquad\qquad\qquad \cdots
\end{aligned}
$$

$$
\begin{aligned}
& 260\ 264\ 260\ 264\ 268\ 268\ 267\ 267\ 268\ 272\ 267\ 271 \\
& 260\ 260\ 260\ 260\ 260\ 264\ 260\ 264\ 268\ 268\ 267\ 267 \\
& 268\ 272\ 267\ 271\ 338\ 338\ 338\ 338\ 340\ 344\ 340\ 344 \\
& 348\ 348\ 347\ 347\ 348\ 352\ 347\ 351\ 276\ 276\ 276\ 276 \\
& 276\ 280\ 276\ 280\ 284\ 284\ 283\ 283\ 284\ 288\ 283\ 287\ ].
\end{aligned}
$$

Next, giving a special set of logical constants, we solve the system of logical equations. Assume $c_1 = 1$, $c_2 = 1$, $c_3 = 1$, $c_4 = 0$, $c_5 = 1$, $c_6 = 1$, $c_7 = 1$, $c_8 = 0$, and $c_9 = 1$. Then,

$$b = \ltimes_{i=1}^{9} c_i = \delta_{2^9}^{35}.$$

Using a computer routine, we can find the set Λ such that for the columns L_λ of L with $L_\lambda = b$, $\lambda \in \Lambda$. It is easy to calculate in this way that

$$\Lambda = \{5, 7, 17, 18, 19, 20, 21, 23, 37, 39\}.$$

According to Theorem 4.2, there are ten corresponding solutions, which can be easily calculated as follows:

1. $x_1 = \delta_{2^9}^{5}$ or

$$
\begin{cases}
p_1 = 1, p_2 = 1, p_3 = 1, \\
p_4 = 1, p_5 = 1, p_6 = 1, \\
p_7 = 0, p_8 = 1, p_9 = 1,
\end{cases}
$$

2. $x_2 = \delta_{2^9}^{7}$ or

$$
\begin{cases}
p_1 = 1, p_2 = 1, p_3 = 1, \\
p_4 = 1, p_5 = 1, p_6 = 1, \\
p_7 = 0, p_8 = 0, p_9 = 1,
\end{cases}
$$

3. $x_3 = \delta_{2^9}^{17}$ or

$$\begin{cases} p_1 = 1, p_2 = 1, p_3 = 1, \\ p_4 = 1, p_5 = 0, p_6 = 1, \\ p_7 = 1, p_8 = 1, p_9 = 1, \end{cases}$$

4. $x_4 = \delta_{2^9}^{18}$ or

$$\begin{cases} p_1 = 1, p_2 = 1, p_3 = 1, \\ p_4 = 1, p_5 = 0, p_6 = 1, \\ p_7 = 1, p_8 = 1, p_9 = 0, \end{cases}$$

5. $x_5 = \delta_{2^9}^{19}$ or

$$\begin{cases} p_1 = 1, p_2 = 1, p_3 = 1, \\ p_4 = 1, p_5 = 0, p_6 = 1, \\ p_7 = 1, p_8 = 0, p_9 = 1, \end{cases}$$

6. $x_6 = \delta_{2^9}^{20}$ or

$$\begin{cases} p_1 = 1, p_2 = 1, p_3 = 1, \\ p_4 = 1, p_5 = 0, p_6 = 1, \\ p_7 = 1, p_8 = 0, p_9 = 0, \end{cases}$$

7. $x_7 = \delta_{2^9}^{21}$ or

$$\begin{cases} p_1 = 1, p_2 = 1, p_3 = 1, \\ p_4 = 1, p_5 = 0, p_6 = 1, \\ p_7 = 0, p_8 = 1, p_9 = 1, \end{cases}$$

8. $x_8 = \delta_{2^9}^{23}$ or

$$\begin{cases} p_1 = 1, p_2 = 1, p_3 = 1, \\ p_4 = 1, p_5 = 0, p_6 = 1, \\ p_7 = 0, p_8 = 0, p_9 = 1, \end{cases}$$

9. $x_9 = \delta_{2^9}^{37}$ or

$$\begin{cases} p_1 = 1, p_2 = 1, p_3 = 1, \\ p_4 = 0, p_5 = 1, p_6 = 1, \\ p_7 = 0, p_8 = 1, p_9 = 1, \end{cases}$$

10. $x_{10} = \delta_{2^9}^{39}$ or

$$\begin{cases} p_1 = 1, p_2 = 1, p_3 = 1, \\ p_4 = 0, p_5 = 1, p_6 = 1, \\ p_7 = 0, p_8 = 0, p_9 = 1. \end{cases}$$

We then consider a general form of logical equation. Consider

$$f(p_1, p_2, \ldots, p_n) = g(q_1, q_2, \ldots, q_m). \tag{4.19}$$

We want to find its algebraic form.

Proposition 4.1 *The algebraic form of logical equation* (4.19) *is*

$$M_e M_f (I_{2^n} \otimes M_g) p_1 \cdots p_n q_1 \cdots q_m = \delta_2^1. \tag{4.20}$$

Proof Define $p := f(p_1, p_2, \ldots, p_n)$ and $q := g(q_1, q_2, \ldots, q_m)$. (4.19) implies that either both p and q are "true" or both p and q are "false". That is, $p \leftrightarrow q$ is a tautology. In algebraic form we have

$$M_e M_f p_1 p_2 \cdots p_n M_g q_1 q_2 \cdots q_m = \delta_2^1.$$

Note that

$$p_1 p_2 \cdots p_n M_g = (I_{2^n} \otimes M_g) p_1 p_2 \cdots p_n.$$

Equation (4.20) follows immediately. □

Finally, we consider a system of logical equations as follows:

$$\begin{cases} f_1(x_1, \ldots, x_n) = g_1(x_1, \ldots, x_n), \\ \quad \vdots \\ f_m(x_1, \ldots, x_n) = g_m(x_1, \ldots, x_n). \end{cases} \tag{4.21}$$

Using vector form and setting $x = \ltimes_{i=1}^n x_i$ we have

$$\begin{cases} M_1^f x = M_1^g x, \\ \quad \vdots \\ M_m^f x = M_m^g x, \end{cases} \tag{4.22}$$

where M_i^f, etc. are the structure matrices of the respective f_i. Multiplying both sides together and using the standard procedure to simplify both sides, we finally have

$$M^f x = M^g x, \tag{4.23}$$

where $M^f, M^g \in \mathscr{L}_{2^m \times 2^n}$. It is easy to verify that $x = \ltimes_{i=1}^n x_i$ is the solution of (4.23) if and only if $(x_1, \ldots, x_n)$ is a set of solutions of (4.20).

The following result is straightforward to verify.

Theorem 4.3 $x = \delta_{2^n}^j$ *is a solution of* (4.23) *if and only if*

$$\mathrm{Col}_j(M^f) = \mathrm{Col}_j(M^g), \tag{4.24}$$

where $\mathrm{Col}_j(M)$ *is the jth column of* M.

In fact we can also perform a computation as an algebraic "transposition of terms".

If $A = (a_{ij}) \in \mathscr{B}_{p \times q}$ and $B = (b_{ij}) \in \mathscr{B}_{p \times q}$, then we define the logical "exclusive or", $\bar{\vee}$, over A and B as

$$A \bar{\vee} B := (a_{ij} \bar{\vee} b_{i,j}) \in \mathscr{B}_{p \times q}.$$

Letting

$$J = \{ j \mid \text{Col}_j(M_f \bar{\vee} M_g) = 0 \},$$

we then have the following result.

Corollary 4.1 $x = \delta_{2^n}^j$ *is a solution of* (4.23) *if and only if* $j \in J$.

We illustrate this with an example.

Example 4.5 Consider the following system of logical equations:

$$\begin{cases} f_1(x_1, x_2, x_3) = g_1(x_1, x_2, x_3), \\ f_2(x_1, x_2, x_3) = g_2(x_1, x_2, x_3), \end{cases} \tag{4.25}$$

where

$$\begin{cases} f_1(x_1, x_2, x_3) = \neg x_1, \\ f_2(x_1, x_2, x_3) = (x_1 \wedge x_2) \vee [\neg x_1 \wedge (x_2 \leftrightarrow x_3)] \end{cases}$$

and

$$\begin{cases} g_1(x_1, x_2, x_3) = x_3, \\ g_2(x_1, x_2, x_3) = (x_1 \wedge \neg x_2) \vee (\neg x_1 \wedge x_2). \end{cases}$$

It is easy to calculate that

$$M^f = \delta_4[3, 3, 4, 4, 1, 2, 2, 1], \qquad M^g = \delta_4[2, 4, 1, 3, 1, 3, 2, 4].$$

It can now be seen that $\text{Col}_5(M^f) = \text{Col}_5(M^g)$ and $\text{Col}_7(M^f) = \text{Col}_7(M^g)$. According to Theorem 4.3, the solutions are $x = \delta_8^5$ and $x = \delta_8^7$.

We conclude that the solutions of (4.25) are

$$\begin{cases} x_1 = 0, \\ x_2 = 1, \\ x_3 = 1, \end{cases} \qquad \begin{cases} x_1 = 0, \\ x_2 = 0, \\ x_3 = 1. \end{cases}$$

4.3 Logical Inference

The purpose of this section is to deduce logical inference by solving logical equations. We will discuss this via several examples.

Example 4.6 A says "B is a liar", B says "C is a liar", C says "Both A and B are liars." Who is a liar?

To solve this problem we define three logical variables:

- p: A is honest,
- q: B is honest,
- r: C is honest.

The three statements can then be expressed in logical version as

$$p \Leftrightarrow \neg q,$$
$$q \Leftrightarrow \neg r, \qquad\qquad\qquad (4.26)$$
$$r \Leftrightarrow \neg p \wedge \neg q.$$

Let $c = \delta_2^1$. The system (4.26) can then be converted into an algebraic form as

$$\begin{cases} M_e p M_n q = c, \\ M_e q M_n r = c, \\ M_e r M_c M_n p M_n q = c. \end{cases} \qquad (4.27)$$

It is easy to convert (4.27) into an algebraic equation as

$$Lx = b, \quad \text{where } x = pqr, b = c^3 = \delta_8^1,$$

and

$$L = \delta_8[8, 5, 2, 3, 4, 1, 5, 8].$$

Since only $\text{Col}_6(L) = b$, we have the unique solution

$$x = \delta_8^6,$$

which implies that

$$p = 0, \qquad q = 1, \qquad r = 0.$$

We conclude that only B is honest.

Example 4.7 A competition between five players took place in a simple-rotating way, which means each player had to play all others. We have the following information about the result:

- C beat E,
- A won three games,
- E won one game,
- among B, C, and D, there is one player who beat the other two,
- each of B, C, and D won two games,
- each of A, C, D, and E won some and lost some.

We use AB to denote "A beat B", and so on. It is clear from the definition that

$$BA = \neg AB, \qquad CA = \neg AC, \ldots.$$

Next, we convert each statement into a logical expression.

1. C beat E:

$$CE = 1.$$

2. A won three games:

$$\begin{cases} (AB \wedge AC \wedge AD) \vee (AB \wedge AC \wedge AE) \\ \quad \vee (AB \wedge AD \wedge AE) \vee (AC \wedge AD \wedge AE) = 1, \\ AB \wedge AC \wedge AD \wedge AE = 0. \end{cases} \qquad (4.28)$$

3. E won one game:

$$\begin{cases} AE \wedge BE \wedge CE \wedge DE = 0, \\ (EA \wedge EB) \vee (EA \wedge EC) \vee (EA \wedge ED) \\ \quad \vee (EB \wedge EC) \vee (EB \wedge ED) \vee (EC \wedge ED) = 0. \end{cases}$$

Since $EC = \neg CE = 0$, it can be removed from the above expression to give the following simplification:

$$\begin{cases} AE \wedge BE \wedge DE = 0, \\ (EA \wedge EB) \vee (EA \wedge ED) \vee (EB \wedge ED) = 0. \end{cases} \qquad (4.29)$$

4. Among B, C, and D, one player beat the other two:

$$(BC \wedge BD) \vee (CB \wedge CD) \vee (DB \wedge DC) = 1. \qquad (4.30)$$

5. Each of B, C, and D won two games:

   • B won two games:

$$\begin{cases} (BA \wedge BC) \vee (BA \wedge BD) \vee (BA \wedge BE) \\ \quad \vee (BC \wedge BD) \vee (BC \wedge BE) \vee (BD \wedge BE) = 1, \\ (BA \wedge BC \wedge BD) \vee (BA \wedge BC \wedge BE) \\ \quad \vee (BA \wedge BD \wedge BE) \vee (BC \wedge BD \wedge BE) = 0. \end{cases} \qquad (4.31)$$

   • D won two games: Note that $CE = 1$ can be used to simplify the expression. We then have

$$\begin{cases} CA \vee CB \vee CD = 1, \\ (CA \wedge CB) \vee (CA \wedge CD) \vee (CB \wedge CD) = 0. \end{cases} \qquad (4.32)$$

- D won two games:

$$\begin{cases} (DA \wedge DB) \vee (DA \wedge DC) \vee (DA \wedge DE) \\ \quad \vee (DB \wedge DC) \vee (DB \wedge DE) \vee (DC \wedge DE) = 1, \\ (DA \wedge DB \wedge DC) \vee (DA \wedge DB \wedge DE) \\ \quad \vee (DA \wedge DC \wedge DE) \vee (DB \wedge DC \wedge DE) = 0. \end{cases} \tag{4.33}$$

6. Each of A, C, D, and E won some and lost some. Obviously, this statement does not contain any additional information.

Next, we convert (4.28)–(4.33) into algebraic form. To save space, we write

$$p = AB, \qquad q = AC, \qquad r = AC, \qquad s = AE, \qquad t = BC$$

$$u = BD, \qquad v = BE, \qquad \alpha = CD, \qquad \beta = DE.$$

Applying De Morgan's law to the second equation of (4.29) and the equations of (4.33), and then combining all the algebraic equations yields

$$\begin{cases} M_d^3 M_c^2 pqr M_c^2 pqs M_c^2 prs M_c^2 qrs = \delta_2^1, \\ M_c^3 pqrs = \delta_2^2, \\ M_c^2 sv\beta = \delta_2^2, \\ M_d^2 M_{ds} v M_{ds}\beta M_{dv}\beta = \delta_2^1, \\ M_d^2 M_c tu M_c M_n t\alpha M_c M_n u M_n \alpha = \delta_2^1, \\ M_d^2 M_c M_n pt M_c M_n pu M_c M_n pv M_c tu M_c tv M_c uv = \delta_2^1, \\ M_d^3 M_c^2 M_n ptu M_c^2 M_n ptv M_c^2 M_n puv M_c^2 tuv = \delta_2^2, \\ M_d^2 M_n q M_n t\alpha = \delta_2^1, \\ M_d^2 M_c M_n q M_n t M_c M_n q\alpha M_c M_n t\alpha = \delta_2^2, \\ M_c^5 M_{dr} u M_{dr}\alpha M_{dr} M_n\beta M_{du}\alpha M_{du} M_n\beta M_{d}\alpha M_n\beta = \delta_2^2, \\ M_c^3 M_d^2 ru\alpha M_d^2 ru M_n\beta M_d^2 r\alpha M_n\beta M_d^2 u\alpha M_n\beta = \delta_2^1. \end{cases} \tag{4.34}$$

Now, multiplying all the equations in (4.34) together and using the standard procedure, we obtain the algebraic form

$$Lx = b, \tag{4.35}$$

where $x = pqrstuv\alpha\beta$. Using (4.5) yields

$$b = \delta_2^1 (\delta_2^2)^2 (\delta_2^1)^3 \delta_2^2 \delta_2^1 (\delta_2^2)^2 \delta_2^1 = \delta_{2^{11}}^{791}.$$

L is a $2^{11} \times 2^9$ matrix. The first and last few columns are

$$\delta_{2^{11}}[5 \quad 261 \quad 15 \quad 269 \quad 277 \quad 405 \quad 287 \quad 413$$
$$\cdots$$
$$1812 \quad 1939 \quad 1812 \quad 1940 \quad 1972 \quad 1971 \quad 1972 \quad 1972].$$

Table 4.2 Solutions of
(4.28)–(4.33)

	p	q	r	s	t	u	v	α	β
x_1	1	1	0	1	1	1	0	1	1
x_2	1	0	1	1	1	1	0	0	1
x_3	1	0	1	1	1	0	1	0	0
x_4	0	1	1	1	0	0	1	0	0

A routine shows that

$$\mathrm{Col}_{69}(L) = \mathrm{Col}_{135}(L) = \mathrm{Col}_{140}(L) = \mathrm{Col}_{284}(L) = b.$$

Therefore, the solutions of (4.28)–(4.33) are

$$x_1 = \delta_{29}^{69}, \qquad x_2 = \delta_{29}^{135}, \qquad x_3 = \delta_{29}^{140}, \qquad x_4 = \delta_{29}^{284}. \tag{4.36}$$

Using formula (4.4) yields the scalar forms of the solutions, as in Table 4.2.

Next, we make the following modification to the last statement "Each of A, C, D, and E won some and lost some.":

- Within the group A, C, D, and E, each won some and lost some.

It is now obvious that the new information is: (i) A can not beat all of C, D, and E, and (ii) E cannot lose to all of A, C, and D (equivalently to A and D). All other items of information have already be implied by previous statements. We then have two more equations:

$$\begin{cases} q \wedge r \wedge s = 0, \\ s \wedge \beta = 0. \end{cases} \tag{4.37}$$

Equivalently, we have algebraic equations as follows:

$$\begin{cases} M_c^2 qrs = \delta_2^2, \\ M_c s\beta = \delta_2^2. \end{cases} \tag{4.38}$$

One way to solve this problem is to add (4.38) to (4.34) and solve this system of equations again. Obviously, this is a computationally intensive task. From Table 4.2, it is easy to check that only x_3 satisfies (4.38). So, in this case, x_3 is the unique solution.

The major disadvantage of the method proposed above is the complexity of computation. We now give an example to illustrate it.

Example 4.8 (Eight queens puzzle) Eight queens are to be placed on an 8×8 chessboard such that none of them is able to capture any other using the standard queen's moves. The queens must be placed in such a way that no two queens are attacking each other. Thus, a solution requires that no two queens share the same row, column, or diagonal.

Fig. 4.1 Eight queens puzzle

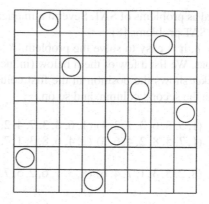

One solution is depicted in Fig. 4.1.

The problem can be extended to the "n queens puzzle" of placing n queens on an $n \times n$ chessboard.

Consider the n queens puzzle. We use P_{ij} to denote the placement of a queen at position (i, j). Then:

- "one row has exactly one queen" can be formulated as

$$\begin{cases} P_{i1} \vee P_{i2} \vee \cdots \vee P_{in} = T, & i = 1, 2, \ldots, n \\ P_{ij} \wedge P_{ik} = F, & j \neq k, i = 1, 2, \ldots, n, \end{cases} \qquad (4.39)$$

- "no two queens share the same column" can be formulated as

$$P_{ji} \wedge P_{ki} = F, \quad j \neq k, \ i = 1, 2, \ldots, n, \qquad (4.40)$$

- "no two queens share the same diagonal" can be formulated as

$$\begin{cases} P_{ij} \wedge P_{i+k\,j+k} = F, & 1 \leq i + k \leq n, 1 \leq j + k \leq n, \\ P_{ij} \wedge P_{i+k\,j-k} = F, & 1 \leq i + k \leq n, 1 \leq j - k \leq n, \\ & i, j = 1, 2, \ldots, n. \end{cases}$$

This can be clarified as

$$\begin{cases} P_{ij} \wedge P_{i+k\,j+k} = F, & 1 - \min\{i, j\} \leq k \leq n - \max\{i, j\}, \\ P_{ij} \wedge P_{i+k\,j-k} = F, & \max\{1 - i, j - n\} \leq k \leq \min\{n - i, j - 1\}, \\ & i, j = 1, 2, \ldots, n. \end{cases} \qquad (4.41)$$

The n queens puzzle is equivalent to solving logical equations (4.39)–(4.41). Since there are n^2 logical unknowns, in general it is impossible to solve the problem by the method proposed earlier. In this case, we may give up the effort of finding all solutions and simply try to find some particular solutions. This kind of problem is called a satisfiability problem (SAT). SAT is an important problem in computer science and its applications. Many decision making problems of intelligent systems can be formulated as problems of SAT. Several numerical methods have been developed to deal with SAT problems [8].

As $n = 8$, it is easy to solve the problem. A simple routine shows that there are 92 solutions. We list a few of them below (in increasing order). The first number in each bracket shows the position in the first column, the second number shows the position in the second column, and so on.

$$(1, 5, 8, 6, 3, 7, 2, 4)\ (1, 6, 8, 3, 7, 4, 2, 5)\ (1, 7, 4, 6, 8, 2, 5, 3)$$
$$(1, 7, 5, 8, 2, 4, 6, 3)\ (2, 4, 6, 8, 3, 1, 7, 5)\ (2, 5, 7, 1, 3, 8, 6, 4)$$
$$\cdots$$
$$(8, 2, 5, 3, 1, 7, 4, 6)\ (8, 3, 1, 6, 2, 5, 7, 4)\ (8, 4, 1, 3, 6, 2, 7, 5).$$

4.4 Substitution

It is well known that in solving a system of linear algebraic equations, a general formula (e.g., using determinants) may be complicated. However, using some unknown substitutions may simplify the calculation substantially. Similarly, in solving logical equations, certain algebraic substitutions may simplify the calculation tremendously. We need some formulas for this and in the following proposition we provide some simple ones which follow directly from the definitions.

Proposition 4.2 *Let* $\mathscr{A}_1, \ldots, \mathscr{A}_s$ *be (possibly compound) logical variables.*

1. *If*

$$M_c^{s-1} \mathscr{A}_1 \cdots \mathscr{A}_s = \delta_2^1, \tag{4.42}$$

   then

$$\mathscr{A}_1 = \cdots = \mathscr{A}_s = \delta_2^1. \tag{4.43}$$

2. *If*

$$M_d^{s-1} \mathscr{A}_1 \cdots \mathscr{A}_s = \delta_2^2, \tag{4.44}$$

   then

$$\mathscr{A}_1 = \cdots = \mathscr{A}_s = \delta_2^2. \tag{4.45}$$

3. *If*

$$M_c^{s-1} \mathscr{A}_1 \cdots \mathscr{A}_s = \delta_2^2 \tag{4.46}$$

   and, for some $1 \leq k \leq s$,

$$\mathscr{A}_k = \delta_2^1, \tag{4.47}$$

   then $\mathscr{A}_k$ *can be removed. That is,* (4.46) *can be reduced to*

$$M_c^{s-2} \mathscr{A}_1 \cdots \mathscr{A}_{k-1} \mathscr{A}_{k+1} \cdots \mathscr{A}_s = \delta_2^2. \tag{4.48}$$

4. *If*

$$M_d^{s-1}\mathscr{A}_1 \cdots \mathscr{A}_s = \delta_2^1 \tag{4.49}$$

and, for some $1 \leq k \leq s$,

$$\mathscr{A}_k = \delta_2^2, \tag{4.50}$$

then $\mathscr{A}_k$ *can be removed. That is,* (4.49) *can be reduced to*

$$M_d^{s-2}\mathscr{A}_1 \cdots \mathscr{A}_{k-1}\mathscr{A}_{k+1} \cdots \mathscr{A}_s = \delta_2^1. \tag{4.51}$$

From distributive laws (1.20) and (1.21), we can obtain the following "factorization" formulas.

Proposition 4.3 *Let* $\mathscr{A}_1, \ldots, \mathscr{A}_s$ *be (possibly compound) logical variables and* p *another logical variable. Then,*

$$M_d^{s-1}(M_c p \mathscr{A}_1 \cdots M_c p \mathscr{A}_s) = M_c p M_d^{s-1}\mathscr{A}_1 \cdots \mathscr{A}_s; \tag{4.52}$$

and

$$M_c^{s-1}(M_d p \mathscr{A}_1 \cdots M_d p \mathscr{A}_s) = M_d p M_c^{s-1}\mathscr{A}_1 \cdots \mathscr{A}_s. \tag{4.53}$$

These formulas are useful for simplifying logical equations. For instance, if

$$M_d^{s-1}M_c p \mathscr{A}_1 \cdots M_c p \mathscr{A}_s = \delta_2^1, \tag{4.54}$$

then we have

$$\begin{cases} p = \delta_2^1, \\ M_d^{s-1}\mathscr{A}_1 \cdots p \mathscr{A}_s = \delta_2^1. \end{cases} \tag{4.55}$$

If

$$M_c^{s-1}M_d p \mathscr{A}_1 \cdots M_d p \mathscr{A}_s = \delta_2^2, \tag{4.56}$$

then we have

$$\begin{cases} p = \delta_2^2, \\ M_c^{s-1}\mathscr{A}_1 \cdots p \mathscr{A}_s = \delta_2^2. \end{cases} \tag{4.57}$$

4.5 *k*-valued Logical Equations

Systems of k-valued logical equations have the same form as the system of Boolean equations (4.1), with f_i as k-valued logical equations, and p_i and c_i as k-valued logical arguments and k-valued constants, respectively. We do not need to repeat the basic concepts discussed in connection with Boolean equations as these can be naturally extended from Boolean logic to k-valued logic.

First, we adapt Algorithm 4.1 to the k-valued case. An argument similar to binary case shows the following. We leave the proof to the reader. Let $x = \ltimes_{s=1}^{n} A_s$, where $A_s \in \Delta_k$. If we assume $x = \delta_{k^n}^{i}$, then the $\{A_s \mid s = 1, \ldots, n\}$ can be calculated by means of the following algorithm.

Algorithm 4.2

- Step 1. Define $b_0 := k^n - i$.
- Step 2. Calculate a_j, b_j, and A_j, $j = 1, 2, \ldots, n$, recursively, by

$$
\begin{cases}
a_j(t) = [\frac{b_{j-1}}{k^{n-j}}], \\
b_j = b_{j-1} - a_j * k^{n-j}, \\
A_j = a_j/(k-1), \quad j = 1, 2, \ldots, n.
\end{cases}
\tag{4.58}
$$

We give an example of this.

Example 4.9 Assume $x = A_1 A_2 A_3 A_4 A_5$ and $x = \delta_{243}^{17}$. Then,

$$
b_0 = 243 - 17 = 226.
$$

It follows that

$$
a_1 = [b_0/3^4] = 2, \qquad A_1 = 1.
$$

Continuing this procedure, we have

$$
\begin{aligned}
&b_1 = b_0 - a_1 * (3^4) = 64, & a_2 = [b_1/3^3] = 2, & \quad A_2 = 1, \\
&b_2 = b_1 - a_2 * 27 = 10, & a_3 = [b_2/3^2] = 1, & \quad A_3 = 0.5, \\
&b_3 = b_2 - a_3 * 3^2 = 1, & a_4 = [b_3/3] = 0, & \quad A_4 = 0, \\
&b_4 = b_3 - a_4 * 3 = 1, & a_5 = [b_4/1] = 1, & \quad A_5 = 0.5.
\end{aligned}
$$

We conclude that $A_1 = 1 \sim \delta_3^1$, $A_2 = 1 \sim \delta_3^1$, $A_3 = 0.5 \sim \delta_3^2$, $A_4 = 0 \sim \delta_3^3$, and $A_5 = 0.5 \sim \delta_3^2$.

Next, we modify Lemma 4.2 for k-valued logic.

Lemma 4.4 *Assume* $z_j = p_1 p_2 \cdots p_j$, *where* $p_i \in \Delta_k$, $i = 1, 2, \ldots, j$. *Then,*

$$
z_j^2 = \Phi_{j,k} z_j,
\tag{4.59}
$$

where

$$
\Phi_{j,k} = \prod_{i=1}^{j} I_{k^{i-1}} \otimes \left[(I_k \otimes W_{[k,k^{j-i}]}) M_{r,k} \right].
\tag{4.60}
$$

Finally, we generalize Theorem 4.2 to the k-valued case. Consider a k-valued logical equation

$$Lx = b, \qquad (4.61)$$

where $L \in \mathcal{L}_{k^m \times k^n}$, $b \in \Delta_{k^m}$, and $k \in \Delta_{k^n}$. Express L in a condensed form as

$$L = \delta_{k^m} [i_1, i_2, \ldots, i_{k^n}]$$

and define the set

$$\Lambda = \left\{ \lambda \mid \delta_{k^m}^{i_\lambda} = b, 1 \leq \lambda \leq k^n \right\}.$$

We then have the following result.

Theorem 4.4 *The solution of* (4.61) *is*

$$x = \delta_{k^n}^\lambda, \quad \lambda \in \Lambda. \qquad (4.62)$$

We now give an example to show how to use k-valued logical equations to deal with logical inference.

Example 4.10 A detective is investigating a murder case. He has the following clues:

- he is 80% sure that either A or B is the murderer,
- if A is the murderer, it is very likely that the murder happened after midnight,
- if B's confession is true, then the light at midnight was on,
- if B's confession is false, it is very likely that the murder happened before midnight,
- there is evidence that the light in the room of the murder was off at midnight.

What conclusion can he draw? First, we must establish the levels of logical values. If we understand "very likely" as more possible that "80%", then we may quantize the logical values into six levels as "T", "very likely", "80%", "$1 - 80\%$", "very unlikely", and "F". Hence, we may consider the problem as one of 6-valued logical inference.

Define the logical variables (unknowns) as

- A: A is the murderer,
- B: B is the murderer,
- M: the murder happened before midnight,
- S: B's confession is true,
- L: the light in the room was on at midnight.

We can then convert the statements into logical equations as follows:

$$A \vee B = \delta_6^3,$$
$$A \to \neg M = \delta_6^2,$$
$$S \to L = \delta_6^1,$$ (4.63)
$$\neg S \to M = \delta_6^2,$$
$$\neg L = \delta_6^1.$$

We may use the general formula provided in Theorem 4.4 to solve this system of logical equations, but substitution will be much easier.

First, from $\neg L = \delta_6^1$ we have

$$L = \neg \delta_6^1 = \delta_6^6.$$

Then, because $S \to L = \delta_6^1$, we have the following matrix form:

$$M_{i,6} SL = M_{i,6} W_{[6]} LS := \Psi_1 S.$$

We then have

$$\Psi_1 S = b,$$

where $b = \delta_6^1$, and it is easy to calculate that

$$\Psi_1 = M_{i,6} W_{[6]} L = \delta_6[6\ 5\ 4\ 3\ 2\ 1].$$

Since only $\mathrm{Col}_6(\Psi_1) = b$, the solution is

$$S = \delta_6^6.$$

Similarly, from $\neg S \to M = \delta_6^2$ we have

$$M_{i,6} M_{n,6} SM = \delta_6^2.$$

We can thus solve for M:

$$M = \delta_6^2.$$

Next, we consider $A \to \neg M = M_{i,6} A M_{n,6} M = \delta_6^2$. Applying some properties of the semi-tensor product, we obtain

$$M_{i,6} A M_{n,6} M = M_{i,6}(I_6 \otimes M_{n,6}) AM = M_{i,6}(I_6 \otimes M_{n,6}) W_{[6]} MA := \psi_2 A.$$

It can be calculated that

$$\psi_2 = M_{i,6}(I_6 \otimes M_{n,6}) W_{[6]} M = \delta_6[5\ 5\ 4\ 3\ 2\ 1].$$

Hence, we have

$$A = \delta_6^5.$$

Finally, from $A \vee B = M_d AB = \delta_6^3$ we can solve for B:

$$B = \delta_6^3.$$

We conclude that

- it is "very unlikely" that A is the murderer,
- it is 80% possible that B is the murderer.

4.6 Failure Location: An Application

As an application of the algebraic expression of logical equations, we consider the failure location problem in networks.

Recently, some methods of quality of service (QoS) degradation locating from observed data on the end-to-end performance of flows have been proposed and investigated by [3–5, 7], etc. The basic idea of this approach can be described as follows. First, set a quality threshold for a network. According to this threshold the flows are classified as good quality flows or bad quality flows. Assume a bad flow is caused by a certain failure link (or several) on the flow path. The location of failure then needs to be specified in order to improve the QoS of the network. The routing information can be obtained by routers. End-to-end verification in a random framework is also a promising method [1].

We assume the route information is not completely known. This is practically reasonable since, for technical reasons, the detected routes between two testing ends may not be as precise as a simple serial line. In this case the routes between two testing ends are allowed to be serial–parallel ones.

It is natural to identify a good link with a through (ON) link and a bad link with a broken (OFF) one. In this way, the problem becomes one of solving a system of Boolean equations.

4.6.1 Matrix Expression of Route Logic

To begin with, we give a rigorous definition of a route network and the logical relationship between an end-to-end path and its links.

Definition 4.2 A route network consists of a finite set of nodes, denoted by $\mathcal{N} = \{A, B, C, \ldots\}$ and a finite set of links, denoted by $\mathcal{S} = \{a, b, c, \ldots\}$. Therefore, a network can be denoted by a pair $(\mathcal{N}, \mathcal{S})$.

Remark 4.2

- A link s is an arc between two nodes. We assume a link to be the smallest possible segment, that is, there are no middle nodes on a link, so a link is an "atom" of the route. A link s could be through (called "ON"), denoted by $s = 1$ (equivalently, in

Fig. 4.2 A network

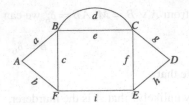

vector form, $s = \delta_2^1$), or broken (called "OFF"), denoted by $s = 0$ (equivalently, $s = \delta_2^2$). Thus, a link is a Boolean variable.

- Unlike in graph theory, there may be several links between two nodes. Also, unlike the Boolean network case, under the route topology a link, not a node, is a Boolean variable.
- A link s has two ending nodes. If s is between A and B, then we write $n(s) = \{A, B\}$.

We need a description for part of a network, which could be considered as the route of an end-to-end testing.

Definition 4.3

1. A network $(\mathcal{N}', \mathcal{S}')$ is said to be a subnet of $(\mathcal{N}, \mathcal{S})$ if $\mathcal{N}' \subset \mathcal{N}$ and $\mathcal{S}' \subset \mathcal{S}$. For such a subnet, we write $(\mathcal{N}', \mathcal{S}') \subset (\mathcal{N}, \mathcal{S})$.
2. A subnet $(\mathcal{N}', \mathcal{S}') \subset (\mathcal{N}, \mathcal{S})$ is said to be complete if $s \in \mathcal{S}$ and $n(s) \subset \mathcal{N}'$ implies that $s \in \mathcal{S}'$.

Definition 4.4

1. A path from a node A to a node B is a set of serial links such that we can get from A to B along the serially connected links. A path without self-intersection is called a legal path, otherwise it is illegal.
2. A route with ending nodes A and B, denoted by $r(A, B)$, is a complete subnet consisting of some nodes $\mathcal{N}' \subset \mathcal{N}$ with $A, B \in \mathcal{N}'$. We denote its node set as $n(r) = \{X \mid X \in \mathcal{N}'\}$.
3. When the logical structure of a route is considered, only legal paths are counted.

We now give an illustrative example.

Example 4.11 Consider the network in Fig. 4.2. We can conclude the following:

1. The set of nodes of the network is $\mathcal{N} = \{A, B, C, D, E, F\}$ and the set of links is $\mathcal{S} = \{a, b, c, d, e, f, g, h, i\}$.
2. Let $\mathcal{N}_1 = \{A, F, E, D\}$, $\mathcal{S}_1 = \{b, i, h\}$, $\mathcal{N}_2 = \{A, B, C, D\}$, and $\mathcal{S}_2 = \{a, e, g\}$. Then $(\mathcal{N}_1, \mathcal{S}_1)$ is a complete subnet and $(\mathcal{N}_2, \mathcal{S}_2)$ is not a complete subnet.
3. $(\mathcal{N}, \mathcal{S})$ can be considered as a route from A to D, denoted by $r(A, D)$.
4. $b - i - f - g$ is a legal path between A and D, denoted by $p(A, D)$.
5. $b - i - f - e - c - i - h$ is an illegal path, so it is not considered as a path in $r(A, D)$.

Fig. 4.3 A network

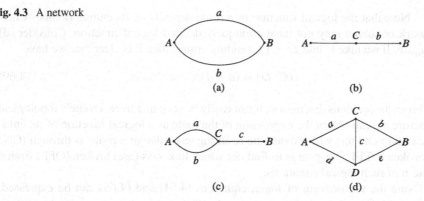

(a)   (b)

(c)   (d)

6. Let $\mathcal{N}_3 = \{A, B, C, D\}$ and $\mathcal{S}_3 = \{a, e, d, g\}$. $(\mathcal{N}_3, \mathcal{S}_3)$ is also a route from A to D, consisting of two paths from A to D: $a - d - g$ and $a - e - g$.

Note that a route can be either ON or OFF, but not both, and it is obvious that a route is also a Boolean variable. Since the logical value of a route is determined by its links, it is a logical function of its links. We must therefore determine the function of a route, where its links act as the arguments of the function. We will explain this by means of the following example.

Example 4.12 Consider the routes in Fig. 4.3.

- Route (a) has a parallel connection. It is clear that $r(A, B)$ is ON if either a or b is ON. Therefore, we have the following logical relation:

$$r(A, B) = a \vee b. \tag{4.64}$$

- Route (b) has a serial connection, hence

$$r(A, B) = a \wedge b. \tag{4.65}$$

- In route (c), a and b are connected in parallel mode and this subnet is then connected with c in serial mode. It is easy to see that

$$r(A, B) = (a \vee b) \wedge c. \tag{4.66}$$

An alternative way to analyze this is as follows: $r(A, B)$ consists of two paths, $a - c$ and $b - c$, so

$$r(A, B) = (a \wedge c) \vee (b \wedge c). \tag{4.67}$$

Obviously, (4.66) and (4.67) are the same.

- Consider route (d). The parallel–serial structure analysis seems complicated. After careful path analysis, it is easy to see that there are four paths: $a - b$, $d - e$, $a - c - e$, and $d - c - b$. Thus, we have

$$r(A, B) = (a \wedge b) \vee (d \wedge e) \vee (a \wedge c \wedge e) \vee (d \wedge c \wedge b). \tag{4.68}$$

Note that the logical function of a route depends on its ending points. A network or subnet may not have a uniquely defined logical function. Consider (d) again. If we take C and D as two ending points, then it is clear that we have

$$r(C, D) = (a \wedge d) \vee c \vee (b \wedge e). \tag{4.69}$$

From the previous discussion, it can easily be seen that from a route's topological structure we can obtain the expression of the route as a logical function of its links. The end-to-end test will provide the resulting test value of a route as through (ON) or broken (OFF). Our goal is to find out which link(s) is (are) broken (OFF) from a system of such logical equations.

Using the vector form of logic, equations (4.64) and (4.65) can be expressed, respectively, as

$$r(A, B) = M_d ab = \delta_2[1\ 1\ 1\ 2]ab \tag{4.70}$$

and

$$r(A, B) = M_c ab = \delta_2[1\ 2\ 2\ 2]ab. \tag{4.71}$$

As for (4.66), we have

$$r(A, B) = M_c(M_d ab)c = M_c M_d abc := M_3 abc. \tag{4.72}$$

The coefficient matrix, M_3, of (4.72) can be calculated as

$$M_3 = M_c M_d = \delta_2[1\ 2\ 1\ 2\ 1\ 2\ 2\ 2].$$

Finally, we consider (4.68). It is easy to calculate that

$$r(A, B) = M_d^3(M_c ab)(M_c de)\left(M_c^2 ace\right)\left(M_c^2 dcb\right) := M_4 abcde, \tag{4.73}$$

where

$$M_4 = M_\alpha M_\beta = \delta_2[1\ 1\ 1\ 1\ 1\ 1\ 1\ 1\ 1\ 2\ 1\ 2\ 1\ 2\ 2\ 2\ 1\ 1\ 2\ 2\ 1\ 2\ 2\ 2\ 1\ 2\ 2\ 2\ 1\ 2\ 2\ 2].$$

Remark 4.3 It can be easily seen that as long as the network structure is known for a route with two fixed ends, its logical value is a known function of its links. Therefore, the method proposed in this section is applicable to networks without routers (routers can only add some information to reduce the computation complexity).

4.6.2   Failure Location

Consider a network $(\mathcal{N}, \mathcal{S})$ where the links are labeled as

$$\mathcal{S} = \{s_1, s_2, \dots, s_n\}.$$

Let $A, B \in \mathcal{N}$. From the network structure, with available information obtained from routers, we can have a route $r(A, B)$. A trivial case is when $r(A, B)$ is a series connection. Here, though, we assume that it can be an arbitrary logical function of its links. When a real network such as the Internet is considered, obtaining a precise description is difficult. This is particularly the case when the network is simplified by approximations. Of course, the end-to-end testing result is known. As discussed earlier, the route $r(A, B)$ is a logical function of its component links, expressed as

$$r(A, B) = f(s_1, s_2, \ldots, s_n), \qquad (4.74)$$

where f is a logical function. (4.74) can be converted into its algebraic form as

$$r(A, B) = M_f \ltimes_{i=1}^n s_i. \qquad (4.75)$$

Now, assuming we have tested m routes, we have m end-to-end testing results:

$$r_i(A_i, B_i) = f_i(x_1, \ldots, x_n) = b_i, \quad i = 1, 2, \ldots, m.$$

Converting them into algebraic equations, we have

$$\begin{cases} M_1 x = b_1, \\ M_2 x = b_2, \\ \quad \vdots \\ M_n x = b_m, \end{cases} \qquad (4.76)$$

where $M_i = M_{f_i}$ is the structure matrix of the logical function f_i of $r_i(A_i, B_i)$, and

$$x = \ltimes_{i=1}^n x_i.$$

Then, (4.76) can be converted into a linear algebraic equation,

$$Lx = b, \qquad (4.77)$$

where $b = \ltimes_{i=1}^n b_i$, and

$$L = M_1 \ltimes_{j=2}^n \left[(I_{2^n} \otimes M_j) \Phi_n \right],$$

$$\Phi_n = \prod_{i=1}^n I_{2^{i-1}} \otimes \left[(I_2 \otimes W_{[2, 2^{j-i}]}) M_r \right].$$

From the previous section it is clear that:

1. Equation (4.77) has solution if and only if $b \in \mathrm{Col}(L)$,
2. $x = \delta_{2^n}^k$ is a solution of (4.76) if and only if $\mathrm{Col}_k(L) = b$.

Next, we give an example to illustrate the above theorem.

Example 4.13 Consider Fig. 4.4, where (a) is the network, and (b)–(e) are four routes, denoted by r_1 to r_4, respectively.

Fig. 4.4 Network with four
routes

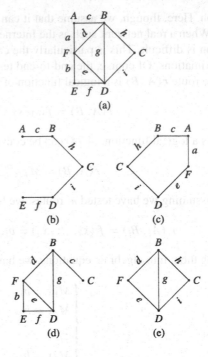

Now, assume the testing results are $r_1(A, E) = 0$, $r_2(C, A) = 1$, $r_3(E, C) = 0$,
and $r_4(F, C) = 1$. We then have the following system of logical equations:

$$\begin{cases} c \wedge h \wedge i \wedge f = 0, \\ (h \wedge c) \vee (i \wedge e \wedge a) = 1, \\ [(b \wedge d) \vee (f \wedge g) \vee (b \wedge e \wedge g) \vee (f \wedge e \wedge d)] \wedge h = 0, \\ (d \wedge h) \vee (e \wedge i) \vee (d \wedge g \wedge i) \vee (e \wedge g \wedge h) = 1. \end{cases} \quad (4.78)$$

Using vector form and matrix expression, we have

$$\begin{cases} M_c^3 chif = \delta_2^2, \\ M_d M_c hc M_c^2 iea = \delta_2^1, \\ M_c M_d M_c bd M_d M_c fg M_d M_c^2 beg M_c^2 fedh = \delta_2^2, \\ M_d M_c dh M_d M_c ei M_c^2 dgi = \delta_2^1. \end{cases} \quad (4.79)$$

After reducing the powers, (4.79) is expressed in normal form as

$$\begin{cases} M_1 cfhi = \delta_2^2, \\ M_2 acehi = \delta_2^1, \\ M_3 bdefgh = \delta_2^2, \\ M_4 deghi = \delta_2^1, \end{cases} \quad (4.80)$$

where

$$M_1 = \delta_2[1\ 2\ 2\ 2\ 2\ 2\ 2\ 2\ 2\ 2\ 2\ 2\ 2\ 2\ 2\ 2],$$

$$M_2 = \delta_2[1\ 1\ 1\ 2\ 1\ 1\ 2\ 2\ 1\ 2\ 1\ 2\ 2\ 2\ 2\ 2\ 1\ 1\ 2\ 2\ 1\ 1\ 2\ 2\ 2\ 2\ 2\ 2\ 2\ 2\ 2],$$

$$M_3 = \delta_2[1\ 2\ 1\ 2\ 1\ 2\ 1\ 2\ 1\ 2\ 1\ 2\ 1\ 2\ 1\ 2\ 1\ 2\ 2\ 2\ 1\ 2\ 2\ 2\ 1\ 2\ 2\ 2\ 2\ 2\ 2\ 2$$
$$\quad\ 1\ 2\ 1\ 2\ 2\ 2\ 2\ 2\ 1\ 2\ 2\ 2\ 2\ 2\ 2\ 2\ 1\ 2\ 2\ 2\ 2\ 2\ 2\ 2\ 1\ 2\ 2\ 2\ 2\ 2\ 2\ 2],$$

$$M_4 = \delta_2[1\ 1\ 1\ 2\ 1\ 1\ 1\ 2\ 1\ 1\ 1\ 2\ 1\ 1\ 2\ 2\ 1\ 2\ 1\ 2\ 1\ 2\ 1\ 2\ 2\ 2\ 2\ 2\ 2\ 2\ 2].$$

Multiplying the four equations in (4.80) together yields

$$Lx = b, \tag{4.81}$$

where $x = abcdefghi$, $b = \delta_2^2\delta_2^1\delta_2^2\delta_2^1 = \delta_{16}^{11}$, and

$$
\begin{aligned}
L = \delta_{16}[\ &1\ \ 9\ 11\ 16\ \ 1\ \ 9\ 11\ 16\ \ 9\ \ 9\ 11\ 16\ \ 9\ \ 9\ 11\ 16 \\
&1\ \ 9\ 15\ \ 1\ \ 1\ \ 9\ 16\ 16\ \ 9\ \ 9\ 15\ 16\ \ 9\ \ 9\ 16\ 16 \\
&1\ 10\ 11\ 16\ \ 3\ 12\ 11\ 16\ \ 9\ 10\ 11\ 16\ 11\ 12\ 11\ 16 \\
&2\ 10\ 16\ 16\ \ 4\ 12\ 16\ 16\ 12\ 12\ 16\ 16\ 12\ 12\ 16\ 16 \\
&9\ 13\ 11\ 16\ \ 9\ 13\ 11\ 16\ \ 9\ 13\ 11\ 16\ \ 9\ 13\ 11\ 16 \\
&13\ 13\ 15\ 16\ 13\ 13\ 16\ 16\ 13\ 13\ 15\ 16\ 13\ 13\ 16\ 16 \\
&9\ 14\ 11\ 16\ 11\ 16\ 11\ 16\ \ 9\ 14\ 11\ 16\ 11\ 16\ 11\ 16 \\
&14\ 14\ 16\ 16\ 16\ 16\ 16\ 16\ 16\ 16\ 16\ 16\ 16\ 16\ 16\ 16 \\
&1\ \ 9\ 11\ 16\ \ 1\ \ 9\ 11\ 16\ 11\ 11\ 11\ 16\ 11\ 11\ 11\ 16 \\
&1\ \ 9\ 15\ 16\ \ 3\ 11\ 16\ 16\ 11\ 11\ 15\ 16\ 11\ 11\ 16\ 16 \\
&1\ 10\ 11\ 16\ \ 3\ 12\ 11\ 16\ 11\ 12\ 11\ 16\ 11\ 12\ 11\ 16 \\
&2\ 10\ 16\ 16\ \ 4\ 12\ 16\ 16\ 12\ 12\ 16\ 16\ 12\ 12\ 16\ 16 \\
&9\ 13\ 11\ 16\ \ 9\ 13\ 11\ 16\ 11\ 15\ 11\ 16\ 11\ 15\ 11\ 16 \\
&13\ 13\ 15\ 16\ 15\ 15\ 16\ 16\ 15\ 15\ 15\ 16\ 15\ 15\ 16\ 16 \\
&9\ 14\ 11\ 16\ 11\ 16\ 11\ 16\ 11\ 16\ 11\ 16\ 11\ 16\ 11\ 16 \\
&14\ 14\ 16\ 16\ 16\ 16\ 16\ 16\ 16\ 16\ 16\ 16\ 16\ 16\ 16\ 16 \\
&1\ \ 9\ 15\ 16\ \ 1\ \ 9\ 15\ 16\ \ 9\ \ 9\ 15\ 16\ \ 9\ \ 9\ 15\ 16 \\
&1\ \ 9\ 15\ 16\ \ 1\ \ 9\ 16\ 16\ \ 9\ \ 9\ 15\ 16\ \ 9\ \ 9\ 16\ 16 \\
&1\ 10\ 15\ 16\ \ 3\ 12\ 15\ 16\ \ 9\ 10\ 15\ 16\ 11\ 12\ 15\ 16 \\
&2\ 10\ 16\ 16\ \ 4\ 12\ 16\ 16\ 12\ 12\ 16\ 16\ 12\ 12\ 16\ 16 \\
&13\ 13\ 15\ 16\ 13\ 13\ 15\ 16\ 13\ 13\ 15\ 16\ 13\ 13\ 15\ 16 \\
&13\ 13\ 15\ 16\ 13\ 13\ 16\ 16\ 13\ 13\ 15\ 16\ 13\ 13\ 16\ 16 \\
&13\ 14\ 15\ 16\ 15\ 16\ 15\ 16\ 13\ 14\ 15\ 16\ 15\ 16\ 15\ 16 \\
&14\ 14\ 16\ 16\ 16\ 16\ 16\ 16\ 16\ 16\ 16\ 16\ 16\ 16\ 16\ 16 \\
&1\ \ 9\ 15\ 16\ \ 1\ \ 9\ 15\ 16\ 11\ 11\ 15\ 16\ 11\ 11\ 15\ 16 \\
&1\ \ 9\ 15\ 16\ \ 3\ 11\ 16\ 16\ 11\ 11\ 15\ 16\ 11\ 11\ 16\ 16 \\
&1\ 10\ 15\ 16\ \ 3\ 12\ 15\ 16\ 11\ 12\ 15\ 16\ 11\ 12\ 15\ 16 \\
&2\ 10\ 16\ 16\ \ 4\ 12\ 16\ 16\ 12\ 12\ 16\ 16\ 12\ 12\ 16\ 16 \\
&13\ 13\ 15\ 16\ 13\ 13\ 15\ 16\ 15\ 15\ 15\ 16\ 15\ 15\ 15\ 16 \\
&13\ 13\ 15\ 16\ 15\ 15\ 16\ 16\ 15\ 15\ 15\ 16\ 15\ 15\ 16\ 16 \\
&13\ 14\ 15\ 16\ 15\ 16\ 15\ 16\ 15\ 16\ 15\ 16\ 15\ 16\ 15\ 16 \\
&14\ 14\ 16\ 16\ 16\ 16\ 16\ 16\ 16\ 16\ 16\ 16\ 16\ 16\ 16\ 16].
\end{aligned}
$$

From L one sees that the columns $\text{Col}_3(L), \text{Col}_7(L), \ldots$ are δ_{16}^{11}. Hence, $x = \delta_{29}^{3}, \delta_{29}^{7}, \ldots$ are the solutions. Now, consider δ_{29}^{3}. Using formula (4.58) it is easy to find the solution

$$x_1 \sim (s_1, s_2, \ldots, s_9) = (1\ 1\ 1\ 1\ 1\ 1\ 1\ 0\ 1).$$

In the following all the solutions are listed:

$$
\begin{array}{ll}
x_1 \sim (1\ 1\ 1\ 1\ 1\ 1\ 1\ 0\ 1), & x_2 \sim (1\ 1\ 1\ 1\ 1\ 1\ 0\ 0\ 1), \\
x_3 \sim (1\ 1\ 1\ 1\ 1\ 0\ 1\ 0\ 1), & x_4 \sim (1\ 1\ 1\ 1\ 1\ 0\ 0\ 0\ 1), \\
x_5 \sim (1\ 1\ 1\ 0\ 1\ 1\ 1\ 0\ 1), & x_6 \sim (1\ 1\ 1\ 0\ 1\ 1\ 0\ 0\ 1), \\
x_7 \sim (1\ 1\ 1\ 0\ 1\ 0\ 1\ 0\ 1), & x_8 \sim (1\ 1\ 1\ 0\ 1\ 0\ 0\ 1\ 1), \\
x_9 \sim (1\ 1\ 1\ 0\ 1\ 0\ 0\ 0\ 1), & x_{10} \sim (1\ 1\ 0\ 1\ 1\ 1\ 1\ 0\ 1), \\
x_{11} \sim (1\ 1\ 0\ 1\ 1\ 1\ 0\ 0\ 1), & x_{12} \sim (1\ 1\ 0\ 1\ 1\ 0\ 1\ 0\ 1), \\
x_{13} \sim (1\ 1\ 0\ 1\ 1\ 0\ 0\ 0\ 1), & x_{14} \sim (1\ 1\ 0\ 0\ 1\ 1\ 1\ 0\ 1), \\
x_{15} \sim (1\ 1\ 0\ 0\ 1\ 1\ 0\ 1\ 1), & x_{16} \sim (1\ 1\ 0\ 0\ 1\ 1\ 0\ 0\ 1), \\
x_{17} \sim (1\ 1\ 0\ 0\ 1\ 0\ 1\ 0\ 1), & x_{18} \sim (1\ 1\ 0\ 0\ 1\ 0\ 0\ 1\ 1), \\
x_{19} \sim (1\ 1\ 0\ 0\ 1\ 0\ 0\ 0\ 1), & x_{20} \sim (1\ 0\ 1\ 1\ 1\ 1\ 1\ 0\ 1), \\
x_{21} \sim (1\ 0\ 1\ 1\ 1\ 1\ 0\ 0\ 1), & x_{22} \sim (1\ 0\ 1\ 1\ 1\ 0\ 1\ 1\ 1), \\
x_{23} \sim (1\ 0\ 1\ 1\ 1\ 0\ 1\ 1\ 0), & x_{24} \sim (1\ 0\ 1\ 1\ 1\ 0\ 1\ 0\ 1), \\
x_{25} \sim (1\ 0\ 1\ 1\ 1\ 0\ 0\ 1\ 1), & x_{26} \sim (1\ 0\ 1\ 1\ 1\ 0\ 0\ 1\ 0), \\
x_{27} \sim (1\ 0\ 1\ 1\ 1\ 0\ 0\ 0\ 1), & x_{28} \sim (1\ 0\ 1\ 1\ 0\ 1\ 0\ 1\ 0), \\
x_{29} \sim (1\ 0\ 1\ 1\ 0\ 0\ 1\ 1\ 1), & x_{30} \sim (1\ 0\ 1\ 1\ 0\ 0\ 1\ 1\ 0), \\
x_{31} \sim (1\ 0\ 1\ 1\ 0\ 0\ 0\ 1\ 1), & x_{32} \sim (1\ 0\ 1\ 1\ 0\ 0\ 0\ 1\ 0), \\
x_{33} \sim (1\ 0\ 1\ 0\ 1\ 1\ 1\ 0\ 1), & x_{34} \sim (1\ 0\ 1\ 0\ 1\ 1\ 0\ 0\ 1), \\
x_{35} \sim (1\ 0\ 1\ 0\ 1\ 0\ 1\ 1\ 1), & x_{36} \sim (1\ 0\ 1\ 0\ 1\ 0\ 1\ 0\ 1), \\
x_{37} \sim (1\ 0\ 1\ 0\ 1\ 0\ 0\ 1\ 1), & x_{38} \sim (1\ 0\ 1\ 0\ 1\ 0\ 0\ 0\ 1), \\
x_{39} \sim (1\ 0\ 0\ 1\ 1\ 1\ 1\ 0\ 1), & x_{40} \sim (1\ 0\ 0\ 1\ 1\ 1\ 0\ 0\ 1), \\
x_{41} \sim (1\ 0\ 0\ 1\ 1\ 0\ 1\ 1\ 1), & x_{42} \sim (1\ 0\ 0\ 1\ 1\ 0\ 1\ 0\ 1), \\
x_{43} \sim (1\ 0\ 0\ 1\ 1\ 0\ 0\ 1\ 1), & x_{44} \sim (1\ 0\ 0\ 1\ 1\ 0\ 0\ 0\ 1), \\
x_{45} \sim (1\ 0\ 0\ 0\ 1\ 1\ 1\ 0\ 1), & x_{46} \sim (1\ 0\ 0\ 0\ 1\ 1\ 0\ 1\ 1), \\
x_{47} \sim (1\ 0\ 0\ 0\ 1\ 1\ 0\ 0\ 1), & x_{48} \sim (1\ 0\ 0\ 0\ 1\ 0\ 1\ 1\ 1), \\
x_{49} \sim (1\ 0\ 0\ 0\ 1\ 0\ 1\ 0\ 1), & x_{50} \sim (1\ 0\ 0\ 0\ 1\ 0\ 0\ 1\ 1), \\
x_{51} \sim (1\ 0\ 0\ 0\ 1\ 0\ 0\ 0\ 1), & x_{52} \sim (0\ 1\ 1\ 0\ 1\ 0\ 0\ 1\ 1), \\
x_{53} \sim (0\ 0\ 1\ 1\ 1\ 0\ 1\ 1\ 1), & x_{54} \sim (0\ 0\ 1\ 1\ 1\ 0\ 1\ 1\ 0), \\
x_{55} \sim (0\ 0\ 1\ 1\ 1\ 0\ 0\ 1\ 1), & x_{56} \sim (0\ 0\ 1\ 1\ 1\ 0\ 0\ 1\ 0), \\
x_{57} \sim (0\ 0\ 1\ 1\ 0\ 1\ 0\ 1\ 0), & x_{58} \sim (0\ 0\ 1\ 1\ 0\ 0\ 1\ 1\ 1), \\
x_{59} \sim (0\ 0\ 1\ 1\ 0\ 0\ 1\ 1\ 0), & x_{60} \sim (0\ 0\ 1\ 1\ 0\ 0\ 0\ 1\ 1), \\
x_{61} \sim (0\ 0\ 1\ 1\ 0\ 0\ 0\ 1\ 0), & x_{62} \sim (0\ 0\ 1\ 0\ 1\ 0\ 1\ 1\ 1), \\
x_{63} \sim (0\ 0\ 1\ 0\ 1\ 0\ 0\ 1\ 1). &
\end{array}
$$

From the above data, one may doubt the value of this approach. What conclusion can we draw from so many solutions? We need the following hypothesis: Failure is unlikely to happen in many places. In other words, the probability of a failure is low. We believe this hypothesis is practically reasonable. Based on this hypothesis, we have the following principle.

Least Side Principle: The most likely failure is the one with the smallest number of ill (broken, or OFF) links.

According to this principle, we conclude that the most likely failure case is x_1 because it is the only one which contains only one broken link. It follows that the ill (broken) link is (very likely) h.

Remark 4.4 In most cases we do not need to solve such an elaborate system of algebraic equations when a practical problem is considered. In fact, if we have a good serial route, meaning that

$$x_{i_1} \wedge x_{i_2} \wedge \cdots \wedge x_{i_k} = 1,$$

then we have

$$x_{i_1} = x_{i_2} = \cdots = x_{i_k} = 1.$$

In a vector-form logical equation we can simply replace x_{i_j} by δ_2^1.

Similarly, if we have a bad parallel route, meaning that

$$x_{i_1} \vee x_{i_2} \vee \cdots \vee x_{i_k} = 0,$$

then we have

$$x_{i_1} = x_{i_2} = \cdots = x_{i_k} = 0.$$

In a vector-form logical equation we can simply replace x_{i_j} by δ_2^2, but this case should be very rare.

4.6.3 Cascading Inference

When a network is not small, the method proposed in last section fails because of the computation complexity. For a large-scale network, [6] proposed a method that logically divides the network into subnets in order to utilize parallelism. Here, we propose an algorithm called cascading inference. This algorithm may considerable reduce the time for inferring.

Definition 4.5 Let $(\mathcal{M}, \mathcal{T}) \subset (\mathcal{N}, \mathcal{S})$ be a complete subnet. The neighborhood degree of $(\mathcal{M}, \mathcal{T})$ is the number of links with one end on $\mathcal{M}$ and the other end on $\mathcal{N} \setminus \mathcal{M}$. Such a link is called a front link of the subnet. A node of the subnet which is attached a front link is called a front node.

We illustrate the notion of cascading inference in Fig. 4.5. If we have a large network, we split it into several subnets. As in Fig. 4.5 we split it into three subnets: $\mathcal{S}_1$, $\mathcal{S}_2$, and $\mathcal{S}_3$. In general we split it in such a way that each subnet has neighborhood degree as small as possible. In our example, $\mathcal{S}_1$, $\mathcal{S}_2$, and $\mathcal{S}_3$ have neighborhood degrees 4, 3, and 3, respectively, and sets of front nodes are $\{A_1, A_2, A_3\}$,

Fig. 4.5 Cascading inference
of a large network

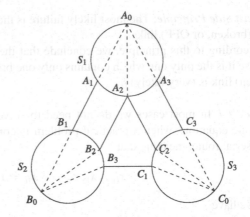

$\{B_1, B_2, B_3\}$, and $\{C_1, C_2, C_3\}$, respectively. In each subnet, we then choose one or more node(s) as the testing end node(s). In the example, we have chosen $A_0 \in \mathscr{S}_1$, $B_0 \in \mathscr{S}_2$, and $C_0 \in \mathscr{S}_3$ as testing end nodes. The test nodes are connected to the front nodes of their own subnets by auxiliary links. In Fig. 4.5 the auxiliary links are drawn as dashed lines. We now obtain a simplified network with all testing nodes and front nodes as its nodes and all front links and auxiliary links as its links. Testing this simplified network, we can determine which subnet contains the failure. We may need to introduce further end-to-end tests to eventually detect which subnet is the troublemaker. Next, we consider the problematic subnet and repeat the same procedure until the failure is located.

Summarizing the above procedure, we propose the following algorithm.

Algorithm 4.3

- Step 1. Split the network into a few subnets according to the principle that each subnet has a low neighborhood degree.
- Step 2. For each subnet simply connect the test point(s) [or end point(s)] with all front nodes to form a new, simplified network.
- Step 3. For the simplified network, use end-to-end testing to find the bad (OFF) link, which could be an auxiliary link.
- Step 4. Replace the original network by the subnet containing the bad link, then go back to Step 1.

We use the following example to illustrate this algorithm.

Example 4.14 Consider the network depicted in Fig. 4.6. The subnet within triangle $\triangle ADE$ has neighborhood degree 2 and its two front nodes are D and E.

We assume that it is a self-similar network, that is, there are three subnets, triangles $\triangle ADE$, $\triangle DBF$, and $\triangle EFC$, which have same structure as $\triangle ABC$. We assume the network contains k layers of such refinement.

We now apply the algorithm to this network. In Step 1 we divide it into three subnets, $\triangle ADE$, $\triangle DBF$, and $\triangle EFC$. For each triangle, in addition to the two

Fig. 4.6 Cascading inference

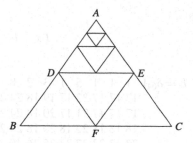

front points, we choose the third vertex as the end point for testing. For ease of statement, we name the segments as follows:

$$x_1 = \overline{AD}, \qquad x_2 = \overline{DB}, \qquad x_3 = \overline{BF}, \qquad x_4 = \overline{FC}, \qquad x_5 = \overline{CE},$$

$$x_6 = \overline{AE}, \qquad x_7 = \overline{DE}, \qquad x_8 = \overline{DF}, \qquad x_9 = \overline{EF}.$$

Assume that we test six end-to-end routes, which are classified into two categories: (I) $r_1 = r(A, B)$, $r_2 = r(A, C)$, and $r_3 = r(B, C)$, with $n(r_1) = \{A, D, B\}$, $n(r_2) = \{A, E, C\}$, and $n(r_3) = \{B, F, C\}$, which are connected as in route (b) of Example 4.11; (II) $r_4 = r(A, F)$, $r_5 = r(B, E)$, and $r_6 = r(C, D)$, with $n(r_4) = \{A, D, E, F\}$, $n(r_5) = \{B, D, E, F\}$, and $n(r_6) = \{C, E, F, D\}$, which are connected as in route (d) of Example 4.11. We have the following system of logical equations:

$$r(A, B) = x_1 \wedge x_2,$$

$$r(B, C) = x_3 \wedge x_4,$$

$$r(A, C) = x_5 \wedge x_6,$$

$$r(A, F) = (x_1 \wedge x_8) \vee (x_6 \wedge x_9) \vee (x_1 \wedge x_7 \wedge x_9) \vee (x_6 \wedge x_7 \wedge x_8),$$

$$r(B, E) = (x_2 \wedge x_7) \vee (x_3 \wedge x_9) \vee (x_2 \wedge x_8 \wedge x_9) \vee (x_3 \wedge x_8 \wedge x_7),$$

$$r(C, D) = (x_5 \wedge x_7) \vee (x_4 \wedge x_8) \vee (x_5 \wedge x_9 \wedge x_8) \vee (x_4 \wedge x_9 \wedge x_7).$$

$$(4.82)$$

Converting these into algebraic equations, we have

$$b_1 = r(A, B) = M_d x_1 x_2,$$

$$b_2 = r(B, C) = M_d x_3 x_4,$$

$$b_3 = r(A, C) = M_d x_5 x_6,$$

$$b_4 = r(A, F) = M_d^3 M_c x_1 x_8 M_c x_6 x_9 M_c^2 x_1 x_7 x_9 M_c^2 x_6 x_7 x_8,$$

$$b_5 = r(B, E) = M_d^3 M_c x_2 x_7 M_c x_3 x_9 M_c^2 x_2 x_8 x_9 M_c^2 x_3 x_8 x_7,$$

$$b_6 = r(C, D) = M_d^3 M_c x_5 x_7 M_c x_4 x_8 M_c^2 x_9 x_8 M_c^2 x_4 x_9 x_7,$$

$$(4.83)$$

where b_i, $i = 1, 2, \ldots, 6$, are measured values. Let $b = b_1 b_2 b_3 b_4 b_5 b_6$ and $x = x_1 x_2 x_3 x_4 x_5 x_6 x_7 x_8 x_9$. The equations can then be converted into a linear algebraic equation as

$$Lx = b, \tag{4.84}$$

where

$$
\begin{aligned}
L = \delta_{64}[\; &1 \;\; 1 \;\; 1 \;\; 5 \;\; 1 \;\; 3 \;\; 2 \;\; 8 \;\; 9 \;\; 9 \;\; 9 \; 13 \;\; 9 \; 11 \; 14 \; 16 \\
&17 \; 17 \; 17 \; 22 \; 17 \; 19 \; 18 \; 24 \; 25 \; 25 \; 25 \; 30 \; 25 \; 27 \; 30 \; 32 \\
&17 \; 17 \; 17 \; 21 \; 17 \; 20 \; 18 \; 24 \; 25 \; 25 \; 25 \; 29 \; 25 \; 28 \; 30 \; 32 \\
&18 \; 18 \; 18 \; 22 \; 18 \; 20 \; 18 \; 24 \; 26 \; 26 \; 26 \; 30 \; 26 \; 28 \; 30 \; 32 \\
&33 \; 33 \; 33 \; 37 \; 33 \; 35 \; 36 \; 40 \; 41 \; 41 \; 41 \; 45 \; 41 \; 43 \; 48 \; 48 \\
&49 \; 49 \; 49 \; 54 \; 49 \; 51 \; 52 \; 56 \; 57 \; 57 \; 57 \; 62 \; 57 \; 59 \; 64 \; 64 \\
&49 \; 49 \; 49 \; 53 \; 49 \; 52 \; 52 \; 56 \; 57 \; 57 \; 57 \; 61 \; 57 \; 60 \; 64 \; 64 \\
&50 \; 50 \; 50 \; 54 \; 50 \; 52 \; 52 \; 56 \; 58 \; 58 \; 58 \; 62 \; 58 \; 60 \; 64 \; 64 \\
&33 \; 33 \; 33 \; 39 \; 33 \; 35 \; 34 \; 40 \; 41 \; 41 \; 41 \; 47 \; 41 \; 43 \; 46 \; 48 \\
&49 \; 49 \; 49 \; 56 \; 49 \; 51 \; 50 \; 56 \; 57 \; 57 \; 57 \; 64 \; 57 \; 59 \; 62 \; 64 \\
&49 \; 49 \; 49 \; 55 \; 49 \; 52 \; 50 \; 56 \; 57 \; 57 \; 57 \; 63 \; 57 \; 60 \; 62 \; 64 \\
&50 \; 50 \; 50 \; 56 \; 50 \; 52 \; 50 \; 56 \; 58 \; 58 \; 58 \; 64 \; 58 \; 60 \; 62 \; 64 \\
&35 \; 35 \; 35 \; 39 \; 35 \; 35 \; 36 \; 40 \; 43 \; 43 \; 43 \; 47 \; 43 \; 43 \; 48 \; 48 \\
&51 \; 51 \; 51 \; 56 \; 51 \; 51 \; 52 \; 56 \; 59 \; 59 \; 59 \; 64 \; 59 \; 59 \; 64 \; 64 \\
&51 \; 51 \; 51 \; 55 \; 51 \; 52 \; 52 \; 56 \; 59 \; 59 \; 59 \; 63 \; 59 \; 60 \; 64 \; 64 \\
&52 \; 52 \; 52 \; 56 \; 52 \; 52 \; 52 \; 56 \; 60 \; 60 \; 60 \; 64 \; 60 \; 60 \; 64 \; 64 \\
&33 \; 33 \; 33 \; 37 \; 33 \; 39 \; 34 \; 40 \; 45 \; 45 \; 45 \; 45 \; 45 \; 47 \; 46 \; 48 \\
&49 \; 49 \; 49 \; 54 \; 49 \; 55 \; 50 \; 56 \; 61 \; 61 \; 61 \; 62 \; 61 \; 63 \; 62 \; 64 \\
&49 \; 49 \; 49 \; 53 \; 49 \; 56 \; 50 \; 56 \; 61 \; 61 \; 61 \; 61 \; 61 \; 64 \; 62 \; 64 \\
&50 \; 50 \; 50 \; 54 \; 50 \; 56 \; 50 \; 56 \; 62 \; 62 \; 62 \; 62 \; 62 \; 64 \; 62 \; 64 \\
&33 \; 33 \; 33 \; 37 \; 33 \; 39 \; 36 \; 40 \; 45 \; 45 \; 45 \; 45 \; 45 \; 47 \; 48 \; 48 \\
&49 \; 49 \; 49 \; 54 \; 49 \; 55 \; 52 \; 56 \; 61 \; 61 \; 61 \; 62 \; 61 \; 63 \; 64 \; 64 \\
&49 \; 49 \; 49 \; 53 \; 49 \; 56 \; 52 \; 56 \; 61 \; 61 \; 61 \; 61 \; 61 \; 64 \; 64 \; 64 \\
&50 \; 50 \; 50 \; 54 \; 50 \; 56 \; 52 \; 56 \; 62 \; 62 \; 62 \; 62 \; 62 \; 64 \; 64 \; 64 \\
&33 \; 33 \; 33 \; 39 \; 33 \; 39 \; 34 \; 40 \; 45 \; 45 \; 45 \; 47 \; 45 \; 47 \; 46 \; 48 \\
&49 \; 49 \; 49 \; 56 \; 49 \; 55 \; 50 \; 56 \; 61 \; 61 \; 61 \; 64 \; 61 \; 63 \; 62 \; 64 \\
&49 \; 49 \; 49 \; 55 \; 49 \; 56 \; 50 \; 56 \; 61 \; 61 \; 61 \; 63 \; 61 \; 64 \; 62 \; 64 \\
&50 \; 50 \; 50 \; 56 \; 50 \; 56 \; 50 \; 56 \; 62 \; 62 \; 62 \; 64 \; 62 \; 64 \; 62 \; 64 \\
&35 \; 35 \; 35 \; 39 \; 35 \; 39 \; 36 \; 40 \; 47 \; 47 \; 47 \; 47 \; 47 \; 47 \; 48 \; 48 \\
&51 \; 51 \; 51 \; 56 \; 51 \; 55 \; 52 \; 56 \; 63 \; 63 \; 63 \; 64 \; 63 \; 63 \; 64 \; 64 \\
&51 \; 51 \; 51 \; 55 \; 51 \; 56 \; 52 \; 56 \; 63 \; 63 \; 63 \; 63 \; 63 \; 64 \; 64 \; 64 \\
&52 \; 52 \; 52 \; 56 \; 52 \; 56 \; 52 \; 56 \; 64 \; 64 \; 64 \; 64 \; 64 \; 64 \; 64 \; 64].
\end{aligned}
$$

It is easy to verify that if the number of layers is k, then the number of total links (unknowns) is $N_s = 3^{k+1}$. The number of triangles we have to test is $N_t = 2k - 1$, which is much smaller than N_s.

References

1. Chen, Y., Bindel, D., Song, H.H., Katz, R.H.: Algebra-based scalable overlay network monitoring: Algorithms evaluation, and applications. IEEE/ACM Trans. Netw. **15**(5), 1084–1097 (2007). doi:10.1109/TNET.2007.896251

2. Cheng, D., Li, Z.: Solving logic equation via matrix expression. Front. Electr. Electron. Eng. China **4**(3), 259–269 (2009)
3. Cheng, D., Takahashi, Y.: Network failure locating via end-to-end verification. Preprint (2009)
4. Duffield, N.: Simple network performance tomography. In: Proc. 3rd ACM SIGCOMM Conference on Internet Measurement, pp. 210–215. ACM, New York (2003). doi:http://doi.acm.org/10.1145/948205.948232
5. Kobayashi, M., Hasegawa, Y., Murase, T.: Estimating points of QoS degradation in the network from the aggregation of per-flow quality information. Tech. rep., Institute of Electronics, Information and Communication Engineers (2005)
6. Kobayashi, M., Murase, T.: Scalable QoS degradation locating from end-to-end quality of flows on various routes. Preprint (2005)
7. Tachibana, A., Ano, S., Hasegawa, T., Tsuru, M., Oie, Y.: Empirical study on locating congested segments over the internet based on multiple end-to-end path measurements. In: Proceedings of the 2005 Symposium on Applications and the Internet pp. 342–351 (2005). doi:http://doi.ieeecomputersociety.org/10.1109/SAINT.2005.26
8. Truemper, K.: Design of Logic-based Intelligent Systems. Wiley, New York (2004)

Chapter 5
Topological Structure of a Boolean Network

5.1 Introduction to Boolean Networks

Inspired by the Human Genome Project, a new view of biology, called systems biology, is emerging. Systems biology does not investigate individual genes, proteins or cells in isolation. Rather, it studies the behavior and relationships of all the cells, proteins, DNA and RNA in a biological system called a cell network. The most active networks may be the genetic regulatory networks, which, reacting to changes of environment, regulate the growth, replication, and death of cells. We refer to [14, 17] for a general introduction to systems biology.

How do genetic regulatory networks function? According to [25], in the early 1960s Jacob and Monod showed that any cell contains a number of "regulatory" genes that act as switches and which can turn one another on and off. This indicates that a genetic network is acting in a Boolean manner. The logical essence of a cell network was also pointed out by Paul Nurse [21], who stated that the cells "then need to be linked and integrated together to define the modules and overall regulatory networks required to bring about the reproduction of the cell. This task will require system analysis that emphasize the logical relationships between elements of the networks, ..."

The Boolean network, first introduced by Kauffman [15], then developed by [1, 2, 4, 7, 10, 16, 23, 24] and many others, has become a powerful tool for describing, analyzing, and simulating cell networks. Hence, it has received much attention, not only from the biology community, but also within physics, systems science, etc. In this model, a gene state is quantized to only two levels: true and false. The state of each gene is then determined by the states of its neighboring genes using logical rules. It has been shown that Boolean networks play an important role in modeling cell regulation because they can represent important features of living organisms [3, 12]. The structure of a Boolean network is described in terms of its cycles and the transient states that lead to them. Two different methods, iteration and scalar form, were developed in [11] to determine the cyclic structure and the transient states that lead to them. In [8], a linear reduced scalar equation was derived from a more rudimentary nonlinear scalar equation to obtain immediate information about

D. Cheng et al., *Analysis and Control of Boolean Networks*,
Communications and Control Engineering,
DOI 10.1007/978-0-85729-097-7_5, © Springer-Verlag London Limited 2011

both cycles and the transient structure of the network. Several useful Boolean networks have been analyzed and their cycles revealed (see, e.g., [8, 11] and references therein). It was pointed out in [26] that finding fixed points and cycles of a Boolean network is an NP-complete problem.

Boolean models have been studied for a long time and many useful tools have been developed to find the solutions of static and dynamic Boolean equations, such as discrete iteration [22] and satisfiability [6, 18]. As pointed out in [21], "Perhaps a proper understanding of the complex regulatory networks making up cellular systems like the cell cycle will require a similar shift from common sense thinking. We might need to move into a strange more abstract world, more readily analyzable in terms of mathematics than our present imaginings of cells operating as a microcosm of our everyday world."

The algorithms developed in this chapter can be used to obtain all the fixed points, cycles, transient periods, and basins of attractors. Theoretically, the algorithms presented in the following section can provide complete solutions, but the algorithms are limited by computational complexity (they can hardly be used for large-scale networks).

5.2 Dynamics of Boolean Networks

Definition 5.1 [8] A Boolean network is a set of nodes, $x_1, x_2, \ldots, x_n$, which simultaneously interact with each other. At each given time $t = 0, 1, 2, \ldots$, a node has only one of two different values: 1 or 0. Thus, the network can be described by a system of equations:

$$
\begin{cases}
x_1(t+1) = f_1(x_1(t), x_2(t), \ldots, x_n(t)), \\
x_2(t+1) = f_2(x_1(t), x_2(t), \ldots, x_n(t)), \\
\vdots \\
x_n(t+1) = f_n(x_1(t), x_2(t), \ldots, x_n(t)),
\end{cases}
\tag{5.1}
$$

where f_i, $i = 1, 2, \ldots, n$, are n-ary logical functions.

In the following, we give a rigorous description of a network graph.

Definition 5.2 A network graph, $\Sigma = \{\mathcal{N}, \mathcal{E}\}$, consists of a set of nodes, $\mathcal{N} = \{x_i \mid i = 1, \ldots, n\}$, and a set of edges, $\mathcal{E} \subset \{x_1, \ldots, x_n\} \times \{x_1, \ldots, x_n\}$. If $(x_i, x_j) \in \mathcal{E}$, there is an edge from $x_i \to x_j$, which means that node x_j is affected by node x_i.

The network graph is also sometimes called the connectivity graph [22].
We now give a simple example to show the structure of a Boolean network.

Fig. 5.1 Network graph of
(5.2)

Example 5.1 Consider a Boolean network, $\Sigma = (\mathcal{N}, \mathcal{E})$, of three nodes, given by

$$\begin{cases} A(t+1) = B(t) \wedge C(t), \\ B(t+1) = \neg A(t), \\ C(t+1) = B(t) \vee C(t). \end{cases} \qquad (5.2)$$

Its set of nodes is $\mathcal{N} = \{x_1 := A,\ x_2 := B,\ x_3 := C\}$ and its set of edges is
$\mathcal{E} = \{(x_1, x_2), (x_2, x_1), (x_2, x_3), (x_3, x_1), (x_3, x_3)\}$. Its network graph is depicted in
Fig. 5.1.

Using mod 2 algebra, it can also be expressed as

$$\begin{cases} A(t+1) = B(t)C(t), \\ B(t+1) = 1 + A(t), \\ C(t+1) = B(t) + C(t) + B(t)C(t). \end{cases} \qquad (5.3)$$

Note that in mod 2 algebra we have addition $\oplus$ and product $*$:

$$a \oplus b = a + b \bmod 2,$$
$$a * b = ab \bmod 2.$$

In most cases we omit "mod 2" and use the conventional addition and product symbols for $\oplus$ and $*$.

For a Boolean network, the number of edges which point to a node i is called the
in-degree of node i and the number of edges which start from node i is called the
out-degree of node i. In Example 5.1 the in-degrees of A and C are both 2, and the
in-degree of B is 1; the out-degree of A is 1, and the out-degrees of B and C are 2.

The network graph can also be expressed by an $n \times n$ matrix, called the incidence
matrix, defined as

$$\mathcal{I} = (b_{ij}), \quad \text{where } b_{ij} = \begin{cases} 1, & (x_i, x_j) \in \mathcal{N}, \\ 0, & \text{otherwise.} \end{cases} \qquad (5.4)$$

Consider the network Σ in Example 5.1. Its incidence matrix is

$$\mathcal{I}(\Sigma) = \begin{bmatrix} 0 & 1 & 1 \\ 1 & 0 & 0 \\ 0 & 1 & 1 \end{bmatrix}. \qquad (5.5)$$

Our first task is to convert the Boolean network dynamics (5.1) into an algebraic form or, more precisely, to express it as a conventional discrete-time linear system. Using the technique developed in the previous chapter, we use vector form $x_i(t) \in \Delta$ and define

$$x(t) = x_1(t)x_2(t) \cdots x_n(t) := \ltimes_{i=1}^n x_i(t).$$

Using Theorem 3.1, there exist structure matrices, $M_i = M_{f_i}, i = 1, \ldots, n$, such that

$$x_i(t+1) = M_i x(t), \quad i = 1, 2, \ldots, n. \tag{5.6}$$

Remark 5.1 Note that the in-degree is usually much less than n, that is, the right-hand side of the ith equation of (5.1) may not involve all x_j, $j = 1, 2, \ldots, n$. For instance, in the previous example, for node A we have

$$A(t+1) = B(t) \wedge C(t).$$

In matrix form this is

$$A(t+1) = M_c B(t) C(t). \tag{5.7}$$

To obtain the form of (5.6), using dummy matrix E_d (4.9), we can rewrite (5.7) as

$$A(t+1) = M_c E_d A(t) B(t) C(t) = M_c E_d x(t).$$

Multiplying the equations in (5.6) together yields

$$x(t+1) = M_1 x(t) M_2 x(t) \cdots M_n x(t). \tag{5.8}$$

Using Theorem 4.1, (5.8) can be expressed as

$$x(t+1) = Lx(t), \tag{5.9}$$

where

$$L = M_1 \prod_{j=2}^n \left[(I_{2^n} \otimes M_j) \Phi_n \right]$$

is called the transition matrix.

The question now is: Is the system (5.9) enough to describe the dynamics? The answer is "yes".

Theorem 5.1 *The dynamics of the Boolean network (5.1) is uniquely determined by the linear dynamical system (5.9).*

Proof From (5.9) one sees that

$$x(t) = L^t x(0), \quad t = 0, 1, 2, \ldots. \tag{5.10}$$

It follows that

$$x_i(t) = M_i L^{t-1} x(0), \quad i = 1, 2, \ldots. \tag{5.11}$$

Hence (5.9) completely determines the dynamics (5.1). □

Definition 5.3 Equation (5.9) is called the algebraic form of the network (5.1). Equation (5.6) is called the componentwise algebraic form of the network (5.1).

In fact, a direct computation using the properties of the semi-tensor product can easily produce the algebraic form. We give a simple example to show how to obtain the algebraic form of the dynamics of a Boolean network.

Example 5.2 Recall the Boolean network in Example 5.1. Its dynamics is given by (5.2). In algebraic form, we have

$$\begin{cases} A(t+1) = M_c B(t) C(t), \\ B(t+1) = M_n A(t), \\ C(t+1) = M_d B(t) C(t). \end{cases} \tag{5.12}$$

Setting $x(t) = A(t)B(t)C(t)$ we can calculate L as

$$\begin{aligned} x(t+1) &= M_c BC M_n A M_d BC \\ &= M_c(I_4 \otimes M_n) BCA M_d BC \\ &= M_c(I_4 \otimes M_n)(I_8 \otimes M_d) BCABC \\ &= M_c(I_4 \otimes M_n)(I_8 \otimes M_d) W_{[2,4]} ABCBC \\ &= M_c(I_4 \otimes M_n)(I_8 \otimes M_d) W_{[2,4]} ABW_{[2]} BCC \\ &= M_c(I_4 \otimes M_n)(I_8 \otimes M_d) W_{[2,4]} (I_4 \otimes W_{[2]}) AM_r BM_r C \\ &= M_c(I_4 \otimes M_n)(I_8 \otimes M_d) W_{[2,4]} \\ &\qquad (I_4 \otimes W_{[2]})(I_2 \otimes M_r)(I_4 \otimes M_r) ABC. \end{aligned} \tag{5.13}$$

Then system (5.2) can be expressed in matrix form as

$$x(t+1) = Lx(t),$$

where the network transition matrix is

$$\begin{aligned} L &= M_c(I_4 \otimes M_n)(I_8 \otimes M_d) W_{[2,4]}(I_4 \otimes W_{[2]})(I_2 \otimes M_r)(I_4 \otimes M_r) \\ &= \delta_8[3\ 7\ 7\ 8\ 1\ 5\ 5\ 6]. \end{aligned}$$

Remark 5.2

- It is obvious that a mod 2 equation such as (5.3) can be converted into a logical equation such as (5.12). A logical equation can also be converted into a mod 2 equation because "$1 +$ (mod 2)" is equivalent to "$\neg$", "$\times$ (mod 2)" is equivalent to "$\wedge$", and $\{\neg, \wedge\}$ is an adequate set. Logical form may provide a clear meaning for the relationship between logical variables, but in numerical computations, e.g., identification, mod 2 algebra is more convenient. Therefore, we use both.
- Equation (5.9) is a standard linear system with L being a square Boolean matrix. Therefore, all classical methods and conclusions for linear systems can be used to analyze the dynamics of the Boolean network.

5.3 Fixed Points and Cycles

Consider the Boolean network equation (5.9). We have the following result.

Lemma 5.1

$$\text{Col}_i(L) \in \Delta_{2^n}, \quad where \ L \in \mathcal{L}_{2^n \times 2^n}. \tag{5.14}$$

Proof We only have to show $\text{Col}(L) \subset \Delta_{2^n}$. Assume there is a j ($1 \leq j \leq 2^n$) such that $\text{Col}_j(L) \notin \Delta_{2^n}$. Then, when $x(t) = \delta_{2^n}^j$, we have

$$x(t+1) = Lx(t) = \text{Col}_j(L) \notin \Delta_{2^n},$$

which is a contradiction.   □

Definition 5.4

1. A state $x_0 \in \Delta_{2^n}$ is called a fixed point of system (5.9) if $Lx_0 = x_0$.
2. $\{x_0, Lx_0, \ldots, L^k x_0\}$ is called a cycle of system (5.9) with length k if $L^k x_0 = x_0$ and the elements in the set $\{x_0, Lx_0, \ldots, L^{k-1} x_0\}$ are pairwise distinct.

Remark 5.3 We use L to denote both the matrix and its corresponding linear mapping. So, x_0 may be in an L-invariant subspace. In this way, a cycle (or a fixed point) can be defined on an L-invariant subspace.

The next two theorems are the main results of this chapter. They show how many fixed points and cycles of different lengths a Boolean network has.

Theorem 5.2 *Consider the Boolean network* (5.1). $\delta_{2^n}^i$ *is its fixed point if and only if, in its algebraic form* (5.9), *the diagonal element* ℓ_{ii} *of the network transition matrix* L *equals* 1. *It follows that the number of fixed points of the network* (5.1), *denoted by* N_e, *equals the number of* i *for which* $\ell_{ii} = 1$. *Equivalently,*

$$N_e = \text{tr}(L). \tag{5.15}$$

Proof Assume that $\delta_{2^n}^i$ is its fixed point. Note that $L\delta_{2^n}^i = \text{Col}_i(L)$. It is clear that $\delta_{2^n}^i$ is its fixed point if and only if $\text{Col}_i(L) = \delta_{2^n}^i$, which completes the proof. □

For ease of statement, if $\ell_{ii} = 1$, then $\text{Col}_i(L)$ is called a diagonal nonzero column of L.

Next, we consider the cycles of the Boolean network system (5.1). Let $k \in \mathbb{Z}_+$. A positive integer $s \in \mathbb{Z}_+$ is called a proper factor of k if $s < k$ and $k/s \in \mathbb{Z}_+$. The set of proper factors of k is denoted by $\mathscr{P}(k)$. For instance, $\mathscr{P}(8) = \{1, 2, 4\}$, $\mathscr{P}(12) = \{1, 2, 3, 4, 6\}$, etc. Using a similar argument as for Theorem 5.2, we can have the following theorem.

Theorem 5.3 *The number of cycles of length s, denoted by N_s, is inductively determined by*

$$\begin{cases} N_1 = N_e, \\ N_s = \dfrac{\text{tr}(L^s) - \sum_{k \in \mathscr{P}(s)} k N_k}{s}, & 2 \le s \le 2^n. \end{cases} \tag{5.16}$$

Proof First, if $\delta_{2^n}^i$ is an element of a cycle of length s, then $L^s \delta_{2^n}^i = \delta_{2^n}^i$. From the proof of Theorem 5.2 the ith column of L^s, denoted by $\text{Col}_i(L^s)$, is a diagonal nonzero column of L^s, which adds 1 to $\text{tr}(L^s)$. Note that if $\delta_{2^n}^k$ is an element of a cycle of length $k \in \mathscr{P}(s)$, then we also have $L^s \delta_{2^n}^k = \delta_{2^n}^k$, and $\text{Col}_k(L^s)$ will also add 1 to $\text{tr}(L^s)$. Such diagonal elements have to be subtracted from $\text{tr}(L^s)$. Taking this into consideration, the second part of formula (5.16) is obvious.

As for the upper boundary of s, note that since $x(t)$ can have at most 2^n possible values, the length of any cycle is less than or equal to 2^n. □

Next, we consider how to find the cycles. If

$$\text{tr}(L^s) - \sum_{k \in \mathscr{P}(s)} k N_k > 0, \tag{5.17}$$

then we call s a nontrivial power.

Assume s is a nontrivial power. Denote by ℓ_{ii}^s the (i, i)th entry of matrix L^s. We then define

$$C_s = \{i \mid \ell_{ii}^s = 1\}, \quad s = 1, 2, \ldots, 2^n,$$

and

$$D_s = C_s \bigcap_{i \in \mathscr{P}(s)} C_i^c,$$

where C_i^c is the complement of C_i.

From the above argument the following result is obvious.

Proposition 5.1 *Let $x_0 = \delta_{2^n}^i$. Then $\{x_0, Lx_0, \ldots, L^s x_0\}$ is a cycle with length s if and only if $i \in D_s$.*

Theorem 5.3 and Proposition 5.1 provide a simple algorithm for constructing cycles. We give an example to illustrate the algorithm.

Example 5.3 Recall Example 5.1. It is easy to check that

$$\text{tr}(L^t) = 0, \quad t \le 3,$$

and

$$\text{tr}(L^t) = 4, \quad t \ge 4.$$

Using Theorem 5.3, we conclude that there is only one cycle of length 4. Moreover, note that

$$L^4 = \delta_8[1\ 3\ 3\ 1\ 5\ 7\ 7\ 3].$$

Each diagonal nonzero column can then generate the cycle. For instance, choosing $Z = \delta_8^1$, we have

$$LZ = \delta_8^3, \qquad L^2 Z = \delta_8^7, \qquad L^3 Z = \delta_8^5, \qquad L^4 Z = Z.$$

Using Algorithm 4.1 to convert the vector forms back to the scalar forms of $A(t)$, $B(t)$, and $C(t)$, we have the cycle as $(1, 1, 1) \to (1, 0, 1) \to (0, 0, 1) \to (0, 1, 1) \to (1, 1, 1)$.

Next, we consider the transient period, i.e., the minimum number of transient steps that leads any point to the limit set, Ω, which consists of all fixed points and cycles. First, note that L has only $r := 2^n \times 2^n$ possible independent values. Hence, if we construct a sequence of $r + 1$ matrices as

$$L^0 = I_{2^n}, L, L^2, \dots, L^r,$$

then there must be two equal matrices. Let $r_0 < r$ be the smallest i such that L^i appears again in the sequence. That is, there exists a $k > i$ such that $L^i = L^k$. More precisely,

$$r_0 = \underset{0 \le i < r}{\text{argmin}} \{ L^i \in \{ L^{i+1}, L^{i+2}, \dots, L^r \} \}. \tag{5.18}$$

Then, such r_0 exists. The following proposition is obvious.

Proposition 5.2 *Let* r_0 *be defined as in* (5.18). *Starting from any state, the trajectory will then enter into a cycle after* r_0 *iterations.*

For a given state x_0, the transient period of x_0, denoted by $T_t(x_0)$, is the smallest k satisfying $x(0) = x_0$ and $x(k) \in \Omega$. The transient period of a Boolean network, denoted by T_t, is defined as

$$T_t = \max_{\forall x \in \Delta_{2^n}} (T_t(x)).$$

In fact, we can show that r_0 is the transient period of the system.

Theorem 5.4 *The r_0 defined in (5.18) is the transient period of the system, that is,*

$$T_t = r_0. \tag{5.19}$$

Proof First, assume that

$$L^{r_0} = L^{r_0+T} \tag{5.20}$$

and that $T > 0$ is the smallest positive number which verifies (5.20). By definition, $r_0 + T \leq r$. We first claim that if there is a cycle of length t, then t is a factor of T. We prove this claim by contradiction, as follows. Assume $T \pmod{t} = s$ and $1 \leq s < t$. Let x_0 be a state on the cycle. $L^{r_0} x_0$ is then also a state on the same cycle. Hence,

$$L^{r_0} x_0 = L^{r_0+T} x_0 = L^T L^{r_0} x_0 = L^s \left(L^{r_0} x_0 \right) \neq L^{r_0} x_0,$$

which is a contradiction.

From (5.20) and the definition of T_t it is obvious that $T_t \leq r_0$. To prove $T_t = r_0$ we assume that $T_t < r_0$. By definition, for any x, $L^{T_t} x$ is on a cycle, the length of which is a factor of T. Hence,

$$L^{T_t} x = L^T L^{T_t} x = L^{T_t+T} x, \quad \forall x. \tag{5.21}$$

It is easy to check that if, for any $x \in \Delta_{2^n}$, (5.21) holds, then $L^{T_t} = L^{T_t+T}$, which contradicts the definition of r_0. □

Remark 5.4

1. According to Theorem 5.4 it is clear that $r_0 \leq 2^n$, because the transient period cannot be larger than 2^n.
2. Let $r_0 = T_t$ be defined as above and $T > 0$ be the smallest positive number which verifies (5.20). It is then easy to see that T is the least common multiple of the lengths of all cycles. For convenience, we call such a T the cycle multiplier.
3. From the first and second items of this remark, it is easily seen that to find r_0 we have only to check L^s for $s \leq r_0 + T$ and hence for

$$s \leq 2^{n+1}.$$

Next, we consider when a network converges to one point.

Definition 5.5 A Boolean network is said to be globally convergent if its limit set, Ω, consists of only one fixed point. Global convergence is also called global stability.

Note that, by definition, the global convergence of a Boolean network means that starting from any state, the trajectory of the network converges to the unique fixed point.

The following result is a consequence of Theorems 5.2 and 5.3.

Corollary 5.1 *The system (5.1) is globally convergent if and only if one of the following equivalent conditions is satisfied:*

1. 1 *is the only nontrivial power and*

$$\text{tr}(L) = 1. \tag{5.22}$$

2.

$$\text{tr}(L^{2^n}) = 1. \tag{5.23}$$

3. *The cycle multiplier* $T = 1$.

Next, we give an example of global convergence.

Example 5.4 [22] Consider a system with state space $\mathscr{X} = \mathscr{D}^7$. Using scalar form (i.e., $x_i \in \{1, 0\}$), the system can be expressed in mod 2 algebra as

$$x(t+1) = Ax, \tag{5.24}$$

where

$$A = \begin{bmatrix} 1 & 0 & 0 & 0 & 1 & 0 & 0 \\ 0 & 1 & 1 & 0 & 0 & 0 & 0 \\ 1 & 1 & 0 & 0 & 0 & 0 & 0 \\ 0 & 0 & 1 & 0 & 0 & 0 & 0 \\ 0 & 1 & 1 & 0 & 0 & 0 & 0 \\ 0 & 0 & 0 & 1 & 1 & 0 & 0 \\ 0 & 1 & 1 & 0 & 1 & 1 & 0 \end{bmatrix}.$$

Equivalently, we have

$$\begin{cases} x_1(t+1) = x_1(t)\bar{\vee}x_5(t), \\ x_2(t+1) = x_2(t)\bar{\vee}x_3(t), \\ x_3(t+1) = x_1(t)\bar{\vee}x_2(t), \\ x_4(t+1) = x_3(t), \\ x_5(t+1) = x_2(t)\bar{\vee}x_3(t), \\ x_6(t+1) = x_4(t)\bar{\vee}x_5(t), \\ x_7(t+1) = x_2(t)\bar{\vee}x_3(t)\bar{\vee}x_5(t)\bar{\vee}x_6(t). \end{cases} \tag{5.25}$$

Recall that the structure matrix of $\bar{\vee}$ is

$$M_p = \begin{bmatrix} 0 & 1 & 1 & 0 \\ 1 & 0 & 0 & 1 \end{bmatrix}.$$

Using this, we obtain the componentwise algebraic form:

$$\begin{cases} x_1(t+1) = M_p x_1(t) x_5(t), \\ x_2(t+1) = M_p x_2(t) x_3(t), \\ x_3(t+1) = M_p x_1(t) x_2(t), \\ x_4(t+1) = x_3(t), \\ x_5(t+1) = M_p x_2(t) x_3(t), \\ x_6(t+1) = M_p x_4(t) x_5(t), \\ x_7(t+1) = M_p^3 x_2(t) x_3(t) x_5(t) x_6(t). \end{cases} \qquad (5.26)$$

Multiplying together yields

$$x(t+1) = Lx(t), \qquad (5.27)$$

where

$$\begin{aligned} L = M_p \big(I_2 \otimes \big(I_2 \otimes M_p \big(I_2 \otimes \big(I_2 \otimes M_p \big(I_2 \otimes \big(I_2 \otimes M_X \big(I_2 \otimes \big(I_2 \otimes M_p \\ \big(I_2 \otimes \big(I_2 \otimes M_p \big(I_2 \otimes \big(I_2 \otimes M_p M_p M_p\big)\big)\big)\big)\big)\big)\big)\big)\big)\big)\big)\big)(I_8 \otimes W_{[2]})(I_4 \otimes W_{[2]}) \\ (I_2 \otimes W_{[2]})(I_4 \otimes W_{[2]})(I_{16} \otimes W_{[2]})(I_8 \otimes W_{[2]})(I_{128} \otimes W_{[2]}) \\ (I_{64} \otimes W_{[2]})(I_{32} \otimes W_{[2]})(I_{16} \otimes W_{[2]})(I_{2048} \otimes W_{[2]})(I_{1024} \otimes W_{[2]}) \\ (I_{512} \otimes W_{[2]})(I_{256} \otimes W_{[2]})(I_{128} \otimes W_{[2]})(I_{64} \otimes W_{[2]})(I_{32} \otimes W_{[2]}) \\ (I_{64} \otimes W_{[2]})(I_{256} \otimes W_{[2]})(I_{128} \otimes W_{[2]})(I_{512} \otimes W_{[2]})(I_{256} \otimes W_{[2]}) \\ (I_{4096} \otimes W_{[2]})(I_{2048} \otimes W_{[2]})(I_{1024} \otimes W_{[2]})(I_{512} \otimes W_{[2]})(I_{2048} \otimes W_{[2]}) \\ (I_{1024} \otimes W_{[2]})(I_{4096} \otimes W_{[2]})(I_{8192} \otimes W_{[2]})(I_{16384} \otimes W_{[2]}) M_r \\ \big(I_2 \otimes M_r M_r M_r \big(I_2 \otimes M_r M_r M_r \big(I_2 \otimes \big(I_2 \otimes M_r M_r\big)\big)\big)\big). \end{aligned}$$

This can be calculated as

$$\begin{aligned} \delta_{128}[&120\ 120\ 119\ 119\ \ 53\ \ 53\ \ 54\ \ 54\ 118\ 118\ 117\ 117 \\ &55\ \ 55\ \ 56\ \ 56\ \ 91\ \ 91\ \ 92\ \ 92\ \ 26\ \ 26\ \ 25\ \ 25 \\ &89\ \ 89\ \ 90\ \ 90\ \ 28\ \ 28\ \ 27\ \ 27\ \ 67\ \ 67\ \ 68\ \ 68 \\ &\ 2\ \ \ 2\ \ \ 1\ \ \ 1\ \ 65\ \ 65\ \ 66\ \ 66\ \ \ 4\ \ \ 4\ \ \ 3\ \ \ 3 \\ &112\ 112\ 111\ 111\ \ 45\ \ 45\ \ 46\ \ 46\ 110\ 110\ 109\ 109 \\ &47\ \ 47\ \ 48\ \ 48\ \ 40\ \ 40\ \ 39\ \ 39\ 101\ 101\ 102\ 102 \\ &38\ \ 38\ \ 37\ \ 37\ 103\ 103\ 104\ 104\ \ 11\ \ 11\ \ 12\ \ 12 \\ &74\ \ 74\ \ 73\ \ 73\ \ \ 9\ \ \ 9\ \ 10\ \ 10\ \ 76\ \ 76\ \ 75\ \ 75 \\ &19\ \ 19\ \ 20\ \ 20\ \ 82\ \ 82\ \ 81\ \ 81\ \ 17\ \ 17\ \ 18\ \ 18 \\ &84\ \ 84\ \ 83\ \ 83\ \ 64\ \ 64\ \ 63\ \ 63\ 125\ 125\ 126\ 126 \\ &62\ \ 62\ \ 61\ \ 61\ 127\ 127\ 128\ 128]. \end{aligned}$$

Fig. 5.2 The dynamic graph
of (5.24)

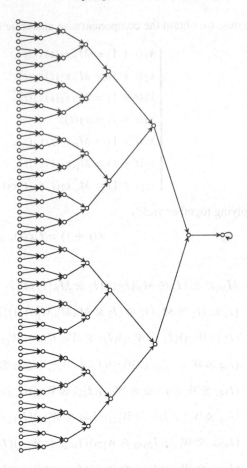

It is then easy to calculate that the smallest repeating power $r_0 = 7$ and that $L^7 = L^8$. That is, the transient period $T_t = 7$ and the cycle multiplier $T = 1$. According to Corollary 5.1 the system (5.24) is globally convergent.

To find the fixed point, which is the global attractor, we can check L to find the ith column $\text{Col}_i(L)$ satisfying $\text{Col}_i(L) = \delta_{128}^i$. The only solution is $i = 128$. $\delta_{128}^{128} \sim (0, 0, 0, 0, 0, 0, 0)$, which is the fixed point.

We will call the state-space graph of a network the dynamic graph. It is also called an iteration graph [22].

The dynamic graph of the system is shown in Fig. 5.2 [22].

Further consideration of Example 5.4 may be rewarding. Recall system (5.27). x_7 does not appear in the dynamics of any variable. We call such a variable the following-up variable. It is easy to see that in a network, the following-up variables do not affect the structure of the attractors. Finding a cycle for the remaining system and calculating the corresponding values of the following-up variables for each state on the cycle, we have the corresponding overall cycle.

Now, for a linear mod 2 system, the coefficient matrix always coincides with its incidence matrix. So, we can check A. If we remove the last row (equation of x_7), we can see that x_6 becomes a following-up variable. If we remove it, then x_4 becomes a following-up matrix. We conclude that the subsystem of x_1, x_2, x_3, x_5 determines the cycles of the original system. This subsystem is

$$\begin{cases} x_1(t+1) = M_p x_1(t) x_5(t), \\ x_2(t+1) = M_p x_2(t) x_3(t), \\ x_3(t+1) = M_p x_1(t) x_2(t), \\ x_5(t+1) = M_p x_2(t) x_3(t). \end{cases} \qquad (5.28)$$

Its network transition matrix is

$$L = M_p \big(I_2 \otimes \big(I_2 \otimes M_p \big(I_2 \otimes \big(I_2 \otimes M_p \big(I_2 \otimes (I_2 \otimes M_p) M_r \big) \big) \big) \big) \big) (I_8 \otimes W_{[2]})$$
$$(I_4 \otimes W_{[2]})(I_2 \otimes W_{[2]})(I_4 \otimes W_{[2]})(I_{16} \otimes W_{[2]})(I_8 \otimes W_{[2]})(I_{16} \otimes W_{[2]})$$
$$(I_{32} \otimes W_{[2]}) M_r \big(I_2 \otimes M_r (I_2 \otimes M_r) \big).$$

L, it follows, is

$$L = \delta_{16}[16 \; 8 \; 11 \; 3 \; 9 \; 1 \; 14 \; 6 \; 6 \; 14 \; 1 \; 9 \; 3 \; 11 \; 8 \; 16].$$

Moreover, the transient period $T_t = 4$ and the cycle multiplier is $T = 1$. We also conclude that the system is globally convergent.

From the above argument we arrive at the following result.

Proposition 5.3 *Assume the incidence matrix of a network Σ (with possible reordering of the variables) is expressed as*

$$\mathscr{I}(\Sigma) = \begin{bmatrix} A_1 & 0 \\ A_2 & A_3 \end{bmatrix},$$

where A_3 is a strictly lower triangular matrix (i.e., lower triangular matrix with zero diagonal elements). The structure of the limit set is then exactly the same as that of the subsystem consisting of first block variables.

Before concluding this section, we consider the basin of each attractor. Define

$$\Omega := \bigcup_{i=1}^{k} C_i,$$

where $\{C_i \,|\, i = 1, \ldots, k\}$ is the set of attractors. We give the following definition.

Definition 5.6

1. S_i is called the basin of attractor C_i if S_i is the set of points which converge to C_i. More precisely, $p \in S_i$ if and only if the trajectory $x(t, p)$ with $x(0, p) = p$ satisfies $x(t, p) \in C_i$ for $t \geq T_t$.
2. q is called the parent state of p if $p = x(1, q)$.

Remark 5.5

• Let $C \subset \Delta_{2^n}$ and let

$$L^{-1}(C) = \{q \mid Lq \in C\}.$$

The set of parent states of p is then $L^{-1}(p)$.

• $\Delta_{2^n} = \bigcup_{i=1}^{k} S_i$. Moreover, since $\{S_i \mid i = 1, \dots, k\}$ are disjoint, it is a partition of the state space $\mathcal{X} = \Delta_{2^n}$.

An alternative way to describe basins is as follows. Consider the system (5.9). Let $p \in \mathcal{X}$. Define the descendant set of p as

$$Des(p) = \{q \mid \text{for some } k \geq 0, q = L^k p\}.$$

Next, if $Des(p) \cap Des(q) \neq \emptyset$, p and q are said to be equivalent, denoted by $p \sim q$. It is then easy to see the following:

• $\sim$ is an equivalence relation. If we denote the equivalence class of p by $[p]$, that is,

$$[p] = \{q \mid q \sim p\},$$

then $\{[p] \mid p \in \mathcal{X}\}$ is a partition of $\mathcal{X}$. That is, either $[p] = [q]$ or $[p] \cap [q] = \emptyset$.
• In this partition, each component contains exactly one attractor. Therefore, each component is the basin of the unique attractor contained in it.
• For an attractor C, let $p \in C$. Its basin is then $[p]$.

What remains to investigate now is how to find S_i. We start from each point $p \in C_i$. If we can find its parent states $L^{-1}(p)$, then, for each point $p_1 \in L^{-1}(p)$, we can also find $L^{-1}(p_1)$. Continuing this process, after T_t iterations, we obtain a tree of states which converge to p. Summarizing the above arguments, we have the following result.

Proposition 5.4

$$S_i = C_i \cup L^{-1}(C_i) \cup L^{-2}(C_i) \cup \cdots \cup L^{-T_t}(C_i). \tag{5.29}$$

Finally, let us consider how to find $L^{-1}(p)$. It is easy to verify the following.

Proposition 5.5

$$\begin{cases} L^{-1}(p) = \{\delta_{2^n}^j \mid \operatorname{Col}_j(L) = p\}, \\ L^{-k}(p) = \{\delta_{2^n}^j \mid \operatorname{Col}_j(L^k) = p\}, \quad k = 2, \dots, T_t. \end{cases} \tag{5.30}$$

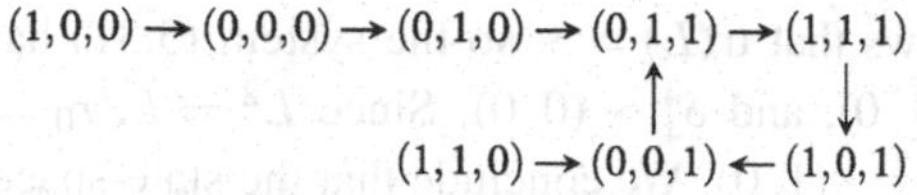

Fig. 5.3 The dynamic graph
of (5.2)

$$(1,0,0) \rightarrow (0,0,0) \rightarrow (0,1,0) \rightarrow (0,1,1) \rightarrow (1,1,1)$$
$$\uparrow \qquad\qquad \downarrow$$
$$(1,1,0) \rightarrow (0,0,1) \leftarrow (1,0,1)$$

Example 5.5 Recall Example 5.1. It is easy to check that $r_0 = 3$ and

$$L^3 = L^7 = \delta_8[5\ 1\ 1\ 5\ 7\ 3\ 3\ 1].$$

We then have the transient period $T_t = 3$. Using Propositions 5.4 and 5.5, we may choose any point $p \in C$, where C is its only cycle, to find $L^{-1}(p)$, $L^{-2}(p)$, and $L^{-3}(p)$.

For instance, suppose we choose $p = (0, 1, 1) \sim \delta_8^5$. We can then see two columns, $\text{Col}_6(L)$ and $\text{Col}_7(L)$, equal to p. Therefore, $\delta_8^6 \sim (0, 1, 0)$ and $\delta_8^7 \sim (0, 0, 1)$ form $L^{-1}(p)$. However, $(0, 0, 1)$ is on the cycle, so we are only interested in $p_1 = \delta_8^6 \sim (0, 1, 0)$. Now, since only $\text{Col}_8(L) = p_1$, we have $L^{-1}(p_1) = \{\delta_8^8\}$. Let $p_2 = \delta_8^8 \sim (0, 0, 0)$. Only $\text{Col}_4(L) = p_2$, so we have $p_3 := \delta_8^4 \sim (1, 0, 0) \in L^{-1}(p_2)$. Thus, we have a chain $p_3 \rightarrow p_2 \rightarrow p_1 \rightarrow p$. If we choose $q = (0, 0, 1) \sim \delta_8^7$, then $\text{Col}_2(L) = \text{Col}_3(L) = q$. Since $\delta_8^3 \sim (1, 0, 1)$ is on the cycle, we choose $q_1 = \delta_8^2 \sim (1, 1, 0)$. It is easy to check that $L^{-1}(q_1) = \emptyset$, and we have no more parent states. Finally, we obtain the dynamical graph of the network in Example 5.1 as in Fig. 5.3. (Note that we only use L^{-1} here. The iterative calculation provides the whole tree. If we need only the basins S_i, then the L^{-k} are convenient.)

In the following we consider some examples from the literature to show that the aforementioned method is universally applicable.

Example 5.6 [26] Consider the following Boolean network:

$$\begin{cases} A(t+1) = A(t) \vee B(t), \\ B(t+1) = A(t) \wedge B(t). \end{cases} \tag{5.31}$$

It is easy to calculate that

$$x(t+1) = M_d A B M_c A B$$
$$= M_d(I_4 \otimes M_c) A W_{[2]} A B^2$$
$$= M_d(I_4 \otimes M_c)(I_2 \otimes W_{[2]}) M_r A M_r B$$
$$= M_d(I_4 \otimes M_c)(I_2 \otimes W_{[2]}) M_r (I_2 \otimes M_r) x(t)$$
$$:= L x(t).$$

L can be calculated as

$$L = \delta_4[1\ 2\ 2\ 4].$$

It follows that $\text{tr}(L) = 3$, so the system (5.31) has three fixed points: $\delta_4^1 \sim (1, 1)$, $\delta_4^2 \sim (1, 0)$, and $\delta_4^4 \sim (0, 0)$. Since $L^2 = L$, $r_0 = 1$. Finally, for $x = \delta_4^3 \sim (0, 1)$, $Lx = \delta_4^2 \sim (1, 0)$. We conclude that the state-space graph of the system (5.31) contains three components: two fixed points, $(1, 1)$ and $(0, 0)$, and a length-2 cycle, $\{(0, 1), (1, 0)\}$.

Example 5.7 [8] Consider the following Boolean network:

$$\begin{cases} A(t+1) = B(t)C(t), \\ B(t+1) = 1 \,\bar\vee\, A(t), \\ C(t+1) = B(t). \end{cases} \qquad (5.32)$$

It is easy to calculate that

$$\begin{aligned} x(t+1) &= M_c BCM_n AB \\ &= M_c(I_4 \otimes M_n)BCAB \\ &= M_c(I_4 \otimes M_n)W_{[2,4]}ABCB \\ &= M_c(I_4 \otimes M_n)W_{[2,4]}ABW_{[2]}BC \\ &= M_c(I_4 \otimes M_n)W_{[2,4]}(I_4 \otimes W_{[2]})AM_rBC \\ &= M_c(I_4 \otimes M_n)W_{[2,4]}(I_4 \otimes W_{[2]})(I_2 \otimes M_r)x(t) \\ &:= Lx(t). \end{aligned}$$

L follows immediately as

$$L = \delta_8[3\ 7\ 8\ 8\ 1\ 5\ 6\ 6].$$

We then have

$$\text{tr}(L^k) = 0, \quad k = 1, 2, 3, 4,$$

and

$$L^5 = \delta_8[1\ 3\ 3\ 3\ 5\ 6\ 8\ 8],$$
$$\text{tr}(L^5) = 5.$$

Choosing any diagonal nonzero column of L^5, e.g., $x = \delta_8^1 \sim (1, 1, 1)$, we can generate a length-5 cycle $x \to Lx \to L^2x \to L^3x \to L^4x \to L^5x = x$, where $Lx = \delta_8^3 \sim (1, 0, 1)$, $L^2x = \delta_8^8 \sim (0, 0, 0)$, $L^3x = \delta_8^6 \sim (0, 1, 0)$, $L^4x = \delta_8^5 \sim (0, 1, 1)$, and $L^5x = \delta_8^1 \sim (1, 1, 1)$.

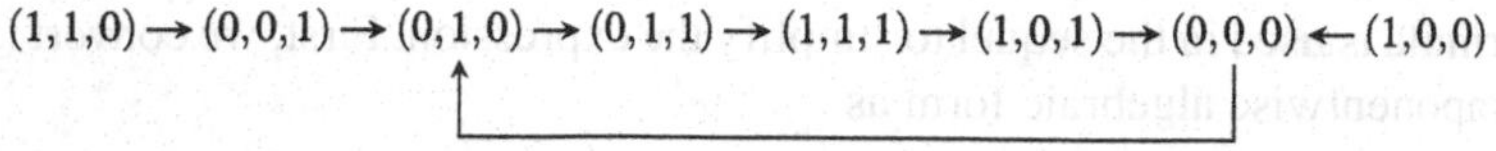

$$(1,1,0) \rightarrow (0,0,1) \rightarrow (0,1,0) \rightarrow (0,1,1) \rightarrow (1,1,1) \rightarrow (1,0,1) \rightarrow (0,0,0) \leftarrow (1,0,0)$$

Fig. 5.4 The dynamic graph of (5.32)

It is easy to check that $r_0 = 2$ and $L^2 = L^7$. That is, $T_t = 2$. Since $T = 5$, there are no cycles of length greater than 5. If we choose $z = \delta_8^2 \sim (1, 1, 0)$, then

$$Lz = \delta_8^7 \sim (0, 0, 1), \qquad L^2 z = \delta_8^6 = L^3 x.$$

If we choose $y = \delta_8^4 \sim (1, 0, 0)$, then

$$Ly = \delta_8^8 = L^2 x.$$

The dynamic graph is shown in Fig. 5.4, which coincides with the one in [8].

5.4 Some Classical Examples

In this section we revisit some examples which have been previously investigated in the literature. Compared with known results, it is evident that the approach introduced in this chapter is universal and precise.

The following example is the Boolean model of cell growth, differentiation, and apoptosis (programmed cell death) introduced in [13] and reinvestigated in [8].

Example 5.8

$$\begin{cases} A(t+1) = K(t) \bar{\vee} K(t) \wedge H(t), \\ B(t+1) = A(t) \bar{\vee} A(t) \wedge C(t), \\ C(t+1) = 1 \bar{\vee} D(t) \bar{\vee} D(t) \wedge I(t), \\ D(t+1) = J(t) \wedge K(t), \\ E(t+1) = 1 \bar{\vee} C(t) \bar{\vee} C(t) \wedge F(t), \\ F(t+1) = E(t) \bar{\vee} E(t) \wedge G(t), \\ G(t+1) = 1 \bar{\vee} B(t) \wedge E(t), \\ H(t+1) = F(t) \bar{\vee} F(t) \wedge G(t), \\ I(t+1) = H(t) \bar{\vee} H(t) \wedge I(t), \\ J(t+1) = J(t), \\ K(t+1) = K(t). \end{cases} \qquad (5.33)$$

Note that

$$A \bar{\vee} (A \wedge B) = \neg (A \rightarrow B).$$

This formula is used in the sequel to simplify the expression. First, we convert (5.33) into componentwise algebraic form as

$$
\begin{cases}
A(t+1) = M_n M_i K(t) H(t), \\
B(t+1) = M_n M_i A(t) C(t), \\
C(t+1) = M_i D(t) I(t), \\
D(t+1) = M_c J(t) K(t), \\
E(t+1) = M_i C(t) F(t), \\
F(t+1) = M_n M_i E(t) G(t), \\
G(t+1) = M_n M_c B(t) E(t), \\
H(t+1) = M_n M_i F(t) G(t), \\
I(t+1) = M_n M_i H(t) I(t), \\
J(t+1) = J(t), \\
K(t+1) = K(t).
\end{cases}
\tag{5.34}
$$

It is easy to calculate the structure matrix L as

$$
L = M_n M_i \big(I_2 \otimes \big(I_2 \otimes M_n M_i \big(I_2 \otimes \big(I_2 \otimes M_i \big(I_2 \otimes \big(I_2 \otimes M_c \big(I_2 \otimes \big(I_2 \otimes M_i \big(I_2 \otimes
$$

$$
\big(I_2 \otimes M_n M_i \big(I_2 \otimes \big(I_2 \otimes M_n M_c \big(I_2 \otimes \big(I_2 \otimes M_n M_i \big(I_2 \otimes \big(I_2 \otimes
$$

$$
M_n M_i \big)\big)\big)\big)\big)\big)\big)\big)\big)\big)\big)\big)\big)\big)\big)\big) (I_2 \otimes W_{[2]}) W_{[2]} (I_{2048} \otimes W_{[2]})(I_{1024} \otimes W_{[2]})
$$

$$
(I_{512} \otimes W_{[2]})(I_{256} \otimes W_{[2]})(I_{128} \otimes W_{[2]})(I_{64} \otimes W_{[2]})(I_{32} \otimes W_{[2]})
$$

$$
(I_{16} \otimes W_{[2]})(I_8 \otimes W_{[2]})(I_4 \otimes W_{[2]})(I_2 \otimes W_{[2]})(I_8 \otimes W_{[2]})(I_4 \otimes W_{[2]})
$$

$$
(I_{256} \otimes W_{[2]})(I_{128} \otimes W_{[2]})(I_{64} \otimes W_{[2]})(I_{32} \otimes W_{[2]})(I_{16} \otimes W_{[2]})(I_8 \otimes W_{[2]})
$$

$$
(I_{32} \otimes W_{[2]})(I_{16} \otimes W_{[2]})(I_{1024} \otimes W_{[2]})(I_{512} \otimes W_{[2]})(I_{256} \otimes W_{[2]})
$$

$$
(I_{128} \otimes W_{[2]})(I_{64} \otimes W_{[2]})(I_{32} \otimes W_{[2]})(I_{4096} \otimes W_{[2]})(I_{2048} \otimes W_{[2]})
$$

$$
(I_{1024} \otimes W_{[2]})(I_{512} \otimes W_{[2]})(I_{256} \otimes W_{[2]})(I_{128} \otimes W_{[2]})(I_{64} \otimes W_{[2]})
$$

$$
(I_{2048} \otimes W_{[2]})(I_{1024} \otimes W_{[2]})(I_{512} \otimes W_{[2]})(I_{256} \otimes W_{[2]})(I_{128} \otimes W_{[2]})
$$

$$
(I_{8192} \otimes W_{[2]})(I_{4096} \otimes W_{[2]})(I_{2048} \otimes W_{[2]})(I_{1024} \otimes W_{[2]})(I_{512} \otimes W_{[2]})
$$

$$
(I_{256} \otimes W_{[2]})(I_{8192} \otimes W_{[2]})(I_{4096} \otimes W_{[2]})(I_{2048} \otimes W_{[2]})(I_{1024} \otimes W_{[2]})
$$

$$
(I_{512} \otimes W_{[2]})(I_{16384} \otimes W_{[2]})(I_{8192} \otimes W_{[2]})(I_{4096} \otimes W_{[2]})(I_{2048} \otimes W_{[2]})
$$

$$
(I_{1024} \otimes W_{[2]})(I_{2048} \otimes W_{[2]})(I_{32768} \otimes W_{[2]})(I_{16384} \otimes W_{[2]})(I_{8192} \otimes W_{[2]})
$$

$$(I_{4096} \otimes W_{[2]})(I_{8192} \otimes W_{[2]})(I_{65536} \otimes W_{[2]})(I_{32768} \otimes W_{[2]})(I_{16384} \otimes W_{[2]})$$

$$(I_{32768} \otimes W_{[2]})(I_{131072} \otimes W_{[2]})(I_{65536} \otimes W_{[2]})\big(I_2 \otimes \big(I_2 \otimes M_r \big(I_2 \otimes \big(I_2 \otimes M_r$$

$$\big(I_2 \otimes M_r \big(I_2 \otimes M_r \big(I_2 \otimes M_r \big(I_2 \otimes M_r \big(I_2 \otimes M_r(I_2 \otimes M_r M_r)\big)\big)\big)\big)\big)\big)\big)\big)\big).$$

Since this is a $2^{11} \times 2^{11}$ matrix, it is too long to display here, even in condensed form. However, it can be easily stored in a computer. It is then easy to calculate that

$$\mathrm{tr}(L) = 3, \qquad \mathrm{tr}(L^9) = 12,$$

and that there are no other nontrivial powers. We conclude that there are only three fixed points and one cycle of length 9. Finding diagonal nonzero columns of L and L^9, respectively, it is easy to deduce that the three fixed points are

$$E_1 = (1, 0, 1, 0, 0, 0, 1, 0, 0, 0, 1),$$

$$E_2 = (0, 0, 1, 0, 0, 0, 1, 0, 0, 1, 0),$$

$$E_3 = (0, 0, 1, 0, 0, 0, 1, 0, 0, 0, 0).$$

The only cycle of length 9 is

$$(1, 1, 0, 1, 1, 1, 0, 1, 0, 1, 1) \to (0, 1, 0, 1, 1, 1, 0, 1, 1, 1, 1) \to$$
$$(0, 0, 1, 1, 1, 1, 0, 1, 0, 1, 1) \to (0, 0, 0, 1, 1, 1, 1, 1, 1, 1, 1) \to$$
$$(0, 0, 1, 1, 1, 0, 1, 0, 0, 1, 1) \to (1, 0, 0, 1, 0, 0, 1, 0, 0, 1, 1) \to$$
$$(1, 1, 0, 1, 1, 0, 1, 0, 0, 1, 1) \to (1, 1, 0, 1, 1, 0, 0, 0, 0, 1, 1) \to$$
$$(1, 1, 0, 1, 1, 1, 0, 0, 0, 1, 1) \to (1, 1, 0, 1, 1, 1, 0, 1, 0, 1, 1).$$

The minimum power for repeating L^k is $L^{10} = L^{19}$, so the transient period $T_t = 10$.

Remark 5.6 It was shown in [13] that a nontrivial growth attractor exists. Our result shows that there are exactly three fixed points and one cycle of length 9. As $J = K = D = 1$, both [13] and [8] showed that the cycle exists. Our result agrees with this. In the case of $J = K = D = 1$, it is easy to check that the transient period is still $T_t = 10$. [8] claimed that $T_t \le 7$, but this is incorrect. Consider $x(0) := x_0 = (0, 1, 1, 1, 1, 0, 0, 0, 0, 1, 1)$. It is easy to calculate that $x(10) = (1, 0, 0, 1, 0, 1, 1, 0, 0, 1, 1)$, which is not in the cycle, and $x(11) = (1, 1, 0, 1, 1, 0, 1, 0, 0, 1, 1)$, which is in the cycle. Thus, $T_t(x_0) = 10$.

The following example is from [9] and was reinvestigated in [11].

Example 5.9 Consider the following system:

$$
\begin{cases}
A(t+1) = 1 \,\bar{\vee}\, C(t) \,\bar{\vee}\, F(t) \,\bar{\vee}\, C(t) \wedge F(t), \\
B(t+1) = A(t), \\
C(t+1) = B(t), \\
D(t+1) = 1 \,\bar{\vee}\, C(t) \,\bar{\vee}\, F(t) \,\bar{\vee}\, I(t) \,\bar{\vee}\, C(t) \wedge F(t) \,\bar{\vee}\, C(t) \wedge I(t) \\
\qquad\quad \bar{\vee}\, F(t) \wedge I(t) \,\bar{\vee}\, C(t) \wedge F(t) \wedge I(t), \\
E(t+1) = D(t), \\
F(t+1) = E(t), \\
G(t+1) = 1 \,\bar{\vee}\, F(t) \,\bar{\vee}\, I(t) \,\bar{\vee}\, F(t) \wedge I(t), \\
H(t+1) = G(t), \\
I(t+1) = H(t).
\end{cases}
\tag{5.35}
$$

The componentwise algebraic form of the above equation is

$$
\begin{cases}
A(t+1) = M_n M_d C F, \\
B(t+1) = A, \\
C(t+1) = B, \\
D(t+1) = M_c^2 M_n I M_n C M_n F, \\
E(t+1) = D, \\
F(t+1) = E, \\
G(t+1) = M_n M_d F I, \\
H(t+1) = G, \\
I(t+1) = H.
\end{cases}
\tag{5.36}
$$

Let $x(t) = A(t)B(t)C(t)D(t)E(t)F(t)G(t)H(t)I(t)$ and $x(t+1) = Lx(t)$. Then,

$$
L = M_n M_d \big(I_2 \otimes \big(I_2 \otimes \big(I_2 \otimes \big(I_2 \otimes M_c M_c M_n \big(I_2 \otimes M_n \big(I_2 \otimes M_n \big(I_2 \otimes \big(I_2 \otimes
$$

$$
(I_2 \otimes M_n M_d)\big)\big)\big)\big)\big)\big)\big)\big)(I_2 \otimes W_{[2]}) W_{[2]} (I_4 \otimes W_{[2]})(I_2 \otimes W_{[2]})(I_{16} \otimes W_{[2]})
$$

$$
(I_8 \otimes W_{[2]})(I_{64} \otimes W_{[2]})(I_{32} \otimes W_{[2]})(I_{16} \otimes W_{[2]})(I_{128} \otimes W_{[2]})(I_{64} \otimes W_{[2]})
$$

$$
(I_{32} \otimes W_{[2]})(I_{128} \otimes W_{[2]})(I_{256} \otimes W_{[2]})(I_{1024} \otimes W_{[2]})(I_{512} \otimes W_{[2]})
$$

$$
(I_{2048} \otimes W_{[2]})(I_{1024} \otimes W_{[2]})\big(I_2 \otimes \big(I_2 \otimes M_r \big(I_2 \otimes \big(I_2 \otimes
$$

$$
(I_2 \otimes M_r M_r \big(I_2 \otimes \big(I_2 \otimes (I_2 \otimes M_r)\big)\big)\big)\big)\big)\big)\big).
$$

The nontrivial powers are $\operatorname{tr}(L^2) = 4$ and $\operatorname{tr}(L^6 = 64)$. It follows from Theorem 5.3 that there are only two cycles of length 2 and ten cycles of length 6. Searching diagonal nonzero columns of L^2 yields

$(1, 0, 1, 1, 0, 1, 1, 0, 1) \rightarrow (0, 1, 0, 0, 1, 0, 0, 1, 0) \rightarrow (1, 0, 1, 1, 0, 1, 1, 0, 1)$,
$(1, 0, 1, 0, 0, 0, 0, 1, 0) \rightarrow (0, 1, 0, 0, 0, 0, 1, 0, 1) \rightarrow (1, 0, 1, 0, 0, 0, 0, 1, 0)$.

Searching diagonal nonzero columns of L^6 yields

$(1, 1, 1, 1, 1, 1, 1, 1, 1) \rightarrow (0, 1, 1, 0, 1, 1, 0, 1, 1) \rightarrow (0, 0, 1, 0, 0, 1, 0, 0, 1) \rightarrow$
$(0, 0, 0, 0, 0, 0, 0, 0, 0) \rightarrow (1, 0, 0, 1, 0, 0, 1, 0, 0) \rightarrow (1, 1, 0, 1, 1, 0, 1, 1, 0) \rightarrow$
$(1, 1, 1, 1, 1, 1, 1, 1, 1)$,

$(1, 1, 1, 1, 1, 0, 1, 1, 0) \rightarrow (0, 1, 1, 0, 1, 1, 1, 1, 1) \rightarrow (0, 0, 1, 0, 0, 1, 0, 1, 1) \rightarrow$
$(0, 0, 0, 0, 0, 0, 0, 0, 1) \rightarrow (1, 0, 0, 0, 0, 0, 0, 0, 0) \rightarrow (1, 1, 0, 1, 0, 0, 1, 0, 0) \rightarrow$
$(1, 1, 1, 1, 1, 0, 1, 1, 0)$,

$(1, 1, 1, 1, 0, 1, 1, 0, 1) \rightarrow (0, 1, 1, 0, 1, 0, 0, 1, 0) \rightarrow (0, 0, 1, 0, 0, 1, 1, 0, 1) \rightarrow$
$(0, 0, 0, 0, 0, 0, 0, 10) \rightarrow (1, 0, 0, 1, 0, 0, 1, 0, 1) \rightarrow (1, 1, 0, 0, 1, 0, 0, 1, 0) \rightarrow$
$(1, 1, 1, 1, 0, 1, 1, 0, 1)$,

$(1, 1, 1, 1, 0, 0, 1, 0, 0) \rightarrow (0, 1, 1, 0, 1, 0, 1, 1, 0) \rightarrow (0, 0, 1, 0, 0, 1, 1, 1, 1) \rightarrow$
$(0, 0, 0, 0, 0, 0, 0, 1, 1) \rightarrow (1, 0, 0, 0, 0, 0, 0, 0, 1) \rightarrow (1, 1, 0, 0, 0, 0, 0, 0, 0) \rightarrow$
$(1, 1, 1, 1, 0, 0, 1, 0, 0)$,

$(1, 1, 1, 0, 1, 1, 0, 1, 1) \rightarrow (0, 1, 1, 0, 0, 1, 0, 0, 1) \rightarrow (0, 0, 1, 0, 0, 0, 0, 0, 0) \rightarrow$
$(0, 0, 0, 0, 0, 0, 1, 0, 0) \rightarrow (1, 0, 0, 1, 0, 0, 1, 1, 0) \rightarrow (1, 1, 0, 1, 1, 0, 1, 1, 1) \rightarrow$
$(1, 1, 1, 0, 1, 1, 0, 1, 1)$,

$(1, 1, 1, 0, 1, 0, 0, 1, 0) \rightarrow (0, 1, 1, 0, 0, 1, 1, 0, 1) \rightarrow (0, 0, 1, 0, 0, 0, 0, 1, 0) \rightarrow$
$(0, 0, 0, 0, 0, 0, 1, 0, 1) \rightarrow (1, 0, 0, 0, 0, 0, 0, 1, 0) \rightarrow (1, 1, 0, 1, 0, 0, 1, 0, 1) \rightarrow$
$(1, 1, 1, 0, 1, 0, 0, 1, 0)$,

$(1, 1, 1, 0, 0, 1, 0, 0, 1) \rightarrow (0, 1, 1, 0, 0, 0, 0, 0, 0) \rightarrow (0, 0, 1, 0, 0, 0, 1, 0, 0) \rightarrow$
$(0, 0, 0, 0, 0, 1, 1, 0) \rightarrow (1, 0, 0, 1, 0, 0, 1, 1, 1) \rightarrow (1, 1, 0, 0, 1, 0, 0, 1, 1) \rightarrow$
$(1, 1, 1, 0, 0, 1, 0, 0, 1)$,

$(1, 1, 1, 0, 0, 0, 0, 0, 0) \rightarrow (0, 1, 1, 0, 0, 0, 1, 0, 0) \rightarrow (0, 0, 1, 0, 0, 0, 1, 1, 0) \rightarrow$
$(0, 0, 0, 0, 0, 0, 1, 1, 1) \rightarrow (1, 0, 0, 0, 0, 0, 0, 1, 1) \rightarrow (1, 1, 0, 0, 0, 0, 0, 0, 1) \rightarrow$
$(1, 1, 1, 0, 0, 0, 0, 0, 0)$,

$(1, 0, 1, 1, 0, 1, 1, 1, 1) \rightarrow (0, 1, 0, 0, 1, 0, 0, 1, 1) \rightarrow (1, 0, 1, 0, 0, 1, 0, 0, 1) \rightarrow$
$(0, 1, 0, 0, 0, 0, 0, 0, 0) \rightarrow (1, 0, 1, 1, 0, 0, 1, 0, 0) \rightarrow (0, 1, 0, 0, 1, 0, 1, 1, 0) \rightarrow$
$(1, 0, 1, 1, 0, 1, 1, 1, 1)$,

$(1, 0, 1, 1, 0, 0, 1, 1, 0) \rightarrow (0, 1, 0, 0, 1, 0, 1, 1, 1) \rightarrow (1, 0, 1, 0, 0, 1, 0, 1, 1) \rightarrow$
$(0, 1, 0, 0, 0, 0, 0, 0, 1) \rightarrow (1, 0, 1, 0, 0, 0, 0, 0, 0) \rightarrow (0, 1, 0, 0, 0, 0, 1, 0, 0) \rightarrow$
$(1, 0, 1, 1, 0, 0, 1, 1, 0)$.

Finally, we can calculate that the first repeating L^k is $L^3 = L^9$. Thus, $T_t = 3$.

Remark 5.7 In [11] it was shown that there are no fixed points and that there are two cycles of length 2. Our results concerning fixed points and cycles of length 2 coincide with those of [11]. Heidel et al. [11] pointed out only six cycles of length 6, but according to our result there are exactly ten cycles of length 6.

5.5 Serial Boolean Networks

The Boolean network defined by (5.1) is called a parallel Boolean network. Most Boolean networks discussed in this book are of this class. Sometimes, though, we may need to update the elements in a serial way. Consider the following system:

$$
\begin{cases}
x_1(t+1) = f_1(x_1(t), x_2(t), \ldots, x_{n-1}(t), x_n(t)), \\
x_2(t+1) = f_2(x_1(t+1), x_2(t), \ldots, x_{n-1}(t), x_n(t)), \\
\vdots \\
x_n(t+1) = f_n(x_1(t+1), \ldots, x_{n-1}(t+1), x_n(t)).
\end{cases} \tag{5.37}
$$

Here we update x_1 first, then use the updated x_1 to update x_2, then use the updated x_1 and x_2 to update x_3, and so on. Such a Boolean network is called a serial Boolean network.

We now give an example of a serial Boolean network.

Example 5.10 Consider a game with n players, denoted by $P_1, \ldots, P_n$. Each player has two possible actions, denoted by $\mathscr{D} = \{0, 1\}$, and the next action of each player depends on the current actions of all players. Denote the strategy of player i by f_i. Then, f_i is a Boolean function of the current actions $x_1, \ldots, x_n$ of the players $P_1, \ldots, P_n$. If the game is played by all players simultaneously, then the dynamics of the strategies can be described by (5.1). However, if it is played one-by-one, that is, P_1 plays first, then P_2, and so on, then the strategy of P_2 depends on an updated x_1 and not updated $x_2, \ldots, x_n$, and so on. The dynamics of the strategies are then described by (5.37).

It is easy to prove the following result.

Proposition 5.6 *Assume that the systems* (5.1) *and* (5.37) *have the same logical functions* $f_i, i = 1, \ldots, n$. *They then have the same fixed points.*

In general, the systems (5.1) and (5.37) have different cycles, as shown by the following example.

Example 5.11 Consider Example 5.2, which is a continuation of Example 5.1. Now, if we convert (5.3) into a serial network, we have

$$\begin{cases} A(t+1) = B(t) \wedge C(t), \\ B(t+1) = \neg A(t+1), \\ C(t+1) = B(t+1) \vee C(t). \end{cases} \qquad (5.38)$$

Plugging the first equation into second yields

$$B(t+1) = \neg\big(B(t) \wedge C(t)\big) = \neg B(t) \vee \neg C(t).$$

Replacing the $B(t+1)$ in the third equation with this expression, we have

$$C(t+1) = \neg B(t) \vee \neg C(t) \vee C(t) = 1.$$

Collecting these together, we have

$$\begin{cases} A(t+1) = B(t) \wedge C(t), \\ B(t+1) = \neg B(t) \vee \neg C(t), \\ C(t+1) = 1. \end{cases} \qquad (5.39)$$

It is easy to deduce that there is a length-2 cycle of (5.39) and hence (5.38), which is $(1, 0, 1) \rightarrow (0, 1, 1) \rightarrow (1, 0, 1)$. Comparing this with the cycle obtained in Example 5.2, we see that they are different.

Note that the above example shows how to convert a serial Boolean network into an equivalent parallel Boolean network. Therefore, if we are only interested in their topological structures, we do not need a special tool to deal with serial Boolean networks. However, for optimization, etc. (in game theory, for instance) they are quite different. The terms "parallel" and "serial" come from their cycle-based realizations in automata theory.

We can, of course, have a serial–parallel model. Usually we use a partition with ordered subsets to describe this. Suppose we have a six-node network. We may have nodes 1, 2, and 5 updated first, then nodes 3 and 6, and, finally, 4. The ordered partition becomes $(\{1, 2, 5\}, \{3, 6\}, \{4\})$. The system of dynamic equations becomes

$$\begin{cases} x_1(t+1) = f_1(x_1(t), x_2(t), x_3(t), x_4(t), x_5(t), x_6(t)), \\ x_2(t+1) = f_2(x_1(t), x_2(t), x_3(t), x_4(t), x_5(t), x_6(t)), \\ x_5(t+1) = f_5(x_1(t), x_2(t), x_3(t), x_4(t), x_5(t), x_6(t)), \\ x_3(t+1) = f_3(x_1(t+1), x_2(t+1), x_3(t), x_4(t), x_5(t+1), x_6(t)), \\ x_6(t+1) = f_6(x_1(t+1), x_2(t+1), x_3(t), x_4(t), x_5(t+1), x_6(t)), \\ x_4(t+1) = f_4(x_1(t+1), x_2(t+1), x_3(t+1), x_4(t), x_5(t+1), x_6(t+1)). \end{cases}$$

It is obvious that Proposition 5.6 remains true for serial–parallel Boolean networks.

5.6 Higher Order Boolean Networks

In this section we consider higher order Boolean networks. This is based on [19].

Definition 5.7 A Boolean network is called a μth order network if the current states depend on length-μ histories. Precisely, its dynamics can be described as

$$
\begin{cases}
x_1(t+1) = f_1(x_1(t-\mu+1), \ldots, x_n(t-\mu+1), \ldots, x_1(t), \ldots, x_n(t)), \\
x_2(t+1) = f_2(x_1(t-\mu+1), \ldots, x_n(t-\mu+1), \ldots, x_1(t), \ldots, x_n(t)), \\
\vdots \\
x_n(t+1) = f_n(x_1(t-\mu+1), \ldots, x_n(t-\mu+1), \ldots, x_1(t), \ldots, x_n(t)), \\
t \geq \mu - 1,
\end{cases}
\tag{5.40}
$$

where $f_i : \mathscr{D}^{\mu n} \to \mathscr{D}, i = 1, \ldots, n$, are logical functions.

Note that, as for higher order discrete-time difference equations, to determine the solution (also called a trajectory) we need a set of initial conditions

$$
x_i(j) = a_{ij}, \quad i = 1, \ldots, n, j = 0, \ldots, \mu - 1.
\tag{5.41}
$$

We give an example to illustrate this kind of system. It is a biochemical network of coupled oscillations in the cell cycle [9].

Example 5.12 Consider the following Boolean network:

$$
\begin{cases}
A(t+3) = \neg(A(t) \wedge B(t+1)), \\
B(t+3) = \neg(A(t+1) \wedge B(t)).
\end{cases}
\tag{5.42}
$$

It can be easily converted into the canonical form (5.40) as

$$
\begin{cases}
A(t+1) = \neg(A(t-2) \wedge B(t-1)), \\
B(t+1) = \neg(A(t-1) \wedge B(t-2)), \quad t \geq 2.
\end{cases}
\tag{5.43}
$$

This is a third order Boolean network.

The second example comes from [20]. We refer to Example 2.5 in Chap. 2 for a detailed description of the Prisoners' Dilemma and to Chap. 18 for background on game theory.

Example 5.13 Consider the infinite Prisoners' Dilemma. Assume that player 1 is a machine and player 2 is a human. Set the strategies as follows:

> 0: the player cooperates with his partner,
> 1: the player betrays his partner.

Denote by $\{x(0), x(1), \dots\}$ the machine's strategy and by $\{y(0), y(1), \dots\}$ the human's strategy. Assume the machine's strategy, f_m, depends on μ-memory. The machine's strategy can then be described as

$$x(t+1) = f_m\big(x(t-\mu+1), y(t-\mu+1), \dots, x(t), y(t)\big). \tag{5.44}$$

It was proven in [20] that the human's best strategy, f_h, can be obtained by also using μ-memory. That is,

$$y(t+1) = f_h\big(x(t-\mu+1), y(t-\mu+1), \dots, x(t), y(t)\big). \tag{5.45}$$

Putting these together, we have a μth order Boolean network:

$$\begin{cases} x(t+1) = f_m(x(t-\mu+1), y(t-\mu+1), \dots, x(t), y(t)), \\ y(t+1) = f_h(x(t-\mu+1), y(t-\mu+1), \dots, x(t), y(t)). \end{cases} \tag{5.46}$$

These two examples will be revisited later.

As with standard Boolean networks, we have to explore the topological structures of higher order Boolean networks. Since the trajectories of a μth order Boolean network depend on μ initial values, we need rigorous definitions for cycles and/or fixed points.

Definition 5.8 Consider the system (5.40). Denote the state space by

$$\mathcal{X} = \big\{ X \mid X = (x_1, \dots, x_n) \in D^n \big\}.$$

1. Let $X^i = (x_1^i, \dots, x_n^i)$, $X^j = (x_1^j, \dots, x_n^j) \in \mathcal{X}$. (X^i, X^j), is said to be a directed edge if there exist $X^{j\alpha}$, $\alpha = 1, \dots, \mu - 1$, such that $X^i, X^j, \{X^{j\alpha}\}$ satisfy (5.40). More precisely,

$$x_k^j = f_k\big(X^{j_1}, X^{j_2}, \dots, X^{j_{\mu-1}}, X^i\big), \quad k = 1, \dots, n.$$

The set of edges is denoted by $\mathcal{E} \subset \mathcal{X} \times \mathcal{X}$.
2. $(X^1, X^2, \dots, X^\ell)$ is called a path if $(X^i, X^{i+1}) \in \mathcal{E}$, $i = 1, 2, \dots, \ell - 1$.
3. A path $(X^1, X^2, \dots)$ is called a cycle if $X^{i+\ell} = X^i$ for all i, the smallest such ℓ being called the length of the cycle. In particular, a cycle of length 1 is called a fixed point.

A standard Boolean network can be expressed formally as a higher order Boolean network with order $\mu = 1$. Hence, Definition 5.4 is a special case of Definition 5.8.

As with a standard Boolean network, to explore the topological structure of a higher order Boolean network we will first try to convert it into its algebraic form. In the following we will discuss two algebraic forms of (5.40).

5.6.1 First Algebraic Form of Higher Order Boolean Networks

Using vector form, we define

$$\begin{cases} x(t) = \ltimes_{i=1}^{n} x_i(t) \in \Delta_{2^n}, \\ z(t) = \ltimes_{i=t}^{t+\mu-1} x(i) \in \Delta_{2^{\mu n}}, \quad t = 0, 1, \ldots. \end{cases}$$

Assume the structure matrix of f_i is $M_i \in \mathscr{L}_{2 \times 2^{\mu n}}$. We can then express (5.40) in its componentwise algebraic form as

$$x_i(t+1) = M_i z(t - \mu + 1), \quad i = 1, \ldots, n, t = \mu - 1, \mu, \mu + 1, \ldots. \quad (5.47)$$

Multiplying the equations in (5.47) together yields

$$x(t+1) = L_0 z(t - \mu + 1), \quad t \geq \mu, \quad (5.48)$$

where

$$L_0 = M_1 \ltimes_{j=2}^{n} \left[(I_{2^{\mu n}} \otimes M_j) \Phi_{\mu n} \right]. \quad (5.49)$$

Note that the L_0 here can be calculated with a standard procedure as was used before, and we refer to (4.6) for the definition of Φ_k. Using some properties of the semi-tensor product of matrices, we have

$$\begin{aligned} z(t+1) &= \ltimes_{i=t+1}^{t+\mu} x(i) \\ &= (E_d)^n \ltimes_{i=t}^{t+\mu-1} x(i) \left(L_0 \ltimes_{i=t}^{t+\mu-1} x(i) \right) \\ &= (E_d)^n (I_{2^{\mu n}} \otimes L_0) \Phi_{\mu n} \ltimes_{i=t}^{t+\mu-1} x(i) \\ &:= Lz(t), \end{aligned} \quad (5.50)$$

where

$$E_d = \delta_2[1\ 2\ 1\ 2], \qquad L = (E_d)^n (I_{2^{\mu n}} \otimes L_0) \Phi_{\mu n}. \quad (5.51)$$

Equation (5.50) is called the first algebraic form of the network (5.40). We now give an example to illustrate it.

Example 5.14 Consider the following Boolean network:

$$\begin{cases} A(t+1) = C(t-1) \vee (A(t) \wedge B(t)), \\ B(t+1) = \neg(C(t-1) \wedge A(t)), \\ C(t+1) = B(t-1) \wedge B(t). \end{cases} \quad (5.52)$$

Using vector form, we have

$$\begin{cases} A(t+1) = M_d C(t-1) M_c A(t) B(t), \\ B(t+1) = M_n M_c C(t-1) A(t), \\ C(t+1) = M_c B(t-1) B(t). \end{cases} \quad (5.53)$$

Let $x(t) = A(t)B(t)C(t)$. Then, (5.53) can be converted into its componentwise algebraic form as

$$\begin{cases} A(t+1) = M_1 x(t-1)x(t), \\ B(t+1) = M_2 x(t-1)x(t), \\ C(t+1) = M_3 x(t-1)x(t), \end{cases} \tag{5.54}$$

where

$$M_1 = \delta_4[1\,1\,1\,1\,1\,1\,1\,1\,1\,1\,2\,2\,2\,2\,2\,2\,1\,1\,1\,1\,1\,1\,1\,1\,1\,1\,2\,2\,2\,2\,2\,2$$
$$1\,1\,1\,1\,1\,1\,1\,1\,1\,1\,2\,2\,2\,2\,2\,2\,1\,1\,1\,1\,1\,1\,1\,1\,1\,1\,2\,2\,2\,2\,2\,2],$$

$$M_2 = \delta_4[2\,2\,2\,2\,1\,1\,1\,1\,1\,1\,1\,1\,1\,1\,1\,1\,2\,2\,2\,2\,1\,1\,1\,1\,1\,1\,1\,1\,1\,1\,1\,1$$
$$2\,2\,2\,2\,1\,1\,1\,1\,1\,1\,1\,1\,1\,1\,1\,1\,2\,2\,2\,2\,1\,1\,1\,1\,1\,1\,1\,1\,1\,1\,1\,1],$$

$$M_3 = \delta_4[1\,1\,2\,2\,1\,1\,2\,2\,1\,1\,2\,2\,1\,1\,2\,2\,2\,2\,2\,2\,2\,2\,2\,2\,2\,2\,2\,2\,2\,2\,2\,2$$
$$1\,1\,2\,2\,1\,1\,2\,2\,1\,1\,2\,2\,1\,1\,2\,2\,2\,2\,2\,2\,2\,2\,2\,2\,2\,2\,2\,2\,2\,2\,2\,2].$$

Multiplying the three equations in (5.54) together yields

$$x(t+1) = L_0 x(t-1)x(t), \tag{5.55}$$

where

$$L_0 = \delta_8[3\,3\,4\,4\,1\,1\,2\,2\,1\,1\,6\,6\,5\,5\,6\,6\,4\,4\,4\,4\,2\,2\,2\,2\,2\,6\,6\,6\,6\,6\,6$$
$$3\,3\,4\,4\,1\,1\,2\,2\,1\,1\,6\,6\,5\,5\,6\,6\,4\,4\,4\,4\,2\,2\,2\,2\,2\,6\,6\,6\,6\,6\,6].$$

Setting $z(t) = x(t)x(t+1), t \geq 1$, we finally have

$$\begin{aligned} z(t+1) &= x(t+1)x(t+2) \\ &= (E_d)^3 x(t)x(t+1)x(t+2) \\ &= (E_d)^3 x(t)x(t+1)L_0 x(t)x(t+1) \\ &= (E_d)^3 (I_{2^6} \otimes L_0)\Phi_6 x(t)x(t+1) \\ &:= Lz(t), \end{aligned} \tag{5.56}$$

where

$$\begin{aligned} L = \delta_{64}[&3\ 11\ 20\ 28\ 33\ 41\ 50\ 58\ 1\ 9\quad 22\ 30\ 37\ 45\ 54\ 62 \\ &4\ 12\ 20\ 28\ 34\ 42\ 50\ 58\ 2\ 10\ 22\ 30\ 38\ 46\ 54\ 62 \\ &3\ 11\ 20\ 28\ 33\ 41\ 50\ 58\ 1\ 9\quad 22\ 30\ 37\ 45\ 54\ 62 \\ &4\ 12\ 20\ 28\ 34\ 42\ 50\ 58\ 2\ 10\ 22\ 30\ 38\ 46\ 54\ 62]. \end{aligned} \tag{5.57}$$

In fact, we can prove the two Boolean networks have the same topological structure, including fixed points, cycles, and the transient time, which is the time for all points to enter the set of cycles. Therefore, the first order Boolean network (5.50) provides all such results for higher order Boolean networks (5.48). We prove this in the form of the following result.

Lemma 5.2 *There is a one-to-one correspondence between the trajectories of* (5.48) *and the trajectories of* (5.50).

Proof Denote the sets of trajectories of (5.48) and (5.50) by T_x and T_z, respectively. Note that a trajectory is completely determined by its initial values. Now, because of the order, each trajectory of (5.48) depends on $\{x(0), x(1), \ldots, x(\mu - 1)\}$, and each trajectory of (5.50) depends on $z(0)$. Setting $z(0) = \ltimes_{i=0}^{\mu-1} x(i)$, we have a one-to-one correspondence between T_x and T_z.   □

Define a mapping $\phi : T_x \to T_z$, which maps each trajectory in T_x with initial value $\{x(0), x(1), \ldots, x(\mu - 1)\}$ to a trajectory in T_z with initial value $z(0) = \ltimes_{i=0}^{\mu-1} x(i)$. Then, ϕ is bijective. It is easy to see that there are 2^{kn} trajectories of each system, and we write

$$T_x = \left\{ \xi_i^x \,\middle|\, i = 1, 2, \ldots, 2^{\mu n} \right\}, \qquad T_z = \left\{ \xi_i^z \,\middle|\, i = 1, 2, \ldots, 2^{\mu n} \right\}.$$

Denote the sets of cycles of (5.48) and (5.50) by Ω_x and Ω_z, respectively. Here, a fixed point is considered as a cycle of length 1. Note that for a Boolean network, each trajectory will eventually converge to a unique cycle. The cycle to which $\xi^x \in T_x$ (resp., $\xi^z \in T_z$) converges is then denoted by $C(\xi^x)$ [resp., $C(\xi^z)$].

Now, for each $\xi^x \in T_x$, we have $C(\xi^x) \in \Omega_x$ and for $\phi(\xi^x) \in T_z$, we have $C(\phi(\xi^x)) \in \Omega_z$. Using this relation, we define $\psi : C(\xi^x) \mapsto C(\phi(\xi^x))$ by

$$\psi\left(C(\xi^x)\right) := C(\phi(\xi^x)). \tag{5.58}$$

Lemma 5.3

1. *Let* $\xi^x \in T_x$ *be* $\xi^x = \{x(0), x(1), \ldots\}$. *Then,*

$$\xi^z := \phi(\xi^x) = \left\{ z(0) = \ltimes_{i=0}^{\mu-1} x(i), z(1) = \ltimes_{i=1}^{\mu} x(i), \ldots \right\}. \tag{5.59}$$

2. *Let* ξ^x *converge to a trajectory* $\{x(t), x(t+1), \ldots, x(t+\alpha) = x(t)\}$. *Then,* $\xi^z = \phi(\xi^x)$ *converges to a trajectory* $\{z(t), z(t+1), \ldots, z(t+\alpha) = z(t)\}$, *where*

$$\begin{cases} z(t) = \ltimes_{i=t}^{t+\mu-1} x(i), \\ z(t+1) = \ltimes_{i=t+1}^{t+\mu} x(i), \\ \quad\vdots \\ z(t+\alpha) = \ltimes_{i=t+\alpha}^{t+\alpha+\mu-1} x(i) = \ltimes_{i=t}^{t+\mu-1} x(i) = z(t). \end{cases} \tag{5.60}$$

3. $\psi : \Omega_x \to \Omega_z$ *is a one-to-one and onto mapping.*

Proof 1. (5.59) follows directly from the definition of $z(t)$.

2. To see that the elements in (5.60) constitute a cycle, since we have $z(t) = z(t+\alpha)$, it is enough to show that the elements of $\{z(t), \ldots, z(t+\alpha-1)\}$ are distinct, i.e., $C_z := \{z(t), \ldots, z(t+\alpha-1), z(t+\alpha) = z(t)\}$ is not a multifold cycle. This follows as it can easily be shown that if the set $\{z(t), \ldots, z(t+\alpha-1)\}$

has duplicated elements, then so does $\{x(t), \ldots, x(t + \alpha - 1)\}$, contradicting the assumption that $C_x = \{x(t), \ldots, x(t + \alpha - 1), x(t + \alpha) = x(t)\}$ is a cycle.

3. Since the mapping ψ is defined via each trajectory, we must first show that it is well defined. That is, if the trajectories ξ^x and $\xi^{x'}$ determine the same cycle in Ω_x, then $\psi(C(\xi^x)) = \psi(C(\xi^{x'}))$.

Let the cycle determined by ξ^x be $\{x(t), x(t+1), \ldots, x(t + \alpha) = x(t)\}$. Using (5.60), we obtain

$$
\begin{cases}
z(t) = \ltimes_{i=t}^{t+k-1} x(i), \\
z(t+1) = \ltimes_{i=t+1}^{t+k} x(i), \\
\vdots \\
z(t + \alpha) = \ltimes_{i=t+\alpha}^{t+\alpha+k-1} x(i) = \ltimes_{i=t}^{t+k-1} x(i) = z(t).
\end{cases}
$$

Since $\xi^{x'}$ determines the same cycle, there is a constant $a \in \mathbb{Z}$, $0 \le a \le \alpha$, such that

$$x'(t) = x(t + a),$$
$$\vdots$$
$$x'(t + \alpha - a) = x(t + \alpha) = x(t),$$
$$\vdots$$
$$x'(t + \alpha) = x(t + a).$$

By (5.60), we have

$$
\begin{cases}
z'(t) = \ltimes_{i=t}^{t+k-1} x'(i) = \ltimes_{i=t+a}^{t+a+k-1} x(i) = z(t + a), \\
\vdots \\
z'(t + \alpha - a) = \ltimes_{i=t+\alpha-a}^{t+\alpha-a+k-1} x'(i) = \ltimes_{i=t+\alpha}^{t+\alpha+k-1} x(i) = z(t), \\
\vdots \\
z'(t + \alpha) = \ltimes_{i=t+\alpha}^{t+\alpha+k-1} x'(i) = \ltimes_{i=t+a}^{t+a+k-1} x(i) = z(t + a).
\end{cases}
$$

That is, $\psi(C(\xi^x)) = \psi(C(\xi^{x'}))$. Hence, $\psi : \Omega_x \to \Omega_z$ is a well-defined mapping.

Next, we will prove that ψ is injective. That is, if the trajectories ξ^x and $\xi^{x'}$ determine the same cycle in Ω_z, i.e., $\psi(C(\xi^x)) = \psi(C(\xi^{x'}))$, then $C(\xi^x) \in \Omega_x$ and $C(\xi^{x'}) \in \Omega_x$ are the same.

Now assume that $\xi^{x'}$ converges to a cycle $\{x'(t), \ldots, x'(t + \alpha) = x'(t)\}$, which determined the same cycle as ξ^x. Precisely, $\phi(C(\xi^{x'})) = \phi(C(\xi^x)) \in \Omega_z$, which has elements as in (5.60). It follows that

$$\ltimes_{i=0}^{k-1} x'(t + i) \in \phi(C(\xi^x)).$$

Hence, there exists a $0 \le \mu < \alpha$ such that

$$\ltimes_{i=1}^{k-1} x'(t + i) = z(t + \mu).$$

According to the definition of ϕ, we have

$$\ltimes_{i=0}^{k-1} x'(t+i) = z(t+\mu),$$

$$\vdots$$

$$\ltimes_{i=0}^{k-1} x'(t+\alpha-\mu+i) = z(t+\alpha) = z(t),$$

$$\vdots \tag{5.61}$$

$$\ltimes_{i=0}^{k-1} x'(t+\alpha-1+i) = z(t+\mu-1),$$

$$\ltimes_{i=0}^{k-1} x'(t+\alpha+i) = z(t+\mu).$$

Since $z(t) = \ltimes_{i=0}^{k-1} x(t+i)$ can uniquely determine all $\{x(t+i) \mid i = 0, \dots, k-1\}$, (5.61) implies that

$$x'(t) = x(t+\mu),$$

$$\vdots$$

$$x'(t+\alpha-\mu) = x(t), \tag{5.62}$$

$$\vdots$$

$$x'(t+\alpha-1) = x(t+\mu-1).$$

That is, $C(\xi^x)$ and $C(\xi^{x'})$ are the same cycle. Therefore, ψ is a one-to-one mapping. Finally, we have to prove that ψ is surjective. To see this let $C_z := \{z(t), z(t+1), \dots, z(t+\alpha) = z(t)\} \in \Omega_z$, where

$$z(t+i) = \ltimes_{j=0}^{k-1} x(t+i+j), \quad i = 0, 1, \dots, \alpha.$$

It follows from $z(t) = z(t+\alpha)$ that $x(t) = x(t+\alpha)$. Moreover, it is easy to see that the elements of $\{x(t), x(t+1), \dots, x(t+\alpha-1)\}$ are distinct. Hence, $C_x := \{x(t), x(t+1), \dots, x(t+\alpha-1), x(t+\alpha) = x(t)\} \in \Omega_x$ and $\psi(C_x) = C_z$. $\qquad\square$

To construct the inverse mapping $\Psi^{-1} : \Omega_z \to \Omega_x$, we define a mapping $\pi : \Delta_{2^{kn}} \to \Delta_{2^n}$ as follows:

$$\pi(z) = \left(I_{2^n} \otimes 1^{\mathrm{T}}_{2^{(k-1)n}}\right) z. \tag{5.63}$$

Some straightforward computations then lead to the following result.

Lemma 5.4

1. *If* $z = \ltimes_{i=0}^{\mu-1} x_i \in \Delta_{2^{\mu n}}$, *where* $x_i \in \Delta_{2^n}$, *then*

$$\pi(z) = x_0. \tag{5.64}$$

2. *Consider* $\phi : T_x \to T_z$. *If* $\{z(0), z(1), \dots\} \in T_z$, *then*

$$\phi^{-1}\left(\{z(0), z(1), \dots\}\right) = \{\pi(z(0)), \pi(z(1)), \dots\} \in T_x. \tag{5.65}$$

3. *Consider* $\psi : \Omega_x \rightarrow \Omega_z$. *If* $\{z(t), z(t+1), \ldots, z(t+\alpha) = z(t)\} \in \Omega_z$, *then*

$$\psi^{-1}(\{z(t), z(t+1), \ldots, z(t+\alpha) = z(t)\})$$

$$= \{\pi(z(t)), \pi(z(t+1)), \ldots, \pi(z(t+\alpha)) = \pi(z(t))\} \in \Omega_x. \quad (5.66)$$

Summarizing Lemmas 5.2–5.4, we have the following result, which shows how to obtain the topological structure of (5.40) from its first algebraic form (5.50).

Theorem 5.5

1. *Each trajectory ξ^x of* (5.48) *can be obtained from a trajectory ξ^z of* (5.50). *More precisely,*

$$T_x = \{\pi(\xi^z) \mid \xi^z \in T_z\}. \quad (5.67)$$

2. *Each cycle C_x of* (5.48) *can be obtained from a cycle of* (5.50). *More precisely,*

$$\Omega_x = \{\pi(C_z) \mid C_z \in \Omega_z\}. \quad (5.68)$$

3. *The transient period of the network* (5.48) *equals the transient period of the network* (5.50).

Theorem 5.5 shows that to find the cycles of (5.48) it is enough to find the cycles of (5.50). Hence, the method developed in the previous sections of this chapter can be applied to the system (5.50).

We now consider some examples.

Example 5.15 Recall Example 5.12. Set $x(t) = A(t)B(t)$. Using vector form, (5.43) can be expressed as

$$x(t+1) = L_0 x(t-2)x(t-1)x(t), \quad (5.69)$$

where

$$L_0 = \delta_4[4\,4\,4\,4\,2\,2\,2\,2\,3\,3\,3\,3\,1\,1\,1\,1$$
$$3\,3\,3\,3\,1\,1\,1\,1\,3\,3\,3\,3\,1\,1\,1\,1$$
$$2\,2\,2\,2\,2\,2\,2\,2\,1\,1\,1\,1\,1\,1\,1\,1$$
$$1\,1\,1\,1\,1\,1\,1\,1\,1\,1\,1\,1\,1\,1\,1\,1].$$

Set $z(t) = x(t)x(t+1)x(t+2)$. Then,

$$z(t+1) = x(t+1)x(t+2)x(t+3)$$
$$= (E_d)^2 x(t)x(t+1)x(t+2)x(t+3)$$
$$= (E_d)^2 x(t)x(t+1)x(t+2)L_0 x(t)x(t+1)x(t+2)$$
$$= (E_d)^2 (I_{2^6} \otimes L_0)\Phi_6 x(t)x(t+1)x(t+2)$$
$$:= Lz(t), \quad (5.70)$$

Fig. 5.5 The cycle of (5.43)
with length 2

where

$$L = \delta_{26}[4\ 8\ 12\ 16\ 18\ 22\ 26\ 30\ 35\ 39\ 43\ 47\ 49\ 53\ 57\ 61$$
$$3\ 7\ 11\ 15\ 17\ 21\ 25\ 29\ 35\ 39\ 43\ 47\ 49\ 53\ 57\ 61$$
$$2\ 6\ 10\ 14\ 18\ 22\ 26\ 30\ 33\ 37\ 41\ 45\ 49\ 53\ 57\ 61$$
$$1\ 5\ 9\ \ \ 13\ 17\ 21\ 25\ 29\ 33\ 37\ 41\ 45\ 49\ 53\ 57\ 61].$$

To find the cycles of (5.69), it is enough to find all the cycles in the system (5.70). We can check $\mathrm{tr}(L^k)$, $k = 1, 2, \ldots, 64$, and look for nontrivial powers. These can be easily calculated as

$$\mathrm{tr}(L^2) = 2, \qquad \mathrm{tr}(L^5) = 5, \qquad \mathrm{tr}(L^{10}) = 17.$$

Using Theorem 5.3, we conclude that the system does not have fixed point, but it has one cycle of length 2, one cycle of length 5, and one cycle of length 10.

To determine the cycles of (5.70), we first consider L^2. It is easy to deduce that the 26th column, $\mathrm{Col}_{26}(L^2)$, is a diagonal nonzero column. We can then use it to generate the cycle of length 2. Since $L\delta_{64}^{26} = \delta_{64}^{29}$ and $L\delta_{64}^{29} = \delta_{64}^{26}$, we have a cycle of length 2.

Now, define $\pi(z) = \Gamma z$, where

$$\Gamma = I_4 \otimes \mathbf{1}_{16}^{\mathrm{T}}.$$

Using Theorem 5.5, the cycle of system (5.43) with length 2 is

$$\pi(\delta_{64}^{26}) \to \pi(\delta_{64}^{39}) \to \pi(\delta_{64}^{26}).$$

Equivalently,

$$\delta_4^2 \to \delta_4^3 \to \delta_4^2.$$

Back in its scalar form, it is shown in Fig. 5.5.

Similarly, since $\mathrm{Col}_1(L^5) = \delta_{64}^1$ is a diagonal nonzero column of L^5, then δ_{64}^1, $L\delta_{64}^1 = \delta_{64}^4$, $L^2\delta_{64}^1 = \delta_{64}^{16}$, $L^3\delta_{64}^1 = \delta_{64}^{61}$ and $L^4\delta_{64}^1 = \delta_{64}^{49}$ form a cycle of length 5. Using Theorem 5.5, the cycle of system (5.43) with length 5 is

$$\pi(\delta_{64}^1) \to \pi(\delta_{64}^4) \to \pi(\delta_{64}^{16}) \to \pi(\delta_{64}^{61}) \to \pi(\delta_{64}^{49}) \to \pi(\delta_{64}^1).$$

Equivalently, it is

$$\delta_4^1 \to \delta_4^1 \to \delta_4^1 \to \delta_4^4 \to \delta_4^4 \to \delta_4^1.$$

Back in its scalar form, it is the cycle depicted in Fig. 5.6.

Since $\mathrm{Col}_2(L^{10}) = \delta_{64}^2$ is a diagonal nonzero column of L^{10}, it follows that δ_{64}^2, $L\delta_{64}^2 = \delta_{64}^8$, $L^2\delta_{64}^2 = \delta_{64}^{30}$, $L^3\delta_{64}^2 = \delta_{64}^{53}$, $L^4\delta_{64}^2 = \delta_{64}^{17}$, $L^5\delta_{64}^2 = \delta_{64}^3$, $L^6\delta_{64}^2 = \delta_{64}^{12}$,

Fig. 5.6 The cycle of (5.43) with length 5

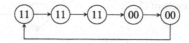

Fig. 5.7 The cycle of (5.43) with length 10

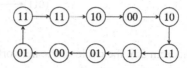

$L^7\delta_{64}^2 = \delta_{64}^{47}$, $L^8\delta_{64}^2 = \delta_{64}^5$, and $L^9\delta_{64}^2 = \delta_{64}^{33}$ form a cycle of length 10.

Using Theorem 5.5, the cycle of the system (5.43) with length 10 is

$$\pi\left(\delta_{64}^2\right) \to \pi\left(\delta_{64}^8\right) \to \pi\left(\delta_{64}^{30}\right) \to \pi\left(\delta_{64}^{53}\right) \to \pi\left(\delta_{64}^{17}\right) \to \pi\left(\delta_{64}^3\right) \to$$
$$\pi\left(\delta_{64}^{12}\right) \to \pi\left(\delta_{64}^{47}\right) \to \pi\left(\delta_{64}^5\right) \to \pi\left(\delta_{64}^{33}\right) \to \pi\left(\delta_{64}^2\right).$$

Equivalently,

$$\delta_4^1 \to \delta_4^1 \to \delta_4^2 \to \delta_4^4 \to \delta_4^2 \to \delta_4^1 \to \delta_4^1 \to \delta_4^3 \to \delta_4^4 \to \delta_4^3 \to \delta_4^1.$$

In scalar form, it is the cycle depicted in Fig. 5.7.

It is easy to calculate the transient period of (5.70), which is 4. From Theorem 5.5 we know that the transient time of the network (5.43) is 4. That is, for any initial state $(A(t_0), B(t_0))$, the trajectory will enter into a cycle after four steps.

The result coincides with the one in [11].

Example 5.16 Recall Example 5.14. To find the cycles of (5.53), it is enough to find all the cycles of the network (5.56). We can check $\mathrm{tr}(L^k)$, $k = 1, 2, \ldots, 64$, and look for nontrivial powers. It can be easily calculated that

$$\mathrm{tr}\left(L^k\right) = \begin{cases} 8, & k = 8i, i = 1, 2, \ldots, \\ 0, & \text{others.} \end{cases}$$

From Theorem 5.3, we conclude that the system (5.56) has only one cycle with length 8. To find this cycle, we consider L^8. It is easy to deduce that the third column, $\mathrm{Col}_3(L^8)$, is a diagonal nonzero column. We can then use it to generate the cycle of length 8. Since $L\delta_{64}^3 = \delta_{64}^{20}$, $L^2\delta_{64}^3 = \delta_{64}^{28}$, $L^3\delta_{64}^3 = \delta_{64}^{30}$, $L^4\delta_{64}^3 = \delta_{64}^{46}$, $L^5\delta_{64}^3 = \delta_{64}^{45}$, $L^6\delta_{64}^3 = \delta_{64}^{37}$, $L^7\delta_{64}^3 = \delta_{64}^{33}$, and $L^8\delta_{64}^3 = \delta_{64}^3$, we obtain a cycle of length 8 of the network (5.56) as

$$\delta_{64}^3 \to \delta_{64}^{20} \to \delta_{64}^{28} \to \delta_{64}^{30} \to \delta_{64}^{46} \to \delta_{64}^{45} \to \delta_{64}^{37} \to \delta_{64}^{33} \to \delta_{64}^3.$$

Define $\pi(z) = \Gamma z$, where

$$\Gamma = I_8 \otimes 1_8^{\mathrm{T}}.$$

Fig. 5.8 The cycle of (5.53)
with length 8

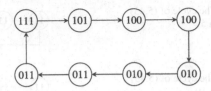

Using Theorem 5.5, the cycle of the network (5.53) can be obtained as

$$\pi\left(\delta_{64}^3\right) \to \pi\left(\delta_{64}^{20}\right) \to \pi\left(\delta_{64}^{28}\right) \to \pi\left(\delta_{64}^{30}\right) \to \pi\left(\delta_{64}^{46}\right) \to \pi\left(\delta_{64}^{45}\right) \to$$
$$\pi\left(\delta_{64}^{37}\right) \to \pi\left(\delta_{64}^{33}\right) \to \pi\left(\delta_{64}^3\right).$$

Equivalently,

$$\delta_8^1 \to \delta_8^3 \to \delta_8^4 \to \delta_8^4 \to \delta_8^6 \to \delta_8^6 \to \delta_8^5 \to \delta_8^5 \to \delta_8^1.$$

Using its scalar form, the cycle of the system (5.53) with length 8 is shown in Fig. 5.8.

It is easy to calculate that the transient period of (5.56) is 7. From Theorem 5.5 we know that the transient period of the network (5.53) is $T_t = 7$. That is, for any initial state $(A(t_0), B(t_0))$, the trajectory will enter into the above cycle after seven steps.

Remark 5.8 In this section, we do not consider the basin of an attractor (fixed point or cycle) as discussed for first order Boolean networks. For higher order Boolean networks, the basin of a cycle is meaningless because an initial point $x(0)$ in the original state space $\mathscr{X}$ may enter into more than one cycle.

To see this, we consider the following example. In Example 5.15 or Example 5.12, there are three cycles, with lengths 2, 5, and 10. We consider only the cycles with lengths 2 and 5. Consider the algebraic form (5.70). From initial states δ_{64}^{60} and δ_{64}^{50}, we have

$$\delta_{64}^{60} \to \delta_{64}^{45} \to \left\{\delta_{64}^{49} \to \delta_{64}^1 \to \delta_{64}^4 \to \delta_{64}^{16} \to \delta_{64}^{61} \to \delta_{64}^{49}\right\},$$
$$\delta_{64}^{50} \to \delta_{64}^5 \to \delta_{64}^{18} \to \delta_{64}^7 \to \left\{\delta_{64}^{26} \to \delta_{64}^{39} \to \delta_{64}^{26}\right\}.$$

For system (5.43), based on Lemma 5.4, from initial states $\pi(\delta_{64}^{60})$ and $\pi(\delta_{64}^{50})$, we have

$$\pi\left(\delta_{64}^{60}\right) \to \pi\left(\delta_{64}^{45}\right) \to \left\{\pi\left(\delta_{64}^{49}\right) \to \pi\left(\delta_{64}^1\right) \to \pi\left(\delta_{64}^4\right) \to \pi\left(\delta_{64}^{16}\right) \to \pi\left(\delta_{64}^{61}\right) \to \pi\left(\delta_{64}^{49}\right)\right\},$$
$$\pi\left(\delta_{64}^{50}\right) \to \pi\left(\delta_{64}^5\right) \to \pi\left(\delta_{64}^{18}\right) \to \pi\left(\delta_{64}^7\right) \to \left\{\pi\left(\delta_{64}^{26}\right) \to \pi\left(\delta_{64}^{39}\right) \to \pi\left(\delta_{64}^{26}\right)\right\}.$$

Equivalently, for system (5.43), from initial states δ_4^4, we have

$$\delta_4^4 \to \delta_4^3 \to \left\{\delta_4^4 \to \delta_4^1 \to \delta_4^1 \to \delta_4^1 \to \delta_4^4 \to \delta_4^4\right\},$$

$$\delta_4^4 \to \delta_4^1 \to \delta_4^2 \to \delta_4^1 \to \{\delta_4^2 \to \delta_4^3 \to \delta_4^2\}.$$

That is, δ_4^4 enters into two different cycles.

5.6.2 Second Algebraic Form of Higher Order Boolean Networks

Define

$$w(\tau) := x(\mu\tau)x(\mu\tau + 1) \cdots x\big(\mu\tau + (\mu - 1)\big) = z(\mu\tau). \qquad (5.71)$$

We then have

$$w(\tau + 1) = z(\mu\tau + \mu) = L^\mu z(\mu\tau) = L^\mu w(\tau),$$

where L is obtained in (5.51). Therefore, we have

$$w(\tau + 1) = \Gamma w(\tau), \qquad (5.72)$$

where

$$\Gamma = \big[(E_d)^n (I_{2^{\mu n}} \otimes L_0)\Phi_{\mu n}\big]^\mu,$$

with initial value $w(0) = \ltimes_{i=0}^{\mu-1} x(i)$. We call (5.72) the second algebraic form of the μth order Boolean network (5.40).

In fact, by rescheduling the sampling time, the second algebraic form provides the state variable, $w(\tau)$, $\tau = 0, 1, \ldots$, as a set of non-overlapping segments of $x(t)$. Hence, there is an obvious one-to-one correspondence between the trajectories of (5.40) and the trajectories of (5.72).

Proposition 5.7 *There is an obvious one-to-one correspondence between the trajectories of (5.40) and the trajectories of its second algebraic form (5.72), given by*

$$w(\tau) := x(\mu\tau)x(\mu\tau + 1) \cdots x\big(\mu\tau + (\mu - 1)\big), \qquad \tau = 0, 1, \ldots.$$

Therefore, it is easier to use the second algebraic form to calculate the trajectories of higher order Boolean networks. Unfortunately, for analyzing the topological structures, it is not as convenient as the first algebraic form.

Proposition 5.8 *Assume that the system (5.40) has a cycle of length α. Let the least common multiple (lcm) of α and μ be $\beta = lcm(\alpha, \mu)$. The system (5.72) then has a cycle of length $\gamma = \beta/\mu$.*

Proof Assume $s > 0$ is sufficiently large so that $x((s-1)\mu + 1)$ is on the cycle of length α. Since β is a multiply of α, we have that

$$\begin{cases} x((s-1)\mu + 1) = x((s-1)\mu + 1 + \beta), \\ x((s-1)\mu + 2) = x((s-1)\mu + 2 + \beta), \\ \vdots \\ x(s\mu) = x(s\mu + \beta). \end{cases} \tag{5.73}$$

Multiplying both sides of (5.73) yields

$$w(s) = w\left(s + \frac{\beta}{\mu}\right) = w(s + \gamma). \tag{5.74}$$

It is also easy to check that γ is the smallest positive integer which satisfies (5.74). The conclusion follows. $\square$

Example 5.17 Consider Example 5.14 again. From (5.56) we know that

$$z(t+1) = Lz(t),$$

where L is the matrix given in (5.57).

If we set $w(\tau) = x(2\tau)x(2\tau + 1) = z(2\tau)$, then the second algebraic form of (5.52) is

$$w(\tau + 1) = z(2\tau + 2) = L^2 z(2\tau) = L^2 w(\tau) = \Gamma w(\tau), \tag{5.75}$$

where

$$\begin{aligned} \Gamma = \delta_{64}[&20\ 22\ 28\ 30\ 3\quad 1\ 12\ 10\ 3\quad 1\ 42\ 46\ 33\ 37\ 42\ 46 \\ &28\ 30\ 28\ 30\ 11\ 9\ 12\ 10\ 11\ 9\ 42\ 46\ 41\ 45\ 42\ 46 \\ &20\ 22\ 28\ 30\ 3\quad 1\ 12\ 10\ 3\quad 1\ 42\ 46\ 33\ 37\ 42\ 46 \\ &28\ 30\ 28\ 30\ 11\ 9\ 12\ 10\ 11\ 9\ 42\ 46\ 41\ 45\ 42\ 46]. \end{aligned}$$

To find the cycles of (5.52), we check $\text{tr}(\Gamma^s)$, $s = 1, 2, \ldots$, and look for nontrivial powers s. These can be easily calculated as

$$\text{tr}(\Gamma^s) = \begin{cases} 8, & s = 4i, i = 1, 2, \ldots, \\ 0, & \text{others.} \end{cases}$$

Using Theorem 5.3 we conclude that the system (5.75) has two cycles of length 4.

Next, we investigate the cycles. Consider Γ^4. It is easy to determine that its third column is a diagonal nonzero column. We can then use it to generate one cycle of length 4. Since $\Gamma\delta_{64}^3 = \delta_{64}^{28}$, $\Gamma\delta_{64}^{28} = \delta_{64}^{46}$, $\Gamma\delta_{64}^{46} = \delta_{64}^{37}$, and $\Gamma\delta_{64}^{37} = \delta_{64}^3$, we have a cycle of length 4. Similarly, since δ_{64}^{20} is a diagonal nonzero column of Γ^4, it follows that δ_{64}^{20}, $\Gamma\delta_{64}^{20} = \delta_{64}^{30}$, $\Gamma\delta_{64}^{30} = \delta_{64}^{45}$, $\Gamma\delta_{64}^{45} = \delta_{64}^{33}$, and $\Gamma\delta_{64}^{33} = \delta_{64}^{20}$ form another cycle of length 4.

Using formula (4.4) to convert δ_{64}^3, δ_{64}^{28}, δ_{64}^{46}, δ_{64}^{37} and δ_{64}^{20}, δ_{64}^{30}, δ_{64}^{45}, δ_{64}^{33} back to binary form, we have

$$\delta_{64}^3 \sim (1,1,1,1,0,1), \qquad \delta_{64}^{20} \sim (1,0,1,1,0,0),$$

$$\delta_{64}^{28} \sim (1,0,0,1,0,0), \qquad \delta_{64}^{30} \sim (1,0,0,0,1,0),$$

$$\delta_{64}^{46} \sim (0,1,0,0,1,0), \qquad \delta_{64}^{45} \sim (0,1,0,0,1,1),$$

$$\delta_{64}^{37} \sim (0,1,1,0,1,1), \qquad \delta_{64}^{33} \sim (0,1,1,1,1,1).$$

Thus, the two cycles of length 4 are

$$(1,1,1,1,0,1) \rightarrow (1,0,0,1,0,0) \rightarrow (0,1,0,0,1,0) \rightarrow (0,1,1,0,1,1) \rightarrow$$
$$(1,1,1,1,0,1) \rightarrow (1,0,0,1,0,0) \rightarrow (0,1,0,0,1,0) \rightarrow (0,1,1,0,1,1) \rightarrow \cdots$$

and

$$(1,0,1,1,0,0) \rightarrow (1,0,0,0,1,0) \rightarrow (0,1,0,0,1,1) \rightarrow (0,1,1,1,1,1) \rightarrow$$
$$(1,0,1,1,0,0) \rightarrow (1,0,0,0,1,0) \rightarrow (0,1,0,0,1,1) \rightarrow (0,1,1,1,1,1) \rightarrow \cdots .$$

Comparing the set of cycles of system (5.52) with that of system (5.75), one sees easily that there is no one-to-one correspondence between them. Of course, in this simple case, we can calculate that $w(\tau) = z(2\tau - 1) = A(2\tau - 1)B(2\tau - 1) \times C(2\tau - 1)A(2\tau)B(2\tau)C(2\tau)$. For $A(t)$, $B(t)$, $C(t)$, we have

$$(1,1,1) \rightarrow (1,0,1) \rightarrow (1,0,0) \rightarrow (1,0,0) \rightarrow (0,1,0) \rightarrow (0,1,0) \rightarrow$$
$$(0,1,1) \rightarrow (0,1,1) \rightarrow (1,1,1) \rightarrow (1,0,1) \rightarrow (1,0,0) \rightarrow (1,0,0) \rightarrow$$
$$(0,1,0) \rightarrow (0,1,0) \rightarrow (0,1,1) \rightarrow (0,1,1) \rightarrow \cdots$$

and

$$(1,0,1) \rightarrow (1,0,0) \rightarrow (1,0,0) \rightarrow (0,1,0) \rightarrow (0,1,0) \rightarrow (0,1,1) \rightarrow$$
$$(0,1,1) \rightarrow (1,1,1) \rightarrow (1,0,1) \rightarrow (1,0,0) \rightarrow (1,0,0) \rightarrow (0,1,0) \rightarrow$$
$$(0,1,0) \rightarrow (0,1,1) \rightarrow (0,1,1) \rightarrow (1,1,1) \rightarrow \cdots .$$

It is easy to see that the two cycles of (5.75) become

$$(1,0,1) \rightarrow (1,0,0) \rightarrow (1,0,0) \rightarrow (0,1,0) \rightarrow (0,1,0) \rightarrow (0,1,1) \rightarrow$$
$$(0,1,1) \rightarrow (1,1,1) \rightarrow (1,0,1) \rightarrow (1,0,0) \rightarrow (1,0,0) \rightarrow (0,1,0) \rightarrow$$
$$(0,1,0) \rightarrow (0,1,1) \rightarrow (0,1,1) \rightarrow (1,1,1) \rightarrow \cdots ,$$

which is the only cycle of (5.52) of length 8.

References

1. Akutsu, T., Miyano, S., Kuhara, S.: Inferring qualitative relations in genetic networks and metabolic pathways. Bioinformatics **16**, 727–734 (2000)

2. Albert, R., Barabási, A.: Dynamics of complex systems: Scaling laws for the period of Boolean networks. Phys. Rev. Lett. **84**(24), 5660–5663 (2000)
3. Albert, R., Othmer, H.: The topology and signature of the regulatory interactions predict the expression pattern of the segment polarity genes in drosophila melanogaster. J. Theor. Biol. **223**(1), 1–18 (2003)
4. Aldana, M.: Boolean dynamics of networks with scale-free topology. Phys. D: Nonlinear Phenom. **185**(1), 45–66 (2003)
5. Cheng, D., Qi, H.: A linear representation of dynamics of Boolean networks. IEEE Trans. Automat. Contr. **55**(10), 2251–2258 (2010)
6. Clarke, E., Kroening, D., Ouaknine, J., Strichman, O.: Completeness and complexity of bounded model checking. In: Verification, Model Checking, and Abstract Interpretation. Lecture Notes in Computer Science, vol. 2937, pp. 85–96. Springer, Berlin/Heidelberg (2004)
7. Drossel, B., Mihaljev, T., Greil, F.: Number and length of attractors in a critical Kauffman model with connectivity one. Phys. Rev. Lett. **94**(8), 88,701 (2005)
8. Farrow, C., Heidel, J., Maloney, J., Rogers, J.: Scalar equations for synchronous Boolean networks with biological applications. IEEE Trans. Neural Netw. **15**(2), 348–354 (2004)
9. Goodwin, B.: Temporal Organization in Cells. Academic Press, San Diego (1963)
10. Harris, S., Sawhill, B., Wuensche, A., Kauffman, S.: A model of transcriptional regulatory networks based on biases in the observed regulation rules. Complexity **7**(4), 23–40 (2002)
11. Heidel, J., Maloney, J., Farrow, C., Rogers, J.: Finding cycles in synchronous Boolean networks with applications to biochemical systems. Int. J. Bifurc. Chaos **13**(3), 535–552 (2003)
12. Huang, S.: Regulation of cellular states in mammalian cells from a genomewide view. In: Collado-Vodes, J., Hofestadt, R. (eds.) Gene Regulation and Metabolism: Post-Genomic Computational Approaches, pp. 181–220. MIT Press, Cambridge (2002)
13. Huang, S., Ingber, D.: Shape-dependent control of cell growth, differentiation, and apoptosis: Switching between attractors in cell regulatory networks. Exp. Cell Res. **261**(1), 91–103 (2000)
14. Ideker, T., Galitski, T., Hood, L.: A new approach to decoding life: systems biology. Annu. Rev. Genom. Hum. Genet. **2**, 343–372 (2001)
15. Kauffman, S.: Metabolic stability and epigenesis in randomly constructed genetic nets. J. Theor. Biol. **22**(3), 437 (1969)
16. Kauffman, S.: The Origins of Order: Self-organization and Selection in Evolution. Oxford University Press, London (1993)
17. Kitano, H.: Systems biology: a brief overview. Science **259**, 1662–1664 (2002)
18. Langmead, C., Jha, S., Clarke, E.: Temporal-logics as query languages for Dynamic Bayesian Networks: Application to D. melanogaster Embryo Development. Tech. rep., School of Computer Science, Carnegie Mellon University, Pittsburgh, PA 15213 (2006)
19. Li, Z., Zhao, Y., Cheng, D.: Structure of higher order Boolean networks. Preprint (2010)
20. Mu, Y., Guo, L.: Optimization and identification in a non-equilibrium dynamic game. In: Proc. CDC-CCC'09, pp. 5750–5755 (2009)
21. Nurse, P.: A long twentieth century of the cell cycle and beyond. Cell **100**(1), 71–78 (2000)
22. Robert, F.: Discrete Iterations: A Metric Study. Springer, Berlin (1986). Translated by J. Rolne
23. Samuelsson, B., Troein, C.: Superpolynomial growth in the number of attractors in Kauffman networks. Phys. Rev. Lett. **90**(9), 98,701 (2003)
24. Shmulevich, I., Dougherty, E., Kim, S., Zhang, W.: Probabilistic Boolean networks: a rule-based uncertainty model for gene regulatory networks. Bioinformatics **18**(2), 261–274 (2002)
25. Waldrop, M.: Complexity: The Emerging Science at the Edge of Order and Chaos. Touchstone, New York (1992)
26. Zhao, Q.: A remark on 'Scalar equations for synchronous Boolean networks with biological capplications' by C. Farrow, J. Heidel, J. Maloney, and J. Rogers. IEEE Trans. Neural Netw. **16**(6), 1715–1716 (2005)

Chapter 6
Input-State Approach to Boolean Control Networks

6.1 Boolean Control Networks

A Boolean control network is defined as

$$\begin{cases} x_1(t+1) = f_1(x_1(t), x_2(t), \ldots, x_n(t), u_1(t), \ldots, u_m(t)), \\ x_2(t+1) = f_2(x_1(t), x_2(t), \ldots, x_n(t), u_1(t), \ldots, u_m(t)), \\ \vdots \\ x_n(t+1) = f_n(x_1(t), x_2(t), \ldots, x_n(t), u_1(t), \ldots, u_m(t)), \end{cases} \quad (6.1)$$

and

$$y_j(t) = h_j\big(x_1(t), x_2(t), \ldots, x_n(t)\big), \quad j = 1, 2, \ldots, p, \quad (6.2)$$

where $f_i : \mathscr{D}^{n+m} \to \mathscr{D}, i = 1, 2, \ldots n$, and $h_j : \mathscr{D}^n \to \mathscr{D}, j = 1, 2, \ldots p$, are logical functions, $x_i \in \mathscr{D}, i = 1, 2, \ldots n$, are states, $y_j \in \mathscr{D}, j = 1, 2, \ldots, p$ are outputs, and $u_\ell \in \mathscr{D}, \ell = 1, 2, \ldots m$, are inputs (or controls).

In this chapter we assume that the controls are logical variables satisfying certain logical rules, called the input network, described as follows:

$$\begin{cases} u_1(t+1) = g_1(u_1(t), u_2(t), \ldots, u_m(t)), \\ u_2(t+1) = g_2(u_1(t), u_2(t), \ldots, u_m(t)), \\ \vdots \\ u_m(t+1) = g_m(u_1(t), u_2(t), \ldots, u_m(t)). \end{cases} \quad (6.3)$$

Let

$$u(t) = \ltimes_{i=1}^{m} u_i(t),$$
$$x(t) = \ltimes_{i=1}^{n} x_i(t),$$
$$y(t) = \ltimes_{i=1}^{p} y_i(t).$$

D. Cheng et al., *Analysis and Control of Boolean Networks*,
Communications and Control Engineering,
DOI 10.1007/978-0-85729-097-7_6, © Springer-Verlag London Limited 2011

Fig. 6.1 A control network

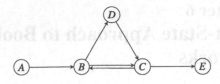

The Boolean control network (6.1)–(6.3) can then be expressed in algebraic form as

$$\begin{cases} u(t+1) = Gu(t), & u \in \Delta_{2^m}, \\ x(t+1) = Lu(t)x(t) := L(u)x(t), & x \in \Delta_{2^n}, \\ y(t) = Hx(t), & y \in \Delta_{2^p}, \end{cases} \quad (6.4)$$

where $G \in \mathcal{L}_{2^m \times 2^m}$, $L \in \mathcal{L}_{2^n \times 2^{n+m}}$, $H \in \mathcal{L}_{2^p \times 2^n}$, and $L(u) = Lu(t)$ is the control-dependent network transition matrix.

Example 6.1 Consider the system depicted in Fig. 6.1.

We consider $u(t) = A(t)$ as the input and $y(t) = E(t)$ as the output. The dynamics of the Boolean network is then described as

$$\begin{cases} B(t+1) = u(t) \to C(t), \\ C(t+1) = B(t) \vee D(t), \\ D(t+1) = \neg B(t), \end{cases} \quad (6.5)$$

the control network is

$$u(t+1) = \varphi\big(u(t)\big), \quad (6.6)$$

and the output is

$$y(t) = h\big(C(t)\big). \quad (6.7)$$

Set $x(t) = B(t) \ltimes C(t) \ltimes D(t)$. Converting this system into its algebraic form, we have

$$\begin{cases} u(t+1) = Gu(t), \\ x(t+1) = Lu(t)x(t) = L(u(t))x(t), \\ y(t) = Hx(t), \end{cases} \quad (6.8)$$

where $L(u(t)) = Lu(t)$ is the control-dependent network transition matrix. First, we assume φ is an identity mapping, that is, $u(t+1) = u(t)$ is a constant control. $L(u)$ can then be easily calculated as

$$L(u) = M_i u (I_2 \otimes M_d)(I_8 \otimes M_n) W_{[2]} W_{[2,8]} M_r$$

$$= \begin{cases} \delta_8[2\,2\,6\,6\,1\,3\,5\,7], & u = \delta_2^1, \\ \delta_8[2\,2\,2\,2\,1\,3\,1\,3], & u = \delta_2^2. \end{cases}$$

Now, both δ_2^1 and δ_2^2 are fixed points of the control network. Using Theorems 5.2 and 5.3, it is easy to deduce that for $u = \delta_2^1$, there is a fixed point for the system, which is $x = \delta_8^2$, or, equivalently, $X = (1, 1, 0)$, and there is also a cycle of length 2, which is $(1, 0, 1) \rightarrow (0, 1, 0) \rightarrow (1, 0, 1)$. When $u = \delta_2^2$, there is only a fixed point $X = (1, 1, 0)$.

In general, we would like to consider the structure of the Boolean control system (6.1), where the controls are varying, according to its own dynamical evolution rule (6.3). It is now clear that the system is evolving on an input-state "product space". We will need some preparatory results concerning this product space.

6.2 Semi-tensor Product Vector Space vs. Semi-tensor Product Space

Definition 6.1 Let M and N be vector spaces of dimensions m and n, respectively, with bases $\{\alpha_1, \alpha_2, \ldots, \alpha_m\}$ and $\{\beta_1, \beta_2, \ldots, \beta_n\}$, respectively. A vector space of dimension mn is called a semi-tensor product (STP) vector space of M and N, denoted by $\mathrm{Span}\{M \ltimes N\}$, if there exists a linear mapping $\ltimes : M \times N \rightarrow W$ such that

$$\gamma_{(i-1)n+j} := \alpha_i \ltimes \beta_j, \quad i = 1, \ldots, m, \ j = 1, \ldots, n,$$

form a basis of W.

Remark 6.1

1. We fix the bases $\alpha = \{\alpha_i \mid i = 1, \ldots, m\}$, $\beta = \{\beta_i \mid i = 1, \ldots, n\}$ and $\gamma = \{\alpha_i \ltimes \beta_j \mid i = 1, \ldots, m; \ j = 1, \ldots, n\}$ as the default bases for M, N, and $W = \mathrm{Span}\{M \ltimes N\}$, respectively.
   If we assume that $X = \sum_{i=1}^{m} a_i \alpha_i$, $Y = \sum_{j=1}^{n} b_j \beta_j$, then

$$X \ltimes Y = \sum_{i=1}^{m} a_i \alpha_i \ltimes \sum_{j=1}^{n} b_j \beta_j = \sum_{i=1}^{m} \sum_{j=1}^{n} a_i b_j \alpha_i \ltimes \beta_j := c\gamma. \quad (6.9)$$

We use a vector form for coefficients:

$$a = [a_1 \quad a_2 \quad \cdots \quad a_m]^T,$$
$$b = [b_1 \quad b_2 \quad \cdots \quad b_n]^T,$$
$$c = [c_1 \quad c_2 \quad \cdots \quad c_{mn}]^T.$$

It is then easy to check that

$$c = a \ltimes b. \quad (6.10)$$

Therefore, as in linear algebra we can ignore the bases and simply express vectors as $X = a$, $Y = b$, and $X \ltimes Y = a \ltimes b$.

2. Note that

$$M \ltimes N = \{u \ltimes v \mid u \in M, v \in N\}$$

is not a vector space. However, it contains a basis of W, which is why we use $W = \text{Span}\{M \ltimes N\}$. We call $M \ltimes N$ the semi-tensor product space.

3. To make Definition 6.1 meaningful, we have to show that it is independent of the choice of bases for M and N.

First, let

$$\tilde{\alpha} = \begin{bmatrix} \tilde{\alpha}_1 \\ \tilde{\alpha}_2 \\ \vdots \\ \tilde{\alpha}_m \end{bmatrix}, \qquad \tilde{\beta} = \begin{bmatrix} \tilde{\beta}_1 \\ \tilde{\beta}_2 \\ \vdots \\ \tilde{\beta}_n \end{bmatrix}$$

be alternate bases for M, N, respectively. We have to show that

$$\tilde{\gamma} = \tilde{\alpha} \ltimes \tilde{\beta} = \begin{bmatrix} \tilde{\gamma}_1 \\ \tilde{\gamma}_2 \\ \vdots \\ \tilde{\gamma}_{mn} \end{bmatrix}$$

is also a basis of W.

To see this let $\tilde{\alpha} = A\alpha$ and $\tilde{\beta} = B\beta$. Using (2.32) we have

$$\tilde{\gamma} = \tilde{\alpha} \ltimes \tilde{\beta} = (A\alpha) \ltimes (B\beta) = (A \otimes B)\alpha \ltimes \beta = (A \otimes B)\gamma.$$

Since both A and B are nonsingular, so is $A \otimes B$, which means that $\tilde{\gamma}$ is a basis of W.

Next, we show that the product is independent of the choice of bases. Let $X = \tilde{a}^{\mathrm{T}}\tilde{\alpha} = a^{\mathrm{T}}\alpha$ and $Y = \tilde{b}^{\mathrm{T}}\tilde{\beta} = b^{\mathrm{T}}\beta$. We then have $\tilde{a}^{\mathrm{T}}A\alpha = a^{\mathrm{T}}\alpha$. That is,

$$a = A^{\mathrm{T}}\tilde{a}.$$

Similarly, we have

$$b = B^{\mathrm{T}}\tilde{b}.$$

Using (6.9), we see that under two different product bases of W we have two product values:

$$X \ltimes Y = (\tilde{a} \ltimes \tilde{b})^{\mathrm{T}}\tilde{\gamma}, \tag{6.11}$$

$$X \ltimes Y = (a \ltimes b)^{\mathrm{T}}\gamma. \tag{6.12}$$

We then have to show that (6.11) and (6.12) are equal. This is true because

$$X \ltimes Y = (a \ltimes b)^{\mathrm{T}}\gamma = \left[A^{\mathrm{T}}\tilde{a} \ltimes B^{\mathrm{T}}\tilde{b}\right]^{\mathrm{T}}\gamma$$
$$= \left[(A^{\mathrm{T}} \otimes B^{\mathrm{T}})\tilde{a} \ltimes \tilde{b}\right]^{\mathrm{T}}\gamma$$

$$= [\tilde{a} \ltimes \tilde{b}]^{\mathrm{T}} (A \otimes B)\gamma$$

$$= [\tilde{a} \ltimes \tilde{b}]^{\mathrm{T}} \tilde{\gamma}.$$

Consider $\ltimes : M \times N \to W = \mathrm{Span}\{M \ltimes N\}$. It is easy to check that the image

$$\ltimes(M \times N) := \{uv \mid u \in M \text{ and } v \in N\} = M \ltimes N \subset W.$$

As mentioned earlier, in general, this is not an onto mapping. Naturally, we would like to know whether or not $\ltimes$ is a one-to-one mapping. It turns out that in general, it is not. Let $Z \in M \ltimes N$. If $Z = 0$ and $XY = Z$, where $X \in M$ and $Y \in N$, then at least one of X and Y should be zero. Assume $Z \neq 0$. By definition, we can find $X_0 \in M$ and $Y_0 \in N$ such that $X_0 Y_0 = Z$. It is then easy to prove that all the solutions of $XY = Z$ are

$$\begin{cases} X = X_0/\mu, \\ Y = \mu Y_0, \quad \mu \neq 0. \end{cases} \tag{6.13}$$

Now, assume that $M = \Delta_{k^m}$ and $N = \Delta_{k^n}$. Note that these are not vector spaces. We may consider them as topological spaces with the discrete topology. Thus, we can also call $M \ltimes N := \{uv \mid u \in M \text{ and } v \in N\}$ the STP topological space of M and N (sometimes just called the STP space of M and N). It is easy to check the following property.

Proposition 6.1 *Let $M = \Delta_{k^m}$, $N = \Delta_{k^n}$, and $W = M \ltimes N$.*

1. *The STP space is*

$$W = \Delta_{k^{m+n}}.$$

2. *Let $w \in W$. There then exist unique $u \in M$ and $v \in N$ such that $w = uv$.*

We define the matrix

$$\mathbf{1}_{p \times q} = I_p \otimes \mathbf{1}_q^{\mathrm{T}}.$$

A straightforward computation then shows that we have the following formulas for the decomposition.

Proposition 6.2 *Let $M = \Delta_{k^m}$, $N = \Delta_{k^n}$, and $w \in W = M \ltimes N$. Decompose $w = uv$, where $u \in M$ and $v \in N$. Then,*

$$\begin{cases} u = \mathbf{1}_{k^m \times k^n} w, \\ v = \mathbf{1}_{k^m}^{\mathrm{T}} w. \end{cases} \tag{6.14}$$

We now give a simple example.

Example 6.2 Let $M = \Delta_3$, $N = \Delta_9$, and $W = M \ltimes N = \Delta_{27}$. Then,

$$\mathbf{1}_{3\times 9} = \begin{bmatrix} \mathbf{1}_9^T & 0 & 0 \\ 0 & \mathbf{1}_9^T & 0 \\ 0 & 0 & \mathbf{1}_9^T \end{bmatrix}.$$

Assume that $w = \delta_{27}^{13} \in W$, $u \in M$, $v \in N$, and $w = uv$. It is then easy to calculate that

$$\begin{cases} u = \mathbf{1}_{3\times 9}w = \delta_3^2 = [0\ 1\ 0]^T, \\ v = \mathbf{1}_3^T w = \delta_9^4 = [0\,0\,0\,1\,0\,0\,0\,0\,0]^T. \end{cases}$$

6.3 Cycles in Input-State Space

Recall system (6.1) with input (6.3). Note that the state space is $X \in \mathcal{D}^n$ (equivalently, in vector form, $x \in \mathcal{X} = \Delta_{2^n}$) and the input space is $U \in \mathcal{D}^m$ (equivalently, in vector form, $u \in \mathcal{U} = \Delta_{2^m}$). The input-state STP space (sometimes just called the input-state space) is $U \times X = \mathcal{D}^{m+n}$ (or $\mathcal{W} = \mathcal{U} \ltimes \mathcal{X} = \Delta_{2^{m+n}}$).

In this section we consider the structure of a cycle in the input-state space. Denote by $C_{\mathcal{W}}^r$ a cycle in the space $\mathcal{W}$ with length r. Assume that there is a cycle of length k in the input-state space $\mathcal{W}$, say,

$$C_{\mathcal{W}}^k: \quad w(0) = w_0 = u_0 x_0 \to w(1) = w_1 = u_1 x_1 \to \cdots \to$$
$$w(k) = w_k = u_k x_k = w_0.$$

First, it is easily seen that since $u_0 = u_k$, in the input space $\mathcal{U}$, the sequence $\{u_0, u_1, \ldots, u_k\}$ contains, say, j folds of a cycle of length ℓ, where $j\ell = k$. Note that "j folds of a cycle" means the cycle is repeated j times. Hence, $u_\ell = u_0$. Now, let us see what conditions the $\{x_i\}$ in the cycle $C_{\mathcal{W}}^k$ should satisfy. Define a network transition matrix

$$\Psi := L(u_{\ell-1})L(u_{\ell-2})\cdots L(u_1)L(u_0). \tag{6.15}$$

Starting from $w_0 = u_0 x_0$, the x component of the cycle $C_{\mathcal{W}}^k$ is

$$x_0 \to x_1 = L(u_0)x_0 \to x_2 = L(u_1)L(u_0)x_0 \to \cdots \to x_\ell = \Psi x_0 \to$$
$$x_{\ell+1} = L(u_0)\Psi x_0 \to x_{\ell+2} = L(u_1)L(u_0)\Psi x_0 \to \cdots \to x_{2\ell} = \Psi^2 x_0 \to \cdots \to$$
$$x_{(j-1)\ell+1} = L(u_0)\Psi^{j-1}x_0 \to x_{(j-1)\ell+2} = L(u_1)L(u_0)\Psi^{j-1}x_0 \to \cdots \to$$
$$x_{j\ell} = \Psi^j x_0 = x_0.$$
$$\tag{6.16}$$

We conclude that $x_0 \in \Delta_{2^n}$ is a fixed point of the equation

$$x(t+1) = \Psi^j x(t). \tag{6.17}$$

For convenience, we assume that $j > 0$ is the smallest positive integer which makes x_0 a fixed point of (6.17). Conversely, we assume that $x_0 \in \Delta_{2^n}$ is a fixed point of (6.17) and that u_0 is a point on a cycle of the control space $C_{\mathcal{U}}^\ell$. It is then obvious that we have the cycle (6.16).

Summarizing the above arguments yields the following theorem.

Theorem 6.1 *Consider the Boolean control network* (6.1)–(6.3). *A set* $C_{\mathcal{W}}^k \subset \Delta_{2^{k(n+m)}}$ *is a cycle in the input-state space* $\mathcal{W}$ *with length* k *if and only if for any point* $w_0 = u_0 x_0 \in C_{\mathcal{W}}^k$, *there exists an* $\ell \leq k$ *which is a factor of* k *such that* $u_0, u_1 = G u_0, u_2 = G^2 u_0, \dots, u_\ell = G^\ell u_0 = u_0$ *is a cycle in the control space, and* x_0 *is a fixed point of equation* (6.17) *with* $j = k/\ell$.

Theorem 6.1 shows how to find all the cycles in the input-state space. First, we can find cycles in the input space. If we pick a cycle in the input space, say $C_{\mathcal{U}}^\ell$, then for each point $u_0 \in C_{\mathcal{U}}^\ell$ we can construct an auxiliary system

$$x(t+1) = \Psi x(t). \tag{6.18}$$

Suppose that $C_{\mathcal{U}}^\ell = (u_0, u_1, \dots, u_\ell = u_0)$ is a cycle in $\mathcal{U}$ and $C_{\mathcal{X}}^j = (x_0, x_1, \dots, x_j = x_0)$ is a cycle of (6.18). There is then a cycle $C_{\mathcal{W}}^k$, $k = \ell j$, in the input-state STP space, which can be constructed as follows:

$$w_0 = u_0 x_0 \to w_1 = u_1 L(u_0) x_0 \to w_2 = u_2 L(u_1) L(u_0) x_0 \to \cdots \to$$

$$w_\ell = u_0 x_1 \to w_{\ell+1} = u_1 L(u_0) x_1 \to w_{\ell+2} = u_2 L(u_1) L(u_0) x_1 \to \cdots \to$$

$$\vdots$$

$$\tag{6.19}$$

$$w_{(j-1)\ell} = u_0 x_{(j-1)} \to w_{(j-1)\ell+1} = u_1 L(u_0) x_{(j-1)} \to$$

$$w_{(j-1)\ell+2} = u_2 L(u_1) L(u_0) x_{(j-1)} \to \cdots \to$$

$$w_{j\ell} = u_0 x_j = u_0 x_0 = w_0.$$

We call this $C_{\mathcal{W}}^k$ the composed cycle of $C_{\mathcal{U}}^\ell$ and $C_{\mathcal{X}}^j$, denoted by $C_{\mathcal{W}}^k = C_{\mathcal{U}}^\ell \circ C_{\mathcal{X}}^j$.

Note that from a cycle $C_{\mathcal{U}}^\ell$ we can choose any point as the starting point u_0. In equation (6.18) we then have different Ψ, which produce different $C_{\mathcal{X}}^j$. It is reasonable to guess that the composed cycle $C_{\mathcal{W}}^k = C_{\mathcal{U}}^\ell \circ C_{\mathcal{X}}^j$ is independent of the choice of u_0. In fact, this is true.

Definition 6.2 Let $C_{\mathcal{W}}^k = \{w(t) \mid t = 0, 1, \dots, k\}$ be a cycle in the input-state space and $C_{\mathcal{U}}^\ell$ be a cycle in the input space. Splitting $w(t) = u(t)x(t)$, we say that $C_{\mathcal{W}}^k$ is attached to $C_{\mathcal{U}}^\ell$ at u_0 if $w(0) = u_0 x_0$ and

1. $u(t) \in C_{\mathcal{U}}^\ell$ with $u(0) = u_0$,
2. $x(0) = x_0$ is a fixed point of (6.17) with $j = \frac{k}{\ell} \in \mathbb{Z}_+$.

Remark 6.2 According to Theorem 6.1, each cycle $C_{\mathscr{W}}^k$ in the input-state space must be attached to exactly one cycle in the input space. In fact, the following argument shows that $C_{\mathscr{W}}^k$ attaches $C_{\mathscr{U}}^\ell$ at u_0 at moment $t = 0$ (where the attaching point of $C_{\mathscr{W}}^k$ is $w_0 = u_0 x_0$) and will attach it at u_1 at moment $t = 1$ (where the attaching point of $C_{\mathscr{W}}^k$ is $w_1 = u_1 x_1$), and so on. So, $C_{\mathscr{W}}^k$ and $C_{\mathscr{U}}^\ell$ are moving as two assembled together gears.

Proposition 6.3 *The sets of cycles in the input-state space which are attached to any point of a given cycle $C_{\mathscr{U}}^\ell$ are the same.*

Proof Let $C_{\mathscr{U}}^\ell = \{u_0, u_1, \ldots, u_\ell = u_0\}$ be the cycle under consideration and let S_0, $S_1, \ldots, S_{\ell-1}$ be the sets of input-state cycles attached to $u_0, u_1, \ldots, u_{\ell-1}$, respectively. First, we show that

$$S_0 \subset S_i, \quad i = 1, 2, \ldots, \ell - 1.$$

Let $C_k^0 = \{w_0, w_1, \ldots, w_k\} \in S_0$, i.e., it is a cycle attached to $C_{\mathscr{U}}^\ell$ at u_0. Using the elements of a control cycle, we can define

$$L_i := L(u_i), \quad i = 0, 1, \ldots, \ell - 1.$$

We then construct ℓ system matrices:

$$\begin{cases} \Psi_0 := L_{\ell-1} L_{\ell-2} \cdots L_0, \\ \Psi_1 := L_0 L_{\ell-1} L_{\ell-2} \cdots L_1, \\ \vdots \\ \Psi_{\ell-1} := L_{\ell-2} L_{\ell-3} \cdots L_0 L_{\ell-1}. \end{cases}$$

Correspondingly, we then construct ℓ auxiliary systems:

$$x(t+1) = \Psi_i x(t), \quad i = 0, 1, \ldots, \ell - 1. \tag{6.20}$$

Since $w_0 = u_0 x_0 \in C_{\mathscr{W}}^k \in S_0$, it follows that x_0 satisfies

$$(\Psi_0)^j x_0 = x_0. \tag{6.21}$$

Note that $w(1) := w_1 = u_1 L_0 x_0$. To see that $w_1 \in C_k^1 \in S_1$ we have to show that $L_0 x_0$ satisfies

$$(\Psi_1)^j L_0 x_0 = L_0 x_0. \tag{6.22}$$

This is true because

$$L_0 x_0 = L_0 (\Psi_0)^j x_0$$
$$= L_0 \left(L_{\ell-1} \cdots L_0 \right)^j x_0$$
$$= L_0 \underbrace{\left(L_{\ell-1} \cdots L_0 \right) \cdots \left(L_{\ell-1} \cdots L_0 \right)}_{j} x_0$$

Fig. 6.2 The structure of a
cycle in the input-state space

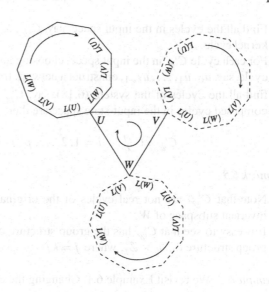

$$= \underbrace{(L_0 L_{\ell-1} \cdots L_1) \cdots (L_0 L_{\ell-1} \cdots L_1)}_{j} L_0 x_0$$

$$= (L_0 L_{\ell-1} \cdots L_1)^j L_0 x_0$$

$$= \Psi_1^j L_0 x_0.$$

Similarly, we can show that

$$u_s L_{s-1} L_{s-2} \cdots L_0 x_0 \in C_k^s \in S_s, \quad s = 1, 2, \ldots, \ell - 1.$$

Note that, precisely speaking, (6.22) can only ensure that there is a cycle of length
$\ell \times j'$ attached to the cycle at u_1, where j' is a factor of j. However, since the above
definition of $\{\Psi_i\}$ is in a rotating form, starting from a point $w_0 = u_1 x_0'$, the same
argument shows $j \le j'$. Therefore, $j' = j$.

The same argument shows that

$$S_j \subset S_i, \quad 0 \le i \ne j \le \ell - 1.$$

We conclude that

$$S_j = S_i, \quad 0 \le i, j \le \ell - 1. \qquad \square$$

Remark 6.3 To see the rolling process, assume $C_{\mathcal{U}}^3 = \{U, V, W\}$. An attached cycle
in the input-state space is then depicted in Fig. 6.2, where the dashed-line cycles are
duplicated ones.

Remark 6.4 From Proposition 6.3 we see that cycles in the input-state space of a
Boolean control network can be found in the following way:

1. Find all the cycles in the input space, say, $C_{\mathcal{U}}^1, \ldots, C_{\mathcal{U}}^p$. We call such a cycle a kernel cycle.
2. For each cycle $C_{\mathcal{U}}^i$ in the input space, choosing any point as starting point of the cycle, say, $u_0, u_1, \ldots, u_{\ell-1}$, construct a network transition matrix (6.15) and then find all the cycles of the system (6.18), say, $C_{\mathcal{X}}^{i,1}, \ldots, C_{\mathcal{X}}^{i,q_i}$. The set of overall composed cycles in the input-state space are then

$$C_{\mathcal{U}}^i \circ C_{\mathcal{X}}^{i,j}, \quad i = 1, 2, \ldots, p, \, j = 1, 2, \ldots, q_i.$$

Remark 6.5

1. Note that $C_{\mathcal{X}}^{i,j}$ are not real cycles of the original system unless X is also an invariant subspace of W.
2. It is easy to see that $C_{\mathcal{U}}^\ell$ has the group structure of $\mathbb{Z}_\ell$ and $C_{\mathcal{W}}^k$ has the product group structure of $\mathbb{Z}_\ell \times \mathbb{Z}_j$, where $j = k/\ell$.

Example 6.3 We revisit Example 6.1. Changing the control to

$$u(t+1) = \neg u(t),$$

we now have an obvious kernel cycle, $0 \to 1 \to 0$. We can then easily calculate that

$$L(0) = \delta_8[2\,2\,2\,2\,1\,3\,1\,3],$$

$$L(1) = \delta_8[2\,2\,6\,6\,1\,3\,5\,7].$$

Hence, we consider an auxiliary system

$$x(t+1) = \Psi x(t), \tag{6.23}$$

where

$$\Psi = L(1)L(0) = \delta_8[2\,2\,2\,2\,2\,6\,2\,6].$$

A routine calculation shows that a nontrivial power of Ψ is 1 and that $\mathrm{tr}(\Psi^1) = 2$. Thus, there are two fixed points, $\delta_8^1 \sim (1, 1, 0)$ and $\delta_8^6 \sim (0, 1, 0)$. The overall composed cycles are depicted in Fig. 6.3, where the dashed lines show the duplicated cycles. Overall, we have a cycle in the input space and two product cycles of length 2 in the input-state space.

Finally, we consider transient periods of product cycles. Assume that $C_{\mathcal{U}}^{l_i}$, $i = 1, \ldots, p$, are the cycles of length l_i in the control space. We can construct Ψ_i and find the smallest r^i such that

$$(\Psi_i)^{r^i} = (\Psi_i)^{r^i + T_i}.$$

It is clear that if a point will eventually enter the cycle attached to this cycle, then after r^i (composed) steps the second component will enter the rotating cycle. Note

Fig. 6.3 Cycles of a control system

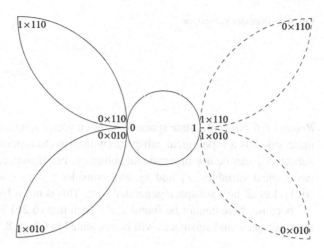

that Ψ_i is a composed mapping, consisting of ℓ_i steps. Taking the first part $(C^\ell_{\mathscr{U}})$ into consideration, it is easily seen that the transient period for cycles attached to $C^{\ell_i}_{\mathscr{U}}$, denoted by $T_t(C^{\ell_i}_{\mathscr{U}})$, satisfies

$$\max\left\{r_0, \ell_i(r^i - 1)\right\} \le T_t(C^i_u) \le \max\left\{r_0, \ell_i(r^i)\right\}, \quad i = 1, \dots, p. \qquad (6.24)$$

Define

$$V_i := \max\left\{r_0, \ell_i(r^i - 1)\right\},$$
$$U_i := \max\left\{r_0, \ell_i(r^i)\right\}, \quad i = 1, \dots, p.$$

The following is then obvious.

Proposition 6.4 *The transient period of the control system satisfies*

$$\max_{1 \le i \le p}\{V_i\} < T_t \le \max_{1 \le i \le p}\{U_i\}. \qquad (6.25)$$

6.4 Cascaded Boolean Networks

The input-state structure proposed in previous sections is very useful for analyzing the structure of Boolean networks with cascading structure.

Definition 6.3 Consider the system (5.1) [or the system (6.1)]. Let $\mathscr{X} = \Delta_{2^n}$ be the state space. $\mathscr{V} = \Delta_{2^k}$ is said to be a k-dimensional subspace of $\mathscr{X}$ if there is a space $\mathscr{V}^c = \Delta_{2^{n-k}}$, called the complement space (or, simply, complement) of $\mathscr{V}$, such that

$$\mathscr{V} \ltimes \mathscr{V}^c = \mathscr{X}. \qquad (6.26)$$

Fig. 6.4 Invariant subspaces

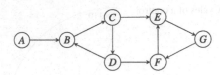

Remark 6.6 Since the state space $\mathscr{X}$ is not a vector space, the subspace can only be understood as a topological subspace (with discrete topology). Note that not every subset Δ_{2^k} can be a k-dimensional subspace. For instance, let $\mathscr{X}$ be generated by two logical variables, x_1 and x_2, and consider $z = x_1 \wedge x_2$. This can take values $\{0, 1\}$. Let $\mathscr{Z}$ be a subspace generated by z. This is not a 1-dimensional subspace of $\mathscr{X}$ because there cannot be found a $\mathscr{Z}^c$ such that (6.26) holds. A general study of the state space and subspaces will be presented in Chap. 8.

Definition 6.4 Consider the system (5.1), where $x \in \mathscr{X} = \Delta_{2^n}$. A k-dimensional subspace $\mathscr{V} = \Delta_{2^k}$ is called an invariant subspace if $x_0 \in \mathscr{V}$ implies that $x(t, x_0) \in \mathscr{V}, \forall t > 0$.

From the last section it is easily seen that the input space $\mathscr{U}$ is an invariant subspace of the input-state STP space $\mathscr{W}$. Conversely, an invariant subspace can also be considered as an input subspace.

To test whether a subspace is an invariant subspace, we can use either a network graph or a network equation. We will use some examples to illustrate this. A more general definition and some verifiable conditions will be discussed in the next chapter.

Let $\{x_{i_1}, \ldots, x_{i_s}\}$ be a subset of the vertices of a network. Define a subspace $\mathscr{V} = \text{Span}\{x_{i_1}, \ldots, x_{i_s}\}$. $\mathscr{V} \subset \mathscr{X}$ is the subspace describing the states of $\{x_{i_1}, \ldots, x_{i_s}\}$.

Note that "Span" is not clearly defined here. In Chap. 8 we will see that "Span" is the same as "$\mathscr{F}_\ell\{\cdots\}$", which means "the set of logical functions of $\{\cdots\}$".

Example 6.4 Consider the network graph shown in Fig. 6.4. It is easily seen that $\mathscr{V}_1 = \text{Span}\{A\}$ and $\mathscr{V}_2 = \text{Span}\{A, B, C, D\}$ are two invariant subspaces. Denote by $\mathscr{X}$ the total space. We then have the nested invariant subspaces

$$\mathscr{V}_1 \subset \mathscr{V}_2 \subset \mathscr{X}.$$

Note that $\mathscr{V} = \text{Span}\{A, B, C\}$ is not an invariant subspace because it is affected by D. (For readers familiar with graph theory, it is easy to see that a subspace is invariant if and only if the subgraph generated by the set of vertices with the inherent edges between them has in-degree zero.)

The structure of nested invariant subspaces can also be determined from systems of equations. Consider the following example.

Example 6.5 Consider the following system:

$$\begin{cases} A_1(t+1) = f_1^1(A_1(t), \dots, A_\ell(t)), \\ \vdots \\ A_\ell(t+1) = f_\ell^1(A_1(t), \dots, A_\ell(t)), \\ B_1(t+1) = f_1^2(A_1(t), \dots, A_\ell(t), B_1(t), \dots, B_m(t)), \\ \vdots \\ B_m(t+1) = f_m^2(A_1(t), \dots, A_\ell(t), B_1(t), \dots, B_m(t)), \\ C_1(t+1) = f_1^3(A_1(t), \dots, A_\ell(t), B_1(t), \dots, B_m(t), C_1(t), \dots, C_n(t)), \\ \vdots \\ C_n(t+1) = f_n^3(A_1(t), \dots, A_\ell(t), B_1(t), \dots, B_m(t), C_1(t), \dots, C_n(t)). \end{cases}$$

$$(6.27)$$

Here we have at least two nested invariant subspaces:

$$\mathscr{V}_1 = \text{Span}\{A_1, \dots, A_\ell\} = \mathscr{D}^\ell,$$

$$\mathscr{V}_2 = \text{Span}\{A_1, \dots, A_\ell, B_1, \dots, B_m\} = \mathscr{D}^{\ell+m},$$

$$\mathscr{V}_1 \subset \mathscr{V}_2 \subset \mathscr{X} = \mathscr{D}^{\ell+m+n}.$$

We consider a cycle, say $U^3 \in \mathscr{X}$. As discussed in the previous section, it must attach to a cycle, say $U^2 \in \mathscr{V}_2$. Similarly, U_2 must attach to a cycle, say $U^1 \in \mathscr{V}_1$. Now, in Fig. 6.5 we assume that cycles $U_1^2, U_2^2 \in \mathscr{V}_2$ are attached to $U_1 \in \mathscr{V}_1$, that $U_1^3, U_2^3 \in \mathscr{X}$ are attached to U_1^2, and that $U_3^3, U_4^3 \in \mathscr{X}$ are attached to U_2^2. We call such connected cycles chained gears.

Chained gears have the following properties:

- The gears in each chain, such as $U^1 \to U_1^2 \to U_1^3$, have multiple perimeters. Here the perimeter of a cycle means the number of states in the cycle. For instance, the perimeter of U_1^3 is a multiple of the perimeter of U_1^2, and the perimeter of U_1^2 is a multiple of the perimeter of U^1.
- In each chain the smaller gears affect the larger gears, and the larger gears do not affect the smaller gears.
- Smallest gears act as leading gears, the other gears will follow them.

Kauffman claimed that in a cellular network the tiny attractors decide the vast order [4]. The "rolling gear" structure may explain why small cycles determine the order of the whole network. We are led to speculate that the structure of "rolling gear" may account for the "hidden order" in human lives!

Finally, you may ask why there should be invariant subspaces. In fact, if a large or, potentially, huge network has small cycles, then the small cycles with the states in their regions of attraction form small invariant subspaces. If there are no such small cycles, then the system is in chaos [4]! Therefore, an ordered large-scale network should have the structure of nested invariant subspaces.

Fig. 6.5 Structure of cycles
in a cascaded Boolean
network

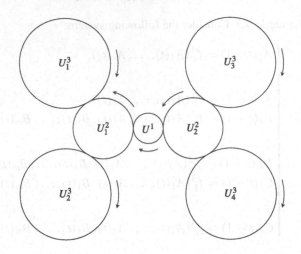

Table 6.1 Truth table of
(6.28)

	f_1	f_2	f_3	f_4	f_5
	1	1	1	1	1
	1	1	0	1	0
	1	1	1	1	0
	0	0	1	0	0
	1	0	0	1	0
	1	1	1	0	0
	1	1	1	0	0
	0	0	0	0	0
j_1	5	3	3	3	5
j_2	2	5	1	4	4
j_3	4	4	5	4	1

6.5 Two Illustrative Examples

The first example is from [5]. It serves two purposes: (1) to illustrate the standard
algorithm, and (2) to demonstrate that the "small cycles" play a decisive role in
determining the overall structure of the network.

Example 6.6 Consider a system with five nodes:

$$A_i = f_i(A_{j_1}, A_{j_2}, A_{j_3}), \quad i = 1, 2, 3, 4, 5, \tag{6.28}$$

where the logical functions $f_i, i = 1, \ldots, 5$, are determined by Table 6.1.

The algebraic form of system (6.28) is then

$$
\begin{cases}
A(t+1) = \delta_2[1\,1\,1\,2\,1\,1\,1\,2]E(t)B(t)D(t), \\
B(t+1) = \delta_2[1\,1\,1\,2\,2\,1\,1\,2]C(t)E(t)D(t), \\
C(t+1) = \delta_2[1\,2\,1\,1\,2\,1\,1\,2]C(t)A(t)E(t), \\
D(t+1) = \delta_2[1\,1\,1\,2\,1\,2\,2\,2]C(t)D(t)D(t), \\
E(t+1) = \delta_2[1\,2\,2\,2\,2\,2\,2\,2]E(t)D(t)A(t).
\end{cases}
\tag{6.29}
$$

To obtain the structure matrices, note that the first row of the structure matrix of f_i is exactly the same as its values in the truth table.

To convert the algebraic form back to logical form, mod 2 algebra is more convenient. Using mod 2 algebra, the system (6.28) can be expressed as

$$
\begin{cases}
A(t+1) = B(t) + D(t) + B(t)D(t), \\
B(t+1) = D(t) + E(t) + C(t)D(t)E(t), \\
C(t+1) = A(t) + C(t) + E(t) + C(t)E(t) + A(t)C(t)E(t), \\
D(t+1) = D(t), \\
E(t+1) = A(t)D(t)E(t).
\end{cases}
\tag{6.30}
$$

In fact, the mod 2 product $A \cdot B = A \wedge B$, and the mod 2 addition $A + B = A \bar{\vee} B$. Hence,

$$
M_\times = M_c = \delta_2[1\,2\,2\,2], \qquad M_+ = M_p = \delta_2[2\,1\,1\,2].
$$

If we let $x(t) = A(t)B(t)C(t)D(t)E(t)$, then

$$
x(t+1) = M_p^2 B D M_c B D M_p^2 D E M_c^2 C D E M_p^4 A C E M_c C E M_c^2 A C E D M_c^2 A D E.
$$

Now, there is a standard procedure to determine L. In fact,

$$
L = \delta_{32}[1\,6\,4\,16\,13\,2\,8\,12\,1\,6\,20\,32\,13\,2\,24\,28
$$
$$
2\,2\,4\,12\,10\,6\,4\,16\,2\,2\,20\,28\,10\,6\,20\,32].
$$

It is then easy to check that the nontrivial powers are 1 and 2, and that

$$
\mathrm{tr}(L) = 4, \qquad \mathrm{tr}(L^2) = 6.
$$

We conclude that there are four fixed points and one cycle of length 2. Using Theorem 5.2, it is easily seen that the fixed points are

$$
E_1 = (1, 1, 1, 1, 1), \qquad E_2 = (1, 0, 0, 1, 1),
$$
$$
E_3 = (0, 0, 1, 0, 0), \qquad E_4 = (0, 0, 0, 0, 0),
$$

and the cycle of length 2 is

$$
(1, 1, 1, 1, 0) \rightarrow (1, 1, 0, 1, 0) \rightarrow (1, 1, 1, 1, 0).
$$

The smallest repeating L^k is $L^3 = L^5$, so the transient period $T_t = 3$.

Finally, we use Proposition 5.5 to obtain the overall picture of the state space.

- Starting from $E_1 = (1, 1, 1, 1, 1)$, we calculate its parent states, its grandparent states, and so on. We have the following retrieval process and results. Note that in the following, $[x]$ is used to show that x is already on the cycle and can thus be removed from the retrieving chain:

$$E_1 = (1, 1, 1, 1, 1) \sim \delta_{32}^1 \Rightarrow [L_1 \to \delta_{32}^1], L_9 \to \delta_{32}^9 \sim (1, 0, 1, 1, 1) \Rightarrow \emptyset.$$

$$E_2 = (1, 0, 0, 1, 1) \sim \delta_{32}^{13} \Rightarrow [L_{13} \to \delta_{32}^{13}], L_5 \to \delta_{32}^5 \sim (1, 1, 0, 1, 1) \Rightarrow \emptyset.$$

$$E_3 = (0, 0, 1, 0, 0) \sim \delta_{32}^{28} \Rightarrow [L_{28} \to \delta_{32}^{28}], L_{16} \to \delta_{32}^{16} \sim (1, 0, 0, 0, 0) \Rightarrow$$

$$\begin{cases} & \begin{cases} L_3 \to \delta_{32}^3 \sim (1, 1, 1, 0, 1) \Rightarrow \emptyset, \\ L_4 \to \delta_{32}^4 \sim (1, 1, 1, 0, 0) \Rightarrow & L_{19} \to \delta_{32}^{19} \sim (0, 1, 1, 0, 1) \Rightarrow \emptyset, \\ & L_{23} \to \delta_{32}^{23} \sim (0, 1, 0, 0, 1) \Rightarrow \emptyset, \end{cases} \\ L_{24} \to \delta_{32}^{24} \sim (0, 1, 0, 0, 0) \Rightarrow L_{15} \to \delta_{32}^{15} \sim (1, 0, 0, 0, 1) \Rightarrow \emptyset. \end{cases}$$

$$E_4 = (0, 0, 0, 0, 0) \sim \delta_{32}^{32} \Rightarrow [L_{32} \to \delta_{32}^{32}], L_{12} \to \delta_{32}^{12} \sim (1, 0, 1, 0, 0) \Rightarrow$$

$$\begin{cases} & \begin{cases} L_{11} \to \delta_{32}^{11} \sim (1, 0, 1, 0, 1) \Rightarrow \emptyset, \\ L_{20} \to \delta_{32}^{20} \sim (0, 1, 1, 0, 0) \Rightarrow & L_{27} \to \delta_{32}^{27} \sim (0, 0, 1, 0, 1) \Rightarrow \emptyset, \\ & L_{31} \to \delta_{32}^{31} \sim (0, 0, 0, 0, 1) \Rightarrow \emptyset, \end{cases} \\ L_8 \to \delta_{32}^8 \sim (1, 1, 0, 0, 0) \Rightarrow L_7 \to \delta_{32}^7 \sim (1, 1, 0, 0, 1) \Rightarrow \emptyset. \end{cases}$$

- Next, we consider two points on the cycle: $C_1 = (1, 1, 0, 1, 0)$ and $C_2 = (1, 1, 1, 1, 0)$. For C_1:

$$C_1 = 11010 \sim \delta_{32}^6 \Rightarrow$$

$$\begin{cases} [L_2 \to \delta_{32}^2], \\ L_{10} \to \delta_{32}^{10} \sim (1, 0, 1, 1, 0) \Rightarrow \begin{cases} L_{21} \to \delta_{32}^{21} \sim (0, 1, 0, 1, 1) \Rightarrow \emptyset, \\ L_{29} \to \delta_{32}^{29} \sim (0, 0, 0, 1, 1) \Rightarrow \emptyset, \end{cases} \\ L_{22} \to \delta_{32}^{22} \sim (0, 1, 0, 1, 0) \Rightarrow \emptyset, \\ L_{31} \to \delta_{32}^{31} \sim (0, 0, 0, 1, 0) \Rightarrow \emptyset. \end{cases}$$

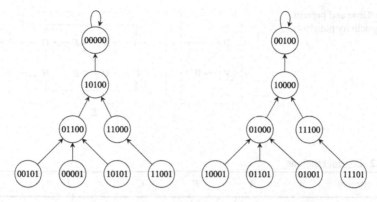

Part 1. $D = 0$

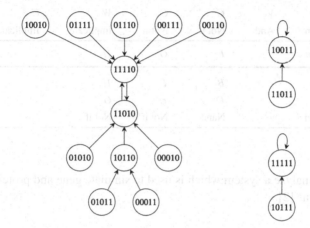

Part 2. $D = 1$

Fig. 6.6 The state-transition diagram

- And for C_2, we have

$$C_2 = (1, 1, 1, 1, 0) \sim \delta_{32}^2 \Rightarrow \begin{cases} [L_6 \to \delta_{32}^6], \\ L_{14} \to \delta_{32}^{14} \sim (1, 1, 0, 1, 0) \Rightarrow \emptyset, \\ L_{17} \to \delta_{32}^{17} \sim (0, 1, 1, 1, 1) \Rightarrow \emptyset, \\ L_{18} \to \delta_{32}^{18} \sim (0, 1, 1, 1, 0) \Rightarrow \emptyset, \\ L_{25} \to \delta_{32}^{25} \sim (0, 0, 1, 1, 1) \Rightarrow \emptyset, \\ L_{26} \to \delta_{32}^{26} \sim (0, 0, 1, 1, 0) \Rightarrow \emptyset. \end{cases}$$

The state transition diagram in Fig. 6.6 from [5] verifies our conclusion.

The significance of this example lies in the following observation: There is a smallest "cycle", the fixed point D. From Fig. 6.6 it is easily seen that for $D = 0$ and $D = 1$ the topological structures of the state-space graphs are completely different.

Fig. 6.7 Gene and protein
signaling activity patterns

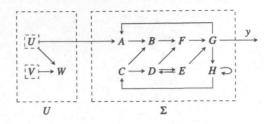

Table 6.2 Logical relations

Network element	W	A	B	C	D	E
Input 1	U	U	A	H	E	D
Input 2	V	G	C	W	C	F
Boolean function	And	Not if	Not if	Implicate	Implicate	Not if

Network element	F	G	H
Input 1	B	F	H
Input 2	D	E	G
Boolean function	Nand	Not if	Not if

Next, we analyze a system which is used to simulate gene and protein signaling activity patterns [2].

Example 6.7 The network depicted in Fig. 6.7 and Table 6.2 is presented in [2] to simulate gene and protein signaling activity patterns within a small model Boolean network. For notational brevity, we use A for "Erk", B for "cyclin D1", C for "p27", D for "cyclin E", E for "E2F", F for "pRb", G for "S phase genes", U for "growth factors", V for "cell shape (spreading)", and W for "X". We refer to [2] for the biological meanings of this notation.

The logical equation is then expressed as

$$\begin{cases} A(t+1) = \neg(U(t) \rightarrow G(t)), \\ B(t+1) = \neg(A(t) \rightarrow C(t)), \\ C(t+1) = H(t) \rightarrow W(t), \\ D(t+1) = E(t) \rightarrow C(t), \\ E(t+1) = \neg(D(t) \rightarrow F(t)), \\ F(t+1) = \neg(B(t) \wedge D(t)), \\ G(t+1) = \neg(F(t) \rightarrow E(t)), \\ H(t+1) = \neg(H(t) \rightarrow G(t)). \end{cases} \qquad (6.31)$$

As for the control network, we have

$$
\begin{cases}
U(t+1) = g_1(U(t)), \\
V(t+1) = g_2(V(t)), \\
W(t+1) = g_3(U(t), V(t)).
\end{cases}
\tag{6.32}
$$

In vector form, we have the componentwise system of algebraic equations

$$
\begin{cases}
A(t+1) = M_n M_i U(t) G(t), \\
B(t+1) = M_n M_i A(t) C(t), \\
C(t+1) = M_i H(t) W(t), \\
D(t+1) = M_i E(t) C(t), \\
E(t+1) = M_n M_i D(t) F(t), \\
F(t+1) = M_n M_c B(t) D(t), \\
G(t+1) = M_n M_i F(t) E(t), \\
H(t+1) = M_n M_i H(t) G(t).
\end{cases}
\tag{6.33}
$$

As in [2], we first set the control network as

$$
\begin{cases}
U(t+1) = \sigma_1(U(t)), \\
V(t+1) = \sigma_2(V(t)), \\
W(t+1) = U(t) \wedge V(t).
\end{cases}
\tag{6.34}
$$

Case 1: $U(0) = V(0) = \delta_2^1$. In this case $\sigma_1 = \sigma_2 = $ identity, i.e., $U(t)$ and $V(t)$ are equal to the constant δ_2^1.

Plugging these into (6.33) yields the system transition matrix

$$
L\big(U(t), W(t)\big) = L\big(\delta_2^1, \delta_2^1\big).
$$

In calculation, a control can be treated as a logical operator, so the procedure for calculating the network transition matrix remains applicable. It is then easy to obtain the following results:

- The only attractor is a fixed point, $(0, 0, 1, 1, 0, 1, 1, 0)$.
- $L^{10} = L^{11}$ and the transient period is $T_t = 10$.

Case 2: $U(0) = \delta_2^1$ and $V(0) = \delta_2^2$. In this case we arrive at the same conclusion as above.

Case 3: $U(0) = \delta_2^2$. In this case we always have $W(t) = \delta_2^2, t \geq 1$. The conclusion is then:

- The only attractor is a fixed point, $(0, 0, 1, 1, 0, 1, 1, 0)$.
- $L^6 = L^7$ and the transient period is $T_t = 6$. [Taking $W(0)$ into consideration, T_t should be 7.]

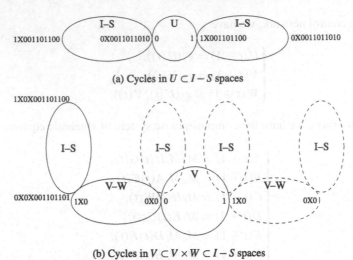

(a) Cycles in $U \subset I - S$ spaces

(b) Cycles in $V \subset V \times W \subset I - S$ spaces

Fig. 6.8 Chained cycles

Next, we assume $\sigma_1 = \sigma_2 = \neg$, and the control network is then

$$\begin{cases} U(t+1) = \neg U(t), \\ V(t+1) = \neg V(t), \\ W(t+1) = V(t) \wedge W(t). \end{cases} \tag{6.35}$$

We then have two sequences of nested invariant subspaces which we consider separately. Consider the first chain, which is

$$\mathscr{V}_1 = \mathrm{Span}\{U\} \subset \mathscr{V}_2 = \mathrm{Span}\{A, B, C, D, E, F, G, H, U, V, W\}.$$

In $\mathscr{V}_1$ we have an obvious cycle: $(0) \to (1) \to (0)$. For $U = 0$ a routine computation shows that there is only a cycle of length 2, which is

$$(0,0,1,1,0,1,1,0,1,0) \to (0,0,1,1,0,1,1,0,0,0) \to (0,0,1,1,0,1,1,0,1,0).$$

$L(0)$ is a 1024×1024 matrix. We omit this here, but we can calculate that $L(0)^7 = L(0)^9$ and $T_t = 7$. For $U = 1$, we have the same cycle, and $L(1)^{11} = L(1)^{13}$ and $T_t = 11$.

Finally, let $\Psi = L(1)L(0)$. Then, Ψ has only one fixed point, $(0,0,1,1,0,1,1,0,1,0)$. We conclude that, overall, in U space we have only one cycle, $0 \to 1 \to 0$, and in the whole space we have only one product cycle,

$$0 \times (0,0,1,1,0,1,1,0,1,0) \to 1 \times (0,0,1,1,0,1,1,0,0,0) \to$$
$$0 \times (0,0,1,1,0,1,1,0,1,0).$$

These are depicted in Fig. 6.8(a), where $I - S$ is the overall input-state space.

Next, we consider the second chain, which is

$$\mathcal{V}_1 = \mathrm{Span}\{V\} \subset \mathcal{V}_2 = \mathrm{Span}\{V, W\} \subset \mathcal{V}_3$$
$$= \mathrm{Span}\{A, B, C, D, E, F, G, H, U, V, W\}.$$

First, there is a trivial cycle in V space: $0 \to 1 \to 0$. Then, in $V \times W$ space, it is easy to calculate that

$$L(0) = \delta_2[2\ 2], \qquad L(1) = \delta_2[1\ 2].$$

Therefore,

$$\Psi = L(1)L(0) = \delta_2[2\ 2],$$

which has unique fixed point $\delta_2^2 \sim 0$. We conclude that in $V \times W$ space we have only one cycle, 0×0 and 1×0. Finally, we consider the space $V \times W \times ABCDEFGHU$. Calculating $\Psi = L(0 \times 0)L(1 \times 0)$, it is easy to show that the only cycle is a fixed point: $(0, 0, 1, 1, 0, 1, 1, 0, 1)^{\mathrm{T}}$. We conclude that there is only one cycle of length 2 in the overall product space, which is $0 \times 0 \times (0, 0, 1, 1, 0, 1, 1, 0, 1)^{\mathrm{T}} \to 1 \times 0 \times (0, 0, 1, 1, 0, 1, 1, 0, 0)^{\mathrm{T}}$. Cycles in different levels are depicted in Fig. 6.8(b).

References

1. Cheng, D.: Input-state approach to Boolean networks. IEEE Trans. Neural Netw. **20**(3), 512–521 (2009)
2. Huang, S., Ingber, D.: Shape-dependent control of cell growth, differentiation, and apoptosis: switching between attractors in cell regulatory networks. Exp. Cell Res. **261**(1), 91–103 (2000)
3. Ideker, T., Galitski, T., Hood, L.: A new approach to decoding life: systems biology. Annu. Rev. Genomics Hum. Genet. **2**, 343–372 (2001)
4. Kauffman, S.: At Home in the Universe. Oxford University Press, London (1995)
5. Paszek, E.: Boolean Networks. http://cnx.org/content/m12394/latest/ (2008)

Chapter 7
Model Construction via Observed Data

7.1 Reconstructing Networks

Recall that the dynamics of a Boolean network can be expressed as

$$\begin{cases} x_1(t+1) = f_1(x_1(t), x_2(t), \ldots, x_n(t)), \\ x_2(t+1) = f_2(x_1(t), x_2(t), \ldots, x_n(t)), \\ \vdots \\ x_n(t+1) = f_n(x_1(t), x_2(t), \ldots, x_n(t)). \end{cases} \tag{7.1}$$

In vector form we have

$$x(t) = \ltimes_{i=1}^{n} x_i(t)$$

and assume that the structure matrices of f_i, $i = 1, \ldots, n$, are $M_i \in \mathscr{L}_{2 \times 2^n}$, $i = 1, \ldots, n$. Then, (7.1) can be converted to

$$\begin{cases} x_1(t+1) = M_1 x(t), \\ x_2(t+1) = M_2 x(t), \\ \vdots \\ x_n(t+1) = M_n x(t). \end{cases} \tag{7.2}$$

The system (7.2) is called the componentwise algebraic form of (7.1). Multiplying the equations in (7.2) together yields a linear representation of (7.1) as

$$x(t+1) = Lx(t), \tag{7.3}$$

where $L \in \mathscr{L}_{2^n \times 2^n}$ is called the network transition matrix. Equation (7.3) is called the algebraic form of (7.1).

It was proven in Chap. 5 that (7.1), (7.2), and (7.3) are equivalent. The advantages of (7.1) are that it provides clear logical relations and that it is easy to realize via circuitry. The advantage of using (7.3) is that it is a conventional linear discrete-time

D. Cheng et al., *Analysis and Control of Boolean Networks*,
Communications and Control Engineering,
DOI 10.1007/978-0-85729-097-7_7, © Springer-Verlag London Limited 2011

dynamical system, so many tools developed in control theory can be used to design and analyze the network.

Chapter 5 provided a standard procedure to calculate (7.3) from (7.1) via (7.2). It consists of an algorithm, which is used to calculate M_i from its logical form, and a formula, which calculates the transition matrix by using structure matrices M_i, $i = 1, \ldots, n$. A direct computation is also convenient to use. For the purposes of review, we give a simple example.

Example 7.1 Consider the following Boolean network (see Fig. 7.1)

Its dynamics can be described as

$$\begin{cases} x_1(t+1) = x_1(t) \to x_4(t), \\ x_2(t+1) = \neg x_1(t), \\ x_3(t+1) = x_2(t) \wedge x_4(t), \\ x_4(t+1) = x_2(t) \leftrightarrow x_3(t). \end{cases} \tag{7.4}$$

We first express the system (7.4) in its componentwise algebraic form as

$$\begin{cases} x_1(t+1) = M_i x_1(t) x_4(t) = \delta_2[1\ 2\ 1\ 1] x_1(t) x_4(t), \\ x_2(t+1) = M_n x_1(t) = \delta_2[2\ 1] x_1(t), \\ x_3(t+1) = M_c x_2(t) x_4(t) = \delta_2[1\ 2\ 2\ 2] x_2(t) x_4(t), \\ x_4(t+1) = M_e x_2(t) x_3(t) = \delta_2[1\ 2\ 2\ 1] x_2(t) x_3(t). \end{cases} \tag{7.5}$$

Setting $x = \ltimes_{i=1}^4 x_i(t)$, the system (7.4) can be expressed in algebraic form as

$$\begin{aligned} x(t+1) &= M_i x_1(t) x_4(t) M_n x_1(t) M_c x_2(t) x_4(t) M_e x_2(t) x_3(t) \\ &= M_i (I_4 \otimes M_n) x_1(t) x_4(t) x_1(t) M_c x_2(t) x_4(t) M_e x_2(t) x_3(t) \\ &\ \ \vdots \\ &= L x(t), \end{aligned} \tag{7.6}$$

where

$$L = \delta_{16}[5\ 15\ 6\ 16\ 8\ 16\ 7\ 15\ 1\ 3\ 2\ 4\ 4\ 4\ 3\ 3]. \tag{7.7}$$

At this point it is pertinent to ask how to reconstruct the Boolean network from its network transition matrix L. This is important because we will work on a state

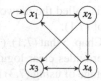

Fig. 7.1 Network graph of (7.4)

space and try to design a network transition matrix. We will then have to convert it back to the network and give its logical relations for design purposes.

Consider a Boolean network with input–output structure. From a set of input–output data we may identify the network transition matrix L. Particularly, in the case of large or, potentially, huge networks, we may find a matrix L to approximate the original system or a particular input–output functional part of the original network. We refer to Chap. 17 for the investigation of Boolean control networks. Since L is the coefficient matrix of a standard discrete-time linear system, it seems that many known methods can be used for this purpose. This makes the deduction from L of the dynamics of network variables more important.

Assuming that L is known, we will try to retrieve (7.1) and the network. First, we have to reconstruct the structure matrices M_i of the logical operators f_i. We define a set of logical matrices $S_i^n \in \mathcal{L}_{2 \times 2^n}$, called retrievers, in the following way. Divide the set of columns, labeled $1, 2, \ldots, 2^n$, into 2^i equal-sized segments, where $1 \le i \le n$. Then, put δ_2^1 into the first segment of columns, put δ_2^2 into the second segment of columns, then δ_2^1 again, and so on, continuing this process to define S_i^n. In this way we have defined

$$S_1^n = \delta_2[\underbrace{1 \cdots 1}_{2^{n-1}} \ \underbrace{2 \cdots 2}_{2^{n-1}}],$$

$$S_2^n = \delta_2[\underbrace{1 \cdots 1}_{2^{n-2}} \ \underbrace{2 \cdots 2}_{2^{n-2}} \ \underbrace{1 \cdots 1}_{2^{n-2}} \ \underbrace{2 \cdots 2}_{2^{n-2}}],$$

$$\vdots$$

$$S_n^n = \delta_2[\underbrace{1 \ 2 \ \cdots \ 1 \ 2}_{2^{n-1}}].$$

(7.8)

The following result shows how to calculate M_i, $i = 1, \ldots, n$, from L.

Theorem 7.1 *The structure matrices M_i of f_i can be retrieved as follows*:

$$M_i = S_i^n L, \quad i = 1, 2, \ldots, n. \tag{7.9}$$

To prove Theorem 7.1, we need the following lemma.

Lemma 7.1

$$S_k^n = S_1^n W_{[2^{k-1}, 2]}, \quad k = 1, 2, \ldots, n. \tag{7.10}$$

Proof First, note that S_k^n can be expressed as

$$S_k^n = \delta_2[\Gamma_k^n],$$

where

$$\Gamma_k^n = \mathbf{1}_{2^{k-1}}^T \otimes [\underbrace{1 \cdots 1}_{2^{n-k}} \ \underbrace{2 \cdots 2}_{2^{n-k}}].$$

We now prove (7.10) by mathematical induction. For $k = 1$, the right-hand side of (7.10) is

$$RHS = S_1^n W_{[1,2]} = S_1^n I_2 = S_1^n,$$

so (7.10) is true.

Now, assume (7.10) is true for $k = j < n$. For $k = j + 1$, using (2.52), we then have

$$
\begin{aligned}
RHS &= S_1^n W_{[2^j,2]} \\
&= S_1^n (W_{[2^{j-1},2]} \otimes I_2)(I_{2^{j-1}} \otimes W_{[2]}) \\
&= S_j^n (I_{2^{j-1}} \otimes W_{[2]}) \\
&= \delta_2 \big[\mathbf{1}_{2^{j-1}}^T \otimes [\underbrace{1 \cdots 1}_{2^{n-j}} \ \underbrace{2 \cdots 2}_{2^{n-j}}] \big] (I_{2^{j-1}} \otimes W_{[2]}) \\
&= \delta_2 \big[\mathbf{1}_{2^{j-1}}^T \otimes I_{2^{j-1}} \big] \otimes \big(\delta_2 [\underbrace{1 \cdots 1}_{2^{n-j}} \ \underbrace{2 \cdots 2}_{2^{n-j}}] W_{[2]} \big) \\
&= \delta_2 \big[\mathbf{1}_{2^{j-1}}^T \big] \otimes \delta_2 [\underbrace{1 \cdots 1}_{2^{n-j-1}} \ \underbrace{2 \cdots 2}_{2^{n-j-1}} \ \underbrace{1 \cdots 1}_{2^{n-j-1}} \ \underbrace{2 \cdots 2}_{2^{n-j-1}}] \\
&= \delta_2 \big[\mathbf{1}_{2^{j-1}}^T \big] \delta_2 [\underbrace{1 \cdots 1}_{2^{n-j-1}} \ \underbrace{2 \cdots 2}_{2^{n-j-1}}] \\
&= S_{j+1}^n.
\end{aligned}
$$

(7.11)

$\square$

Proof of Theorem 7.1 If $x = \ltimes_{i=1}^n x_i$, where $x_i \in \Delta$, then $\ltimes_{i=2}^n x_i \in \Delta_{2^{n-1}}$. We therefore denote it as $\ltimes_{i=2}^n x_i = \delta_{2^{n-1}}^j$.

Assume $x_1 = \delta_2^1$. Then,

$$x = \delta_2^1 \delta_{2^{n-1}}^j = \begin{bmatrix} \delta_{2^{n-1}}^j \\ 0 \end{bmatrix}.$$

A straightforward computation shows that

$$S_1^n x = x_1.$$

(7.12)

Next, we also have

$$x = W_{[2,2^{k-1}]} x_k \ltimes_{i \neq k} x_i$$

or

$$W_{[2,2^{k-1}]}x = x_k \ltimes_{i \neq k} x_i.$$

Using (7.12), we have

$$S_1^n W_{[2,2^{k-1}]}x = x_k.$$

Using Lemma 7.1, the conclusion follows.   □

Note that the number of neighborhood nodes of node i (equivalently, edges, starting from other nodes, toward i), called the in-degree of node i, is usually much smaller than n. We have to determine which node is connected to i. We have the following result.

Proposition 7.1 *Consider the system* (7.2). *If M_i satisfies*

$$M_i W_{[2,2^{j-1}]}(M_n - I_2) = 0, \tag{7.13}$$

then j is not in the neighborhood of i. In other words, the edge $j \to i$ does not exist. Moreover, the equation of A_i can be replaced by

$$x(t+1) = M_i' x_1(t) \cdots x_{j-1}(t)x_{j+1}(t) \cdots x_n(t), \tag{7.14}$$

where

$$M_i' = M_i W_{[2,2^{j-1}]}\delta_2^1.$$

Proof Note that we can rewrite the ith equation of (7.2) as

$$x_i(t+1) = M_i W_{[2,2^{j-1}]}x_j(t)x_1(t) \cdots x_{j-1}(t)x_{j+1}(t) \cdots x_n(t).$$

We now replace $x_j(t)$ by $\neg x_j(t)$. If this does not affect the overall structure matrix, it means that $x_i(t+1)$ is independent of $x_j(t)$. The invariance of replacement is illustrated by (7.13). As for (7.14), since $x_j(t)$ does not affect $x_i(t+1)$, we can simply set $x_j(t) = \delta_2^1$ [equivalently, we could set $x_j(t) = \delta_2^2$] to simplify the expression.   □

Remark 7.1 Repeating the verification of (7.13), all the fabricated variables can be removed from the equation and we can finally obtain the network expression with clean logical dynamics. This is called a clean form.

We now illustrate this with an example.

Example 7.2 Assume we have a Boolean network with five nodes and that its network matrix $L \in \mathscr{L}_{32 \times 32}$ is

$$L = \delta_{32}[3\ 6\ 7\ 6\ 19\ 22\ 31\ 30\ 19\ 22\ 23\ 22\ 3\ 6\ 15\ 14$$
$$3\ 5\ 7\ 5\ 19\ 21\ 31\ 29\ 19\ 21\ 23\ 21\ 3\ 5\ 15\ 13].$$

Using the retriever S_i^5, we have

$$M_i = S_i^5 L, \quad i = 1, 2, 3, 4, 5,$$

which are

$M_1 = \delta_2[1\ 1\ 1\ 1\ 2\ 2\ 2\ 2\ 2\ 2\ 2\ 2\ 1\ 1\ 1\ 1\ 1\ 1\ 1\ 1\ 2\ 2\ 2\ 2\ 2\ 2\ 2\ 2\ 1\ 1\ 1\ 1]$,
$M_2 = \delta_2[1\ 1\ 1\ 1\ 1\ 1\ 2\ 2\ 1\ 1\ 1\ 1\ 1\ 1\ 2\ 2\ 1\ 1\ 1\ 1\ 1\ 1\ 2\ 2\ 1\ 1\ 1\ 1\ 1\ 1\ 2\ 2]$,
$M_3 = \delta_2[1\ 2\ 2\ 2\ 1\ 2\ 2\ 2\ 1\ 2\ 2\ 2\ 1\ 2\ 2\ 2\ 1\ 2\ 2\ 2\ 1\ 2\ 2\ 2\ 1\ 2\ 2\ 2\ 1\ 2\ 2\ 2]$,
$M_4 = \delta_2[2\ 1\ 2\ 1\ 2\ 1\ 2\ 1\ 2\ 1\ 2\ 1\ 2\ 1\ 2\ 1\ 2\ 1\ 2\ 1\ 2\ 1\ 2\ 1\ 2\ 1\ 2\ 1\ 2\ 1\ 2\ 1]$,
$M_5 = \delta_2[1\ 2\ 1\ 2\ 1\ 2\ 1\ 2\ 1\ 2\ 1\ 2\ 1\ 2\ 1\ 2\ 1\ 1\ 1\ 1\ 1\ 1\ 1\ 1\ 1\ 1\ 1\ 1\ 1\ 1\ 1\ 1]$.

Next, considering M_1, it is easy to verify that

$$\begin{cases} M_1 M_n = M_1, \\ M_1 W_{[2]} M_n \neq M_1, \\ M_1 W_{[2,2^2]} M_n \neq M_1, \\ M_1 W_{[2,2^3]} M_n = M_1, \\ M_1 W_{[2,2^4]} M_n = M_1. \end{cases}$$

We conclude that $x_1(t+1)$ depends only on $x_2(t)$ and $x_3(t)$. We can then remove the fabricated variables $x_1(t)$, $x_4(t)$, and $x_5(t)$ from the first equation

$$x_1(t+1) = M_1 x_1(t) x_2(t) x_3(t) x_4(t) x_5(t) \tag{7.15}$$

by replacing $x_1(t)$, $x_4(t)$ and $x_5(t)$ in (7.15) with any constant logical values. If, say, we let $x_1(t) = x_4(t) = x_5(t) = \delta_2^1$, then we get

$$\begin{aligned} x_1(t+1) &= M_1 \delta_2^1 x_2(t) x_3(t) \delta_2^1 \delta_2^1 \\ &= M_1 W_{[4,8]}(\delta_2^1)^3 x_2(t) x_3(t) \\ &= \delta_2[1\ 2\ 2\ 1] x_2(t) x_3(t). \end{aligned} \tag{7.16}$$

We can similarly remove the fabricated variables from other equations. Skipping the mechanical verification process, we finally have

$$\begin{cases} x_1(t+1) = \delta_2[1\ 2\ 2\ 1] x_2(t) x_3(t), \\ x_2(t+1) = \delta_2[1\ 1\ 1\ 2] x_3(t) x_4(t), \\ x_3(t+1) = \delta_2[1\ 2\ 2\ 2] x_4(t) x_5(t), \\ x_4(t+1) = \delta_2[2\ 1] x_5(t), \\ x_5(t+1) = \delta_2[1\ 2\ 1\ 1] x_1(t) x_5(t). \end{cases} \tag{7.17}$$

We can then reconstruct the network as Fig. 7.2.

Fig. 7.2 Reconstructed
graph from network matrix

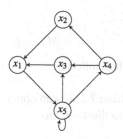

Moreover, from the above algebraic equations of the network it is easy to obtain the logical equations as follows:

$$\begin{cases} x_1(t+1) = x_2(t) \leftrightarrow x_3(t), \\ x_2(t+1) = x_3(t) \vee x_4(t), \\ x_3(t+1) = x_4(t) \wedge x_5(t), \\ x_4(t+1) = \neg x_5(t), \\ x_5(t+1) = x_1(t) \rightarrow x_5(t). \end{cases} \tag{7.18}$$

Remark 7.2 Unlike an algebraic function, for a logical function it might be very difficult to verify that an argument is fabricated. Converting a logical function into its algebraic form and using Proposition 7.1, the fabricated arguments can be eliminated. We then convert it back to logical form. We will call such a logical equation a clean form. When we construct the incidence matrix of a Boolean network from its dynamics, the latter should be in its clean form.

In Example 7.2 we convert the algebraic form (7.17) to logical form (7.18) by observation. In general, converting a logical function from its algebraic form back to logical form is not easy. We now describe a procedure for doing this.

Proposition 7.2 *Assume that a logical variable E has an algebraic expression as*

$$E = f(x_1, x_2, \ldots, x_n) = M_f x_1 x_2 \cdots x_n, \tag{7.19}$$

where $M_f \in \mathcal{L}_{2 \times 2^n}$ is the structure matrix of f. Then,

$$E = \left[x_1 \wedge f_1(x_2, \ldots, x_n) \right] \vee \left[\neg x_1 \wedge f_2(x_2, \ldots, x_n) \right], \tag{7.20}$$

where

$$M_f = (M_{f_1} \mid M_{f_2}),$$

i.e., the structure matrix of f_1 (f_2) is the first (last) half of M_f.

Proof Using (7.19), when $x_1 = 1$,

$$E = M_f \delta_2^1 x_2 \cdots x_n = M_{f_1} x_2 \cdots x_n$$

and when $x_1 = 0$,

$$E = M_f \delta_2^2 x_2 \cdots x_n = M_{f_2} x_2 \cdots x_n.$$

Equation (7.20) then follows.   □

    Using Proposition 7.2 we can obtain the logical expression of E recursively. We give an example to illustrate this.

Example 7.3 Assume

$$E = \delta_2[1\,2\,2\,1\,2\,1\,2\,1\,1\,1\,2\,2\,2\,1\,1\,2]x_1x_2x_3x_4. \tag{7.21}$$

Then,

$$E = [x_1 \wedge f_1(x_2, x_3, x_4)] \vee [\neg x_1 \wedge f_2(x_2, x_3, x_4)],$$

and

$$M_{f_1} = \delta_2[1\,2\,2\,1\,2\,1\,2\,1],$$

$$M_{f_2} = \delta_2[1\,1\,2\,2\,2\,1\,1\,2].$$

Next,

$$f_1(x_2, x_3, x_4) = [x_2 \wedge f_{11}(x_3, x_4)] \vee [\neg x_2 \wedge f_{12}(x_2, x_4)],$$

where

$$M_{f_{11}} = \delta_2[1\,2\,2\,1] \implies f_{11}(x_3, x_4) = x_3 \leftrightarrow x_4,$$

$$M_{f_{12}} = \delta_2[2\,1\,2\,1] \implies f_{12}(x_3, x_4) = \neg x_4.$$

$$f_2(x_2, x_3, x_4) = [x_2 \wedge f_{21}(x_3, x_4)] \vee [\neg x_2 \wedge f_{22}(x_3, x_4)],$$

where

$$M_{f_{21}} = \delta_2[1\,1\,2\,2] \implies f_{21}(x_3, x_4) = x_3,$$

$$M_{f_{22}} = \delta_2[2\,1\,1\,2] \implies f_{22}(x_3, x_4) = \neg(x_3 \leftrightarrow x_4).$$

Combining all of this, we have

$$E = [x_1 \wedge x_2 \wedge (x_3 \leftrightarrow x_4)] \vee [x_1 \wedge (\neg x_2) \wedge (\neg x_4)] \vee [(\neg x_1) \wedge x_2 \wedge x_3]$$
$$\vee [(\neg x_1) \wedge (\neg x_2) \wedge (\neg(x_3 \leftrightarrow x_4))].$$

Remark 7.3 Consider a Boolean control network in its algebraic form as

$$\begin{cases} x(t+1) = Lu(t)x(t), & x(t) \in \Delta_{2^n}, u(t) \in \Delta_{2^m}, \\ y(t) = Hx(t), & y(t) \in \Delta_{2^p}. \end{cases} \tag{7.22}$$

Theorem 7.1 and Proposition 7.2 can then be applied to convert the algebraic forms of state equation and output equations back to their logical forms. These will be used frequently in the sequel.

7.2 Model Construction for General Networks

Assume a Boolean network consists of n nodes. Let $X(t) = \{x_1(t), \ldots, x_n(t)\}$. Denote the observed data as $\{X(0), X(1), \ldots, X(N)\}$. We give a rigorous definition for the model construction.

Definition 7.1 Assume a set of observed data $\{X(0), X(1), \ldots, X(N)\}$ is given, where $X(t) = \{x_1(t), \ldots, x_n(t)\}$. The model construction problem is the problem of finding a logical dynamical system (7.1) such that the given data verify the dynamical equation.

A model which is verified by the given data is called a realization of the data.

The model construction problem is also called the identification problem. It has been investigated by several authors. For instance, in [5] a reverse engineering algorithm was proposed for inference of genetic network architecture. Identification by using a small number of gene expression patterns was proposed in [1] and another identification algorithm based on matrix multiplication and the "fingerprint function" was later proposed by the same authors [2]. Nam et al. [6] presented a randomized network search algorithm, which requires less time on average.

From the definition we have the following, immediate, result.

Proposition 7.3 *The system is uniquely identifiable if and only if the data* $\{X(0), X(1), \ldots, X(N-1)\}$ *contain all possible states.*

Proof Convert the data into vector form by using $x(t) := \ltimes_{i=1}^{n} x_i(t)$. In algebraic form, we then have that $x(t) = \delta_{2^n}^i$ and $x(t+1) = \delta_{2^n}^j$ if and only if the ith column of L is

$$\mathrm{Col}_i(L) = \delta_{2^n}^j. \tag{7.23}$$

It follows that L is identifiable if and only if, in vector form,

$$\{x(0), x(1), \ldots, x(N-1)\} = \Delta_{2^n}.$$

The conclusion then follows. $\square$

If the procedure has been carried out more than once, the following result is obvious.

Corollary 7.1 *Assume the observed data consists of k groups, as*

$$\{X^i(0), X^i(1), \ldots, X^i(N_i)\},$$

where $i = 1, \ldots, k$. *The system is then uniquely identifiable if and only if (in vector form)*

$$\{x^i(0), \ldots, x^i(N_i - 1) \mid i = 1, 2, \ldots, k\} = \Delta_{2^n}. \tag{7.24}$$

Remark 7.4

1. From Proposition 7.3 one sees that to identify a Boolean network of n nodes, at least $2^n + 1$ data are necessary.
2. If the data are not sufficient or do not satisfy the condition of Proposition 7.3, we still can use (7.23) to identify some columns. The model is then not unique. Uncertain columns of L can be chosen arbitrarily.

Example 7.4 Assume a set of five cells is considered. The 12 groups of experimental data are demonstrated in Fig. 7.3, where a white disc, labeled 1, represents a healthy cell, and a black disc, labeled 0, represents an infected cell. Our goal is to build a dynamical model for the process of infection.

From the first experimental data we have (where the nodes are ordered from left to right and then from top to bottom),

$$X^1(0) = (0, 0, 1, 0, 0), \quad X^1(1) = (0, 1, 1, 1, 0), \quad X^1(2) = (1, 1, 0, 1, 1),$$
$$X^1(3) = (0, 1, 1, 0, 0), \quad X^1(4) = (1, 1, 1, 1, 1), \quad X^1(5) = (1, 1, 1, 0, 0),$$
$$X^1(6) = (1, 1, 1, 1, 0), \quad X^1(7) = (1, 1, 0, 1, 0), \quad X^1(8) = (0, 1, 0, 1, 0),$$
$$X^1(9) = (0, 1, 0, 1, 1), \quad X^1(10) = (0, 1, 1, 0, 1), \quad X^1(11) = (1, 1, 0, 0, 1),$$
$$X^1(12) = (0, 0, 0, 0, 0), \quad X^1(13) = (0, 0, 1, 1, 0), \quad X^1(14) = (0, 1, 0, 1, 0).$$

In vector form we now have $X^1(0) = \delta_2\{2, 2, 1, 2, 2\}$ and

$$x^1(0) = \delta_2^2 \ltimes \delta_2^2 \ltimes \delta_2^1 \ltimes \delta_2^2 \ltimes \delta_2^2 = \delta_{32}^{28}.$$

Similarly, we can calculate that

$$x^1(0) = \delta_{32}^{28}, \quad x^1(1) = \delta_{32}^{18}, \quad x^1(2) = \delta_{32}^{5}, \quad x^1(3) = \delta_{32}^{20},$$
$$x^1(4) = \delta_{32}^{1}, \quad x^1(5) = \delta_{32}^{4}, \quad x^1(6) = \delta_{32}^{2}, \quad x^1(7) = \delta_{32}^{6},$$
$$x^1(8) = \delta_{32}^{22}, \quad x^1(9) = \delta_{32}^{21}, \quad x^1(10) = \delta_{32}^{19}, \quad x^1(11) = \delta_{32}^{7},$$
$$x^1(12) = \delta_{32}^{32}, \quad x^1(13) = \delta_{32}^{26}, \quad x^1(14) = \delta_{32}^{22}.$$

Using Proposition 7.3 [or, more precisely, equation (7.23)], we know that

$$\text{Col}_{28}(L) = \delta_{32}^{18}, \quad \text{Col}_{18}(L) = \delta_{32}^{5}, \quad \text{Col}_5(L) = \delta_{32}^{20}, \ldots.$$

The 14 columns of L have thus been determined.

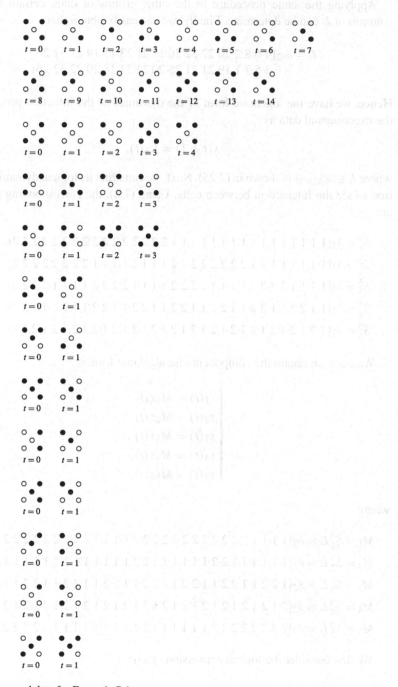

Fig. 7.3 Observed data for Example 7.4

Applying the same procedure to the other groups of data, certain values of columns of L can be determine. Finally, we can easily obtain that

$$L = \delta_{32}[4\ 6\ 8\ 2\ 20\ 22\ 32\ 26\ 19\ 21\ 23\ 17\ 19\ 21\ 31\ 25$$
$$3\ 5\ 7\ 1\ 19\ 21\ 31\ 25\ 20\ 22\ 24\ 18\ 20\ 22\ 32\ 26]. \tag{7.25}$$

Hence, we have the algebraic form of the dynamics of the infection process from the experimental data as

$$x(t+1) = Lx(t), \tag{7.26}$$

where $L \in \mathscr{L}_{32 \times 32}$ is shown in (7.25). Next, we construct its logical dynamical equation to see the interaction between cells. Using (7.8), the corresponding retrievers are

$$S_1^5 = \delta_2[1\ 1\ 1\ 1\ 1\ 1\ 1\ 1\ 1\ 1\ 1\ 1\ 1\ 1\ 1\ 1\ 2\ 2\ 2\ 2\ 2\ 2\ 2\ 2\ 2\ 2\ 2\ 2\ 2\ 2\ 2\ 2],$$

$$S_2^5 = \delta_2[1\ 1\ 1\ 1\ 1\ 1\ 1\ 1\ 2\ 2\ 2\ 2\ 2\ 2\ 2\ 2\ 1\ 1\ 1\ 1\ 1\ 1\ 1\ 1\ 2\ 2\ 2\ 2\ 2\ 2\ 2\ 2],$$

$$S_3^5 = \delta_2[1\ 1\ 1\ 1\ 2\ 2\ 2\ 2\ 1\ 1\ 1\ 1\ 2\ 2\ 2\ 2\ 1\ 1\ 1\ 1\ 2\ 2\ 2\ 2\ 1\ 1\ 1\ 1\ 2\ 2\ 2\ 2], \tag{7.27}$$

$$S_4^5 = \delta_2[1\ 1\ 2\ 2\ 1\ 1\ 2\ 2\ 1\ 1\ 2\ 2\ 1\ 1\ 2\ 2\ 1\ 1\ 2\ 2\ 1\ 1\ 2\ 2\ 1\ 1\ 2\ 2\ 1\ 1\ 2\ 2],$$

$$S_5^5 = \delta_2[1\ 2\ 1\ 2\ 1\ 2\ 1\ 2\ 1\ 2\ 1\ 2\ 1\ 2\ 1\ 2\ 1\ 2\ 1\ 2\ 1\ 2\ 1\ 2\ 1\ 2\ 1\ 2\ 1\ 2\ 1\ 2].$$

We can then obtain the componentwise algebraic form

$$\begin{cases} x_1(t) = M_1 x(t), \\ x_2(t) = M_2 x(t), \\ x_3(t) = M_3 x(t), \\ x_4(t) = M_4 x(t), \\ x_5(t) = M_5 x(t), \end{cases} \tag{7.28}$$

where

$$M_1 = S_1^5 L = \delta_2[1\ 1\ 1\ 1\ 2\ 2\ 2\ 2\ 2\ 2\ 2\ 2\ 2\ 2\ 2\ 2\ 1\ 1\ 1\ 1\ 2\ 2\ 2\ 2\ 2\ 2\ 2\ 2\ 2\ 2\ 2\ 2],$$

$$M_2 = S_2^5 L = \delta_2[1\ 1\ 1\ 1\ 1\ 1\ 2\ 2\ 1\ 1\ 1\ 1\ 1\ 1\ 2\ 2\ 1\ 1\ 1\ 1\ 1\ 1\ 2\ 2\ 1\ 1\ 1\ 1\ 1\ 1\ 2\ 2],$$

$$M_3 = S_3^5 L = \delta_2[1\ 2\ 2\ 1\ 1\ 2\ 2\ 1\ 1\ 2\ 2\ 1\ 1\ 2\ 2\ 1\ 1\ 2\ 2\ 1\ 1\ 2\ 2\ 1\ 1\ 2\ 2\ 1\ 1\ 2\ 2\ 1],$$

$$M_4 = S_4^5 L = \delta_2[2\ 1\ 2\ 1\ 2\ 1\ 2\ 1\ 2\ 1\ 2\ 1\ 2\ 1\ 2\ 1\ 2\ 1\ 2\ 1\ 2\ 1\ 2\ 1\ 2\ 1\ 2\ 1\ 2\ 1\ 2\ 1],$$

$$M_5 = S_5^5 L = \delta_2[2\ 2\ 2\ 2\ 2\ 2\ 2\ 2\ 1\ 1\ 1\ 1\ 1\ 1\ 1\ 1\ 1\ 1\ 1\ 1\ 1\ 1\ 1\ 1\ 2\ 2\ 2\ 2\ 2\ 2\ 2\ 2]. \tag{7.29}$$

We first consider the logical expression of $x_1(t)$:

$$x_1(t+1) = M_1 x(t) = \delta_2[1\ 1\ 1\ 1\ 2\ 2\ 2\ 2\ 2\ 2\ 2\ 2\ 2\ 2$$
$$1\ 1\ 1\ 1\ 2\ 2\ 2\ 2\ 2\ 2\ 2\ 2\ 2\ 2]x(t).$$

Fig. 7.4 Network graph of system (7.31)

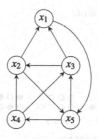

We use Proposition 7.1 to remove the fabricated variables. It is easy to verify that

$$M_1(M_n - I_2) = 0,$$

$$M_1 W_{[2,2]}(M_n - I_2) \neq 0,$$

$$M_1 W_{[2,4]}(M_n - I_2) \neq 0,$$ (7.30)

$$M_1 W_{[2,8]}(M_n - I_2) = 0,$$

$$M_1 W_{[2,16]}(M_n - I_2) = 0.$$

Therefore, $x_1(t)$, $x_4(t)$, and $x_5(t)$ are fabricated variables in the dynamical equation of $x_1(t+1)$. Setting $x_1(t) = x_4(t) = x_5(t) = \delta_2^1$ yields

$$x_1(t+1) = M_1 x_1(t) x_2(t) x_3(t) x_4(t) x_5(t)$$

$$= M_1 W_{[4,8]} x_4(t) x_5(t) x_1(t) x_2(t) x_3(t)$$

$$= M_1 W_{[4,8]} (\delta_2^1)^3 x_2(t) x_3(t)$$

$$= \delta_2[1\ 2\ 2\ 2] x_2(t) x_3(t).$$

Hence, its logical expression is

$$x_1(t+1) = x_2(t) \wedge x_3(t).$$

The same process can be used to reconstruct the logical expression of $x_2(t)$, $x_3(t)$, $x_4(t)$, and $x_5(t)$. Finally, we can obtain the logical expression of the dynamics of the group of cells as

$$\begin{cases} x_1(t+1) = x_2(t) \wedge x_3(t), \\ x_2(t+1) = x_3(t) \vee x_4(t), \\ x_3(t+1) = x_4(t) \leftrightarrow x_5(t), \\ x_4(t+1) = \neg x_5(t), \\ x_5(t+1) = x_1(t) \bar{\vee} x_2(t). \end{cases}$$ (7.31)

Figure 7.4 is its network graph.

Fig. 7.5 Network graph of system (7.33)

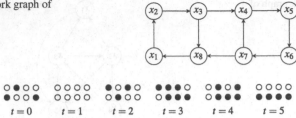

Fig. 7.6 Observed data for Example 7.5

7.3 Construction with Known Network Graph

In previous section a general method was given to construct the dynamical model of a Boolean network from its experimental data. As was pointed out, in general at least $2^n + 1$ data are necessary to uniquely determine the model. As n is not very small, this is a large or, potentially, huge number and in practical experiments such an amount of data cannot easily be obtained. In this section, we consider the case where the network graph is known. In this case, the data required can be reduced considerably.

Note that when the number of a network nodes is large, drawing its network graph is a demanding task. An alternative expression of the dynamical connection of nodes is the incidence matrix [7]. The incidence matrix was defined in Chap. 5 and we now recall it. Consider an n-node network. An $n \times n$ matrix, $\mathscr{I} = (r_{i,j}) \in \mathscr{M}_{n \times n}$, is called its incidence matrix, where $r_{i,j} = 1$, if $x_i(t+1)$ depends on $x_j(t)$ directly; otherwise, $r_{i,j} = 0$. For instance, recall Example 7.1. Its incidence matrix is

$$\mathscr{I}|_{(7.4)} = \begin{pmatrix} 1 & 0 & 0 & 1 \\ 1 & 0 & 0 & 0 \\ 0 & 1 & 0 & 1 \\ 0 & 1 & 1 & 0 \end{pmatrix}. \tag{7.32}$$

We consider the following example.

Example 7.5 Consider a network with eight nodes. Its network graph is depicted in Fig. 7.5.

Assume that for this network we have the experimental data as in Fig. 7.6.

To construct its dynamical model, we use the componentwise algebraic form. That is,

$$\begin{cases} x_1(t+1) = M_1 x_8(t), \\ x_2(t+1) = M_2 x_1(t), \\ x_3(t+1) = M_3 x_2(t), \\ x_4(t+1) = M_4 x_3(t) x_7(t), \\ x_5(t+1) = M_5 x_4(t), \\ x_6(t+1) = M_6 x_5(t), \\ x_7(t+1) = M_7 x_6(t), \\ x_8(t+1) = M_8 x_3(t) x_7(t). \end{cases} \tag{7.33}$$

From the data, it is easy to see that

$$x_8(0) = 0 \implies x_1(1) = 1,$$
$$x_8(1) = 1 \implies x_1(2) = 0, \ldots.$$

In vector form we then have

$$\text{Col}_2(M_1) = \delta_2^1, \qquad \text{Col}_1(M_1) = \delta_2^2, \ldots.$$

We conclude that $M_1 = \delta_2[2\ 1]$ and hence

$$x_1(t+1) = \neg x_8(t).$$

Similarly, the other M_i, $i = 2, 3, \ldots, 8$, can be calculated. Finally, we have the dynamics as

$$\begin{cases} x_1(t+1) = \neg x_8(t), \\ x_2(t+1) = x_1(t), \\ x_3(t+1) = \neg x_2(t), \\ x_4(t+1) = x_3(t) \vee x_7(t), \\ x_5(t+1) = x_4(t), \\ x_6(t+1) = \neg x_5(t), \\ x_7(t+1) = x_6(t), \\ x_8(t+1) = x_3(t) \wedge x_7(t). \end{cases} \tag{7.34}$$

Comparing this example with Example 7.4, it is obvious that when the network graph (equivalently, the incidence matrix) is known, then much less data will be needed to build the model.

7.4 Least In-degree Model

Consider a network with n nodes. The in-degree of node k, denoted by $d_i(k)$, is the number of edges which end at node k. Consider the incidence matrix of the network. Then,

$$d_i(k) = \sum_{j=1}^{n} r_{kj}, \quad k = 1, \ldots, n. \tag{7.35}$$

For instance, consider Example 7.1. Its in-degrees are $d_i(1) = 2$, $d_i(2) = 1$, $d_i(3) = 2$, and $d_i(4) = 2$. Consider Example 7.4. Its in-degrees are $d_i(1) = d_i(2) = d_i(3) = d_i(5) = 2$, and $d_i(4) = 1$.

It is well known that in ordered networks the in-degree is much less than the number of nodes [4]. In an experiment involving random light bulb networks, it was assumed that the number of nodes $n = 100,000$ and the in-degree $d_i = 2$. In this section we consider the least in-degree model.

Definition 7.2 Given a set of experimental data, for an n-node Boolean network, a realization with the in-degree $d_i^*(k)$, $k = 1, \ldots, n$, is called the least in-degree model if, for any other realization with in-degree $d_i(k)$, $k = 1, \ldots, n$, we have

$$d_i^*(k) \le d_i(k), \quad k = 1, \ldots, n.$$

It is obvious that a least in-degree model requires much less data to identify the model. Moreover, a real practical network should be of least in-degree. In the following we consider how to obtain a least in-degree realization. We start from the componentwise algebraic form (7.2). Denote a set of experimental data by $\{X(0), X(1), \ldots, X(N)\}$. Consider the ith subsystem

$$x_i(t + 1) = M_i x(t), \quad \text{where } M_i \in \mathcal{L}_{2 \times 2^n}. \tag{7.36}$$

Using this set of data, some columns of the structure matrix M_i can be determined. Say,

$$M_i = [* \cdots * \ c_{i_1} * \cdots * \ c_{i_2} * \cdots * \cdots * \ c_{i_s} * \cdots *], \tag{7.37}$$

where c_{i_j}, $j = 1, \ldots, s$, are identified columns and $*$ denotes an uncertain column. We call (7.37) the uncertain structure matrix. Next, we construct a set of matrices as

$$M_{i,j} := M_i W_{[2, 2^{j-1}]}, \quad j = 1, 2, \ldots, n,$$

and then split it into two equal-sized parts as

$$M_{i,j} = \begin{bmatrix} M_{i,j}^1 & M_{i,j}^2 \end{bmatrix}. \tag{7.38}$$

We then have the following result.

Proposition 7.4 f_i *has a realization which is independent of* x_j *if and only if*

$$M_{i,j}^1 = M_{i,j}^2 \tag{7.39}$$

has a solution for uncertain columns.

Proof When $j = 1$, we have $M_{i,j} = M_i$, so

$$M_i = \begin{bmatrix} M_i^1 & M_i^2 \end{bmatrix}. \tag{7.40}$$

If

$$M_i^1 = M_i^2 \tag{7.41}$$

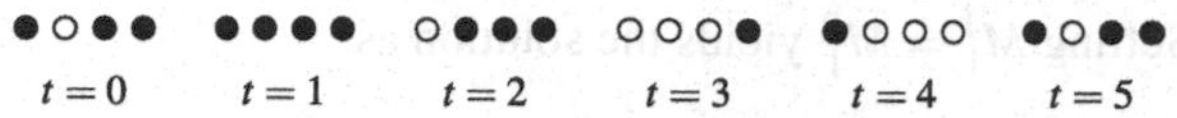

Fig. 7.7 Observed data for
Example 7.6

$t=0$ $t=1$ $t=2$ $t=3$ $t=4$ $t=5$

has a solution for uncertain elements, then the solution makes $M_i^1 = M_i^2$. Using
equation (7.13), it is easy to see that this realization is independent of x_1. Considering x_j, we rewrite (7.36) as

$$x_i(t+1) = M_{i,j} x_j \ltimes_{k=1}^{j-1} x_k \ltimes_{k=j+1}^{n} x_k$$

and the same argument as for x_1 leads to the general conclusion.   □

Next, we give an algorithm for producing a least in-degree realization.

Algorithm 7.1

Step 1. For each componentwise algebraic equation, we use the observed data to
identify some of its columns as (7.37). Moreover, we define the incidence set as
$S_i = \{1, 2, \ldots, n\}$, $i = 1, \ldots, n$.

Step 2. Construct (7.40) to check whether (7.41) has a solution. If it does, then fix
some uncertain columns and update the system to

$$x_i(t+1) = M_i^1 \ltimes_{j=2}^{n} x_j(t).$$

Go to the next step.

Step j. (Repeat this step for $3 \le j \le n$.) Check whether (7.39) has solution. If it
does, then fix some uncertain columns and update the system to

$$x_i(t+1) = M_{i,j}^1 \ltimes_{1 \le k \le j-1, k \in S_i} x_k \ltimes_{k=j+1}^{n} x_k. \tag{7.42}$$

Replace S_i by $S_i \setminus \{j\}$.

The following conclusion follows from the design of the algorithm.

Proposition 7.5 *Algorithm* 7.1 *yields least in-degree realizations.*

We now give an example to illustrate this.

Example 7.6 Consider the experimental data in Fig. 7.7.
The vector form of the data is as follows:

$$x(0) = \delta_{16}^{12}, \qquad x(1) = \delta_{16}^{16}, \qquad x(2) = \delta_{16}^{8},$$
$$x(3) = \delta_{16}^{2}, \qquad x(4) = \delta_{16}^{10}, \qquad x(5) = \delta_{16}^{12}.$$

Using the technique developed in previous sections, we can identify some columns
of M_1 via the known data as

$$M_1 = \delta_2[* \, 2 \, * \, * \, * \, * \, * \, 1 \, * \, 2 \, * \, 2 \, * \, * \, * \, 1].$$

Setting $M_1^1 = M_1^2$ yields the solution as

$$M_1^1 = M_1^2 = \delta_2[*\,2 * 2 * * * 1].$$

Therefore, the system can be simplified as

$$x_1(t+1) = \delta_2[*\,2 * 2 * * * 1]x_2(t)x_3(t)x_4(t).$$

Splitting M_1^1 into two parts and considering the equation

$$\delta_2[*\,2 * 2] = \delta_2[* * * 1],$$

we have no solution. Thus, the equation depends on x_2.

Consider

$$M_{1,2} = M_1^1 W_{[2,2]} = \delta_2[*\,2 * * * 2 * 1].$$

Then,

$$\delta_2[*\,2 * *] = \delta_2[* 2 * 1]$$

has solution as

$$\delta_2[*\,2 * 1].$$

The original equation can then be updated as

$$x_1(t+1) = \delta_2[*\,2 * 1]x_2(t)x_4(t).$$

Finally, we check $x_4(t)$. Since

$$\delta_2[*\,2 * 1]W_{[2,2]} = \delta_2[* * 2\,1],$$

and

$$\delta_2[* *] = \delta_2[2\,1]$$

has solution

$$\delta_2[2\,1],$$

we finally have

$$x_1(t+1) = \delta_2[2\,1]x_2(t).$$

That is,

$$x_1(t+1) = \neg x_2(t).$$

Fig. 7.8 Network graph of
system (7.43)

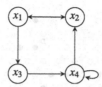

Applying the same procedure to three other equations, we can finally obtain the least
in-degree realization as

$$\begin{cases} x_1(t+1) = \neg x_2(t), \\ x_2(t+1) = x_4(t) \vee x_1(t), \\ x_3(t+1) = x_1(t), \\ x_4(t+1) = x_3(t) \,\bar{\vee}\, x_4(t). \end{cases} \tag{7.43}$$

Figure 7.8 is its network graph.

In fact, if

$$d_i(k) \leq \mu, \quad k = 1, \ldots, n,$$

then it is easy to see that the smallest number of data required to identify the system
is $2^\mu + 1$, which is, in general, much less than $2^n + 1$.

7.5 Construction of Uniform Boolean Network

In this section we consider the case where the network has uniform dynamical structure. Physically, this would correspond to something like the following situation.
Assume we have a set of cells where each cell can be infected only by its neighbors
and, moreover, the rule for a cell being infected is the same for all cells. Each cell
then has the same logical dynamical interactive pattern with its neighbors.

Example 7.7 Let a set of experimental data be given as in Fig. 7.9. We assume that
the infection process is uniform and that each cell x_0 is affected only by its neighboring cells, x_1, x_2, x_3, x_4, x_5, and x_6. Moreover, it is also reasonable to assume that
the infection is isotropic. That is, if we label the six neighboring cells of a cell in
clockwise order (starting from any of them), then the dynamical equation becomes

$$x_0(t+1) = M \ltimes_{i=0}^{6} x_i(t), \quad \text{where } M \in \mathcal{L}_{2 \times 2^7}. \tag{7.44}$$

Our purpose is to identify M.

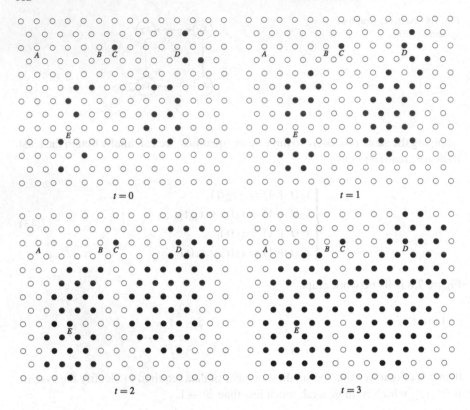

Fig. 7.9 Observed data for Example 7.7

We now consider some special points in order to demonstrate how to find M.

1. Consider $x_0 = A$. Note that $x_0(0) = 1$ and, on its neighborhood, we have $x_1(0) = 1$, $x_2(0) = 1$, $x_3(0) = 1$, $x_4(0) = 1$, $x_5(0) = 1$, and $x_6(0) = 1$, that is, $X(0) = (1\,1\,1\,1\,1\,1\,1) \sim \delta_{128}^1$. Since $x_0(1) = 1$, we conclude that $\mathrm{Col}_1(M) = \delta_2^1$.

2. Consider $x_0 = B$. We have $x_0(0) = 1$ and, on its neighborhood, we have $x_1(0) = 1$, $x_2(0) = 1$, $x_3(0) = 1$, $x_4(0) = 1$, $x_5(0) = 1$, and $x_6(0) = 0$, that is, $X(0) = (1\,1\,1\,1\,1\,1\,0) \sim \delta_{128}^2$. Since $x_0(1) = 1$, we conclude that $\mathrm{Col}_2(M) = \delta_2^1$. Moreover, by isotropy we can also assume that $X(0)$ is obtained by rotating the subscripts of $x_1(0), x_2(0), x_3(0), x_4(0), x_5(0)$, and $x_6(0)$, and all such $X(0)$ yields the same $x_0(1)$. That is,

$$X(0) = (1\,1\,1\,1\,1\,0\,1) \sim \delta_{128}^3 \implies \mathrm{Col}_3(M) = \delta_2^1,$$

$$X(0) = (1\,1\,1\,1\,0\,1\,1) \sim \delta_{128}^5 \implies \mathrm{Col}_5(M) = \delta_2^1,$$

$$X(0) = (1\,1\,1\,0\,1\,1\,1) \sim \delta_{128}^9 \implies \mathrm{Col}_9(M) = \delta_2^1,$$

$$X(0) = (1\,1\,0\,1\,1\,1\,1) \sim \delta_{128}^{17} \implies \mathrm{Col}_{17}(M) = \delta_2^1,$$

$$X(0) = (1\,0\,1\,1\,1\,1\,1) \sim \delta_{128}^{33} \implies \text{Col}_{33}(M) = \delta_2^1.$$

3. Consider $x_0 = C$. Since $x_0(0) = 0$ and $x_0(1) = 0$, we have $X(0) = (0, 1, 1, 1, 1, 1, 1) \sim \delta_{128}^{65}$, which implies that $\text{Col}_{65}(M) = \delta_2^2$.

4. Consider $x_0 = D$. Since $x_0(0) = 1$ and $x_0(1) = 0$, using isotropy, we have

$$X(0) = (1\,0\,1\,0\,1\,1\,1) \sim \delta_{128}^{41} \implies \text{Col}_{41}(M) = \delta_2^2,$$

$$X(0) = (1\,1\,0\,1\,1\,1\,0) \sim \delta_{128}^{18} \implies \text{Col}_{18}(M) = \delta_2^2,$$

$$X(0) = (1\,0\,1\,1\,1\,0\,1) \sim \delta_{128}^{35} \implies \text{Col}_{35}(M) = \delta_2^2,$$

$$X(0) = (1\,1\,1\,1\,0\,1\,0) \sim \delta_{128}^{6} \implies \text{Col}_{6}(M) = \delta_2^2,$$

$$X(0) = (1\,1\,1\,0\,1\,0\,1) \sim \delta_{128}^{11} \implies \text{Col}_{11}(M) = \delta_2^2,$$

$$X(0) = (1\,1\,0\,1\,0\,1\,1) \sim \delta_{128}^{21} \implies \text{Col}_{21}(M) = \delta_2^2.$$

5. Consider $x_0 = E$. Since $x_0(0) = 1$, $x_0(1) = 1$, and $x_0(2) = 0$, we have

$$X(0) = (1\,0\,1\,1\,0\,1\,1) \sim \delta_{128}^{37} \implies \text{Col}_{37}(M) = \delta_2^1,$$

$$X(0) = (1\,1\,1\,0\,1\,1\,0) \sim \delta_{128}^{10} \implies \text{Col}_{10}(M) = \delta_2^1,$$

$$X(0) = (1\,1\,0\,1\,1\,0\,1) \sim \delta_{128}^{19} \implies \text{Col}_{19}(M) = \delta_2^1,$$

and

$$X(0) = (1\,0\,1\,0\,0\,1\,1) \sim \delta_{128}^{45} \implies \text{Col}_{45}(M) = \delta_2^2,$$

$$X(0) = (1\,1\,0\,0\,1\,1\,0) \sim \delta_{128}^{26} \implies \text{Col}_{26}(M) = \delta_2^2,$$

$$X(0) = (1\,0\,0\,1\,0\,0\,1) \sim \delta_{128}^{51} \implies \text{Col}_{51}(M) = \delta_2^2,$$

$$X(0) = (1\,0\,1\,1\,0\,1\,0) \sim \delta_{128}^{38} \implies \text{Col}_{38}(M) = \delta_2^2,$$

$$X(0) = (1\,1\,1\,0\,1\,0\,0) \sim \delta_{128}^{12} \implies \text{Col}_{12}(M) = \delta_2^2,$$

$$X(0) = (1\,1\,0\,1\,0\,0\,1) \sim \delta_{128}^{23} \implies \text{Col}_{23}(M) = \delta_2^2.$$

6. Continuing this process, all the columns of M can finally be identified as

$$\begin{aligned}
M = \delta_2[&1\,1\,1\,2\,1\,2\,2\,2\,1\,1\,2\,2\,2\,2\,2\,1\,2\,1\,2\,2\,2\,2\,2\,2\,2\,2\,2\,2\,2\,2 \\
&1\,2\,2\,2\,1\,2\,2\,2\,2\,2\,2\,2\,2\,2\,2\,2\,2\,2\,2\,2\,2\,2\,2\,2\,2\,2\,2\,2\,2\,2 \\
&2\,2\,2\,2\,2\,2\,2\,2\,2\,2\,2\,2\,2\,2\,2\,2\,2\,2\,2\,2\,2\,2\,2\,2\,2\,2\,2\,2\,2\,2 \\
&2\,2\,2\,2\,2\,2\,2\,2\,2\,2\,2\,2\,2\,2\,2\,2\,2\,2\,2\,2\,2\,2\,2\,2\,2\,2\,2\,2\,2\,2].
\end{aligned}$$

Skipping the standard (and tedious) process, we finally can convert the algebraic form back to logical form as

$$x_0(t+1) = x_0(t) \wedge (x_1(t) \vee x_2(t)) \wedge (x_2(t) \vee x_3(t)) \wedge (x_3(t) \vee x_4(t))$$

$$\wedge (x_4(t) \vee x_5(t)) \wedge (x_5(t) \vee x_6(t)) \wedge (x_6(t) \vee x_1(t))$$

$$\wedge (x_1(t) \vee x_3(t)) \wedge (x_3(t) \vee x_5(t)) \wedge (x_5(t) \vee x_1(t))$$

$$\wedge (x_2(t) \vee x_4(t)) \wedge (x_4(t) \vee x_6(t)) \wedge (x_6(t) \vee x_2(t)). \qquad (7.45)$$

7.6 Modeling via Data with Errors

Until now the data we have considered are precisely correct. In dealing with real-world data, certain numerical methods should be used, so in this section we propose some basic ideas for dealing with imperfect data.

Data can contain errors caused by measurement, for example. In identification, we may have conflicting data. Suppose, with some data we have obtained, that

$$\text{Col}_i(L) = \begin{cases} \delta_{2^n}^p, & k \text{ times,} \\ \delta_{2^n}^q, & s \text{ times.} \end{cases} \qquad (7.46)$$

1. If $k \gg s$, then we can ignore $\delta_{2^n}^q$ and let $\text{Col}_i(L) = \delta_{2^n}^p$.
2. If $k \ll s$, then we can ignore $\delta_{2^n}^p$ and let $\text{Col}_i(L) = \delta_{2^n}^q$.
3. If $k \approx s$, then we may need more data or (when we already have enough data) conclude that the model is not acceptable.

A similar judgment can be applied to each M_i in the componentwise model.

Consider the least in-degree model. Let $\{I, J\}$ be a partition of $\{1, 2, \ldots, 2^n\}$. If

$$\begin{cases} \text{Col}_j(M) = \delta_2^{s_j}, & k_j \geq k, j \in J, \\ \text{Col}_i(M) = \delta_2^{s_i}, & k_i \ll k, i \in I, \end{cases} \qquad (7.47)$$

then we may consider $\text{Col}_i(M)$, $i \in I$, as error columns and set $\text{Col}_i(M) = *$, i.e., consider them as uncertain columns.

Roughly speaking, for model construction, we use only the data about which we are confident. Many statistical testing methods can be used to tell whether particular data is reliable. We give a simple example to illustrate this.

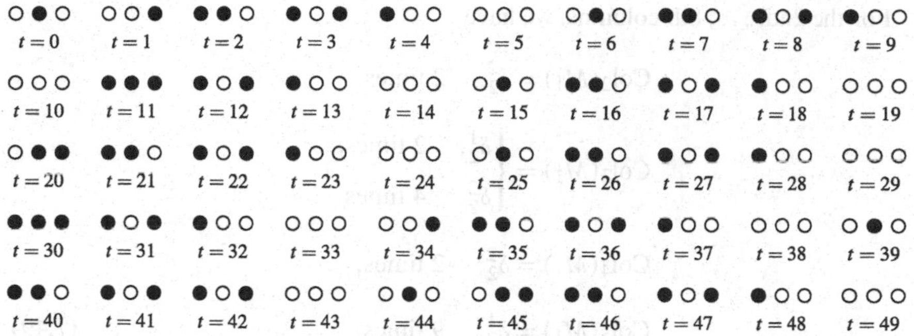

t=0	t=1	t=2	t=3	t=4	t=5	t=6	t=7	t=8	t=9
○●○	○○●	●●○	●○●	●○○	○○○	○●○	●●○	●○●	●○○

t=10	t=11	t=12	t=13	t=14	t=15	t=16	t=17	t=18	t=19
○○○	●●●	●○●	●○○	○○○	○●○	●●○	●○●	●○○	○○○

t=20	t=21	t=22	t=23	t=24	t=25	t=26	t=27	t=28	t=29
○●●	●●○	●○●	●○○	○○○	○●○	●●○	●○●	●○○	○○○

t=30	t=31	t=32	t=33	t=34	t=35	t=36	t=37	t=38	t=39
●●●	●○●	●○○	○○○	○○●	●●○	●○●	●○○	○○○	○●○

t=40	t=41	t=42	t=43	t=44	t=45	t=46	t=47	t=48	t=49
●●○	●○●	●○●	○○○	○●○	○●●	●●○	●○●	●○○	○○○

Fig. 7.10 Observed data for Example 7.8

Example 7.8 Suppose we have a Boolean network with three nodes. The experimental data are depicted in Fig. 7.10.

The 50 experiment data can be converted into vector form as

$$
\begin{array}{lllll}
x(0) = \delta_8^3, & x(1) = \delta_8^2, & x(2) = \delta_8^7, & x(3) = \delta_8^6, & x(4) = \delta_8^5, \\
x(5) = \delta_8^1, & x(6) = \delta_8^3, & x(7) = \delta_8^7, & x(8) = \delta_8^6, & x(9) = \delta_8^5, \\
x(10) = \delta_8^1, & x(11) = \delta_8^8, & x(12) = \delta_8^6, & x(13) = \delta_8^5, & x(14) = \delta_8^1, \\
x(15) = \delta_8^3, & x(16) = \delta_8^7, & x(17) = \delta_8^6, & x(18) = \delta_8^5, & x(19) = \delta_8^1, \\
x(20) = \delta_8^4, & x(21) = \delta_8^7, & x(22) = \delta_8^6, & x(23) = \delta_8^5, & x(24) = \delta_8^1, \\
x(25) = \delta_8^3, & x(26) = \delta_8^7, & x(27) = \delta_8^6, & x(38) = \delta_8^5, & x(29) = \delta_8^1, \\
x(30) = \delta_8^8, & x(31) = \delta_8^6, & x(32) = \delta_8^5, & x(33) = \delta_8^1, & x(34) = \delta_8^2, \\
x(35) = \delta_8^7, & x(36) = \delta_8^6, & x(37) = \delta_8^5, & x(38) = \delta_8^1, & x(39) = \delta_8^3, \\
x(40) = \delta_8^7, & x(41) = \delta_8^6, & x(42) = \delta_8^6, & x(43) = \delta_8^1, & x(44) = \delta_8^3, \\
x(45) = \delta_8^4, & x(46) = \delta_8^7, & x(47) = \delta_8^6, & x(48) = \delta_8^5, & x(49) = \delta_8^1.
\end{array}
$$

Suppose the componentwise algebraic form of $x_1(t)$ is

$$
x_1(t+1) = M_1 x(t), \quad M_1 \in \mathcal{L}_{2 \times 8}.
$$

From the data, using the technique developed in Section 7.2, we have

$$
\text{Col}_1(M_1) = \begin{cases} \delta_2^1, & 8 \text{ times,} \\ \delta_2^2, & 2 \text{ times.} \end{cases} \tag{7.48}
$$

Hence, we set $\text{Col}_1(M_1) = \delta_2^1$.

For the 2nd, ..., 8th columns, we have

$$\text{Col}_2(M_1) = \delta_2^2, \quad 2 \text{ times,}$$

$$\text{Col}_3(M_1) = \begin{cases} \delta_2^1, & 2 \text{ times,} \\ \delta_2^2, & 4 \text{ times,} \end{cases}$$

$$\text{Col}_4(M_1) = \delta_2^2, \quad 2 \text{ times,}$$

$$\text{Col}_5(M_1) = \delta_2^1, \quad 9 \text{ times,} \tag{7.49}$$

$$\text{Col}_6(M_1) = \begin{cases} \delta_2^1, & 1 \text{ time,} \\ \delta_2^2, & 10 \text{ times,} \end{cases}$$

$$\text{Col}_7(M_1) = \delta_2^2, \quad 6 \text{ times,}$$

$$\text{Col}_8(M_1) = \delta_2^2, \quad 2 \text{ times.}$$

Hence, we can obtain the matrix M_1 as

$$M_1 = \delta_2[1\ 2\ 2\ 2\ 1\ 2\ 2\ 2].$$

Splitting M_1 as $M_1 = [M_{11}\ M_{12}]$, we have $M_{11} = M_{12}$. The algebraic form of $x_1(t)$ is

$$x_1(t+1) = \delta_2[1\ 2\ 2\ 2]x_2(t)x_3(t).$$

Converting this into its logical form, we get

$$x_1(t+1) = x_2(t) \wedge x_3(t).$$

Using the same technique for $x_2(t)$ and $x_3(t)$, we obtain the logical expression from data as

$$\begin{cases} x_1(t+1) = x_2(t) \wedge x_3(t), \\ x_2(t+1) = \neg x_1(t), \\ x_3(t+1) = x_1(t) \vee x_2(t). \end{cases} \tag{7.50}$$

Returning to the data, it is easy to check that eight of them are wrong. The method seems relatively robust.

Finally, we note that if a model is constructed and, later, additional data become available, then we can update the model as follows. If the kth equation verifies new data, it remains available. Otherwise, we can add newly identified columns to the existing set and use them to construct a new structure matrix M_k. The new kth equation can then be updated.

References

1. Akutsu, T., Miyano, S., Kuhara, S.: Identification of genetic networks from a small number of gene expression patterns under the Boolean network model. In: Proc. Pacific Symposium on Biocomputing, pp. 17–28. World Scientific, Singapore (1999)
2. Akutsu, T., Miyano, S., Kuhara, S.: Algorithms for identifying Boolean networks and related biological networks based on matrix multiplication and fingerprint function. J. Comput. Biol. 7(3/4), 331–343 (2000)
3. Cheng, D., Qi, H., Li, Z.: Model construction of Boolean network via observed data (2010, submitted)
4. Kauffman, S.: At Home in the Universe. Oxford University Press, London (1995)
5. Liang, S., Fuhrman, S., Somogyi, R.: Reveal, a general reverse engineering algorithm for inference of genetic network architectures. In: Proc. Pacific Symposium on Biocomputing, vol. 3, pp. 18–29 (1998)
6. Nam, D., Seo, S., Kim, S.: An efficient top-down search algorithm for learning Boolean networks of gene expression. Mach. Learn. 65, 229–245 (2006)
7. Robert, F.: Discrete Iterations: A Metric Study. Springer, Berlin (1986). Translated by J. Rolne

Chapter 8
State Space and Subspaces

8.1 State Spaces of Boolean Networks

One of the fundamental pillars of modern control theory is the state-space description of control systems, first proposed by Kalman [7]. Consider a linear system

$$\begin{cases} \dot{x} = Ax + Bu, & x \in \mathbb{R}^n, u \in \mathbb{R}^m, \\ y = Cx, & y \in \mathbb{R}^p. \end{cases} \tag{8.1}$$

Many subspaces of the state space $\mathbb{R}^n$ then play important roles in system analysis and control design, e.g., the controllable subspace, observable subspace and (A, B)-invariant subspace. Consider an affine nonlinear system

$$\begin{cases} \dot{x} = f(x) + \sum_{i=1}^{m} g_i(x)u_i, & x \in M, u \in U, \\ y = h(x), & y \in N, \end{cases} \tag{8.2}$$

where M, U, and N are n-, m-, and p-dimensional manifolds, respectively. The vector fields on M, denoted by $V(M)$, form a vector space, called the Lie algebra, and $f(x), g_i(x) \in V(M)$. Like the subspaces of $V(M)$, the accessibility Lie algebra, (f, g)-invariant distribution, etc. also play important roles in the control of affine nonlinear systems.

Consider a Boolean network,

$$\begin{cases} x_1(t+1) = f_1(x_1(t), \dots, x_n(t)), \\ \vdots \\ x_n(t+1) = f_n(x_1(t), \dots, x_n(t)), & x_i \in \mathscr{D}, \end{cases} \tag{8.3}$$

D. Cheng et al., *Analysis and Control of Boolean Networks*,
Communications and Control Engineering,
DOI 10.1007/978-0-85729-097-7_8, © Springer-Verlag London Limited 2011

or a Boolean control network,

$$
\begin{cases}
x_1(t+1) = f_1(x_1(t), \ldots, x_n(t), u_1(t), \ldots, u_m(t)), \\
\vdots \\
x_n(t+1) = f_n(x_1(t), \ldots, x_n(t), u_1(t), \ldots, u_m(t)), \\
y_j(t) = h_j\big(x_1(t), \ldots, x_n(t)\big), \quad x_i, u_i, y_j \in \mathscr{D}.
\end{cases}
\tag{8.4}
$$

Unlike quantity-based dynamical (control) systems, the logic-based dynamical (control) systems do not have a natural vector space structure. To use the state-space approach, we have to define the state space and its various subspaces. In fact, in Chap. 6 we have already used the concepts of state space and subspaces based on the coordinates of the system. In this chapter, the general coordinate-independent definitions will be given. These will play a similar role as their counterparts in modern control theory.

Definition 8.1 Consider the Boolean network (8.3) or the Boolean control network (8.4).

1. The state space $\mathscr{X}$ is defined as the set of all logical functions of $x_1, \ldots, x_n$, denoted by

$$
\mathscr{X} = \mathscr{F}_\ell\{x_1, \ldots, x_n\}.
\tag{8.5}
$$

2. Let $z_1, \ldots, z_k \in \mathscr{X}$. The subspace $\mathscr{Z}$ generated by $z_1, \ldots, z_k$ is the set of logical functions of $z_1, \ldots, z_k$, denoted by

$$
\mathscr{Z} = \mathscr{F}_\ell\{z_1, \ldots, z_k\}.
\tag{8.6}
$$

Remark 8.1

1. Let $\xi \in \mathscr{X}$. Then, ξ is a logical function of $x_1, \ldots, x_n$, say

$$
\xi = g(x_1, \ldots, x_n).
$$

It can then be uniquely expressed in algebraic form as

$$
\xi = M_g \ltimes_{i=1}^n x_i,
$$

where $M_g \in \mathscr{L}_{2 \times 2^n}$. Now, M_g can be expressed as

$$
\delta_2[i_1 \, i_2 \, \cdots \, i_{2^n}],
$$

where i_s can be either 1 or 2. It follows that there are 2^{2^n} different functions. That is,

$$
|\mathscr{X}| = 2^{2^n}.
$$

2. Using a set of functions to define a (sub)space is reasonable. For instance, in the linear space $\mathbb{R}^n$ with the coordinate frame $\{x_1, \ldots, x_n\}$, we consider all the linear functions over $x_{i_1}, \ldots, x_{i_k}$, that is,

$$L_k = \left\{ \sum_{j=1}^{k} c_j x_{i_j} \,\middle|\, c_1, \ldots, c_k \in \mathbb{R} \right\},$$

which is obviously a k-dimensional subspace. In fact, we can identify L_k with its domain, which is a k-dimensional subspace of the state space $\mathbb{R}^n$, called the dual space of L_k.

The logical space (subspace) defined here is also in the dual sense and we consider its domain as a subspace of the state space.

8.2 Coordinate Transformation

From modern control theory, we know that in the state-space approach the coordinate transformation plays a fundamental role. To apply this approach to logical dynamical systems, we also need to define a coordinate transformation (or change of coordinates) on the state space $\mathscr{X}$.

Definition 8.2 Let $Z = \{z_1, \ldots, z_n\} \subset \mathscr{X}$. For notational ease, we also consider $Z = (z_1, \ldots, z_n)^T$ as a column vector. The mapping $G : \mathscr{D}^n \to \mathscr{D}^n$ defined by $X = (x_1, \ldots, x_n)^T \mapsto Z = (z_1, \ldots, z_n)^T$ is called a coordinate transformation if T is one-to-one and onto.

To obtain the verifiable condition for the coordinate transformation, we consider its algebraic form. Suppose the mapping is determined by

$$G : \begin{cases} z_1 = g_1(x_1, \ldots, x_n), \\ \vdots \\ z_n = g_n(x_1, \ldots, x_n). \end{cases} \tag{8.7}$$

Setting $x = \ltimes_{i=1}^{n} x_i$ and $z = \ltimes_{i=1}^{n} z_i$, the algebraic form of G is described as

$$z = T_G x, \tag{8.8}$$

where $T_G \in \mathscr{L}_{2^n \times 2^n}$ is the structure matrix of G. Since there is a one-to-one correspondence between Δ_{2^n} and $\mathscr{D}^n$, that is, a one-to-one correspondence between $X = (x_1, \ldots, x_n)$ and $x = \ltimes_{i=1}^{n} x_i$, the following result is obvious.

Theorem 8.1 *G is a coordinate transformation if and only if its structure matrix T_G is nonsingular.*

Remark 8.2 If a matrix $T \in \mathscr{L}_{s \times s}$ is nonsingular, then it is an orthogonal matrix. Hence, if (8.8) is a coordinate transformation, then

$$x = T_G^{\mathrm{T}} z. \tag{8.9}$$

Remark 8.3 Let a_i, $i = 1, \dots, n$, and b_i, $i = 1, \dots, n$, be sets of n vectors. From linear algebra it is well known that if

$$\begin{bmatrix} b_1 \\ b_2 \\ \vdots \\ b_n \end{bmatrix} = G \begin{bmatrix} a_1 \\ a_2 \\ \vdots \\ a_n \end{bmatrix}$$

and G is singular, then $b_1, b_2, \dots, b_n$ are linearly dependent, so at least one b_i can be determined by the others. This is not true in the logical case. For suppose

$$B_1 B_2 = \delta_4 [1\ 1\ 3\ 4] A_1 A_2 := G A_1 A_2.$$

Here, G is singular and

$$B_1 = S_1^2 G A_1 A_2 = \delta_2 [1\ 1\ 2\ 2] A_1 A_2 \sim A_1,$$

$$B_2 = S_2^2 G A_1 A_2 = \delta_2 [1\ 1\ 1\ 2] A_1 A_2 \sim A_1 \vee A_2,$$

but neither one can be determined by the other.

This example shows that even if n logical functions are not dependent (no one can be determined by others), they may not be valid as a set of coordinate variables.

Next, we give an example to illustrate coordinate transformation in the logical setting.

Example 8.1 Let

$$\begin{aligned} B_1 &= \neg A_2, \\ B_2 &= A_1 \leftrightarrow A_2, \\ B_3 &= \neg A_3. \end{aligned} \tag{8.10}$$

Define $x = A_1 A_2 A_3$, $y = B_1 B_2 B_3$. Then

$$\begin{aligned} y &= B_1 B_2 B_3 \\ &= M_n A_2 M_e A_1 A_2 M_n A_3 \\ &= M_n (I_2 \otimes M_e) W_{[2]} A_1 A_2^2 M_n A_3 \\ &= M_n (I_2 \otimes M_e) W_{[2]} (I_2 \otimes M_r) A_1 A_2 M_n A_3 \\ &= M_n (I_2 \otimes M_e) W_{[2]} (I_2 \otimes M_r)(I_4 \otimes M_n) A_1 A_2 A_3 \\ &:= T x, \end{aligned} \tag{8.11}$$

where $T \in \mathcal{L}_{8 \times 8}$ is

$$T = M_n(I_2 \otimes M_e)W_{[2]}(I_2 \otimes M_r)(I_4 \otimes M_n)$$
$$= \delta_8[6\,5\,4\,3\,8\,7\,2\,1]. \tag{8.12}$$

Since T is nonsingular, (8.10) is a logical coordinate transformation.

Note that since the T in (8.12) is an orthogonal matrix, we have

$$x = T^{-1}y = T^T y.$$

Then,

$$A_1 = S_1^3 T^T y := M_1 y = \delta_2[2\,2\,1\,1\,1\,1\,2\,2]B_1 B_2 B_3,$$
$$A_2 = S_2^3 T^T y := M_2 y = \delta_2[2\,2\,2\,2\,1\,1\,1\,1]B_1 B_2 B_3,$$
$$A_3 = S_3^3 T^T y := M_3 y = \delta_2[2\,1\,2\,1\,2\,1\,2\,1]B_1 B_2 B_3.$$

It is easy to check that

$$M_1 W_{[2,4]}(M_n - I_2) = 0.$$

Therefore, A_1 is independent on B_3. To eliminate it, we replace M_1 by

$$M_1 W_{[2,4]}\delta_2^1 = \delta_2[2\,1\,1\,2].$$

Hence,

$$A_1 = \delta_2[2\,1\,1\,2]B_1 B_2.$$

Similarly, we can check that A_2 is independent of B_2 and B_3, and can be expressed as

$$A_2 = \delta_2[2\,1]B_1.$$

A_3 is independent of B_1 and B_2, and can be expressed as

$$A_3 = \delta_2[2\,1]B_3.$$

Converting these into logical form, we have

$$A_1 = \neg(B_1 \leftrightarrow B_2) = B_1 \,\bar{\vee}\, B_2,$$
$$A_2 = \neg B_1, \tag{8.13}$$
$$A_3 = \neg B_3.$$

Next, we consider the logical coordinate transformation of the dynamics of a Boolean network. Consider a Boolean network in algebraic form as

$$x(t+1) = Lx(t), \quad x \in \Delta_{2^n}. \tag{8.14}$$

Let $z = Tx : \Delta_{2^n} \to \Delta_{2^n}$ be a logical coordinate transformation. Then,

$$z(t+1) = Tx(t+1) = TLx(t) = TLT^{-1}z(t).$$

That is, under the z coordinate frame, the Boolean network dynamics (8.14) becomes

$$z(t+1) = \tilde{L}z(t), \tag{8.15}$$

where

$$\tilde{L} = TLT^{\mathrm{T}}. \tag{8.16}$$

Consider a Boolean control system in algebraic form as

$$\begin{cases} x(t+1) = Lu(t)x(t), & x \in \Delta_{2^n}, u \in \Delta_{2^m}, \\ y(t) = Hx(t), & y \in \Delta_{2^p}. \end{cases} \tag{8.17}$$

Let $z = Tx : \Delta_{2^n} \to \Delta_{2^n}$ be a logical coordinate transformation. A straightforward computation shows that (8.17) can be expressed as

$$\begin{aligned} z(t+1) &= \tilde{L}u(t)z(t), & z \in \Delta_{2^n}, u \in \Delta_{2^m}, \\ y(t) &= \tilde{H}z(t), & y \in \Delta_{2^p}, \end{aligned} \tag{8.18}$$

where

$$\begin{aligned} \tilde{L} &= TL\big(I_{2^m} \otimes T^{\mathrm{T}}\big), \\ \tilde{H} &= HT^{\mathrm{T}}. \end{aligned} \tag{8.19}$$

We give an example to describe this.

Example 8.2 Consider the following system:

$$\begin{cases} A_1(t+1) = \neg(A_1(t) \leftrightarrow A_2(t)), \\ A_2(t+1) = \neg(A_2(t) \leftrightarrow A_3(t)), \\ A_3(t+1) = u(t) \wedge A_1(t), \\ y(t) = A_1(t) \leftrightarrow A_2(t), \\ u(t+1) = \neg u(t). \end{cases} \tag{8.20}$$

In algebraic form, it becomes

$$\begin{cases} A_1(t+1) = M_p A_1 A_2(t), \\ A_2(t+1) = M_p A_2(t) A_3(t), \\ A_3(t+1) = M_c u(t) A_1(t), \\ y(t) = M_e A_1(t) A_2(t), \\ u(t+1) = M_n u(t). \end{cases} \tag{8.21}$$

Let $x(t) = A_1(t)A_2(t)A_3(t)$. Then,

$$x(t+1) = M_p A_1 A_2 M_p A_2 A_3 M_{cu} A_1 := Lu(t)x(t),$$

where $L \in \mathcal{L}_{8 \times 16}$ can be easily calculated as

$$L = M_p(I_4 \otimes M_p)(I_2 \otimes M_r)(I_8 \otimes M_c)W_{[4,8]}(I_2 \otimes M_r)$$
$$= \delta_8[7\,5\,1\,3\,4\,2\,6\,8\,8\,6\,2\,4\,4\,2\,6\,8].$$

Using the dummy matrix E_d, y can be expressed as

$$y(t) = M_e A_1(t) E_d M_n A_2(t) A_3(t)$$
$$= M_e(I_2 \otimes E_d M_n)x(t)$$
$$= \delta_2[1\,2\,1\,2\,2\,1\,2\,1]x(t).$$

Assume that we use the change of coordinates

$$\begin{cases} B_1 = A_1 \,\bar{\vee}\, A_2, \\ B_2 = \neg A_1, \\ B_3 = \neg A_3, \end{cases}$$

which is the inverse of the coordinate transformation in (8.10). Its transfer matrix is then T^T, where T is as in (8.12).

Using logical coordinate transformation (8.18), we have

$$\tilde{L} = TL(I_2 \otimes T^T)$$
$$= \delta_8[7\,3\,4\,8\,5\,1\,2\,6\,7\,3\,3\,7\,5\,1\,1\,5],$$
$$\tilde{H} = HT^T = \delta_2[1\,2\,2\,1\,1\,2\,2\,1].$$

We also have

$$\tilde{M}_1 = \delta_2[2\,1\,1\,2\,2\,1\,1\,2\,2\,1\,1\,2\,2\,1\,1\,2],$$
$$\tilde{M}_2 = \delta_2[2\,2\,2\,2\,1\,1\,1\,1\,2\,2\,2\,2\,1\,1\,1\,1],$$
$$\tilde{M}_3 = \delta_2[1\,1\,2\,2\,1\,1\,2\,2\,1\,1\,1\,1\,1\,1\,1\,1].$$

It is easy to check that

$$\tilde{M}_1 M_n = \tilde{M}_1, \qquad \tilde{M}_1 W_{[2]} M_n = \tilde{M}_1 W_{[2]}.$$

Hence, $B_1(t+1)$ is independent on $u(t)$ and $B_1(t)$, so we can replace it by

$$\tilde{M}_1 \left(\delta_2^1\right)^2 = \delta_2[2\,1\,1\,2].$$

Thus, the first equation becomes

$$B_1(t+1) = B_2(t) \bar{\vee} B_3(t).$$

A similar argument applied to the other equations yields

$$\begin{cases} B_1(t+1) = B_2(t) \bar{\vee} B_3(t), \\ B_2(t+1) = \neg B_1(t), \\ B_3(t+1) = u(t) \rightarrow B_2(t), \\ y(t) = B_2(t) \leftrightarrow B_3(t), \\ u(t+1) = \neg u(t). \end{cases} \tag{8.22}$$

8.3 Regular Subspaces

In Definition 8.1 a subspace of the state space $\mathscr{X}$ was defined as $\mathscr{Z} = \mathscr{F}_\ell\{z_1, \ldots, z_k\}$. Recall that in the previous chapter a subspace was defined as $\mathscr{X}_0 = \mathscr{F}_\ell\{x_{i_1}, \ldots, x_{i_k}\}$. This subspace corresponds to a set of state variables. We can check whether this subspace is invariant, etc., according to their corresponding logical dynamic equations. Now, a fundamental problem is: for a general subspace, can we always consider its basis as part of the coordinate variables? It turns out that we cannot. Let us consider the following simple example.

Example 8.3 Assume $\mathscr{X} = \mathscr{F}_\ell\{x_1, x_2\}$. Let $\mathscr{Z}$ be a subspace generated by $z_1 = x_1 \wedge x_2$, that is, $\mathscr{Z} = \mathscr{F}_\ell\{z_1\}$. Can we then find $z_2 \in \mathscr{X}$ such that $Z = (z_1, z_2)$ is a coordinate frame? Since $z_2 = f(x_1, x_2)$, there are $2^4 = 16$ different functions. Checking them one by one, it can be seen that there is no z_2 which makes (z_1, z_2) a coordinate frame.

We therefore need to introduce the following definition.

Definition 8.3 A subspace $\mathscr{Z} = \mathscr{F}_\ell\{z_1, \ldots, z_k\} \subset \mathscr{X}$ is called a regular subspace of dimension k if there are $z_{k+1}, \ldots, z_n$ such that $Z = (z_1, \ldots, z_n)$ is a coordinate frame. Moreover, $\{z_1, \ldots, z_k\}$ is called a regular basis of $\mathscr{Z}$.

Example 8.4 Consider the state space $\mathscr{X} = \mathscr{F}_\ell\{x_1, x_2, x_3\}$ and the subspace $\mathscr{Z} = \mathscr{F}_\ell\{z_1, z_2\} \subset \mathscr{X}$.

1. Assume that

$$\begin{cases} z_1 = x_1 \leftrightarrow x_2, \\ z_2 = x_2 \bar{\vee} x_3. \end{cases} \tag{8.23}$$

Its algebraic form can then be expressed as

$$\begin{cases} z_1 = \delta_2[1\ 1\ 2\ 2\ 2\ 2\ 1\ 1]x_1x_2x_3, \\ z_2 = \delta_2[2\ 1\ 1\ 2\ 2\ 1\ 1\ 2]x_1x_2x_3. \end{cases} \tag{8.24}$$

We claim that $\mathscr{L}$ is a regular subspace. To see this, we choose $z_3 = (x_1 \wedge (x_2 \leftrightarrow x_3)) \vee (\neg x_1 \wedge (x_2 \bar{\vee} x_3))$ or, equivalently,

$$z_3 = \delta_2[1\,2\,2\,1\,2\,1\,1\,2]x_1x_2x_3. \tag{8.25}$$

Setting $z = z_1z_2z_3$ and $x = x_1x_2x_3$, it is easy to calculate that

$$z = Lx = \delta_8[3\,2\,6\,7\,8\,5\,1\,4]x. \tag{8.26}$$

Since $L \in \mathscr{L}_{8\times8}$ is nonsingular, z is a coordinate frame, and hence $\mathscr{L}$ is a regular subspace.

2. Assume that

$$\begin{cases} z_1 = x_1 \to x_2, \\ z_2 = x_2 \bar{\vee} x_3. \end{cases} \tag{8.27}$$

Its algebraic form can then be expressed as

$$\begin{cases} z_1 = \delta_2[1\,1\,2\,2\,1\,1\,1\,1]x, \\ z_2 = \delta_2[2\,1\,1\,2\,2\,1\,1\,2]x. \end{cases} \tag{8.28}$$

Let $z_3 \in \mathscr{X}$. There are then 2^8 different z_3's and it is then straightforward to check that there is no z_3 which makes $\{z_1, z_2, z_3\}$ a coordinate frame. Therefore, $\mathscr{L} = \mathscr{F}_\ell\{z_1, z_2\}$ is not a regular subspace.

From the above example one sees that using the definition to check whether a subspace is regular is a very demanding task. Therefore, we have to find an efficient way to verify regularity.

Consider the set of functions

$$z_i = g_i(x_1, \ldots, x_n), \quad i = 1, \ldots, k, \tag{8.29}$$

and let $\mathscr{L} = \mathscr{F}_\ell\{z_1, \ldots, z_k\}$. We would like to know when $\mathscr{L}$ is a regular subspace with $\{z_1, \ldots, z_k\}$ as its regular sub-basis. Set $z = \ltimes_{i=1}^k z_i$ and $x = \ltimes_{i=1}^n x_i$. From (8.29) we can get its algebraic form as

$$z = Lx := \begin{bmatrix} \ell_{1,1} & \ell_{1,2} & \cdots & \ell_{1,2^n} \\ \vdots & & & \\ \ell_{2^r,1} & \ell_{2^r,2} & \cdots & \ell_{2^r,2^n} \end{bmatrix} x. \tag{8.30}$$

Recalling Theorem 4.1, we have

$$L = M_1 \prod_{j=2}^k [(I_{2^n} \otimes M_j)\, \Phi_n], \tag{8.31}$$

where

$$\Phi_s = \prod_{i=1}^{s} I_{2^{i-1}} \otimes \left[(I_2 \otimes W_{[2,2^{s-i}]}) M_r \right], \quad s = 1, 2, \ldots,$$

and $M_r = \delta_4[1\ 4]$.

We provide another method to calculate L using M_i. It is very convenient in numerical calculation and, moreover, it reveals certain relations between the elements of L and M_i.

Proposition 8.1 *Consider* (8.29) *and* (8.30). *Assume that the structure matrix of g_i is*

$$M_i = \left[\xi_1^i\ \xi_2^i\ \cdots\ \xi_{2^n}^i \right], \quad i = 1, \ldots, k,$$

and

$$L = [\ell_1\ \ell_2\ \cdots\ \ell_{2^n}].$$

Then,

$$\ell_r = \ltimes_{i=1}^{k} \xi_r^i, \quad r = 1, \ldots, 2^n.$$

Proof If we assume $x_1 = x_2 = \cdots = x_n = \delta_2^1 \sim 1$, then $x = \ltimes_{i=1}^{n} x_i = \delta_{2^n}^1$, and

$$z = Lx = \mathrm{Col}_1(L) = \ell_1.$$

On the other hand,

$$z_i = M_i x = \mathrm{Col}_1(M_i) = \xi_1^i, \quad i = 1, \ldots, k.$$

Hence, $z = \ltimes_{i=1}^{k} \xi_1^i$. We then have

$$\ell_1 = \ltimes_{i=1}^{k} \xi_1^i.$$

Similarly, let $x_i = \alpha_i \in \{0, 1\}$, $i = 1, \ldots, n$, and set $r = 2^n - [\alpha_1 \times 2^{n-1} + \alpha_2 \times 2^{n-2} + \cdots + \alpha_n]$. Then, $z_i = \xi_r^i$, $i = 1, \ldots, k$. Hence,

$$\ell_r = z = \ltimes_{i=1}^{k} \xi_r^i. \qquad \square$$

The following corollary is easily verifiable.

Corollary 8.1 *Assume that $y_1, \ldots, y_p$ and $z_1, \ldots, z_q$ are sets of logical functions of $x_1, \ldots, x_n$. Let $y = \ltimes_{i=1}^{p} y_i$, $z = \ltimes_{i=1}^{q} z_i$, $w = yz$, and $x = \ltimes_{i=1}^{n} x_i$. Moreover, we have*

$$y = Mx, \qquad z = Nx, \qquad w = Lx,$$

where $M \in \mathscr{L}_{2^p \times 2^n}$, $N \in \mathscr{L}_{2^q \times 2^n}$, and $L \in \mathscr{L}_{2^{p+q} \times 2^n}$. We then have

$$\mathrm{Col}_i(L) = \mathrm{Col}_i(M) \mathrm{Col}_i(N), \quad i = 1, \ldots, 2^n. \tag{8.32}$$

To convert a componentwise algebraic form to (overall) algebraic form, (8.32) is very convenient. The following theorem shows when $\mathscr{L} = \mathscr{F}_\ell\{z_1, \ldots, z_k\}$ is a regular subspace with regular sub-basis $\{z_1, \ldots, z_k\}$.

Theorem 8.2 *Assume there is a set of logical variables* $z_1, \ldots, z_k$ $(k \leq n)$ *satisfying* (8.30). $\mathscr{L} = \mathscr{F}_\ell\{z_1, \ldots, z_k\}$ *is a regular subspace with regular sub-basis* $\{z_1, \ldots, z_k\}$ *if and only if the corresponding coefficient matrix L satisfies*

$$\sum_{i=1}^{2^n} \ell_{j,i} = 2^{n-k}, \quad j = 1, 2, \ldots, 2^k, \tag{8.33}$$

where $\ell_{j,i}$ *are defined in* (8.30).

Proof (Sufficiency) Note that condition (8.33) means that there are 2^{n-k} different x's which make $z = \delta_{2^k}^j$, $j = 1, 2, \ldots, 2^k$. We can now choose z_{k+1} as follows. Set

$$S_k^j = \{x \mid Lx = \delta_{2^k}^j\}, \quad j = 1, 2, \ldots, 2^k.$$

Then, the cardinal number $|S_k^j| = 2^{n-k}$. For half of the elements of S_k^j, define $z_{k+1} = 0$, and for the other half, set $z_{k+1} = 1$. It is then easy to see that for $\tilde{z} = \ltimes_{i=1}^{k+1} z_i$, the corresponding $\tilde{L}$ satisfies (8.33) with k being replaced by $k + 1$. Continue this process until $k = n$. Then, (8.33) becomes

$$\sum_{i=1}^{2^n} \ell_{j,i} = 1, \quad j = 1, 2, \ldots, 2^n. \tag{8.34}$$

Equation (8.34) means that the corresponding L contains all the columns of I_{2^n}, i.e., it is obtained from I_{2^n} via a column permutation. It is, therefore, a coordinate transformation.

(Necessity) Note that using the swap matrix, it is easy to see that the order of z_i does not affect the property of (8.33). First, we claim that if $\{z_i \mid i = 1, \ldots, k\}$ satisfies (8.33), then any of its subsets $\{z_{i_t}\} \subset \{z_i \mid i = 1, \ldots, k\}$ will also satisfy (8.33) with k replaced by $|\{z_{i_t}\}|$. Since the order does not affect this property, it is enough to show that a $(k-1)$-element subset $\{z_i \mid i = 2, \ldots, k\}$ is a proper sub-basis, because from $k - 1$ we can proceed to $k - 2$, and so on. Assume that $z^2 = \ltimes_{i=2}^k z_i = Qx$ and $z_1 = Px$. Using Corollary 8.1, we have

$$\text{Col}_i(L) = \text{Col}_i(P)\,\text{Col}_i(Q), \quad i = 1, \ldots, 2^n. \tag{8.35}$$

Next, we split L into two equal-sized blocks as

$$L = \begin{bmatrix} L_1 \\ L_2 \end{bmatrix}.$$

Note that either $\mathrm{Col}_i(P) = \delta_2^1$ or $\mathrm{Col}_i(P) = \delta_2^2$. Applying this fact to (8.35), one easily sees that either $\mathrm{Col}_i(L) = \begin{bmatrix} \mathrm{Col}_i(Q) \\ 0 \end{bmatrix}$ [as $\mathrm{Col}_i(P) = \delta_2^1$] or $\mathrm{Col}_i(L) = \begin{bmatrix} 0 \\ \mathrm{Col}_i(Q) \end{bmatrix}$ [as $\mathrm{Col}_i(P) = \delta_2^2$]. Hence, $\mathrm{Col}_i(Q) = \mathrm{Col}_i(L_1) + \mathrm{Col}_i(L_2)$. It follows that

$$Q = L_1 + L_2. \tag{8.36}$$

Since L satisfies (8.33), (8.36) ensures that Q also satisfies (8.33).

Now, since $\{z_i \,|\, i = 1, \ldots, k\}$ is a proper sub-basis, there exists $\{z_i \,|\, i = k + 1, \ldots, n\}$ such that $\{z_i \,|\, i = 1, \ldots, n\}$ is a coordinate transformation of x, which satisfies (8.33). [More precisely, it satisfies (8.34) with row sum equal to 1.] According to the claim, the subset $\{z_i \,|\, i = 1, \ldots, k\}$ also satisfies (8.34).   □

The constructive proof of the sufficiency of Theorem 8.2 provides a way to construct a basis (equivalently, a coordinate transformation) from a regular sub-basis. To make this process easier, we introduce another concept.

Definition 8.4 Let $z_1, \ldots, z_k \in \mathscr{X}$ with their algebraic forms given by

$$z_i = \delta_2[\ell_{i,1}\ \ell_{i,2}\ \cdots\ \ell_{i,2^n}], \quad i = 1, \ldots, k.$$

The characteristic matrix of $\{z_1, \ldots, z_k\}$ is defined as

$$E(z_1, \ldots, z_k) = \begin{bmatrix} \ell_{1,1} & \ell_{1,2} & \cdots & \ell_{1,2^n} \\ \ell_{2,1} & \ell_{2,2} & \cdots & \ell_{2,2^n} \\ \vdots & & & \\ \ell_{k,1} & \ell_{k,2} & \cdots & \ell_{k,2^n} \end{bmatrix} \in \mathscr{B}_{k \times 2^n}. \tag{8.37}$$

Proposition 8.2 *Let $z_i \in \mathscr{X}$ and*

$$z_i = \delta_2[\ell_{i,1}\ \cdots\ \ell_{i,2^n}]x, \quad i = 1, \ldots, n,$$

where $x = \ltimes_{i=1}^n x_i$. Then, $Z = (z_1, \ldots, z_n)$ is a coordinate frame if and only if the columns of its characteristic matrix

$$E := (\ell_{i,j}) \in \mathscr{B}_{n \times 2^n}, \tag{8.38}$$

denoted by $\mathrm{Col}(E)$, are all distinct.

Proof Let $z = Tx$, where $z = \ltimes_{i=1}^n z_i$. Assume there are two columns of E, which are the same, say, $\mathrm{Col}_p(E) = \mathrm{Col}_q(E)$. Using Proposition 8.1, we know that if the pth column and qth column of T are the same, then T is not a coordinate change. That is, $\{z_i\}$ is not a coordinate frame. The necessity is thus proved. To prove the sufficiency, note that

$$\ltimes_{s=1}^n \delta_2^{i_s} = \delta_{2^n}^k,$$

where

$$k = (i_1 - 1)2^{n-1} + (i_2 - 1)2^{n-2} + \cdots + (i_{n-1} - 1)2 + i_n.$$

It is clear that if $(i_1, \ldots, i_n) \neq (j_1, \ldots, j_n)$, then

$$\ltimes_{s=1}^{n} \delta_2^{i_s} \neq \ltimes_{s=1}^{n} \delta_2^{j_s}.$$

Hence, the assumption ensures that the 2^n columns of T are of the form $\delta_{2^n}^{k}$ with 2^n different k's. It follows that T is nonsingular.   □

The above proposition can easily be used to construct a coordinate frame from a regular subspace. Moreover, it can also be used to test a regular subspace.

Corollary 8.2 *Let* $z_i \in \mathscr{X}$, $i = 1, \ldots, k$. *Then,* $\{z_1, \ldots, z_k\}$ *is a regular sub-basis (equivalently,* $\mathscr{L} = \mathscr{F}_\ell\{z_1, \ldots, z_k\}$ *is a regular subspace of dimension k) if and only if its characteristic matrix* $E \in \mathscr{B}_{k,2^n}$ *has the same number (that is,* 2^{n-k}*) of distinct columns.* $E \in \mathscr{B}_{k \times 2^n}$ *contains equal* (2^{n-k}) *distinct columns* $\mathrm{Col}_i(E) \in \mathscr{B}_{k \times 1}, \forall i$, *is called a regular characteristic matrix. That is, for any* $\xi \in \mathscr{B}_{k \times 1}$,

$$\left| \{i \mid \mathrm{Col}_i(E) = \xi\} \right| = 2^{n-k}. \tag{8.39}$$

To illustrate this we recall Example 8.4.

Example 8.5 Recall Example 8.4.

1. Consider case 1. Using (8.24) we can easily calculate that

$$z_1 z_2 = Mx = \delta_4[2\ 1\ 3\ 4\ 4\ 3\ 1\ 2]x. \tag{8.40}$$

That is,

$$M = \begin{bmatrix} 0 & 1 & 0 & 0 & 0 & 0 & 1 & 0 \\ 1 & 0 & 0 & 0 & 0 & 0 & 0 & 1 \\ 0 & 0 & 1 & 0 & 0 & 1 & 0 & 0 \\ 0 & 0 & 0 & 1 & 1 & 0 & 0 & 0 \end{bmatrix}.$$

We then have

$$\sum_{j=1}^{8} m_{ij} = 2, \quad i = 1, \ldots, 4.$$

According to Theorem 8.2, $\mathscr{L}$ is a regular subspace. If we use Corollary 8.2, it will be more simple. From (8.24) we can construct the characteristic matrix E as

$$E = \begin{bmatrix} 1 & 1 & 2 & 2 & 2 & 2 & 1 & 1 \\ 2 & 1 & 1 & 2 & 2 & 1 & 1 & 2 \end{bmatrix}.$$

Now, the numbers of distinct columns are the same (i.e., 2), so $\mathscr{L}$ is a regular subspace. In this form, finding a z_3 to form a coordinate frame also becomes much easier. We need to add a row to E such that the extended characteristic matrix has no equal columns, say

$$E_e = \begin{bmatrix} 1 & 1 & 2 & 2 & 2 & 2 & 1 & 1 \\ 2 & 1 & 1 & 2 & 2 & 1 & 1 & 2 \\ c_1 & c_2 & c_3 & c_4 & c_5 & c_6 & c_7 & c_8 \end{bmatrix}.$$

To make all the columns distinct, we need

$$c_1 \neq c_8, \qquad c_2 \neq c_7, \qquad c_3 \neq c_6, \qquad c_4 \neq c_5. \tag{8.41}$$

We can therefore choose

$$z_3 = \delta_2[c_1 \ c_2 \ c_3 \ c_4 \ c_5 \ c_6 \ c_7 \ c_8]x,$$

where $c_i \in \{1, 2\}$, $i = 1, \ldots, 8$, and satisfy (8.41). It is easy to see that the z_3 obtained in (8.25) is a particular one which satisfies (8.41).

2. Consider case 2. From (8.28), we can calculate that

$$z_1 z_2 = Mx = \delta_4[2 \ 1 \ 3 \ 4 \ 2 \ 1 \ 1 \ 2]x.$$

Then,

$$\sum_{j=1}^{8} m_{1j} = \sum_{j=1}^{8} m_{2j} = 3, \qquad \sum_{j=1}^{8} m_{3j} = \sum_{j=1}^{8} m_{4j} = 1.$$

According to Theorem 8.2, $\mathscr{Z}$ is not a regular subspace.

We can also use Corollary 8.2. From (8.28) the characteristic matrix

$$E = \begin{bmatrix} 1 & 1 & 2 & 2 & 1 & 1 & 1 & 1 \\ 2 & 1 & 1 & 2 & 2 & 1 & 1 & 2 \end{bmatrix}.$$

We now have three of columns $(1, 2)^T$ and $(1, 1)^T$, and only one of columns $(2, 1)^T$ and $(2, 2)^T$. We conclude that $\mathscr{Z}$ is not a regular subspace.

Next, we consider a set of nested regular sub-bases.

Theorem 8.3 *Let $y_1, \ldots, y_s$ and $z_1, \ldots, z_t$ be regular sub-bases of the state space $\mathscr{X} = \mathscr{F}_\ell\{x_1, \ldots, x_n\}$. If we assume that*

$$y_i \in \mathscr{F}_\ell\{z_1, \ldots, z_t\}, \qquad i = 1, \ldots, s,$$

then $y_1, \ldots, y_s$ is also a regular sub-basis of $z_1, \ldots, z_t$.

Proof Choose $z_{t+1}, \ldots, z_n$ such that $\tilde{z} = \ltimes_{i=t+1}^{n} z_i \ltimes_{i=1}^{t} z_i$ is a coordinate transformation of x. First, we claim that if $y = \ltimes_{i=1}^{s} y_i$ is a regular sub-basis with respect to $x = \ltimes_{i=1}^{n} x_i$, then it is also a regular sub-basis with respect to $\tilde{z}$, i.e., "regularity" is independent of a particular choice of the coordinates.

To prove this claim, let $x = T\tilde{z}$. Since T is a coordinate transformation, it is a permutation of I_n. Note that $y = Hx$ and H satisfies (8.33), and that

$$y = HT\tilde{z}.$$

Since H^T is obtained from H by column permutation, H^T satisfies (8.33). Therefore, we have

$$y = H\tilde{z} := [H_1, H_2]\tilde{z}, \tag{8.42}$$

where H satisfies (8.33), and H_1 and H_2 are two equal-sized blocks of H. Setting $z_{t+1} = \delta_2^1$ we have $H_1 z'$, and setting $z_{t+1} = \delta_2^2$ we have $H_2 z'$, where $z' = \ltimes_{i=t+2}^n z_i \ltimes_{i=1}^t z_i$. Now, since y is independent of z_{t+1}, we conclude that $H_1 = H_2$. Removing the fabricated variable z_{t+1} from (8.42) yields

$$y = [H_1]z'. \tag{8.43}$$

Since $H_1 = H_2$, one sees that H_1 satisfies (8.33). Continuing this procedure, we can finally have

$$y = H_0 z, \tag{8.44}$$

where $z = \ltimes_{i=1}^t z_i$ and H_0 satisfies (8.33). The conclusion follows from Theorem 8.2. $\qquad\square$

Using Theorem 8.3, we can construct a universal coordinate frame for a set of nested regular sub-bases. The following corollary is obvious.

Corollary 8.3 *Let* $\{z_1^i, \ldots, z_{n_i}^i\}, i = 1, \ldots, k$, *be a set of regular sub-bases of* $\mathcal{X} = \mathcal{F}_\ell\{x_1, \ldots, x_n\}$. *If we assume that*

$$\{z_1^i, \ldots, z_{n_i}^i\} \subset \mathcal{F}_\ell\{z_1^{i+1}, \ldots, z_{n_{i+1}}^{i+1}\}, \quad i = 1, \ldots, k-1,$$

then there exists a coordinate frame $w_1, \ldots, w_n$, *such that*

$$\mathcal{F}_\ell\{z_1^i, \ldots, z_{n_i}^i\} = \mathcal{F}_\ell\{w_1, \ldots, w_{n_i}\}, \quad i = 1, \ldots, k.$$

Corollary 8.4 *Let* $\mathcal{Y}$ *and* $\mathcal{Z}$ *be regular subspaces of* $\mathcal{X}$ *such that* $\mathcal{Y} \subset \mathcal{Z}$. *There then exists a regular subspace* $\mathcal{W}$ *such that* $\mathcal{F}_\ell(\mathcal{W}, \mathcal{Y}) = \mathcal{Z}$, *which is denoted by*

$$\mathcal{W} \oplus \mathcal{Y} = \mathcal{Z}. \tag{8.45}$$

Remark 8.4

1. If (8.45) holds, then $\mathcal{W}$ is called the complement space of $\mathcal{Y}$ in $\mathcal{Z}$, denoted by $\mathcal{W} = \mathcal{Z} \backslash \mathcal{Y}$.
2. The complement space was defined in Chap. 6 (Definition 6.3). This alternative definition is essentially the same as the old one.

The following example shows that the complement space $\mathscr{W}$ is, in general, not unique.

Example 8.6 Let $\mathscr{L} = \mathscr{F}_\ell\{x_1, x_2\} \subset \mathscr{X}$ and $\mathscr{Y} = \mathscr{F}_\ell\{x_1 \leftrightarrow x_2\}$. Set $\mathscr{W}_1 = \mathscr{F}_\ell\{x_1\}$ and $\mathscr{W}_2 = \mathscr{F}_\ell\{x_2\}$. Then,

$$\mathscr{Y} \oplus \mathscr{W}_1 = \mathscr{L},$$

and

$$\mathscr{Y} \oplus \mathscr{W}_2 = \mathscr{L}.$$

Corollary 8.5 *Let*

$$\mathscr{L}_1 \subset \mathscr{L}_2 \subset \cdots \subset \mathscr{L}_k = \mathscr{X}$$

be a set of nested regular subspaces. There then exists a coordinate frame $Z = \{z_{1,1}, \ldots z_{1,n_1}, z_{2,1}, \ldots, z_{2,n_2}, \ldots, z_{k,1}, \ldots, z_{k,n_k}\}$ *such that*

$$\mathscr{L}_s = \mathscr{F}_\ell\{z_{11}, \ldots, z_{1,n_1}, \ldots, z_{s,1}, \ldots z_{s,n_s}\}, \quad s = 1, \ldots, k.$$

8.4 Invariant Subspaces

Consider the system (8.3) again. If it can be expressed (under a suitable coordinate frame) as

$$\begin{cases} z^1(t+1) = F^1(z^1(t)), & z^1 \in \mathscr{D}^s, \\ z^2(t+1) = F^2(z(t)), & z^2 \in \mathscr{D}^{n-s}, \end{cases} \tag{8.46}$$

then $\mathscr{L} = \mathscr{F}_\ell\{z^1\} = \mathscr{F}_\ell\{z_1^1, \ldots, z_s^1\}$ is called an invariant subspace of (8.3).

Remark 8.5

1. In a general sense, a subspace $\mathscr{L}$ is invariant with respect to the system (8.46) if, starting from any point $z_0 \in \mathscr{L}$, the trajectory of (8.46) will remain in $\mathscr{L}$.
2. It follows from the definition that an invariant subspace is a regular subspace.
3. In Chap. 6 the invariant subspace was defined in a similar way, but under the original coordinate frame x. Obviously, this new definition is more general than the previous one because it allows for a change of coordinates. This generalization reveals the essence of invariant subspaces.

From Chap. 6 (or [2]) one sees that invariant subspaces are very important for investigating the topological structure of a network. By means of coordinate transformations, we have generalized the concept of an invariant subspace. Obviously, in a general sense, the invariant subspaces play the same role in determining the topological structure of a network.

Next, we consider how to check whether a subspace is an invariant subspace. Let $z_1, \ldots, z_s \in \mathcal{X}$ and $\mathcal{Z} = \mathscr{F}_\ell\{z_1, \ldots, z_s\}$, and set $z = \ltimes_{i=1}^s z_i$. We then have the following result.

Theorem 8.4 *Consider the system* (8.3) *with its algebraic form* (8.14). *Assume that a regular subspace* $\mathcal{Z} = \mathscr{F}_\ell\{z_1, \ldots, z_s\}$ *with* $z = \ltimes_{i=1}^s z_i$ *has the algebraic form*

$$z = Qx, \tag{8.47}$$

where $Q \in \mathscr{L}_{2^s \times 2^n}$. *Then,* $\mathcal{Z} = \mathscr{F}_\ell\{z_1, \ldots, z_s\}$ *is an invariant subspace of the system* (8.3) *if and only if*

$$\mathrm{Row}(QL) \subset \mathrm{Span}_\mathscr{B} \, \mathrm{Row}(Q), \tag{8.48}$$

where $\mathrm{Span}_\mathscr{B}$ *means that the coefficients are in* $\mathscr{D}$ *and where L is as in* (8.14), *i.e., it is the transition matrix of the algebraic form of the system* (8.3).

Proof Since $\mathcal{Z}$ is a regular subspace, there is a set $\{w_1, \ldots, w_{n-s}\}$ such that the elements of $\{z_1, \ldots, z_s, w_1, \ldots, w_{n-s}\}$ form a new coordinate frame.
   (Sufficiency) From (8.47) we have

$$z(t+1) = Qx(t+1) = QLx(t). \tag{8.49}$$

Since $\mathrm{Row}(QL) \subset \mathrm{Span}_\mathscr{B} \, \mathrm{Row}(Q)$ there exists $\eta \in \mathscr{B}_{2^s \times 2^s}$ such that $QL = \eta Q$. Hence,

$$z(t+1) = \eta Qx(t) = \eta z(t). \tag{8.50}$$

Note that from (8.50) we know that $\eta \in \mathscr{L}_{2^s \times 2^s} \subset \mathscr{B}_{2^s \times 2^s}$.
   Converting the algebraic form (8.49) back to logical form (say, F^1 is the logical form of η), we have

$$\begin{cases} z(t+1) = F^1(z(t)), \\ w(t+1) = F^2(z(t), w(t)). \end{cases}$$

(Necessity) Converting $z(t+1) = F^1(z(t))$ into algebraic form, we have

$$z(t+1) = \eta z(t) = \eta Qx(t). \tag{8.51}$$

Comparing (8.51) with (8.49), we have $QL = \eta Q$, which implies (8.48).   □

From the proof of the above theorem, it is easy to see the following.

Corollary 8.6 *Using the notation of Theorem* 8.4, $\mathcal{Z}$ *is an invariant subspace if and only if there exists an* $H \in \mathscr{L}_{2^s \times 2^s}$ *such that*

$$QL = HQ. \tag{8.52}$$

Note that checking (8.48) is not a straightforward computation; it is easier to use (8.52). If (8.52) holds, as we know that Q is of full row rank, we have

$$H = QLQ^{\mathrm{T}}(QQ^{\mathrm{T}})^{-1}.$$

Hence, we have the following result.

Corollary 8.7 $\mathscr{L}$ *is an invariant subspace if and only if*

$$QL = QLQ^{\mathrm{T}}(QQ^{\mathrm{T}})^{-1}Q. \tag{8.53}$$

It is straightforward to verify (8.53).

Example 8.7 Consider the following Boolean network:

$$\begin{cases} x_1(t+1) = (x_1(t) \wedge x_2(t) \wedge \neg x_4(t)) \vee (\neg x_1(t) \wedge x_2(t)), \\ x_2(t+1) = x_2(t) \vee (x_3(t) \leftrightarrow x_4(t)), \\ x_3(t+1) = (x_1(t) \wedge \neg x_4(t)) \vee (\neg x_1(t) \wedge x_2(t)) \\ \qquad\qquad \vee (\neg x_1(t) \wedge \neg x_2(t) \wedge x_4(t)), \\ x_4(t+1) = x_1(t) \wedge \neg x_2(t) \wedge x_4(t). \end{cases} \tag{8.54}$$

Let $\mathscr{L} = \mathscr{F}_\ell\{z_1, z_2, z_3\}$, where

$$\begin{cases} z_1 = x_1 \bar{\vee} x_4, \\ z_2 = \neg x_2, \\ z_3 = x_3 \leftrightarrow \neg x_4. \end{cases} \tag{8.55}$$

Set $x = \ltimes_{i=1}^{4} x_i$, $z = \ltimes_{i=1}^{3} z_i$. We then have

$$z = Qx,$$

where

$$Q = \delta_8[8\,3\,7\,4\,6\,1\,5\,2\,4\,7\,3\,8\,2\,5\,1\,6],$$

and the algebraic form of (8.54) is

$$x(t+1) = Lx(t),$$

where

$$L = \delta_{16}[11\,1\,11\,1\,11\,13\,15\,9\,1\,2\,1\,2\,9\,15\,13\,11].$$

It is easy to calculate that

$$QL = \delta_8[3\,8\,3\,8\,3\,2\,1\,4\,8\,3\,8\,3\,4\,1\,2\,3],$$

which satisfies (8.48). Hence, $\mathscr{L}$ is an invariant subspace of (8.54).

In fact we can choose $z_4 = x_4$ such that

$$\begin{cases} z_1 = x_1 \bar{\vee} x_4, \\ z_2 = \neg x_2, \\ z_3 = x_3 \leftrightarrow \neg x_4, \\ z_4 = x_4 \end{cases} \tag{8.56}$$

is a coordinate transformation. Moreover, under coordinate frame z, the system (8.54) can be expressed in the cascading form (8.46) as

$$\begin{cases} z_1(t+1) = z_1(t) \rightarrow z_2(t), \\ z_2(t+1) = z_2(t) \wedge z_3(t), \\ z_3(t+1) = \neg z_1(t), \\ z_4(t+1) = z_1(t) \vee z_2(t) \vee z_4(t). \end{cases} \tag{8.57}$$

8.5 Indistinct Rolling Gear Structure

Consider the system (8.3). Assume its algebraic form (in decomposed form) is

$$\begin{cases} z^1(t+1) = L_1 z^1(t), \\ z^2(t+1) = L_2 z^1(t) z^2(t). \end{cases} \tag{8.58}$$

Let $\mathscr{Z}_1 = \mathscr{F}_\ell(z_1^1, \dots, z_s^1)$ and $\mathscr{Z}_2 = \mathscr{F}_\ell(z_1^2, \dots, z_{n-s}^2)$. It was proven in Chap. 6 that the cycle of (8.58) is composed of the cycle in $\mathscr{Z}_1$ and a "formal cycle" in $\mathscr{Z}_2$. More precisely, let $C_z^k = (z_0, z_1, \dots, z_k = z_0)$ be a cycle of length k with $z_i = z_i^1 z_i^2$, $i = 0, \dots, k$. Then, for any $z \in C_z^k$, without loss of generality set $z = z_0$, and $z_0 = z_0^1 z_0^2 \in C_z^k$, there exists an $\ell \leq k$, a factor of k such that

$$C_{z^1}^\ell = \left(z_0^1, z_1^1 = (L_1) z_0^1, z_2^1 = (L_1)^2 z_0^1, \dots, z_\ell^1 = (L_1)^\ell z_0^1 = z_0^1 \right)$$

is a cycle in the $\mathscr{Z}_1$ subspace. Moreover, if we define

$$\Psi := L_2 z_{\ell-1}^1 L_2 z_{\ell-2}^1 \cdots L_2 z_1^1 L_2 z_0^1,$$

then we can construct an auxiliary system

$$z^2(t+1) = \Psi z^2(t). \tag{8.59}$$

Then,

$$C_{z^2}^j = \left(z_0^2, z_1^2 = \Psi z_0^2, \dots, z_j^2 = \Psi^j z_0^2 = z_0^2 \right)$$

is a cycle of (8.59), where $j = k/\ell$. Finally, the cycle C_z^k is decomposed as

$$z_0 = z_0^1 z_0^2 \to z_1 = z_1^1 L_2 z_0^1 z_0^2 \to z_2 = z_2^1 L_2 z_1^1 L_2 z_0^1 z_0^2 \to \cdots \to$$

$$z_\ell = z_0^1 z_1^2 \to z_{\ell+1} = z_1^1 L_2 z_0^1 z_1^2 \to z_{\ell+2} = z_2^1 L_2 z_1^1 L_2 z_0^1 z_1^2 \to \cdots \to$$

$$\vdots$$

$$z_{(j-1)\ell} = z_0^1 z_{(j-1)}^2 \to z_{(j-1)\ell+1} = z_1^1 L_2 z_0^1 z_{(j-1)}^2 \to$$

$$z_{(j-1)\ell+2} = z_2^1 L_2 z_1^1 L_2 z_0^1 z_{(j-1)}^2 \to {}'z_{j\ell} = z_0^1 z_j^2 = z_0^1 z_0^2 = z_0. \tag{8.60}$$

We call this C_z^k the composed cycle of $C_{z^1}^\ell$ and $C_{z^2}^j$, denoted by $C_z^k = C_{z^1}^\ell \circ C_{z^2}^j$.

Remark 8.6

1. As long as the dynamics of a Boolean network has a cascading structure as (8.58), its cycles have such a "composed structure", which is called the rolling gear structure, described in Chap. 6.
2. $C_{z^1}^\ell$ is a real cycle, which involves only some of the nodes (precisely, s nodes). $C_{z^2}^j$ is not a real cycle; it is a cycle of the auxiliary system (8.59).
3. To the best of the authors' knowledge, in the current literature (for instance, [1, 3, 5, 6, 8] and the references therein) only cycles and fixed points involving all nodes are considered. Cycles and fixed points involving only some nodes, such as $C_{z^1}^\ell$, are ignored. They can be found only in the cascading form. Furthermore, cycles such as $C_{z^1}^\ell$ can only be found under a coordinate transformation and in the cascading form.

If a system is not originally in cascading form but has cascading form under a suitable coordinate frame, then the system still has the cycles and/or fixed points involving some of the state variables. Moreover, the rolling gear structure still exists, which will be called the indistinct rolling gear structure. We investigate it via the following example.

Example 8.8 Consider the following Boolean network:

$$\begin{cases} x_1(t+1) = [x_5(t) \wedge (x_3(t) \bar{\vee} x_4(t))] \leftrightarrow (x_5(t) \bar{\vee} x_3(t)), \\ x_2(t+1) = x_5(t) \bar{\vee} x_3(t), \\ x_3(t+1) = (x_3(t) \bar{\vee} x_4(t)) \bar{\vee} x_2(t), \\ x_4(t+1) = [\neg(x_1(t) \leftrightarrow x_2(t))] \bar{\vee} [(x_3(t) \bar{\vee} x_4(t)) \bar{\vee} x_2(t)], \\ x_5(t+1) = x_5(t) \vee (x_3(t) \bar{\vee} x_4(t)), \\ x_6(t+1) = [(x_1(t) \leftrightarrow x_2(t)) \leftrightarrow (x_2(t) \bar{\vee} x_6(t))] \bar{\vee} (x_5(t) \bar{\vee} x_3(t)). \end{cases} \tag{8.61}$$

Setting $x = \ltimes_{i=1}^{6} x_i$, the algebraic form of system (8.61) is

$$x(t+1) = Lx(t), \qquad (8.62)$$

where

$$
\begin{aligned}
L = \delta_{64}[&18\ 17\ 35\ 36\ 62\ 61\ 45\ 46\ 13\ 14\ 30\ 29\ 33\ 34\ 20\ 19 \\
&26\ 25\ 43\ 44\ 54\ 53\ 37\ 38\ 5\ \ 6\ \ 22\ 21\ 41\ 42\ 28\ 27 \\
&21\ 22\ 40\ 39\ 57\ 58\ 42\ 41\ 10\ 9\ \ 25\ 26\ 38\ 37\ 23\ 24 \\
&29\ 30\ 48\ 47\ 49\ 50\ 34\ 33\ 2\ \ 1\ \ 17\ 18\ 46\ 45\ 31\ 32].
\end{aligned}
$$

Using the method presented in Chap. 5, it is easy to calculate that the attractive set of (8.61) consists of four cycles of length 8. These are:

C_1: $(1\,1\,1\,1\,1\,1) \rightarrow (1\,0\,1\,1\,1\,0) \rightarrow (1\,0\,0\,1\,1\,1) \rightarrow (1\,1\,1\,0\,1\,1) \rightarrow$
$(0\,0\,0\,0\,1\,0) \rightarrow (0\,1\,0\,0\,1\,1) \rightarrow (0\,1\,1\,0\,1\,0) \rightarrow (0\,0\,0\,1\,1\,0) \rightarrow$
$(1\,1\,1\,1\,1\,1)$,

C_2: $(1\,1\,1\,1\,1\,0) \rightarrow (1\,0\,1\,1\,1\,1) \rightarrow (1\,0\,0\,1\,1\,0) \rightarrow (1\,1\,1\,0\,1\,0) \rightarrow$
$(0\,0\,0\,0\,1\,1) \rightarrow (0\,1\,0\,0\,1\,0) \rightarrow (0\,1\,1\,0\,1\,1) \rightarrow (0\,0\,0\,1\,1\,1) \rightarrow$
$(1\,1\,1\,1\,1\,0)$,

C_3: $(1\,1\,0\,1\,1\,1) \rightarrow (1\,1\,0\,0\,1\,1) \rightarrow (0\,1\,1\,1\,1\,1) \rightarrow (1\,0\,1\,0\,1\,1) \rightarrow$
$(0\,0\,1\,0\,1\,0) \rightarrow (0\,0\,1\,1\,1\,0) \rightarrow (1\,0\,0\,0\,1\,0) \rightarrow (0\,1\,0\,1\,1\,0) \rightarrow$
$(1\,1\,0\,1\,1\,1)$,

C_4: $(1\,1\,0\,1\,1\,0) \rightarrow (1\,1\,0\,0\,1\,0) \rightarrow (0\,1\,1\,1\,1\,0) \rightarrow (1\,0\,1\,0\,1\,0) \rightarrow$
$(0\,0\,1\,0\,1\,1) \rightarrow (0\,0\,1\,1\,1\,1) \rightarrow (1\,0\,0\,0\,1\,1) \rightarrow (0\,1\,0\,1\,1\,1) \rightarrow$
$(1\,1\,0\,1\,1\,0)$.

Under this coordinate frame, we are not able to find cycles which are contained in smaller invariant subspaces. Therefore, we are not able to reveal the rolling gear structure for the network.

To find very small cycles and the rolling gear structure of the network, we try to convert (8.61), if possible, into a cascading form in order to investigate its indistinct rolling gear structure. Note that Theorem 8.4 says that $\mathrm{Span}\{\mathrm{Col}(Q)^{T}\}$ is a standard L^{T} invariant subspace. Therefore, standard tools from linear algebra can be used to find the invariant subspaces. We skip the tedious and straightforward computation and consider the following two nested spaces:

$$\mathscr{Z}_1 = \mathscr{F}_{\ell}\{z_1 = x_1 \leftrightarrow x_2;\ z_2 = x_5;\ z_3 = x_3 \bar{\vee} x_4\},$$
$$\mathscr{Z}_2 = \mathscr{F}_{\ell}\{z_1 = x_1 \leftrightarrow x_2;\ z_2 = x_5;\ z_3 = x_3 \bar{\vee} x_4;\ z_4 = x_2 \bar{\vee} x_6\}.$$

Set $z^1 = z_1 \ltimes z_2 \ltimes z_3$. It is easy to calculate that

$$z^1 = Q_1 x,$$

where

$$
\begin{aligned}
Q_1 = \delta_8[&2\,2\,4\,4\,1\,1\,3\,3\,1\,1\,3\,3\,2\,2\,4\,4\,6\,6\,8\,8\,5\,5\,7\,7\,5\,5\,7\,7\,6\,6\,8\,8 \\
&6\,6\,8\,8\,5\,5\,7\,7\,5\,5\,7\,7\,6\,6\,8\,8\,2\,2\,4\,4\,1\,1\,3\,3\,1\,1\,3\,3\,2\,2\,4\,4].
\end{aligned}
$$

Similarly, set $z^2 = z_1 \ltimes z_2 \ltimes z_3 \ltimes z_4$. We have

$$z^2 = Q_2 x,$$

where

$$Q_2 = \delta_{16}[\ 4\ \ 3\ \ 8\ \ 7\ \ 2\ \ 1\ \ 6\ \ 5\ \ 2\ \ 1\ \ 6\ \ 5\ \ 4\ \ 3\ \ 8\ 7$$
$$11\ 12\ 15\ 16\ \ 9\ 10\ 13\ 14\ \ 9\ 10\ 13\ 14\ 11\ 12\ 15\ 16$$
$$12\ 11\ 16\ 15\ 10\ \ 9\ 14\ 13\ 10\ \ 9\ 14\ 13\ 12\ 11\ 16\ 15$$
$$3\ \ 4\ \ 7\ \ 8\ \ 1\ \ 2\ \ 5\ \ 6\ \ 1\ \ 2\ \ 5\ \ 6\ \ 3\ \ 4\ \ 7\ 8\ ].$$

Using Theorem 8.3, it is easy to check that $\mathscr{Z}_1 \subset \mathscr{Z}_2$ are nested regular subspaces. To see they are invariant subspaces of the system (8.61), it suffices to find H_i, $i = 1, 2$, such that (8.52) holds, that is, $Q_i L = H_i Q_i$. It is easy to calculate that

$$H_1 = \delta_8[2\ 6\ 6\ 8\ 1\ 5\ 5\ 7],$$

$$H_2 = \delta_{16}[3\ 4\ 11\ 12\ 11\ 12\ 15\ 16\ 2\ 1\ 10\ 9\ 10\ 9\ 14\ 13].$$

It is not difficult to find $z_5 = x_2$ and $z_6 = x_3$ such that $\pi : (x_1, \ldots, x_6) \mapsto (z_1, \ldots, z_6)$ is a coordinate transformation:

$$\pi : \begin{cases} z_1 = x_1 \leftrightarrow x_2, \\ z_2 = x_5, \\ z_3 = x_3 \,\bar{\vee}\, x_4, \\ z_4 = x_2 \,\bar{\vee}\, x_6, \\ z_5 = x_2, \\ z_6 = x_3. \end{cases}$$

The algebraic form of π is

$$z = \ltimes_{i=1}^{6} z_i = Tx, \tag{8.63}$$

where

$$T = \delta_{64}[13\ \ 9\ 29\ 25\ \ 5\ \ 1\ 21\ 17\ \ 6\ \ 2\ 22\ 18\ 14\ 10\ 30\ 26$$
$$43\ 47\ 59\ 63\ 35\ 39\ 51\ 55\ 36\ 40\ 52\ 56\ 44\ 48\ 60\ 64$$
$$45\ 41\ 61\ 57\ 37\ 33\ 53\ 49\ 38\ 34\ 54\ 50\ 46\ 42\ 62\ 58$$
$$11\ 15\ 27\ 31\ \ 3\ \ 7\ 19\ 23\ \ 4\ \ 8\ 20\ 24\ 12\ 16\ 28\ 32].$$

Now, under the coordinate frame $z = Tx$ we have the algebraic form of the system (8.61) as

$$z(t+1) = Tx(t+1) = TLx(t) = TLT^{-1}z(t) := \tilde{L}z(t), \tag{8.64}$$

where

$$\tilde{L} = \delta_{64}[12\ 10\ 11\ \ 9\ 16\ 14\ 15\ 13\ 43\ 41\ 44\ 42\ 47\ 45\ 48\ 46$$
$$42\ 44\ 41\ 43\ 46\ 48\ 45\ 47\ 57\ 59\ 58\ 60\ 61\ 63\ 62\ 64$$
$$8\ \ 6\ \ 7\ \ 5\ \ 4\ \ 2\ \ 3\ \ 1\ 39\ 37\ 40\ 38\ 35\ 33\ 36\ 34$$
$$38\ 40\ 37\ 39\ 34\ 36\ 33\ 35\ 53\ 55\ 54\ 56\ 49\ 51\ 50\ 52].$$

Using the method proposed in Proposition 7.2, we can convert (8.64) into a logical form as (omitting the mechanical procedure)

$$
\begin{cases}
z_1(t+1) = z_2(t) \wedge z_3(t), \\
z_2(t+1) = z_2(t) \vee z_3(t), \\
z_3(t+1) = \neg z_1(t), \\
z_4(t+1) = z_1(t) \leftrightarrow z_4(t), \\
z_5(t+1) = z_2(t) \barwedge z_6(t), \\
z_6(t+1) = z_3(t) \barwedge z_5(t).
\end{cases}
\tag{8.65}
$$

From this cascading form one easily sees that $\mathscr{Z}_1 = \mathscr{F}_\ell\{z_1, z_2, z_3\}$ and $\mathscr{Z}_2 = \mathscr{F}_\ell\{z_1, z_2, z_3, z_4\}$ are invariant subspaces.

The subsystem with respect to $\mathscr{Z}_1$ has one cycle of length 4, which is

$$(1\,1\,1) \to (1\,1\,0) \to (0\,1\,0) \to (0\,1\,1) \to (1\,1\,1),$$

and the subsystem with respect to $\mathscr{Z}_2$ has two cycles of length 4, which are

$$(\underline{1\,1\,1}\,1) \to (\underline{1\,1\,0}\,1) \to (\underline{0\,1\,0}\,1) \to (\underline{0\,1\,1}\,0) \to (\underline{1\,1\,1}\,1),$$
$$(\underline{1\,1\,1}\,0) \to (\underline{1\,1\,0}\,0) \to (\underline{0\,1\,0}\,0) \to (\underline{0\,1\,1}\,1) \to (\underline{1\,1\,1}\,0).$$

The corresponding cycles of system (8.61) become

$\tilde{C}_1$: $(1\,1\,\underline{0\,0}\,1\,1) \to (0\,1\,\underline{0\,0}\,0\,1) \to (0\,1\,\underline{1\,1}\,0\,0) \to (1\,1\,\underline{1\,0}\,1\,1) \to$
$\phantom{\tilde{C}_1:\ }(1\,1\,\underline{0\,0}\,0\,0) \to (0\,1\,\underline{0\,0}\,1\,0) \to (0\,1\,\underline{1\,1}\,1\,1) \to (1\,1\,\underline{1\,0}\,0\,0) \to$
$\phantom{\tilde{C}_1:\ }(1\,1\,\underline{0\,0}\,1\,1),$

$\tilde{C}_2$: $(1\,1\,\underline{0\,1}\,1\,1) \to (0\,1\,\underline{0\,1}\,0\,1) \to (0\,1\,\underline{1\,0}\,0\,0) \to (1\,1\,\underline{1\,1}\,1\,1) \to$
$\phantom{\tilde{C}_2:\ }(1\,1\,\underline{0\,1}\,0\,0) \to (0\,1\,\underline{0\,1}\,1\,0) \to (0\,1\,\underline{1\,0}\,1\,1) \to (1\,1\,\underline{1\,1}\,0\,0) \to$
$\phantom{\tilde{C}_2:\ }(1\,1\,\underline{0\,1}\,1\,1),$

$\tilde{C}_3$: $(1\,1\,\underline{1\,0}\,1\,0) \to (1\,1\,\underline{0\,0}\,1\,0) \to (0\,1\,\underline{0\,0}\,1\,1) \to (0\,1\,\underline{1\,1}\,0\,1) \to$
$\phantom{\tilde{C}_3:\ }(1\,1\,\underline{1\,0}\,0\,1) \to (1\,1\,\underline{0\,0}\,0\,1) \to (0\,1\,\underline{0\,0}\,0\,0) \to (0\,1\,\underline{1\,1}\,1\,0) \to$
$\phantom{\tilde{C}_3:\ }(1\,1\,\underline{1\,0}\,1\,0),$

$\tilde{C}_4$: $(1\,1\,\underline{1\,1}\,1\,0) \to (1\,1\,\underline{0\,1}\,1\,0) \to (0\,1\,\underline{0\,1}\,1\,1) \to (0\,1\,\underline{1\,0}\,0\,1) \to$
$\phantom{\tilde{C}_4:\ }(1\,1\,\underline{1\,1}\,0\,1) \to (1\,1\,\underline{0\,1}\,0\,1) \to (0\,1\,\underline{0\,1}\,0\,0) \to (0\,1\,\underline{1\,0}\,1\,0) \to$
$\phantom{\tilde{C}_4:\ }(1\,1\,\underline{1\,1}\,1\,0).$

It is easy to see that the cycle of $\mathscr{Z}_1$ is implicitly contained in the cycles of $\mathscr{Z}_2$ (marked by underlining) and, similarly, the cycles of $\mathscr{Z}_2$ are implicitly contained in the cycles of (8.64). The latter form several groups of three assembled gears, which form the so-called indistinct rolling gear structure.

Note that cycles C_i and $\tilde{C}_i$, $i = 1, 2, 3, 4$, are exactly the same. (They are point-to-point correspondent. The only difference is caused by the different coordinate frames.)

References

1. Aracena, J., Demongeot, J., Goles, E.: On limit cycles of monotone functions with symmetric connection graph. Theor. Comput. Sci. **322**, 237–244 (2004)
2. Cheng, D.: Input-state approach to Boolean networks. IEEE Trans. Neural Netw. **20**(3), 512–521 (2009)
3. Cheng, D., Qi, H.: A linear representation of dynamics of Boolean networks. IEEE Trans. Automat. Contr. **55**(10), 2251–2258 (2010)
4. Cheng, D., Qi, H.: State-space analysis of Boolean networks. IEEE Trans. Neural Netw. **21**(4), 584–594 (2010)
5. Farrow, C., Heidel, J., Maloney, J., Rogers, J.: Scalar equations for synchronous Boolean networks with biological applications. IEEE Trans. Neural Netw. **15**(2), 348–354 (2004)
6. Heidel, J., Maloney, J., Farrow, C., Rogers, J.: Finding cycles in synchronous Boolean networks with applications to biochemical systems. Int. J. Bifurc. Chaos **13**(3), 535–552 (2003)
7. Kalman, R.E.: On the general theory of control systems. In: Automatic and Remote Control, Proc. First Internat. Congress, International Federation of Automatic Control, (IFAC), Moscow, 1960, vol. 1, pp. 481–492. Butterworth, Stoneham (1961)
8. Shih, M.H., Dong, J.L.: A combinatorial analogue of the Jacobian problem in automata networks. Adv. Appl. Math. **34**, 30–46 (2005)

Chapter 9
Controllability and Observability of Boolean Control Networks

9.1 Control via Input Boolean Network

Controllability is a fundamental topic in investigating Boolean control networks, but there are few known results on control design [1, 3, 4]. Using the algebraic form, the dynamics of a Boolean control network can be converted into a discrete-time conventional dynamical system and the analysis method in modern control theory can then be used to investigate the controllability of Boolean control networks.

We will discuss controllability via two types of inputs. In this section we assume the controls are generated by an input Boolean network. In the following sections, we will then consider the problem for controls of free Boolean sequences.

Consider a Boolean control network:

$$
\begin{cases}
x_1(t+1) = f_1(x_1(t), \ldots, x_n(t), u_1(t), \ldots, u_m(t)), \\
\vdots \\
x_n(t+1) = f_n(x_1(t), \ldots, x_n(t), u_1(t), \ldots, u_m(t)), \\
y_j(t) = h_j(x_1(t), \ldots, x_n(t)), \quad j = 1, \ldots, n.
\end{cases}
\tag{9.1}
$$

Assume the control is generated from a control Boolean network,

$$
\begin{cases}
u_1(t+1) = g_1(u_1(t), \ldots, u_m(t)), \\
\vdots \\
u_m(t+1) = g_m(u_1(t), \ldots, u_m(t)).
\end{cases}
\tag{9.2}
$$

Letting $X = (x_1, \ldots, x_n)^{\mathrm{T}}$, $Y = (y_1, \ldots, y_p)^{\mathrm{T}}$, $U = (u_1, \ldots, u_m)^{\mathrm{T}}$, $F = (f_1, \ldots, f_n)^{\mathrm{T}}$, $H = (h_1, \ldots, h_p)^{\mathrm{T}}$, and $G = (g_1, \ldots, g_m)^{\mathrm{T}}$, (9.1) and (9.2) can be simply expressed by the following (9.3) and (9.4), respectively:

$$
\begin{cases}
X(t+1) = F(X(t), U(t)), \\
Y(t) = H(X(t)),
\end{cases}
\tag{9.3}
$$

$$
U(t+1) = G(U(t)).
\tag{9.4}
$$

D. Cheng et al., *Analysis and Control of Boolean Networks*,
Communications and Control Engineering,
DOI 10.1007/978-0-85729-097-7_9, © Springer-Verlag London Limited 2011

Definition 9.1 Consider the control system (9.1)–(9.2). Given initial state $X(0) = X_0$ and destination state X_d, X_d is said to be controllable (or reachable) from X_0 (at the sth step) with fixed (designable) input structure (G) if we can find U_0 (and G) such that $X(U, 0) = X_0$ and $X(U, s) = X_d$ for some $s \geq 1$.

Since, in this section, the control is generated by a control network, the controllability will here be called controllability by networked control. Note that, according to the above definition, we may consider four cases: (1) fixed s and fixed G, (2) fixed s and designable G, (3) free $s > 0$ and fixed G, (4) free $s > 0$ and designable G.

In the following, we use vector form. As a convention, $x = \ltimes_{i=1}^{n} x_i$, etc.

Definition 9.2 For a fixed G, the input-state transfer matrix $\Theta^G(t, 0)$ is defined as follows. For any $u_0 \in \Delta_{2^m}$ and any $x_0 \in \Delta_{2^n}$, we have

$$x(t) = \Theta^G(t, 0)u_0 x_0, \quad x(t) \in \Delta_{2^n}, \ t > 0.$$

It is obvious that $\Theta^G(t, 0)$ depends on G. In the following we will find the input-state transfer matrix. Since

$$x_1 = Lu_0 x_0,$$

we have $\Theta^G(1, 0) = L$. Next, we calculate $x_2 = x(2)$, which is

$$x_2 = Lu_1 x_1 = LGu_0 Fu_0 Fu_0 x_0 = FG(I_{2^m} \otimes F)\Phi_m u_0 x_0,$$

where Φ_m is defined in Chap. 4 (4.6) as

$$\Phi_m = \prod_{i=1}^{m} I_{2^{i-1}} \otimes \left[(I_2 \otimes W_{[2, 2^{m-i}]}) M_r \right].$$

We then have

$$\Theta^G(2, 0) = LG(I_{2^m} \otimes L)\Phi_m.$$

Using mathematical induction, it is easy to prove that

$$\Theta^G(t, 0) = LG^{t-1}\left(I_{2^m} \otimes LG^{t-2}\right)\left(I_{2^{2m}} \otimes LG^{t-3}\right) \cdots (I_{2^{(t-1)m}} \otimes L)$$
$$(I_{2^{(t-2)m}} \otimes \Phi_m)(I_{2^{(t-3)m}} \otimes \Phi_m) \cdots (I_{2^m} \otimes \Phi_m)\Phi_m. \tag{9.5}$$

We start from case (1). From the above argument the following result is obvious.

Theorem 9.1 *Consider the system (9.1) with control (9.2), where G is fixed. x_d is the sth step reachable from x_0 if and only if*

$$x_d \in \mathrm{Col}\left\{\Theta^G(s, 0)W_{[2^n, 2^m]}x_0\right\}. \tag{9.6}$$

Proof Since

$$x(s) = \Theta^G(s,0)u_0x_0 = \Theta^G(s,0)W_{[2^n,2^m]}x_0u_0$$

the conclusion is obvious.   □

We here give an example to describe this result.

Example 9.1 Consider the system

$$\begin{cases} A(t+1) = B(t) \leftrightarrow C(t), \\ B(t+1) = C(t) \vee u_1(t), \\ C(t+1) = A(t) \wedge u_2(t), \end{cases} \tag{9.7}$$

with controls satisfying

$$\begin{cases} u_1(t+1) = g_1(u_1(t), u_2(t)), \\ u_2(t+1) = g_2(u_1(t), u_2(t)). \end{cases} \tag{9.8}$$

Assume g_1 and g_2 are fixed as

$$\begin{cases} g_1(u_1(t), u_2(t)) = \neg u_2(t), \\ g_2(u_1(t), u_2(t)) = u_1(t), \end{cases} \tag{9.9}$$

and assume $A(0) = 1$, $B(0) = 0$, $C(0) = 1$, and $s = 5$. If we let $u(t) = u_1(t)u_2(t)$, then

$$u(t+1) = u_1(t+1)u_2(t+1) = M_n u_2(t)u_1(t) = M_n W_{[2]}u(t).$$

Therefore,

$$G = M_n W_{[2]} = \delta_4[3\ 1\ 4\ 2],$$

$$x(t+1) = M_e B(t)C(t)M_d C(t)u_1(t)M_c A(t)u_2(t) = Lx(t),$$

where $L \in \mathcal{L}_{8 \times 32}$, which is

$$L = \delta_8[1\ 5\ 5\ 1\ 2\ 6\ 6\ 2\ 2\ 6\ 6\ 2\ 2\ 6\ 6\ 2\ 1\ 7\ 5\ 3\ 2\ 8\ 6\ 4\ 2\ 8\ 6\ 4\ 2\ 8\ 6\ 4].$$

$$\Phi_2 = (I_2 \otimes W_{[2]})M_r(I_2 \otimes M_r) = \delta_{16}[1\ 6\ 11\ 16].$$

Finally, using formula (9.5) yields $\Theta(5,0) \in \mathcal{L}_{8 \times 32}$ as

$$\Theta(5,0) = LG^4\big(I_{2^6} \otimes LG^3\big)(I_{2^4} \otimes LG^2)(I_{2^6} \otimes LG)(I_{2^8} \otimes L)$$

$$(I_{2^6} \otimes \Phi_2)(I_{2^4} \otimes \Phi_2)(I_{2^2} \otimes \Phi_2)(I_2 \otimes \Phi_2)\Phi_2$$

$$= \delta_8[6\ 5\ 5\ 6\ 6\ 5\ 5, 6\ 2\ 2\ 2\ 2\ 2\ 2\ 2\ 2$$

$$8\ 8\ 8\ 8\ 2\ 2\ 2\ 2\ 4\ 8\ 4\ 8\ 4\ 8\ 4\ 8].$$

Now, assume that $X(0) = (A(0), B(0), C(0)) = (1, 0, 1)$. Then, in vector form,

$$x_0 = A(0)B(0)C(0) = \delta_2^1 \delta_2^2 \delta_2^1 = \delta_8^3.$$

Using Theorem 9.1, we have

$$\Theta(5, 0)W_{[8,4]}x_0 = \delta_8[5\ 2\ 8\ 4].$$

Note that in the above equation and hereafter we use the following notation:

$$\delta_k\{i_1, \ldots, i_s\} := \left\{\delta_k^{i_1}, \ldots, \delta_k^{i_s}\right\}.$$

We conclude that the reachable set starting from $X(0)$ and at step 5, denoted by $R_5(X(0))$, is

$$R_5\big((1, 0, 1)^T\big) = \mathrm{Col}\big(\Theta(5, 0)W_{[8,4]}x_0\big) = \delta_8\{5\ 2\ 8\ 4\}.$$

Converting to binary form, we have

$$R_5\big((1, 0, 1)\big) = \big\{(0, 1, 1), (1, 1, 0), (0, 0, 0), (1, 0, 0)\big\}.$$

Finally, we have to find the initial control u_0 which drives the trajectory to the assigned x_d. Since

$$x_d = \Theta(5, 0)W_{[8,4]}x_0u_0 = \delta_8[5\ 2\ 8\ 4]u_0$$

it is obvious that to reach, say, $\delta_8^5 \sim (0, 1, 0)$, the $u_0 = \delta_4^1$, i.e., in scalar form, $u_1(0) = 1$ and $u_2(0) = 0$.

Similarly, to reach the other four points

$$\big\{(0, 1, 1), (1, 1, 0), (0, 0, 0), (1, 0, 0)\big\},$$

the corresponding initial controls should be

$$\big(u_1(0), u_2(0)\big) = \big\{(1, 1), (1, 0), (0, 1), (0, 0)\big\},$$

respectively.

Remark 9.1 The $\Theta^G(s, 0)$ can be calculated inductively. For this purpose, we define

$$\Theta_{LG}(t, 0) := LG^{t-1}\big(I_{2^m} \otimes LG^{t-2}\big)\big(I_{2^{2m}} \otimes LG^{t-3}\big) \cdots \big(I_{2^{(t-1)m}} \otimes L\big),$$

$$\Theta_\Phi(t, 0) := (I_{2^{(t-2)m}} \otimes \Phi_m)(I_{2^{(t-3)m}} \otimes \Phi_m) \cdots (I_{2^m} \otimes \Phi_m)\Phi_m. \tag{9.10}$$

Then,

$$\Theta^G(t, 0) = \Theta_{LG}(t, 0)\Theta_\Phi(t, 0). \tag{9.11}$$

We give inductive formulas for these two factors. For $\Theta_\Phi(t, 0)$ we simply have

$$\Theta_\Phi(t + 1, 0) = (I_{2^{(t-1)m}} \otimes \Theta_m)\Theta_\Phi(t, 0). \tag{9.12}$$

As for $\Theta_{LG}(t,0)$, we first convert the semi-tensor product to a conventional matrix product as

$$\Theta_{LG}(t,0) = \prod_{i=0}^{t-1}\left(I_{2^{im}} \otimes LG^{t-1-i} \otimes I_{2^{(t-1-i)m}}\right).$$

If we express it in right semi-tensor product form, we have (referring to Chap. 2 for the right semi-tensor product)

$$\Theta_{LG}(t,0) = \ltimes_{i=0}^{t-1}\left(LG^{t-1-i} \otimes I_{2^{(t-1-i)m}}\right).$$

It is then clear that

$$\Theta_{LG}(t+1,0) = \left(LG^{t} \otimes I_{2^{tm}}\right) \ltimes \Theta_{LG}(t,0). \qquad (9.13)$$

Next, we consider case (2). Since there are $m_0 = (2^m)^{2^m}$ possible distinct G's, we may express each G in condensed form and arrange them in an "increasing order". Say, when $m = 2$, we have $G_1 = \delta_4[1\ 1\ 1\ 1]$, $G_2 = \delta_4[1\ 1\ 1\ 2]$, ..., $G_{256} = \delta_4[4\ 4\ 4\ 4]$. In general, we may consider a subset $\Lambda \subset \{1, 2, \dots, m_0\}$ and allow G to be chosen from the admissible set $\{G_\lambda \mid \lambda \in \Lambda\}$. The following result is an immediate consequence of Theorem 9.1.

Corollary 9.1 *Consider the system* (9.1) *with control* (9.2), *where* $G \in \{G_\lambda \mid \lambda \in \Lambda\}$. *Then, x_d is reachable from x_0 at the sth step if and only if*

$$x_d \in \mathrm{Col}\left\{\Theta^{G_\lambda}(s,0)W_{[2^n,2^m]}x_0 \mid \lambda \in \Lambda\right\}. \qquad (9.14)$$

Example 9.2 Consider the system (9.7) again. We still assume that $A(0) = 1$, $B(0) = 0$, and $C(0) = 1$ [equivalently, $x(0) = \delta_8^3$] and let the step be $s = 5$. Assume the admissible set of G consists of nonsingular G's. There are 24 such G's:

$$\mathscr{G} = \{G \in \mathscr{L}_{4\times4} \mid G \text{ is nonsingular}\} := \{G_i \mid i = 1, 2, \dots, 24\},$$

where $G_1 = \delta_4[1\ 2\ 3\ 4]$, $G_2 = \delta_4[1\ 2\ 4\ 3]$, $G_3 = \delta_4[1\ 3\ 2\ 4]$, ..., $G_{24} = \delta_4[4\ 3\ 2\ 1]$. The corresponding

$$R^i := (R_5)^i = \mathrm{Col}\left\{\Theta^i(5,0)W_{[2^n,2^m]}x_0\right\}$$

are

$$
\begin{array}{lll}
R^1 = \delta_8\{5,6,8,4\}, & R^2 = \delta_8\{5,6,8,6\}, & R^3 = \delta_8\{5,6,8,4\}, \\
R^4 = \delta_8\{5,7,4,2\}, & R^5 = \delta_8\{5,8,2,4\}, & R^6 = \delta_8\{5,2,8,8\}, \\
R^7 = \delta_8\{5,6,8,4\}, & R^8 = \delta_8\{5,6,8,6\}, & R^9 = \delta_8\{6,8,2,4\}, \\
R^{10} = \delta_8\{6,2,7,4\}, & R^{11} = \delta_8\{1,2,4,8\}, & R^{12} = \delta_8\{6,8,8,2\}, \\
R^{13} = \delta_8\{8,5,6,4\}, & R^{14} = \delta_8\{5,2,8,4\}, & R^{15} = \delta_8\{5,6,8,4\}, \\
R^{16} = \delta_8\{7,6,8,1\}, & R^{17} = \delta_8\{5,2,8,8\}, & R^{18} = \delta_8\{2,6,7,8\}, \\
R^{19} = \delta_8\{6,2,7,2\}, & R^{20} = \delta_8\{8,5,8,2\}, & R^{21} = \delta_8\{8,6,1,4\}, \\
R^{22} = \delta_8\{5,6,8,8\}, & R^{23} = \delta_8\{2,2,7,8\}, & R^{24} = \delta_8\{5,6,8,8\}.
\end{array}
$$

Therefore, the reachable set at the fifth step is

$$R_5^{\mathscr{G}}((1,0,1)) = \delta_8\{1, 2, 4, 5, 6, 7, 8\}.$$

It is interesting that starting from $(A(0), B(0), C(0)) = (1, 0, 1)$, the only unreachable point at the fifth step is δ_8^3, which is the starting point. Now, assume we want to reach $(A(5), B(5), C(5)) = (1, 1, 1)$, which is δ_8^1. Since the first component of R^{11} is δ_8^1 (we have some other choices such as R^{16} and R^{21}), we can choose G_{11} and $u_1(0)u_2(0) = \delta_4^1$ to drive $(1, 0, 1)$ to $\delta_8^1 \sim (1, 1, 1)$ at the fifth step. It is easy to show that $G_{11} = \delta_4[2\ 4\ 1\ 3]$.

From $u_1(0)u_2(0) = \delta_4^1$, we have $u_1(0) = 1$ and $u_2(0) = 1$.

To reconstruct the control dynamics, we need retrievers

$$S_1^2 = \delta_2[1\ 1\ 2\ 2], \qquad S_2^2 = \delta_2[1\ 2\ 1\ 2].$$

We then have the structure matrices of g_1 and g_2 as

$$M_1 = S_1^2 G = \delta_2[1\ 2\ 1\ 2], \qquad M_2 = S_2^2 G = \delta_2[2\ 2\ 1\ 1].$$

It follows that the control dynamics is

$$u_1(t+1) = M_1 u_1(t) u_2(t) = u_2(t),$$

$$u_2(t+1) = M_2 u_1(t) u_2(t) = \neg u_1(t).$$

Next, we consider the reachable set for free s. The reachable set is divided into two classes: the steady-state reachable set and the transient reachable set. Inclusion in the steady-state reachable set means that destination points x_d can be reached after any $T > 0$ (equivalently, at infinite times). Its complement is the transient reachable set. Note that for Boolean networks, a state will eventually enter an attractor, so we are interested in the attractor, to which a point will enter under certain controls.

First we give a lemma, which is of independent interest.

Lemma 9.1 *For a Boolean network, if its transition matrix is nonsingular, then every point is on a cycle.*

Proof According to Theorem 5.4 the transient period T_t is the smallest $k \geq 0$ such that there exists a $T > 0$ satisfying

$$L^k = L^{k+T}. \tag{9.15}$$

To prove the lemma it suffices to show that the transient period T_t is zero. Let the network matrix be L. Consider the sequence $L, L^2, \ldots$. Since there are only finitely many distinct logical matrices in $\mathscr{L}_{2^n \times 2^n}$, there must be two integers $p < q$ such that $L^p = L^q$. Since L is nonsingular, it follows that $L^{p-q} = I$, which, in (9.15), means that $k = 0$ and $T = p - q$. That is, the transient period is zero. □

In the following, we require an assumption.

Assumption 1 G is nonsingular.

According to Lemma 9.1, starting from u_0, we can find a minimum $T_0 > 0$ such that $G^{T_0}u_0 = u_0$. Hence, $u_0, Gu_0, \ldots, G^{T_0}u_0$ is a cycle of length T_0. Following the procedure in Chap. 6, we can construct a mapping

$$\Psi := (LG^{T_0-1}u_0)(LG^{T_0-2}u_0)\cdots(LGu_0)(Lu_0). \tag{9.16}$$

For x_0 we then consider the sequence $x_0, \Psi x_0, \ldots$ and find the transient period r_1 and a minimum $T_1 > 0$ such that

$$\Psi^{r_1}x_0 = \Psi^{r_1+T_1}x_0. \tag{9.17}$$

The reachable set starting from x_0 with u_0 can then be easily constructed. We give the following algorithm:

Step 1. Find T_0 such that $u_0, Gu_0, \ldots, G^{T_0}u_0$ is a cycle in the input space.
Step 2. Find the transient period r_1 and minimum $T_1 > 0$ satisfying (9.17).
Step 3. Construct a sequence

$$x_0^i = \Psi^i x_0, \quad i = 0, 1, 2, \ldots, r_1 + T_1 - 1. \tag{9.18}$$

Step 4. For each x_0^i, inductively construct a sequence

$$x_j^i = LG^{j-1}u_0 x_{j-1}^i, \quad j = 1, \ldots, T_0 - 1. \tag{9.19}$$

Note that the above construction is a special case of the general one discussed in Chap. 6 for constructing input-state composed cycles. Thus, it is easily seen that $\{x_j^i\}$ is the set of reachable points starting from x_0 using u_0 and a fixed G. We now present this as a theorem.

Theorem 9.2 *Consider the system* (9.1) *with control* (9.2). *If we assume Assumption 1 and use the above algorithm, then:*

1. *For given u_0 and G_k, the set of reachable states is*

$$R_{u_0}^k = \left\{x_j^i \mid i = 0, 1, \ldots, r_1 + T_1 - 1; \ j = 0, 1, \ldots, T_0 - 1\right\},$$

where $\{x_j^i\}$ are constructed by (9.18)–(9.19) *and the steady-state reachable set is*

$$RS_{u_0}^k = \left\{x_j^i \in R_{u_0}^k \mid i \geq r_1\right\}.$$

2. *For fixed $G = G_k$, the reachable set from x_0 is*

$$R^k = \bigcup_{u_0} R_{u_0}^k.$$

Table 9.1 Reachable set for $G_1 = \delta_4[1\,2\,3\,4]$

$u(0)$	T_0	r_1	T_1	R^{G_1}
1	1	2	2	$\delta_8\{2, 3, 5\}$
2	1	2	1	$\delta_8\{3, 6\}$
3	1	1	7	$\delta_8\{3, 4, 8\}$
4	1	4	1	$\delta_8\{3, 4, 6, 8\}$

Table 9.2 Reachable set for $G_2 = \delta_4[2\,4\,3\,1]$

$u(0)$	T_0	r_1	T_1	R^{G_2}
1	3	2	1	$\delta_8\{1, 2, 3, 4, 5, 8\}$
2	3	2	1	$\delta_8\{2, 3, 5, 6, 8\}$
3	1	1	7	$\delta_8\{2, 3, 4, 5, 6, 7, 8\}$
4	3	2	1	$\delta_8\{3, 6, 8\}$

3. *For admissible $\{G_\lambda \mid \lambda \in \Lambda\}$, the reachable set is*

$$R = \bigcup_{\lambda \in \Lambda} \bigcup_{u_0} R_{u_0}^{\lambda}.$$

Example 9.3 Consider system (9.7) again with $x(0) = \delta_8^3$. It is easy to obtain the reachable set for each G and each $u(0)$. We give two special G's:

- $G_1 = \delta_4[1\,2\,3\,4]$. The reachable sets for the first 4 steps are listed in Table 9.1. Therefore, the overall reachable set for G_1 is $\delta_8\{2, 3, 4, 5, 6, 8\}$.
- $G_2 = \delta_4[2\,4\,3\,1]$. The reachable sets for the first 4 steps are listed in Table 9.2. Therefore, the overall reachable set for G_2 is $\Delta_8 \sim \mathcal{D}^3$, which means that the system is G_2-controllable from $(1, 0, 1)$ [or, equivalently, $x(0) = \delta_8^3$].

9.2 Subnetworks

In this section we consider the controller nodes and controlled nodes. To make the related topological structure clear, we need to discuss the corresponding subnetworks.

Definition 9.3 Let $\Sigma = (\mathcal{N}, \mathcal{E})$ be a network. $\Sigma_s = (\mathcal{N}_s, \mathcal{E}_s)$ is called a subnetwork of Σ if (i) $\mathcal{N}_s \subset \mathcal{N}$, and (ii) $(i, j) \in \mathcal{E}_s$ if and only if $(i, j) \in \mathcal{E}$ and $i, j \in \mathcal{N}_s$. A subnetwork is denoted by $\Sigma_s \subset \Sigma$.

Definition 9.4 Let $\Sigma_s \subset \Sigma$.

1. The in-degree of Σ_s is the number of edges starting from $\mathcal{N}_s^C$ and ending at $\mathcal{N}_s$, where $\mathcal{N}_s^C$ is the complement of $\mathcal{N}_s$. The out-degree of Σ_s is the number of edges starting from $\mathcal{N}_s$ and ending at $\mathcal{N}_s^C$.

Fig. 9.1 Subnetwork

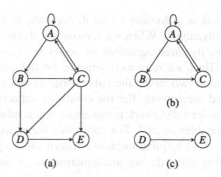

(a)

(b)

(c)

2. If the in-degree of Σ_s is 0, Σ_s is called a controller subnetwork. If the out-degree of Σ_s is 0, Σ_s is called a controlled subnetwork. If Σ is a controller subnetwork, then its complement Σ_s^C is a controlled subnetwork, and vice versa.

We give an example to illustrate these definitions.

Example 9.4 Consider a network Σ consists of five nodes $\mathcal{N} = \{A, B, C, D, E\}$, depicted in Fig. 9.1(a). Its subnetworks Σ_1 and Σ_2, consisting of $\mathcal{N} = \{A, B, C\}$ and $\mathcal{N} = \{D, E\}$, are depicted in Fig. 9.1(b) and (c), respectively.

It is easy to show the following:

- The in-degree of Σ_1 is 0 and its out-degree is 3.
- Σ_1 is an invariant subnetwork.
- The out-degree of Σ_2 is 0 and its in-degree is 3.
- Σ_2 is a controlled subnetwork and its control subnetwork is $\Sigma_2^C = \Sigma_1$.

The incidence matrix of the Σ in Example 9.4 is

$$\mathscr{I}(\Sigma) = \begin{bmatrix} 1 & 0 & 1 & 0 & 0 \\ 1 & 0 & 0 & 0 & 0 \\ 1 & 1 & 0 & 0 & 0 \\ 0 & 1 & 1 & 0 & 0 \\ 0 & 0 & 1 & 1 & 0 \end{bmatrix}. \tag{9.20}$$

Observing (9.20), one sees that the incidence matrix has block lower triangular form, which has two diagonal blocks corresponding to Σ_1 and Σ_2 respectively. In fact, it is easy to prove that this is generally true.

Proposition 9.1 *The subnetwork $\Sigma_s \subset \Sigma$ is invariant if and only if the incidence matrix $\mathscr{I}(\Sigma)$ has block lower triangular form, where the upper part of the matrix corresponds to the subnetwork nodes $\mathcal{N}_s$.*

From a graph theoretical point of view, for a network $\Sigma = (\mathcal{N}, \mathscr{E})$ we can choose a subset of nodes $\mathcal{N}_s \subset \mathcal{N}$, then remove the nodes in $\mathcal{N}_s^c := \mathcal{N} \setminus \mathcal{N}_s$ and all the edges connected to $\mathcal{N}_s^c$. What then remains is a subnetwork. However, if the network is a Boolean (control) network, we may have a problem with the sub-network dynamics. When we remove the dynamic equations of nodes in $\mathscr{N}_s^c$, the remaining dynamic equations of nodes in $\mathscr{N}_s$ may still depend on the variables of $\mathscr{N}_s^c$. Thus, we have to determine the dynamics of a subnetwork. We are only interested in two cases: the subnetwork is either the controller subnetwork or the controlled subnetwork. For the controller subnetwork there is no problem because the controller subnetwork forms an invariant subspace. Its dynamics is independent of the variables of $\mathscr{N}_s^c$. For controlled subnetwork $\mathscr{N}_s$ we assume, for each state variable $x_\lambda \in \mathscr{N}_s^c$, that there is a frozen value x_λ^e such that the dynamics of $\mathscr{N}_s$ is obtained by using the original equations in $\mathscr{N}$ and replacing the variable $x_\lambda \in \mathscr{N}_s^c$ by x_λ^e.

Note that this dynamics is physically realizable if and only if

$$\{x_\lambda^e \,|\, x_\lambda \in \mathscr{N}_s^c\}$$

is a fixed point of the dynamics of $\mathscr{N}_s^c$.

Example 9.5 Consider Example 9.4. Assume the dynamics of Σ is

$$\begin{cases} A(t+1) = A(t) \wedge C(t), \\ B(t+1) = \neg A(t), \\ C(t+1) = B(t) \leftrightarrow A(t), \\ D(t+1) = B(t) \rightarrow C(t), \\ E(t+1) = \neg D(t) \,\bar{\vee}\, C(t). \end{cases} \tag{9.21}$$

It is then obvious that $(0, 1, 0)^{\mathrm{T}}$ is a fixed point of the subnetwork Σ_1, which is a controller subnetwork. Its complement, Σ_2, is a controlled subnetwork. Define $X_1(t) := (A(t), B(t), C(t))^{\mathrm{T}}$, and $X_2(t) := (D(t), E(t))^{\mathrm{T}}$. If we set the frozen value as $X_1^e = (0, 1, 0)^{\mathrm{T}}$, then the dynamics of Σ_2 is

$$\begin{cases} D(t+1) = 1 \rightarrow 0 = 0, \\ E(t+1) = \neg D(t) \,\bar{\vee}\, 0 = \neg D(t). \end{cases} \tag{9.22}$$

9.3 Controllability via Free Boolean Sequence

In the following we consider the case where the control is a free Boolean sequence. Such a control is called an open-loop control. We refer to [1] for an initial description of this kind of controllability.

Definition 9.5 Consider the Boolean control network (9.1) and suppose we are given $x_0, x_d \in \Delta_{2^n}$. The system (9.1) is said to be controllable from x_0 to x_d (by a free Boolean sequence) at the sth step if we can find control $u(t) \in \mathscr{D}^m$, $t = 0, 1, \ldots, s - 1$, such that the initial state $\ltimes_{i=1}^n x_i(0) = x_0$ can be driven to the destination state $\ltimes_{i=1}^n x_i(s) = x_d$.

Fig. 9.2 A Boolean control network

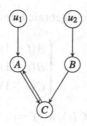

Recall that the algebraic form of (9.1) is

$$x(t+1) = Lu(t)x(t), \quad x \in \Delta_{2^n}, \ u \in \Delta_{2^m}. \tag{9.23}$$

If we define $\tilde{L} = LW_{[2^n,2^m]}$, then (9.23) can be expressed as

$$x(t+1) = \tilde{L}x(t)u(t). \tag{9.24}$$

Using it repetitively yields

$$x(s) = \tilde{L}^s x(0)u(0)u(1)\cdots u(s-1). \tag{9.25}$$

Therefore, the solution to this kind of control problem can easily be deduced, as follows.

Theorem 9.3 x_d *is reachable from* x_0 *at the sth step by controls of Boolean sequences of length s if and only if*

$$x_d \in \text{Col}\{\tilde{L}^s x_0\}. \tag{9.26}$$

Remark 9.2 Note that (9.26) means that x_d is equal to a column of $\tilde{L}^s x_0$. If, say, x_d is equal to the kth column of $\tilde{L}^s x_0$, then the controls should be

$$u(0)u(1)\cdots u(s-1) = \delta_{2^{ms}}^k, \tag{9.27}$$

which uniquely determines all u_i, $i = 0, 1, \ldots, s-1$.

The following example is from [1].

Example 9.6 Consider the Boolean control system depicted in Fig. 9.2.
Its logical equation is

$$\begin{cases} A(t+1) = C(t) \wedge u_1(t), \\ B(t+1) = \neg u_2(t), \\ C(t+1) = A(t) \vee B(t). \end{cases} \tag{9.28}$$

Its componentwise algebraic form is

$$\begin{cases} A(t+1) = M_c C(t)u_1(t), \\ B(t+1) = M_n u_2(t), \\ C(t+1) = M_d A(t)B(t). \end{cases} \tag{9.29}$$

Let $x(t) = A(t)B(t)C(t)$, $u(t) = u_1(t)u_2(t)$. We can then express the system in its algebraic form as

$$x(t+1) = \tilde{L}x(t)u(t), \tag{9.30}$$

where $\tilde{L} \in \mathscr{L}_{8 \times 32}$ is

$$\tilde{L} = \delta_8[3\ 1\ 7\ 5\ 3\ 1\ 7\ 5\ 7\ 5\ 7\ 5\ 7\ 5\ 7\ 5$$
$$3\ 1\ 7\ 5\ 3\ 1\ 7\ 5\ 8\ 6\ 8\ 6\ 8\ 6\ 8\ 6].$$

As in [1] we assume that $X_0 = (A(0), B(0), C(0)) = (0, 0, 0)$. We want to know if a destination state can be reached at the sth step. If, say, $s = 3$, then using Theorem 9.3, we can calculate $\tilde{L}^3 x_0 \in \mathscr{L}_{8 \times 64}$ as

$$\tilde{L}^3 x_0 = \delta_8[8\ 6\ 8\ 6\ 3\ 1\ 7\ 5\ 8\ 6\ 8\ 6\ 3\ 1\ 7\ 5$$
$$7\ 5\ 7\ 5\ 3\ 1\ 7\ 5\ 8\ 6\ 8\ 6\ 3\ 1\ 7\ 5$$
$$8\ 6\ 8\ 6\ 3\ 1\ 7\ 5\ 8\ 6\ 8\ 6\ 3\ 1\ 7\ 5$$
$$7\ 5\ 7\ 5\ 3\ 1\ 7\ 5\ 8\ 6\ 8\ 6\ 3\ 1\ 7\ 5].$$

It is clear that at the third step all states except δ_{16}^2, δ_{16}^4 can be reached. We now choose one state, say $\delta_8^5 \sim (0, 1, 1)$. Note that in the 8th, 16th, 18th, 20th, ... columns we have δ_8^5, which means that any of the controls δ_{64}^8, δ_{64}^{16}, δ_{64}^{18}, δ_{64}^{20}, ... can drive the initial state $(0, 0, 0)$ to the destination state $(0, 1, 1)$. We choose, for example,

$$u_1(0)u_2(0)u_1(1)u_2(1)u_1(2)u_2(2) = \delta_{64}^8.$$

Converting $64 - 8 = 56$ to binary form yields 111000, which means that the corresponding controls are

$$u_1(0) = 1, \qquad u_2(0) = 1, \qquad u_1(1) = 1, \qquad u_2(1) = 0, \qquad u_1(2) = 0,$$

$$u_2(2) = 0.$$

It is easy to check directly that this set of controls works. We may check some others. Choosing, say, δ_{64}^{24} and converting $64 - 24 = 40$ to binary form as 101000, we have

$$u_1(0) = 1, \qquad u_2(0) = 0, \qquad u_1(1) = 1, \qquad u_2(1) = 0, \qquad u_1(2) = 0,$$

$$u_2(2) = 0.$$

This also works.

In general, it is easy to calculate that when $s = 1$, the reachable set from $X_0 = (0, 0, 0)$, denoted by $R_1(X_0)$, is

$$R_1(X_0) = \{(0, 1, 0), (0, 0, 0)\}.$$

When $s > 1$ the reachable set is

$$R_s(X_0) = \{(1, 1, 1), (1, 0, 1), (0, 1, 1), (0, 1, 0), (0, 0, 1), (0, 0, 0)\}, \quad s > 1.$$

A generalization of controllability via control of Boolean sequences is when the length of the sequences, s, is free. An immediate consequence of Theorem 9.3 is the following result.

Corollary 9.2 x_d *is reachable from* x_0 *if and only if*

$$x_d \in \text{Col}\left\{\bigcup_{i=1}^{\infty} \tilde{L}^i x_0\right\}. \tag{9.31}$$

Denote by $R_s(x_0)$ the reachable set from x_0 at time s and let $R(x_0) = \bigcup_{s \geq 0} R_s(x_0)$. The following proposition makes (9.31) verifiable.

Proposition 9.2

1. *The reachable set,* $R(x_0)$, *is a subset of* $\text{Col}\{\tilde{L}\}$.
2. *If we assume that* k^* *is the smallest* $k > 0$ *such that*

$$\text{Col}\{\tilde{L}^{k+1} x_0\} \subset \text{Col}\{\tilde{L}^s x_0 \,|\, s = 1, 2, \ldots, k\},$$

then the reachable set

$$R(x_0) = \text{Col}\left\{\bigcup_{i=1}^{k^*} \tilde{L}^i x_0\right\}. \tag{9.32}$$

Proof 1. A straightforward computation shows that $\tilde{L}^k x_0 \in \mathscr{L}_{2^n \times 2^{km}}$. Since $\tilde{L} \in \mathscr{L}_{2^n \times 2^{n+m}}$, by a property of the semi-tensor product we have

$$\tilde{L}^{k+1} x_0 = \tilde{L} \ltimes \tilde{L}^k x_0 = \tilde{L} \times [\tilde{L}^k x_0 \otimes I_{2^m}],$$

where $\times$ is the conventional matrix product. The conclusion follows immediately.
   2. We use the notation

$$\text{Col}\{\tilde{L}^k\} \otimes I_m := \{X \otimes I_m \,|\, X \in \text{Col}\{\tilde{L}^k\}\}.$$

If we assume

$$\text{Col}\{\tilde{L}^{k+1} x_0\} \subset \text{Col}\{\tilde{L}^s x_0 \,|\, s = 1, 2, \ldots, k\},$$

then

$$\text{Col}\{\tilde{L}^{k+2}x_0\}$$
$$= \{\tilde{L}\eta \mid \eta \in \text{Col}\{\tilde{L}^{k+1}x_0\} \otimes I_m\}$$
$$\subset \{\tilde{L}\eta \mid \eta \in \text{Col}\{\tilde{L}^s x_0\} \otimes I_m, \ s = 1, 2, \ldots, k\}$$
$$= \text{Col}\{\tilde{L}^s x_0 \mid s = 2, 3, \ldots, k+1\}$$
$$\subset \text{Col}\{\tilde{L}^s x_0 \otimes I_m \mid s = 1, 2, 3, \ldots, k\}.$$

This inequality shows that after k there are no more new columns. From part 1 we know that such a k^* does exist. $\square$

Example 9.7 We reconsider Example 9.6. Denote the eight possible initial points by (in decreasing order) $X_0^1 = (1, 1, 1)$, $X_0^2 = (1, 1, 0)$, ..., $X_0^8 = (0, 0, 0)$. It is then easy to see that for all of them, the first degenerate step is the same, that is, $s_0 = 3$. For X_0^1, X_0^2, X_0^5, X_0^6, the first-step reachable set is

$$R_1(X_0^1) = R_1(X_0^2) = R_1(X_0^5) = R_1(X_0^6)$$
$$= \{(1, 1, 1)^T, (1, 0, 1)^T, (0, 1, 1)^T, (0, 0, 1)^T\}.$$

For X_0^3, X_0^4, the first-step reachable set is

$$R_1(X_0^3) = R_1(X_0^4) = \{(0, 1, 1)^T, (0, 0, 1)^T\}.$$

For X_0^7, X_0^8, the first-step reachable set is

$$R_1(X_0^7) = R_1(X_0^8) = \{(0, 1, 0)^T, (0, 0, 0)^T\}.$$

These have same second-step reachable set (which is also the reachable set for any $k > 2$ steps)

$$R_2(X_0^i) = \{(1, 1, 1)^T, (1, 0, 1)^T, (0, 1, 1)^T, (0, 1, 0)^T, (0, 0, 1)^T, (0, 0, 0)^T\},$$

where $i = 1, 2, \ldots, 8$. Note that since $R_2(x_0^i) = \text{Col}\{\tilde{L}\}$, according to part 1 of Proposition 9.2, no more states can be reached.

From above argument, it is reasonable to give the following definition.

Definition 9.6 The system (9.1) is said to be globally reachable from $X_0 \sim x_0$ (by the control of a free length Boolean sequence) if

$$R(x_0) = \text{Col}\{\tilde{L}^k x_0 \mid k = 0, 1, \ldots, 2^n - 1\} = \Delta_{2^n}. \tag{9.33}$$

The system (9.1) is called globally controllable (by the control of a free length Boolean sequence) if

$$\text{Col}\{\tilde{L}^k x_0 \mid k = 0, 1, \ldots, 2^n - 1\} = \Delta_{2^n}, \quad \forall x_0 \in \Delta_{2^n}. \tag{9.34}$$

Example 9.8 Consider the following system:

$$\begin{cases} A(t+1) = B(t) \wedge u_1(t), \\ B(t+1) = C(t) \leftrightarrow (\neg u_2(t)), \\ C(t+1) = A(t) \vee u_2(t). \end{cases} \tag{9.35}$$

It is easy to check that from point $X_0 = (1, 0, 0)$ the first-, second- and third-step reachable sets are

$$R_1(X_0) = \{(0, 1, 1), (0, 0, 1)\},$$

$$R_2(X_0) = \{(1, 1, 0), (1, 0, 1), (0, 1, 0), (0, 0, 1)\},$$

$$R_3(X_0) = \{(1, 1, 1), (1, 0, 1), (1, 0, 0), (0, 1, 1), (0, 1, 0), (0, 0, 1), (0, 0, 0)\}.$$

Therefore, the system (9.35) is globally reachable from $X_0 = (1, 0, 0)$.

Remark 9.3 Unlike the controllability of linear control systems, for system (9.1), $x \in R(y)$ does not mean $y \in R(x)$. A trivial example is as follows. Assume the state equations of (9.1) have algebraic form

$$x(t+1) = Lu(t)x(t),$$

where

$$\mathrm{Col}_i(L) = \delta_{2^n}^i := x_d, \quad \forall 1 \le i \le 2^{n+m}.$$

For any $x_0 \in \Delta_{2^n}$, the reachable set is then $R(x_0) = \{x_d\}$, but if $x_0 \neq x_d$, then $x_0 \neq R(x_d)$.

It is obvious that control by free length Boolean sequences is the strongest form of control. It has been pointed out in the literature that in some Boolean network problems, the controls can only be generated by a Boolean system of controls. The control of free length Boolean sequences could destroy the cycle structure of the system, which could be very important for, e.g, deciding the type of cells.

The controllability of a Boolean control network considered thus far has been solved by means of some entirely theoretical results. The disadvantage of the approach taken here is the computational complexity. We refer to [5] for some sufficient conditions which can be used for larger Boolean control networks. We also refer to Chap. 16, where the verifying condition is a matrix which has fixed size with respect to each step s.

9.4 Observability

It is obvious that for a Boolean network, observability is control-dependent. We first give a definition.

Definition 9.7 Consider the system (9.1).

1. X_1^0 and X_2^0 are said to be distinguishable if there exists a control sequence $\{U(0), U(1), \ldots, U(s)\}$, where $s \geq 0$, such that

$$Y^1(s+1) = y^{s+1}\big(U(s), \ldots, U(0), X_1^0\big) \neq Y^2(s+1)$$

$$= y^{s+1}\big(U(s), \ldots, U(0), X_2^0\big). \tag{9.36}$$

2. The system is said to be observable if any two initial points $X^0, Y^0 \in \mathscr{D}_{2^n}$ are distinguishable.

We now give an algorithm for observability.

Step 1. Construct a sequence $\Gamma_i, i = 1, 2, \ldots$, of sets of $2^p \times 2^n$ matrices as follows:

$$\Gamma_1 = \big\{ L\delta_{2^m}^i \mid i = 1, 2, \ldots, 2^m \big\},$$

$$\Gamma_{k+1} = \big\{ L\delta_{2^m}^i \gamma \mid \gamma \in \Gamma_k; \ i = 1, 2, \ldots, 2^m \big\}, \quad k \geq 1.$$

If $\mathrm{Col}\{\Gamma_{k^*+1}\} \subset \mathrm{Col}\{\Gamma_i \mid i \leq k^*\}$, then $k^* + 1$ is called the degenerate step. If $k^* > 0$ is the last nondegenerate step, then the sequence will stop at k^*. (Since there are at most 2^n different columns, $k^* \leq 2^n$.)

Step 2. Construct a sequence of sets of $2^p \times 2^n$ matrices as $H_0 = H$, $H_i = H\Gamma_i = \{H\gamma \mid \gamma \in \Gamma_i\}$.

Step 3. Using condensed form, each matrix in H_i becomes a 2^n-dimensional row. Choose $h^0 \sim H$ and linearly independent rows $h_j^i \in H_i, i = 1, 2, \ldots, k^*$, to form a matrix as

$$\mathscr{M} = \begin{bmatrix} h^0 \\ h_1^1 \\ \vdots \\ h_{i_1}^1 \\ \vdots \\ h_1^{k^*} \\ \vdots \\ h_{i_{k^*}}^{k^*} \end{bmatrix}. \tag{9.37}$$

We call $\mathscr{M}$ the observability matrix.

Theorem 9.4 *Assuming that the system (9.1) is globally controllable, it is observable, if and only if all columns of $\mathscr{C}$ are distinct.*

Proof Starting from one point x_0 we can observe Hx_0. Using different controls $\delta_{2^n}^i$, we can observe $HL\delta_{2^n}^i$. Using different $\delta_{2^n}^i$ is allowed because the system is globally controllable. Hence, we can start from the same point as many times as we wish. Continuing this process, we see that

$$HL\delta_{2^n}^{i_1} L\delta_{2^n}^{i_2} \cdots L\delta_{2^n}^{i_s} x_0, \quad s \geq 0,$$

are observable. Since $s \geq k_0$ adds no linearly independent rows to the previous set, and a linearly dependent row is useless in distinguishing initial values, the initial values can be distinguished if and only if all columns of $\mathscr{C}$ are distinct. $\qquad\square$

Next, we consider controllability and observability with control of sequence of $1 - 0 - \varnothing$, where $\varnothing$ means the input channel is disconnected. This is reasonable. For instance, in a cellular network the active cycles determine the types of cells. Now, the genetic regulatory network can change the active cycles in a cellular network to change the types of cells, but it acts only over a very short time period, like a pulse. Thus, the control becomes a sequence of $1 - 0 - \varnothing$.

When an input u_l is disconnected, we should ask: What is the *nominal network dynamics*? Principally, it is reasonable to ask that the network graph be a subgraph of the original one by removing u_i related edges. In this way the nominal network graph is unique, but the nominal network dynamics could be different. To specify it, we assume that it has a network matrix $L_\varnothing$. For convenience, we assume that there is a *frozen control* $u_i^\varnothing = $ constant such that the ith input-disconnected system has the form $u_i = u_i^\varnothing$. When $u_i = u_i^\varnothing$, $\forall i$, the control-free system is the nominal network of the original Boolean control network. That is,

$$L_\varnothing = Lu_1^\varnothing u_2^\varnothing \cdots u_m^\varnothing.$$

In many cases we are only interested in the steady-state case. For the nominal Boolean network let C^i, $i = 1, 2, \ldots, k$, be its cycles (attractors) and denote by $\Omega = \bigcup_{i=1}^k C^i$ its set of steady states, by B^i the region of attraction of C^i.

Definition 9.8 A Boolean network is globally steady-state controllable by control sequence of $1 - 0 - \varnothing$ if, for any two points $x, y \in \Omega$, there is a control sequence of $1 - 0 - \varnothing$, which drives the trajectory from x to y. A Boolean network is steady-state observable if, for any $x_0, y_0 \in \Omega$, there is a control sequence of $1 - 0 - \varnothing$ such that x_0, y_0 are distinguished from outputs.

We will need the following assumption.

Assumption 2 $\varnothing$ is a frozen control, which is a fixed point of the input network.

For the rest of this chapter, Assumption 2 is assumed.

The following result is a direct consequence of the last definition and Theorem 9.4.

Proposition 9.3

1. *Consider a Boolean control network such that its nominal system has cycles* C^i, *$1, 2, \ldots, k$. The system is globally steady-state controllable if and only if, for any $1 \le i, j \le k$, there exist at least one $x \in C^i$, one $y \in C^j$ and a $1 - 0 - \varnothing$ sequence of control which drives x to y.*
2. *If a Boolean control network is steady-state controllable, then it is steady-state observable if and only if $\mathcal{M}$, defined in (9.37), has all distinct columns.*

Proof Note that a point on a cycle of the nominal system can be reached infinitely many times as $\varnothing$ is used. The conclusions are then trivial.   □

We now give an example.

Example 9.9 Consider the system (9.7) in Example 9.1. It is natural to assume its nominal system to be (by using frozen controls $u_1^{\varnothing} = 0$ and $u_2^{\varnothing} = 1$)

$$\begin{cases} A(t+1) = B(t) \leftrightarrow C(t), \\ B(t+1) = C(t), \\ C(t+1) = A(t). \end{cases} \tag{9.38}$$

Using the technique developed in Chap. 5 it is easy to calculate that there are two cycles: equilibrium $C^1 : (1, 1, 1)$ and length-7 cycle

$$\begin{aligned} C^2: \quad &(1, 1, 0) \to (0, 0, 1) \to (0, 1, 0) \to (0, 0, 0) \to \\ &(1, 0, 0) \to (1, 0, 1) \to (0, 1, 1) \to (1, 1, 0). \end{aligned}$$

Since there are no transient states, "globally steady-state controllable" is the same as "globally controllable". To prove global steady-state controllability, we have to find a control to drive a point in one cycle to the other and vice versa.

If we let $(A(0), B(0), C(0)) = (1, 1, 1) \in C^1$ and use $u_1(0) = 0$, $u_2(0) = 0$, then $(A(1), B(1), C(1)) = (1, 1, 0) \in C^2$. If we let $(A(0), B(0), C(0)) = (1, 0, 0) \in C^2$ and use $u_1(0) = 1$, $u_2(0) = 1$, then $(A(1), B(1), C(1)) = (1, 1, 1) \in C^1$. By Proposition 9.3, the system (9.7) is globally steady-state controllable.

We now assume that the outputs are

$$\begin{aligned} y_1(t) &= A(t), \\ y_2(t) &= B(t) \vee C(t). \end{aligned} \tag{9.39}$$

We then have

$$y(t) := y_1(t) y_2(t) = A(t) M_d B(t) C(t) = H x(t),$$

where $H \in \mathcal{M}_{4 \times 8}$ is

$$H = \delta_4[1\,1\,1\,2\,3\,3\,3\,4].$$

For the system (9.7), it is easy to calculate that

$$L = \delta_8[1\ 5\ 5\ 1\ 2\ 6\ 6\ 2\ 2\ 6\ 6\ 2\ 2\ 6\ 6\ 2$$
$$1\ 7\ 5\ 3\ 2\ 8\ 6\ 4\ 2\ 8\ 6\ 4\ 2\ 8\ 6\ 4].$$

We can then calculate that

$$HL\delta_4^1 = \delta_4[1\ 3\ 3\ 1\ 1\ 3\ 3\ 1],$$
$$HL\delta_4^2 = \delta_4[1\ 3\ 3\ 1\ 1\ 3\ 3\ 1],$$
$$HL\delta_4^3 = \delta_4[1\ 3\ 3\ 1\ 1\ 4\ 3\ 2],$$
$$HL\delta_4^4 = \delta_4[1\ 4\ 3\ 2\ 1\ 4\ 3\ 2].$$

We only need to construct part of $\mathcal{M}$. Choosing linearly independent rows, we have the observability matrix as

$$\mathcal{M} = \begin{bmatrix} H \\ HL\delta_4^1 \\ HL\delta_4^2 \\ HL\delta_4^3 \\ HL\delta_4^4 \\ \vdots \end{bmatrix} = \begin{bmatrix} 1 & 1 & 1 & 2 & 3 & 3 & 3 & 4 \\ 1 & 3 & 3 & 1 & 1 & 3 & 3 & 1 \\ 1 & 3 & 3 & 1 & 1 & 4 & 3 & 2 \\ 1 & 4 & 3 & 2 & 1 & 4 & 3 & 2 \\ \vdots & & & & & & & \end{bmatrix}.$$

From part of $\mathcal{M}$ it is enough to see that it has no identical columns. Therefore, the system is observable.

References

1. Akutsu, T., Hayashida, M., Ching, W., Ng, M.: Control of Boolean networks: hardness results and algorithms for tree structured networks. J. Theor. Biol. **244**(4), 670–679 (2007)
2. Cheng, D., Qi, H.: Controllability and observability of Boolean control networks. Automatica **45**(7), 1659–1667 (2009)
3. Datta, A., Choudhary, A., Bittner, M., Dougherty, E.: External control in Markovian genetic regulatory networks. Mach. Learn. **52**, 169–191 (2003)
4. Datta, A., Choudhary, A., Bittner, M., Dougherty, E.: External control in Markovian genetic regulatory networks: the imperfect information case. Bioinformatics **20**, 924–930 (2004)
5. Kobayashi, K., Imura, J.I., Hiraishi, K.: Polynomial-time algorithm for controllability test of Boolean networks. In: IEICE Tech. Rep., vol. 109, pp. 13–18 (2009)

Chapter 10
Realization of Boolean Control Networks

10.1 What Is a Realization?

Consider a control system. Overall, it can be considered as a mapping from input(s) to output(s), which is depicted as a black box in Fig. 10.1.

The state-space approach, proposed by Kalman, is one of the cornerstones of modern control theory. It proposes the use of a set of dynamical equations for certain state variables to describe the black box. The dynamical equations of the state variables can represent the input–output mapping, and hence it is called a realization of the system Σ. In general, the realization is not unique. Obtaining a minimum realization involves finding a smallest size of Σ (equivalently, a smallest number of state variables) to realize the required input–output mapping. For a Boolean control network (BCN) with input(s) and output(s), we can consider the same problem. This chapter is devoted to the realization of BCNs.

A dynamical system may have two formally different forms under two different coordinate frames, but it is obvious that any such pair of different forms are equivalent. Because of this coordinate transformation, we may choose a model to be the representation of the equivalence class. This is the so-called normal form.

A BCN can also be considered as a mapping from input space, say $\mathscr{D}^m$, to output space, say $\mathscr{D}^p$. In Chap. 8 we considered coordinate transformations of Boolean (control) systems. It is now clear that if one BCN is obtained from another by a state-space coordinate transformation, then these two BCNs realize the same input–output mapping. If two systems are related by a coordinate transformation, we naturally say that they are equivalent. A question which then naturally arises is: Can two nonequivalent BCNs realize the same input–output mapping? More generally, is it possible for two BCNs with different sizes to realize the same input–output mapping? The answer is "Yes". We give a heuristic example.

D. Cheng et al., *Analysis and Control of Boolean Networks*,
Communications and Control Engineering,
DOI 10.1007/978-0-85729-097-7_10, © Springer-Verlag London Limited 2011

Chapter 10
Realization of Boolean Control Networks

Fig. 10.1 A control system

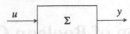

Example 10.1 Consider the following BCN:

$$\Sigma_1 : \begin{cases} x_1(t+1) = u \leftrightarrow \neg(x_1(t) \to x_2(t)), \\ x_2(t+1) = (u \wedge (\neg x_1(t) \wedge x_2(t))) \vee (\neg u \wedge \neg(x_1(t) \to x_1(t))), \\ y(t) = x_1(t) \leftrightarrow x_2(t). \end{cases} \quad (10.1)$$

Using a coordinate transformation

$$\begin{cases} z_1 = x_1 \leftrightarrow x_2, \\ z_2 = \neg x_1, \end{cases} \quad (10.2)$$

the system becomes

$$\Sigma_2 : \begin{cases} z_1(t+1) = z_1(t) \wedge u, \\ z_2(t+1) = (z_1(t) \vee z_2(t)) \leftrightarrow u, \\ y(t) = z_1(t). \end{cases} \quad (10.3)$$

It is not difficult to verify that as the initial values satisfy

$$\begin{cases} z_1(0) = x_1(0) \leftrightarrow x_2(0), \\ z_2(0) = \neg x_1(0), \end{cases} \quad (10.4)$$

the input–output mappings of Σ_1 and Σ_2 are exactly the same.

Moreover, we can see from (10.3) that the output of Σ_2 in fact depends only on z_1, and z_1 is independent of z_2. Therefore, z_2 is a redundant state variable regarding the realization of the input–output mapping and we can remove it to obtain the following:

$$\Sigma_3 : \begin{cases} z(t+1) = z(t) \wedge u, \\ y(t) = z(t). \end{cases} \quad (10.5)$$

Thus, as long as the initial conditions of Σ_1 and Σ_3 satisfy the condition $z(0) = x_1(0) \leftrightarrow x_2(0)$, they realize the same input–output mapping.

From this example one sees that, as with conventional control systems, the realization of BCNs is also an interesting and practically important problem. Moreover, we would like to emphasize the following points:

- To consider the realization of BCNs, coordinate transformations are fundamental.
- It is very likely that a "minimum realization" of a BCN can be found under a suitable coordinate frame.
- Corresponding initial values of different realizations should be taken into consideration.

10.2 Controllable Normal Form

Consider a logical mapping $F : \mathcal{D}^n \to \mathcal{D}^m$, described as

$$F : z_i = f_i(x_1, \ldots, x_n), \quad i = 1, \ldots, m. \tag{10.6}$$

f_i is said to be in a clean form if f_i has no fabricated arguments. That is, if f_i is independent of x_j, then x_j will not appear in f_i. Note that in a logical function, it is not obvious how to identify whether an argument is fabricated or not. In Chap. 7 a procedure is provided to obtain the clean form of an arbitrary logical function f.

The incidence matrix of a dynamic network was defined in Chap. 5. We recall it now. For a mapping F with f_i in clean form, its incidence matrix $\mathscr{I}(F) := (b_{ij}) \in \mathscr{B}_{m \times n}$ is constructed as follows:

$$b_{ij} = \begin{cases} 1, & \text{if } x_j \text{ appears in } f_i, \\ 0, & \text{otherwise.} \end{cases}$$

Consider the BCN

$$\begin{cases} x_1(t+1) = f_1(x_1(t), x_2(t), \ldots, x_n(t), u_1(t), \ldots, u_m(t)), \\ x_2(t+1) = f_2(x_1(t), x_2(t), \ldots, x_n(t), u_1(t), \ldots, u_m(t)), \\ \quad \vdots \\ x_n(t+1) = f_n(x_1(t), x_2(t), \ldots, x_n(t), u_1(t), \ldots, u_m(t)); \\ y_1(t) = h_1(x_1(t), x_2(t), \ldots, x_n(t)), \\ \quad \vdots \\ y_p(t) = h_p(x_1(t), x_2(t), \ldots, x_n(t)). \end{cases} \tag{10.7}$$

Hereafter and throughout this chapter we assume that the logical equations f_i are in clean form. Denote the incidence matrices for x and y by $\mathscr{I}(F) \in \mathscr{B}_{n \times (n+m)}$ and $\mathscr{I}(H) \in \mathscr{B}_{p \times n}$, respectively. For convenience we arrange $\mathscr{I}(F)$ as follows: The first n columns correspond to x and the last m columns correspond to u, that is, $b_{ij} = 1$, $j \leq n$, means x_j appears in $f_i(x, u)$ and $b_{ij} = 1$, $j > n$, means u_{j-n} appears in $f_i(x, u)$.

Definition 10.1 Let z be a coordinate frame. A subspace $\mathscr{V} = \mathscr{F}_\ell\{z_1, z_2, \ldots, z_k\} \subset \mathscr{X}$ is said to be an invariant subspace of the system (10.7) if, for any $z_0 \in \mathscr{V}$ and any control $u(t)$, the trajectory

$$z(t) = z(z_0, u, t) \in \mathscr{V}.$$

A subspace $\mathscr{V} = \mathscr{F}_\ell\{z_1, z_2, \ldots, z_k\} \subset \mathscr{X}$ is said to be a control-independent invariant subspace of the system (10.7) if it is an invariant subspace and, for any $z_0 \in \mathscr{V}$, the trajectory is independent of control u, i.e.,

$$z(t) = z(z_0, t) \in \mathscr{V}.$$

Note that the invariant subspace of a control system is not exactly the same as the invariant subspace of a free system, which is defined in Chaps. 6 and 8.

The following, easily verifiable, results are useful for testing invariant subspaces.

Proposition 10.1

1. $\mathcal{V} = \mathcal{F}_\ell\{z_1, z_2, \ldots, z_k\}$ is an invariant subspace if and only if one of the following two equivalent conditions is satisfied:
   (i) Under coordinate frame z (in clean form)

$$f_i(z_1, z_2, \ldots, z_n, u_1, \ldots, u_m)$$
$$= f_i(z_1, z_2, \ldots, z_k, u_1, \ldots, u_m), \quad i = 1, 2, \ldots, k. \quad (10.8)$$

   (ii) Under coordinate frame z (in clean form) the incidence matrix becomes

$$\mathcal{I}(F) = \begin{bmatrix} B_{11} & 0 & G_1 \\ B_{21} & B_{22} & G_2 \end{bmatrix}. \quad (10.9)$$

2. $\mathcal{V} = \mathcal{F}_\ell\{z_1, z_2, \ldots, z_k\}$ is a control-independent invariant subspace, iff one of the following two equivalent conditions is satisfied:
   (i) Under coordinate frame z (in clean form)

$$f_i(z_1, z_2, \ldots, z_n, u_1, \ldots, u_m)$$
$$= f_i(z_1, z_2, \ldots, z_k), \quad i = 1, 2, \ldots, k. \quad (10.10)$$

   (ii) Under coordinate frame z (in clean form) the incidence matrix becomes

$$\mathcal{I}(F) = \begin{bmatrix} B_{11} & 0 & 0 \\ B_{21} & B_{22} & G_2 \end{bmatrix}. \quad (10.11)$$

Next, we consider how to verify whether a regular subspace $\mathcal{Z} \subset \mathcal{X}$ is a (control-independent) invariant subspace. Let the algebraic form of (10.7) be

$$\begin{cases} x(t+1) = Lx(t)u(t), & x(t) \in \Delta_{2^n}, u(t) \in \Delta_{2^m}, \\ y(t) = Hx(t), & y(t) \in \Delta_{2^p}, \end{cases} \quad (10.12)$$

where $L \in \mathcal{L}_{2^n \times 2^{m+n}}$ and $H \in \mathcal{L}_{2^p \times 2^n}$. Assume that

$$\mathcal{Z} = \mathcal{F}_\ell\{z_1, \ldots, z_k \mid z_i \in \mathcal{X}\} \quad (10.13)$$

is a k-dimensional regular subspace, let $z = \ltimes_{i=1}^k z_i$, and let the algebraic form of $\mathcal{Z}$ be

$$z = Gx, \quad (10.14)$$

where $G \in \mathcal{L}_{2^k \times 2^n}$. We then have the following result.

Proposition 10.2

1. $\mathscr{L}$ *is an invariant subspace if and only if one of the following two equivalent conditions is satisfied:*
   (i)

$$\text{Row}(GL) \subset \text{Span}_{\mathscr{B}} \text{Row}(G). \tag{10.15}$$

   (ii) *There exists an* $E \in \mathscr{L}_{2^k \times 2^{k+m}}$ *such that*

$$GL = EG. \tag{10.16}$$

2. *Define*

$$GLW_{[2^m, 2^n]} := [B_1, B_2, \ldots, B_{2^m}],$$

   where $B_i \in \mathscr{L}_{2^k \times 2^n}$, $i = 1, 2, \ldots, 2^m$. *Then,* $\mathscr{L}$ *is a control-independent invariant subspace if and only if:*
   (i)

$$B_i = B \in \mathscr{L}_{2^k \times 2^n}, \quad i = 1, 2, \ldots, 2^m. \tag{10.17}$$

   (ii) *There exists an* $E \in \mathscr{L}_{2^k \times 2^k}$ *such that*

$$B = EG. \tag{10.18}$$

Proof 1. The proof is the same as that of Theorem 8.4.

2. By definition we have

$$z(t+1) = Ez(t),$$

where $E \in \mathscr{L}_{2^k \times 2^k}$. A similar argument as in the proof of Theorem 8.4 yields that

$$GLW_{[2^m, 2^n]} u(t) x(t) = EGx(t).$$

It follows that

$$GLW_{[2^m, 2^n]} u(t) = EG. \tag{10.19}$$

Since the right-hand side is independent of $u(t)$, taking $u(t) = \delta_{2^m}^1, \ldots, \delta_{2^m}^{2^m}$, the left-hand side becomes $B_1, \ldots, B_{2^m}$. Hence,

$$B_1 = B_2 = \cdots = B_{2^m} := B.$$

Substituting any $u(t)$ into (10.19) yields (10.18). □

Definition 10.2 A regular subspace $\mathscr{V}$ is said to be uncontrollable if it is a control-independent invariant subspace.

Definition 10.3 Consider the system (10.7).

1. A logical variable $\xi \in \mathscr{X}$ is said to be uncontrollable if there is an uncontrollable subspace $\mathscr{V}$ such that $\xi \in \mathscr{V}$.

2.

$$\mathscr{C}_c := \{\xi \in \mathscr{X} \mid \xi \text{ is uncontrollable}\}$$

is called the largest uncontrollable subspace of the system (10.7).

Remark 10.1 By definition any uncontrollable subspace is a subset of $\mathscr{C}_c$, hence it is the largest one. Moreover, it is easy to prove its uniqueness.

Theorem 10.1 *Assume $\mathscr{C}_c$ is a regular subspace with $\{z_{k+1}, \ldots, z_n\}$ as its regular sub-basis. The state equations of the system (10.7) can then be expressed as*

$$\begin{cases} z^1(t+1) = F_1(z(t), u(t)), \\ z^2(t+1) = F_2(z^2(t)), \end{cases} \tag{10.20}$$

where $z^2 = (z_{k+1}, \ldots, z_n)^T$. The system (10.20) is called the controllable normal form of (10.7).

Proof Consider z_s, where $s > k$. Since $z_s \in \mathscr{C}_c$, by definition there is an uncontrollable subspace $\mathscr{W} = \mathscr{F}_\ell\{w_1, \ldots, w_\ell\}$ such that $z_s \in \mathscr{W}$. Let $w = \ltimes_{i=1}^\ell w_i$. In algebraic form, we have

$$z_s = Mw, \quad \text{where } M \in \mathscr{L}_{2 \times 2^k}. \tag{10.21}$$

Since $\mathscr{W}$ is uncontrollable, we have

$$w(t+1) = L_w w(t), \quad \text{where } L_w \in \mathscr{L}_{2^\ell \times 2^\ell}. \tag{10.22}$$

Since both $\mathscr{W}$ and $\mathscr{C}_c$ are regular subspaces of $\mathscr{X}$ and $\mathscr{W} \subset \mathscr{C}_c$, according to Theorem 8.3, $\mathscr{W}$ is a regular subspace of $\mathscr{C}_c$. Hence, we have

$$w = Nz^2, \quad \text{where } N \in \mathscr{L}_{2^\ell \times 2^k}. \tag{10.23}$$

Using (10.21)–(10.23), we have

$$z_s(t+1) = Mw(t+1) = ML_w w(t) = ML_w Nz^2(t). \tag{10.24}$$

Since $s > k$ is arbitrary, (10.24) implies (10.20). □

We now give an example.

Example 10.2 Consider the following system:

$$\begin{cases} x_1(t+1) = x_1(t) \to u(t), \\ x_2(t+1) = (x_1(t) \to u(t)) \leftrightarrow ([(x_1(t) \land x_2(t)) \lor (\neg x_1(t) \land \neg x_2(t))] \lor x_3(t)), \\ x_3(t+1) = \neg(x_1(t) \land x_2(t)) \land (x_1(t) \lor x_2(t)). \end{cases} \tag{10.25}$$

Choose a logical coordinate transformation as follows:

$$\begin{cases} z_1(t) = x_1(t), \\ z_2(t) = (x_1(t) \wedge x_2(t)) \vee (\neg x_1(t) \wedge \neg x_2(t)), \\ z_3(t) = x_3(t). \end{cases}$$

Its inverse mapping is

$$\begin{cases} x_1(t) = z_1(t), \\ x_2(t) = z_1(t) \leftrightarrow z_2(t), \\ x_3(t) = z_3(t). \end{cases}$$

Letting $\mathscr{C}_c = \{z_2, z_3\}$, and setting $x(t) = \ltimes_{i=1}^3 x_i(t)$ and $z^2(t) = z_2(t) \ltimes z_3(t)$, we then have

$$z^2(t) = \delta_4[1\,2\,3\,4\,3\,4\,1\,2]x(t).$$

From Theorem 8.2, $\mathscr{C}_c$ is a regular subspace. Under the coordinates $\{z_i\}$, (10.25) can be expressed as

$$\begin{cases} z_1(t+1) = z_1(t) \rightarrow u(t), \\ z_2(t+1) = z_2(t) \vee z_3(t), \\ z_3(t+1) = \neg z_2(t). \end{cases} \qquad (10.26)$$

The incidence matrix of (10.26) is

$$\mathscr{I}(F) = \begin{pmatrix} 1 & 0 & 0 & 1 \\ 0 & 1 & 1 & 0 \\ 0 & 1 & 0 & 0 \end{pmatrix},$$

so (10.26) is a controllable normal form of (10.25) with $\mathscr{C}_c = \mathscr{F}_\ell\{z_2, z_3\}$.

10.3 Observable Normal Form

To investigate the observable normal form, we have to consider the complement subspace, which was discussed in Chap. 8.

Definition 10.4 Consider the system (10.7). A regular subspace $\mathscr{V}$ is said to be unobservable if there is a complement space $\mathscr{W}$ of $\mathscr{V}$ satisfying:

(i) $\mathscr{W}$ is an invariant subspace of (10.7).
(ii)

$$h_j \in \mathscr{W}, \quad j = 1, \ldots, p.$$

Remark 10.2 Let $Z = (Z^1, Z^2)$ be the coordinate frame such that $\mathcal{W} = \mathcal{F}_\ell\{Z^1\}$ and $\mathcal{V} = \mathcal{F}_\ell\{Z^2\}$. Under coordinate frame Z, the system (10.7) can be expressed as

$$\begin{cases} Z^1(t+1) = F_1(Z^1(t), U(t)), \\ Z^2(t+1) = F_2(Z(t), U(t)), \\ \quad Y(t) = H\big(Z^1(t)\big). \end{cases}$$

Equivalently, under this coordinate frame the incidence matrices of F and H are

$$\mathcal{I}(F) = \begin{bmatrix} B_{11} & 0 & B_{13} \\ B_{21} & B_{22} & B_{23} \end{bmatrix}, \qquad \mathcal{I}(H) = [C \quad 0].$$

Definition 10.5

1. Consider the system (10.7). A logical variable $\xi \in \mathcal{X}$ is called an unobservable variable if there exists an unobservable subspace $\mathcal{V}$ such that $\xi \in \mathcal{V}$.
2.

$$\mathcal{O}_c := \{\xi \mid \xi \text{ is unobservable}\}$$

Now, let $\xi \in \mathcal{O}_c$. There then exist a $\xi \in \mathcal{V}_\xi$ and its complement, denoted by $\mathcal{V}_\xi^c$, which is invariant and such that $Y \subset \mathcal{V}_\xi^c$.

Theorem 10.2 *Consider the system (10.7). Assume that:*

(i) *$\mathcal{O}_c$ is a regular subspace.*
(ii) *For each $\xi \in \mathcal{O}_c$, there exist a $\xi \in \mathcal{V}_\xi$ and its complement, denoted by $\mathcal{V}_\xi^c$, which is invariant, with $Y \subset \mathcal{V}_\xi^c$ and such that*

$$\mathcal{O} := \bigcap_{\xi \in \mathcal{O}_c} \mathcal{V}_\xi^c$$

is a complement of $\mathcal{O}_c$.

There then exists a coordinate frame $Z = (Z^1, Z^2)$ such that:

(1) *$\mathcal{O} = \mathcal{F}_\ell\{Z^1\}$ and $\mathcal{O}_c = \mathcal{F}_\ell\{Z^2\}$.*
(2) *Under coordinate frame Z, the system (10.7) is expressed as*

$$\begin{cases} Z^1(t+1) = F_1(Z^1(t), U(t)), \\ Z^2(t+1) = F_2(Z(t), U(t)), \\ \quad Y(t) = H\big(Z^1(t)\big). \end{cases} \tag{10.27}$$

Proof First, we claim that

$$\mathcal{O}_c = \bigcup_{\xi \in \mathcal{O}_c} \mathcal{V}_\xi. \tag{10.28}$$

Since $\xi \in \mathcal{V}_\xi$, we have

$$\mathcal{O}_c = \bigcup_{\xi \in \mathcal{O}_c} \xi \subset \bigcup_{\xi \in \mathcal{O}_c} \mathcal{V}_\xi.$$

Next, let $\xi \in \mathcal{O}_c$. By definition, there then exists $\mathcal{V}_\xi$ and its complement $\mathcal{V}_\xi^c$ such that (i) and (ii) of Theorem 10.2 hold. Now, for any $\eta \in \mathcal{V}_\xi$, by definition we have $\eta \in \mathcal{O}_c$. Hence, $\mathcal{V}_\xi \subset \mathcal{O}_c$ and then

$$\bigcup_{\xi \in \mathcal{O}_c} \xi \subset \mathcal{O}_c.$$

Equation (10.28) follows. Now, letting $\zeta \in \mathcal{O}$, we express

$$\zeta(t+1) = f(Z(t), U(t)). \tag{10.29}$$

We then assume $z_0 \in \mathcal{O}_c$. According to (10.28), there exists $\xi \in \mathcal{O}_c$ such that $\xi \in \mathcal{V}_\xi$. By definition, $\zeta \in \mathcal{V}_\xi^c$ and hence f is independent of z_0 in (10.29). Since $z_0 \in \mathcal{O}_c$ is arbitrary,

$$\zeta(t+1) = f(Z^1(t), U(t)). \tag{10.30}$$

Since $\zeta \in \mathcal{O}$ is arbitrary, (10.27) follows. □

Remark 10.3

1. Assuming that the conditions of Theorem 10.2 hold, (10.27) is called the observable normal form.
2. Assuming that the conditions of Theorem 10.2 hold, it is then obvious that $\mathcal{O}_c$ is the unique largest unobservable subspace.

We give an example.

Example 10.3 Consider the following system:

$$\begin{cases} x_1(t+1) = x_3(t) \vee u(t), \\ x_2(t+1) = (x_1(t) \wedge \neg x_3(t)) \vee (\neg x_1(t) \wedge (x_3(t) \leftrightarrow u(t))), \\ x_3(t+1) = x_3(t) \to u(t), \\ y = (x_1(t) \leftrightarrow x_3(t)) \to (x_2(t) \bar{\vee} x_3(t)). \end{cases} \tag{10.31}$$

Choose the logical coordinate transformation as follows:

$$\begin{cases} z_1(t) = x_1(t) \leftrightarrow x_3(t), \\ z_2(t) = x_2(t) \bar{\vee} x_3(t), \\ z_3(t) = x_3(t). \end{cases}$$

Its inverse mapping is

$$\begin{cases} x_1(t) = (z_1(t) \wedge z_3(t)) \vee (\neg z_1(t) \wedge \neg z_3(t)), \\ x_2(t) = z_2(t) \bar{\vee} z_3(t), \\ x_3(t) = z_3(t). \end{cases}$$

Letting $\mathcal{O}_c = \{z_3\}$ and setting $x(t) = \ltimes_{i=1}^{3} x_i(t)$, we then have

$$z_3(t) = \delta_2[1\,2\,1\,2\,1\,2\,1\,2]x(t).$$

From Theorem 8.2, $\mathcal{C}_c$ is a regular subspace. Under the coordinates $\{z_i\}$, (10.31) can be expressed as

$$\begin{cases} z_1(t) = u(t), \\ z_2(t) = z_1(t) \wedge u(t), \\ z_3(t) = z_3(t) \to u(t), \\ \quad y = z_1(t) \to z_2(t). \end{cases} \tag{10.32}$$

The incidence matrices of (10.32) are

$$\mathscr{I}(F) = \begin{pmatrix} 0 & 0 & 0 & 1 \\ 1 & 0 & 0 & 1 \\ 0 & 0 & 1 & 1 \end{pmatrix}, \qquad \mathscr{I}(H) = (1 \quad 1 \quad 0).$$

It follows that $\mathcal{O}_c = \mathscr{F}_\ell\{z_3\}$.

10.4 Kalman Decomposition

Combining the controllable and observable normal forms, we may look for a more general form. Given a Boolean control system (10.7), we can find $\mathcal{C}_c$ and $\mathcal{O}_c$. Then, $\mathcal{C}$ and $\mathcal{O}$, as the respective complements of $\mathcal{C}_c$ and $\mathcal{O}_c$, can also be obtained. Note that the uncontrollable subspace $\mathcal{C}_c$ and the unobservable subspace $\mathcal{O}_c$ are uniquely determined, while the controllable subspace $\mathcal{C}$ and the observable subspace $\mathcal{O}$ are not even unique. Consider the subspaces

$$\mathscr{V}_1 = \mathcal{C} \cap \mathcal{O}, \qquad \mathscr{V}_2 = \mathcal{C} \cap \mathcal{O}_c, \qquad \mathscr{V}_3 = \mathcal{C}_c \cap \mathcal{O}, \qquad \mathscr{V}_4 = \mathcal{C}_c \cap \mathcal{O}_c.$$

Using (10.20), (10.27), and the above notation, the following theorem is clear.

Theorem 10.3 *Assume $\mathcal{C}_c$, $\mathcal{O}_c$, $\mathcal{C}_c \cup \mathcal{O}_c$, and $\mathcal{C}_c \cap \mathcal{O}_c$ are regular subspaces. The system (10.7) then has the following Kalman decomposition:*

$$\begin{cases} z^1(t+1) = F^1(z^1(t), z^3(t), u(t)), \quad x^1 \in V_1, \\ z^2(t+1) = F^2(z^1(t), z^2(t), z^3(t), z^4(t), u(t)), \quad x^2 \in V_2, \\ z^3(t+1) = F^3(z^3(t)), \quad z^3 \in V_3, \\ z^4(t+1) = F^4(z^3(t), z^4(t)), \quad z^4 \in V_4, \\ \quad y_s(t) = h_s(z^1(t), z^3(t)), \quad s = 1, 2, \ldots, p, \end{cases} \tag{10.33}$$

where

$$z^1(t) = \left(z_1(t), z_2(t), \ldots, z_{n_1}(t)\right)^{\mathrm{T}},$$
$$z^2(t) = \left(z_{n_1+1}(t), z_{n_1+2}(t), \ldots, z_{n_2}(t)\right)^{\mathrm{T}},$$

$$\vdots$$

$$z^4(t) = \big(z_{n_1+n_2+n_3+1}(t), z_{n_1+n_2+n_3+2}(t), \ldots, z_n(t)\big)^{\mathrm{T}},$$

$$F^1 = (f_1, f_2, \ldots, f_{n_1})^{\mathrm{T}},$$

$$F^2 = (f_{n_1+1}, f_{n_1+2}, \ldots, f_{n_2})^{\mathrm{T}},$$

$$\vdots$$

$$F^4 = (f_{n_1+n_2+n_3+1}, f_{n_1+n_2+n_3+2}, \ldots, f_n)^{\mathrm{T}}.$$

n_i, $i = 1, 2, 3, 4$, *are the respective dimensions of* $\mathscr{V}_i$, $i = 1, 2, 3, 4$.

Proof By assumption, we have a sequence of nested regular subspaces:

$$\mathscr{C}_c \cap \mathscr{O}_c \subset \mathscr{C}_c \subset \mathscr{C}_c \cup \mathscr{O}_c \subset \mathscr{X}.$$

According to Corollary 8.5, we can find a coordinate frame $Z = (Z^4, Z^3, Z^2, Z^1)$ such that $\mathscr{C}_c \cap \mathscr{O}_c = \mathscr{F}\{Z^4\}$, $\mathscr{C}_c = \mathscr{F}\{Z^4, Z^3\}$, $\mathscr{C}_c \cup \mathscr{O}_c = \mathscr{F}\{Z^3, Z^3, Z^2\}$, and $\mathscr{X} = \mathscr{F}\{Z^4, Z^3, Z^2, Z^1\}$. Expressing system (10.7) in coordinate frame Z, we have $\mathscr{V}_i = \mathscr{F}\{Z^i\}$, $i = 1, 2, 3, 4$. Note that since $\mathscr{V}_3 \cup \mathscr{V}_4$ is the largest uncontrollable subspace, Z^1, Z^2, and u will not appear in the equations of Z^3 and Z^4. Since $\mathscr{V}_2 \cup \mathscr{V}_4$ is the largest unobservable subspace, the variables Z^2 and Z^4 will not appear in the equations of Z^1 and Z^3. Moreover, the outputs depend only on Z^1 and Z^3. (10.33) then follows. □

Next, we give an example to illustrate this.

Example 10.4 Consider the following system:

$$\begin{cases} x_1(t+1) = u, \\ x_2(t+1) = \neg x_2(t), \\ x_3(t+1) = [x_3(t) \wedge x_4(t) \wedge (x_5(t) \leftrightarrow x_6(t))] \\ \qquad\qquad \vee [x_3(t) \wedge (\neg(x_4(t)) \wedge x_5(t)] \vee (\neg x_3(t)), \\ x_4(t+1) = \neg(x_1(t) \leftrightarrow x_2(t)), \\ x_5(t+1) = [x_1(t) \wedge (x_2(t) \leftrightarrow x_3(t))] \vee [(\neg x_1(t)) \\ \qquad\qquad \wedge (\neg(x_2(t) \leftrightarrow x_3(t)))], \\ x_6(t+1) = [x_1(t) \leftrightarrow x_2(t)] \wedge \{[x_4(t) \\ \qquad\qquad \wedge (x_5(t) \leftrightarrow x_6(t))] \vee [(\neg x_4(t)) \wedge x_5(t)]\}, \\ y_1(t) = \neg x_4(t), \\ y_2(t) = (x_1(t) \leftrightarrow x_2(t)) \rightarrow (\neg x_2(t)). \end{cases} \qquad (10.34)$$

We skip the tedious process of finding the subspaces by using coordinate transformations and give the logical coordinate transformation as follows:

$$
\begin{cases}
z_1(t) = x_1(t) \leftrightarrow x_2(t), \\
z_2(t) = x_4(t), \\
z_3(t) = x_6(t), \\
z_4(t) = \neg x_2(t), \\
z_5(t) = \neg x_3(t), \\
z_6(t) = [x_4(t) \wedge (x_5(t) \leftrightarrow x_6(t))] \vee [(\neg x_4(t)) \wedge x_5(t)].
\end{cases}
\tag{10.35}
$$

Its inverse mapping is

$$
\begin{cases}
x_1(t) = \neg(z_1(t) \leftrightarrow z_4(t)), \\
x_2(t) = \neg z_4(t), \\
x_3(t) = z_5(t), \\
x_4(t) = z_2(t), \\
x_5(t) = [z_2(t) \wedge (z_3(t) \leftrightarrow z_6(t))] \vee [(\neg z_2(t)) \wedge z_6(t)], \\
x_6(t) = z_3(t).
\end{cases}
\tag{10.36}
$$

Using (10.35)–(10.36), it is easy to show that under coordinate frame $\{z_i\}$, the system (10.34) can be converted into the following form:

$$
\begin{cases}
z_1(t+1) = z_4(t) \leftrightarrow u, \\
z_2(t+1) = \neg z_1(t), \\
z_3(t+1) = z_1(t) \wedge z_6(t), \\
z_4(t+1) = \neg z_4(t), \\
z_5(t+1) = z_5(t) \vee z_6(t), \\
z_6(t+1) = \neg z_5(t), \\
y_1(t) = \neg z_2(t), \\
y_2(t) = z_1(t) \rightarrow z_4(t).
\end{cases}
\tag{10.37}
$$

A straightforward computation verifies that (10.37) is the Kalman decomposition form of system (10.34), with

$$
\mathscr{C} \cap \mathscr{O} = \mathscr{F}\{z_1(t), z_2(t)\}, \qquad \mathscr{C} \cap \mathscr{O}_c = \mathscr{F}\{z_3(t)\},
$$
$$
\mathscr{C}_c \cap \mathscr{O} = \mathscr{F}\{z_4(t)\}, \qquad \mathscr{C}_c \cap \mathscr{O}_c = \mathscr{F}\{z_5(t), z_6(t)\}.
$$

Remark 10.4 In Kalman decomposition (10.33) we assume that $\mathscr{V}_1 = \mathscr{C} \cap \mathscr{O}$ is regular. If $\mathscr{V}_1$ is not regular, then we propose two ways to replace it:

- Replace it by $\mathcal{V}_1^O$, a smallest regular subspace containing $\mathcal{V}_1$. This can be used for a minimum realization. We refer to Chap. 11 (Sect. 11.2) for calculating a regular subspace containing a given subspace.
- Replace it by $\mathcal{V}_1^o$, a largest regular subspace contained in $\mathcal{V}_1$. This can be used to ensure that the regular subspace, $\mathcal{V}_1^o$, is controllable and observable. In the following, we will discuss how to construct a regular subspace contained in a given subspace.

Definition 10.6 Let $H \in \mathcal{L}_{p \times q}$, where H_i is the ith column of H. $\{J_1, \ldots, J_s\}$ is called an s-partition of $J = \{1, 2, \ldots, q\}$ with respect to H if:

(i)

$$\bigcup_{i=1}^{s} J_i = J.$$

(ii)

$$\text{Col}_{i_1}(H) \neq \text{Col}_{i_2}(H), \quad \forall i_1 \in J_i, i_2 \in J_j, 1 \leq i < j \leq s.$$

Example 10.5 If we assume that

$$H = \delta_3[1\ 2\ 3\ 2\ 3\ 2],$$

then

$$J_1 = \{1\}, \qquad J_2 = \{2, 4, 6\}, \qquad J_3 = \{3, 5\}$$

form a 3-partition of $\{1, \ldots, 6\}$ with respect to H and

$$J_1 = \{1, 3, 5\}, \qquad J_2 = \{2, 4, 6\}$$

form a 2-partition of $\{1, \ldots, 6\}$ with respect to H.

Let $\mathcal{Y} \subset \mathcal{X}$ be a subspace. In the following we consider how to find a largest regular subspace $\mathcal{Z} \subset \mathcal{Y}$. Note that "largest" here means that $\mathcal{Z}$ has largest dimension. Let $\mathcal{X} = F_\ell(x_1, \ldots, x_n)$ and $\mathcal{Y} = F_\ell(y_1, \ldots, y_s)$, where $y_i = h_i(x_1, \ldots, x_n)$, $i = 1, \ldots, s$, are logical functions. Assume the algebraic form of $\mathcal{Y}$ is

$$y = Hx, \quad \text{where } H \in \mathcal{L}_{2^s \times 2^n}.$$

Theorem 10.4 $\mathcal{Y}$ *has a regular subspace of dimension k if and only if $\{J_1, J_2, \ldots, J_{2^k}\}$ is a 2^k-partition of $J = \{1, \ldots, 2^s\}$ with respect to $\text{Col}(H)$, satisfying*

$$|J_1| = \cdots = |J_{2^k}| = 2^{n-k}.$$

Proof Denote the algebraic forms of $\mathcal{Z}$ and $\mathcal{Y}$ by $z = Tx$ and $y = Hx$, respectively, where $T \in \mathcal{L}_{2^k \times 2^n}$, $H \in \mathcal{L}_{2^s \times 2^n}$. Since $\mathcal{Z} \subset \mathcal{Y}$ is a subspace of dimension

k, there exists a logical matrix $W \in \mathcal{L}_{2^k \times 2^s}$ such that

$$z = Wy = WHx = Tx.$$

We let $W = \delta_{2^k}[w_1 \cdots w_{2^s}]$ and $H = \delta_{2^s}[h_1 \cdots h_{2^n}]$, and define

$$I_i = \{p \mid w_p = i\},$$

$$J_i = \{q \mid h_q = p, p \in I_i\}, \quad i = 1, \ldots, 2^k.$$

$\{J_1, J_2, \ldots, J_{2^k}\}$ is then a 2^k-partition of $J = \{1, \ldots, 2^s\}$ with respect to H. A straightforward computation shows that

$$T = \delta_{2^k}\big[\mathrm{Col}_{h_1}(W)\,\mathrm{Col}_{h_2}(W), \ldots, \mathrm{Col}_{h_{2^k}}(W)\big].$$

It follows from the definition that

$$\mathrm{Col}_{h_q}(W) = i \quad \Longleftrightarrow \quad q \in J_i.$$

It follows that $\mathcal{Z}$ is a regular subspace if and only if

$$|J_i| = 2^{n-k}, \quad i = 1, \ldots, 2^k. \qquad \square$$

As a convention, we set

$$I_0 = \big\{p \mid \mathrm{Col}_p(W) \text{ is free}\big\}.$$

Example 10.6 Let $\mathcal{X} = F_\ell(x_1, x_2, x_3)$ and $\mathcal{Y} = F_\ell(x_1 \rightarrow x_2, x_1 \vee x_2)$. The algebraic form of $y_1 = x_1 \rightarrow x_2$, $y_2 = x_1 \vee x_2$ is

$$y = Hx = \delta_4[1\ 1\ 3\ 3\ 1\ 1\ 2\ 2]x.$$

Let $J_1 = \{1, 2, 5, 6\}$ and $J_2 = \{3, 4, 7, 8\}$. We then have $I_1 = \{1\}$, $I_2 = \{2, 3\}$, and $I_0 = \{4\}$. Construct the logical matrix $W = \delta_2[w_1\ w_2\ w_3\ w_4]$, where

$$\begin{cases} w_i = 1, & \text{when } i \in I_1 = \{1\}, \\ w_i = 2, & \text{when } i \in I_2 = \{2, 3\}, \\ w_i = 1 \text{ or } 2, & \text{when } i \in I_0 = \{4\}. \end{cases}$$

We choose $W = \delta_2[1\ 2\ 2\ 2]$. Thus, $z = Wy = WHx = \delta_2[1\ 1\ 2\ 2\ 1\ 1\ 2\ 2]x$ is a regular subspace. According to Theorem 10.4, $\mathcal{Z} \in \mathcal{Y}$.

10.5 Realization

Definition 10.7 Two Boolean control networks are said to be equivalent if, for any point x_0 of one network, there is a point $\tilde{x}_0$ of the other network such that for the same inputs $u(t)$, $t = 0, 1, 2, \ldots$, with initial values x_0 and $\tilde{x}_0$, respectively, the outputs $\{y(t)\}$ are the same.

Consider a linear control system [2]:

$$\begin{cases} \dot{x} = Ax + Bu, & x \in \mathbb{R}^n, u \in \mathbb{R}^m, \\ y = Cx, & y \in \mathbb{R}^p. \end{cases} \qquad (10.38)$$

Its Kalman decomposition form is

$$\begin{bmatrix} \dot{z}^1 \\ \dot{z}^2 \\ \dot{z}^3 \\ \dot{z}^4 \end{bmatrix} = \begin{bmatrix} A_{11} & 0 & A_{13} & 0 \\ A_{21} & A_{22} & A_{23} & A_{24} \\ 0 & 0 & A_{33} & 0 \\ 0 & 0 & A_{33} & A_{34} \end{bmatrix} \begin{bmatrix} z^1 \\ z^2 \\ z^3 \\ z^4 \end{bmatrix}, \qquad (10.39)$$

$$y(t) = \begin{bmatrix} C_1 & 0 & C_3 & 0 \end{bmatrix} z.$$

Its minimum realization is then

$$\begin{cases} \dot{z}^1 = A_{11}z^1, \\ y = C_1 z^1. \end{cases} \qquad (10.40)$$

We define the minimum realization of the system (10.7) in an analogous way.

Definition 10.8 Consider the Boolean control network (10.7) with Kalman decomposition (10.33). Given a fixed (frozen) value $z^3 = z_0^3$, the minimum realization of the system (10.7) with frozen $z^3 = z_0^3$ is defined by

$$\begin{cases} z^1(t+1) = F^1(z^1(t), A_3^t z_0^3, u(t)), \\ y_s(t) = h_s(z^1(t), A_3^t z_0^3), & s = 1, 2, \ldots, p, \end{cases} \qquad (10.41)$$

where A_3, as the structure matrix of F^3, is a known $n_3 \times n_3$ logical matrix, and z_0^3 is an adjustable parameter.

Note that, in general, the minimum realization depends on A_3 and z_0^3. In the following two cases the minimum realization is unique:

- Case 1. z^3 does not appear in the dynamical equation of z^1.
- Case 2. The subsystem of z^3 globally converges to ξ. In (10.41), we can then replace $A_3^t z_0^3$ by ξ and call (10.41) the stationary state realization.

Example 10.7 Recall Example 10.4. To obtain the minimum realization of (10.34), we write the first block equation by using its Kalman decomposition form (10.37):

$$\begin{cases} z_1(t+1) = z_4(t) \leftrightarrow u, \\ y_1(t) = \neg z_4(t), \\ y_2(t) = z_1(t) \to z_4(t). \end{cases} \qquad (10.42)$$

Note that in (10.37) the third block variable is $z^3 = z_4$. Since $z_4 = M_n^t z_4^0$ we have the minimum realization

$$\begin{cases} z_1(t+1) = M_e M_n^t z_4^0 u, \\ y_1(t) = M_n^{t+1} z_4^0, \\ y_2(t) = M_i z_1(t) M_n^t z_4^0. \end{cases} \tag{10.43}$$

It is easy to verify that the input–output mapping of the system (10.34) with initial value $(z_1^0, \ldots, z_6^0)$ is exactly the same as that of (10.43) with initial value z_1^0 and parameter z_4^0.

References

1. Cheng, D., Li, Z., Qi, H.: Realization of Boolean control networks. Automatica **46**(1), 62–69 (2010)
2. Wonham, W.: Linear Multivariable Control: A Geometric Approach, 2nd edn. Springer, Berlin (1979)

Chapter 11
Stability and Stabilization

11.1 Boolean Matrices

Let $\mathscr{D} = \{1, 0\}$. We recall the definition of a Boolean matrix and define its operators first.

Definition 11.1

1. A Boolean matrix $X = (x_{ij})$ is an $m \times n$ matrix with entries $x_{ij} \in \mathscr{D}$. When $n = 1$ it is called a Boolean vector. The set of $m \times n$ Boolean matrices is denoted by $\mathscr{B}_{m \times n}$.
2. Let $X = (x_{ij}) \in \mathscr{B}_{m \times n}$. If σ is an unary logical operator, then $\sigma X = (\sigma x_{ij})$.
3. Let $X = (x_{ij}), Y = (y_{ij}) \in \mathscr{B}_{m \times n}$. If σ is a binary logical operator, then $X \sigma Y := (x_{ij} \sigma y_{ij})$.
4. Let $\alpha \in \mathscr{D}$ and $X = (x_{ij}) \in \mathscr{B}_{m \times n}$. If σ is a binary logical operator, then $\alpha \sigma X := (\alpha \sigma x_{ij})$.

The follow example illustrates the operations between Boolean matrices.

Example 11.1

1. Let

$$X = \begin{bmatrix} 1 & 0 & 1 \\ 0 & 1 & 1 \end{bmatrix}.$$

Then,

$$\neg X = \begin{bmatrix} 0 & 1 & 0 \\ 1 & 0 & 0 \end{bmatrix}.$$

2. Let X be as in part 1 and

$$Y = \begin{bmatrix} 1 & 1 & 0 \\ 0 & 0 & 1 \end{bmatrix}.$$

D. Cheng et al., *Analysis and Control of Boolean Networks*,
Communications and Control Engineering,
DOI 10.1007/978-0-85729-097-7_11, © Springer-Verlag London Limited 2011

Then,

$$X \vee Y = \begin{bmatrix} 1 & 1 & 1 \\ 0 & 1 & 1 \end{bmatrix}, \qquad X \bar{\vee} Y = \begin{bmatrix} 0 & 1 & 1 \\ 0 & 1 & 0 \end{bmatrix}.$$

3. Let X and Y be as in part 2, $a = 1$ and $b = 0$. Then,

$$(a \bar{\vee} X) \vee (b \leftrightarrow Y) = \begin{bmatrix} 1 & 0 & 1 \\ 0 & 1 & 1 \end{bmatrix} \vee \begin{bmatrix} 0 & 0 & 1 \\ 1 & 1 & 0 \end{bmatrix} = \begin{bmatrix} 1 & 0 & 1 \\ 1 & 1 & 1 \end{bmatrix}.$$

Next, we consider the scalar product and matrix product.

Definition 11.2

1. Let $\alpha \in \mathscr{D}$. The scalar product of α with $X \in \mathscr{B}_{m \times n}$ is

$$\alpha X = X\alpha := \alpha \wedge X. \tag{11.1}$$

In particular, let $\alpha, \beta \in \mathscr{D}$. Then, $\alpha\beta = \alpha \wedge \beta$, which is the same as the conventional real number product.
2. Let $X = (x_{ij}) \in \mathscr{B}_{p \times q}$ and $Y \in \mathscr{B}_{m \times n}$ be two Boolean matrices. Then,

$$X \otimes Y = (x_{ij}Y) \in \mathscr{B}_{pm \times qn}. \tag{11.2}$$

3. Let $\alpha, \beta, \alpha_i \in \mathscr{D}$, $i = 1, 2, \dots, n$. Boolean addition is defined as follows:

$$\begin{cases} \alpha +_{\mathscr{B}} \beta := \alpha \vee \beta, \\ \sum_{\mathscr{B}}{}_{i=1}^{n} \alpha_i := \alpha_1 \vee \alpha_2 \vee \cdots \vee \alpha_n. \end{cases} \tag{11.3}$$

4. Let $X = (x_{ij}) \in \mathscr{B}_{m \times n}$ and $Y = (y_{ij}) \in \mathscr{B}_{n \times p}$. The Boolean product of Boolean matrices is then defined as

$$X \ltimes_{\mathscr{B}} Y := Z \in \mathscr{B}_{m \times p}, \tag{11.4}$$

where

$$z_{ij} = \sum_{k=1}^{n}{}_{\mathscr{B}} x_{ik} y_{kj}, \quad i = 1, \dots, m; \, j = 1, \dots, p.$$

5. Let $A \prec_t B$ $(A \succ_t B)$. The Boolean product of A, B is then defined as

$$A \ltimes_{\mathscr{B}} B := (A \otimes I_t) \ltimes_{\mathscr{B}} B \quad \left[A \ltimes_{\mathscr{B}} B := A \ltimes_{\mathscr{B}} (B \otimes I_t) \right]. \tag{11.5}$$

6. Assume that $A \ltimes_{\mathscr{B}} A$ is well defined. Boolean powers are then defined as follows:

$$A^{(k)} := \underbrace{A \ltimes_{\mathscr{B}} A \ltimes_{\mathscr{B}} \cdots \ltimes_{\mathscr{B}} A}_{k}.$$

Note that $\ltimes_{\mathscr{B}}$ may be omitted when there is no possible confusion.

We give an example.

Example 11.2 Let

$$A = \begin{bmatrix} 1 & 1 \\ 0 & 1 \\ 0 & 0 \end{bmatrix}, \qquad B = \begin{bmatrix} 0 & 1 \\ 1 & 0 \\ 0 & 1 \end{bmatrix}, \qquad C = \begin{bmatrix} 0 & 1 \\ 1 & 0 \\ 1 & 0 \\ 0 & 1 \end{bmatrix}.$$

Then,

$$A +_{\mathscr{B}} B = A \vee B = \begin{bmatrix} 1 & 1 \\ 1 & 1 \\ 0 & 1 \end{bmatrix}, \qquad A \to B = \begin{bmatrix} 0 & 1 \\ 1 & 0 \\ 1 & 1 \end{bmatrix},$$

$$A \leftrightarrow B = \begin{bmatrix} 0 & 1 \\ 0 & 0 \\ 1 & 0 \end{bmatrix}, \qquad A \bar{\vee} B = \begin{bmatrix} 1 & 0 \\ 1 & 1 \\ 0 & 1 \end{bmatrix},$$

$$A \ltimes_{\mathscr{B}} C = \begin{bmatrix} 1 & 1 \\ 1 & 1 \\ 1 & 0 \\ 0 & 1 \\ 0 & 0 \\ 0 & 0 \end{bmatrix}, \qquad B \ltimes_{\mathscr{B}} C = \begin{bmatrix} 1 & 0 \\ 0 & 1 \\ 1 & 0 \\ 0 & 1 \\ 1 & 0 \\ 0 & 1 \end{bmatrix}.$$

Next, we define some relations between matrices in $\mathscr{B}_{m \times n}$.

Definition 11.3 Let $X = (x_{ij})$, $Y = (y_{ij}) \in \mathscr{B}_{m \times n}$.

1. We write $X \leq Y$ if $x_{ij} \leq y_{ij}$, $\forall i, j$.
2. The vector distance between X and Y, denoted by $D_v(X, Y)$, is defined as

$$D_v(X, Y) = X \bar{\vee} Y. \tag{11.6}$$

Since both the Boolean product and Boolean addition are order-preserving, it is easy to verify the following properties, which generalize the corresponding results (for the vector case) in [2].

Proposition 11.1 *Assume $A \geq B$ and $C \geq E$. Then:*

1.

$$A +_{\mathscr{B}} C \geq B +_{\mathscr{B}} E. \tag{11.7}$$

2. (*As long as the product is well defined*)

$$A \ltimes_{\mathscr{B}} C \geq B \ltimes_{\mathscr{B}} E. \tag{11.8}$$

Proposition 11.2 *Let $X, Y, Z \in \mathcal{B}_{m \times n}$. Vector distance $D_v(X, Y)$ satisfies*

$$\begin{cases} D_v(X, Y) = 0 \Leftrightarrow X = Y, \\ D_v(X, Y) = D_v(Y, X), \\ D_v(X, Z) \leq D_v(X, Y) +_{\mathcal{B}} D_v(Y, Z). \end{cases} \tag{11.9}$$

Finally, we consider the Boolean product of matrices. For simplicity, the "$\ltimes_{\mathcal{B}}$" is omitted.

Proposition 11.3 *Let $A, B \in \mathcal{B}_{m \times n}, C \in \mathcal{B}_{n \times p}$, and $E \in \mathcal{B}_{q \times m}$. Then:*

1.

$$D_v(AC, BC) \leq D_v(A, B)C. \tag{11.10}$$

2.

$$D_v(EA, EB) \leq E D_v(A, B). \tag{11.11}$$

Proof We prove (11.10) only, the proof of (11.11) being identical. By definition, we only have to prove that for any $1 \leq i \leq n$ and $1 \leq j \leq n$,

$$\sum_{k=1}^{n}{}_{\mathcal{B}} a_{ik} c_{kj} \bar{\vee} \sum_{k=1}^{n}{}_{\mathcal{B}} b_{ik} c_{kj} \leq \sum_{k=1}^{n}{}_{\mathcal{B}} (a_{ik} \bar{\vee} b_{ik}) c_{kj}. \tag{11.12}$$

Now, the right-hand side equals zero ($RHS = 0$) if and only if

$$\text{either } a_{ik} = b_{ik} \text{ or } c_{kj} = 0, \quad \forall k. \tag{11.13}$$

However, when either one of (11.13) holds, it is easy to check that the left-hand side is also zero ($LHS = 0$). The conclusion follows. □

Note that when C is a vector, that is, when $X \in \mathcal{B}_{n \times 1}$, equation (11.10) becomes

$$D_v(AX, BX) \leq D_v(A, B)X, \tag{11.14}$$

which is particularly useful. The follow example shows that the inequality is sometimes a strict inequality.

Example 11.3 Let

$$A = \begin{bmatrix} 1 & 0 & 1 & 0 \end{bmatrix}, \qquad B = \begin{bmatrix} 1 & 1 & 1 & 1 \end{bmatrix},$$

and

$$X = \begin{bmatrix} 1 & 0 & 0 & 1 \end{bmatrix}^{\mathrm{T}}.$$

Then,

$$AX \bar{\vee} BX = \left(\begin{bmatrix} 1 & 0 & 1 & 0 \end{bmatrix} \begin{bmatrix} 1 \\ 0 \\ 0 \\ 1 \end{bmatrix} \right) \bar{\vee} \left(\begin{bmatrix} 1 & 1 & 1 & 1 \end{bmatrix} \begin{bmatrix} 1 \\ 0 \\ 0 \\ 1 \end{bmatrix} \right) = 1 \bar{\vee} 1 = 0,$$

$$(A \bar{\vee} B)X = \begin{bmatrix} 0 & 1 & 0 & 1 \end{bmatrix} \begin{bmatrix} 1 \\ 0 \\ 0 \\ 1 \end{bmatrix} = 1,$$

that is,

$$D_{\mathrm{v}}(AX, BX) < D_{\mathrm{v}}(A, B)X.$$

One may ask when the inequalities (11.10) and (11.11) become equalities. In fact, we have the following result, which will be useful in the sequel.

Proposition 11.4 *If $A, B \in \mathcal{B}_{m \times n}$, $C \in \mathcal{L}_{n \times p}$, and $E^t \in \mathcal{L}_{m \times q}$, then*

$$D_{\mathrm{v}}(AC, BC) = D_{\mathrm{v}}(A, B)C \tag{11.15}$$

and

$$D_{\mathrm{v}}(EA, EB) = ED_{\mathrm{v}}(A, B). \tag{11.16}$$

Proof We prove (11.15) only. By definition, it is enough to prove it for $C \in \mathcal{L}_{n \times 1}$. That is, we can assume $C = \delta_n^i$. Then, (11.15) becomes

$$\mathrm{Col}_i(A) \bar{\vee} \mathrm{Col}_i(B) = \mathrm{Col}_i(A \bar{\vee} B),$$

which is obviously true.   □

11.2 Global Stability

In this section we investigate the global stability of a Boolean network, that is, the existence of a fixed point as a global attractor. Equivalently, we consider when a Boolean dynamics converges globally. Our basic tool will be the vector distance. This section is a generalization of the corresponding results in [2].

Consider a Boolean network,

$$\begin{cases} x_1(t+1) = f_1(x_1, \ldots, x_n), \\ x_2(t+1) = f_2(x_1, \ldots, x_n), \\ \vdots \\ x_n(t+1) = f_n(x_1, \ldots, x_n), \quad x_i \in \mathcal{D}, \end{cases} \tag{11.17}$$

or a Boolean control network,

$$\begin{cases} x_1(t+1) = f_1(x_1, \ldots, x_n, u_1, \ldots, u_m), \\ x_2(t+1) = f_2(x_1, \ldots, x_n, u_1, \ldots, u_m), \\ \vdots \\ x_n(t+1) = f_n(x_1, \ldots, x_n, u_1, \ldots, u_m). \end{cases} \tag{11.18}$$

Denote by $\mathscr{X} = \mathscr{D}^n$ their state space. A point $X \in \mathscr{X}$ is expressed as $X = (x_1, \ldots, x_n)^{\mathrm{T}}$. We consider a logical mapping $F : \mathscr{X} \to \mathscr{X}$, which is described as

$$\begin{cases} z_1 = f_1(x_1, \ldots, x_n), \\ \vdots \\ z_n = f_n(x_1, \ldots, x_n). \end{cases} \tag{11.19}$$

We may also express this compactly as

$$Z = F(X), \quad \text{where } X, Z \in \mathscr{X}. \tag{11.20}$$

This mapping may come from the Boolean network (11.17), that is, we may have

$$X_{t+1} = F(X_t). \tag{11.21}$$

Theorem 11.1 *If $X, Y \in \mathscr{X}$, $F : \mathscr{X} \to \mathscr{X}$, then*

$$D_{\mathrm{v}}\big(F(X), F(Y)\big) \le \mathscr{I}(F) \ltimes_{\mathscr{B}} D_{\mathrm{v}}(X, Y), \tag{11.22}$$

where $\mathscr{I}(F)$ is the incidence matrix of F.

Proof Let

$$\mathscr{I}(F) = (b_{ij}) \in \mathscr{B}_{n \times n}.$$

Using the triangle inequality, we have

$$\begin{aligned} D_{\mathrm{v}}\big(f_i(X), f_i(Y)\big) &\le D_{\mathrm{v}}\big(f_i(x_1, \ldots, x_n), f_i(y_1, x_2, \ldots, x_n)\big) \\ &\quad +_{\mathscr{B}} D_{\mathrm{v}}\big(f_i(y_1, x_2, \ldots, x_n), f_i(y_1, y_2, x_3, \ldots, x_n)\big) \\ &\quad +_{\mathscr{B}} \cdots \\ &\quad +_{\mathscr{B}} D_{\mathrm{v}}\big(f_i(y_1, \ldots, y_{n-1}, x_n), f_i(y_1, \ldots, y_n)\big) \\ &\le \sum_{k=1}^{n} {}_{\mathscr{B}} b_{i,k} D_{\mathrm{v}}(x_k, y_k). \end{aligned}$$

The conclusion then follows.   □

Note that the same argument shows that (11.22) is also true for the general case $F : \mathscr{D}^n \to \mathscr{D}^m$.

Theorem 11.2 *For a mapping* $F : \mathscr{X} \to \mathscr{X}$, *where* $\mathscr{X} = \mathscr{D}^n$, *if there exists a matrix* $M \in \mathscr{B}_{n \times n}$ *such that*

$$D_v\big(F(X), F(Y)\big) \leq M \ltimes_{\mathscr{B}} D_v(X, Y), \quad \forall X, Y \in \mathscr{X}, \tag{11.23}$$

then

$$\mathscr{I}(F) \leq M.$$

Proof We prove this by contradiction. Suppose that there exists an M satisfying (11.23) and that there is an entry $m_{ij} < b_{ij}$. It follows that $m_{ij} = 0$ and $b_{ij} = 1$. Now, since f_i depends on x_j, we can find $X = (x_1, \ldots, x_j, \ldots, x_n)$ and $Y = (x_1, \ldots, y_j, \ldots, x_n)$ such that $f_i(X) \neq f_i(Y)$. That is,

$$D_v\big(f_i(X), f_i(Y)\big) = 1.$$

However, using (11.23) we have

$$D_v\big(f_i(X), f_i(Y)\big) \leq \sum_{k \neq j} {}_{\mathscr{B}} m_{ik} D_v(x_k, x_k) +_{\mathscr{B}} m_{ij} D_v(x_j, y_j) = 0,$$

which is absurd. □

Theorem 11.3 *If* $E, F : \mathscr{X} \to \mathscr{X}$ *are logical mappings, then*

$$\mathscr{I}(E \circ F) \leq \mathscr{I}(E) \ltimes_{\mathscr{B}} \mathscr{I}(F). \tag{11.24}$$

Proof For any $X, Y \in \mathscr{X}$

$$D_v\big(E \circ F(X), E \circ F(Y)\big) \leq \mathscr{I}(E) \ltimes_{\mathscr{B}} d\big(F(X), F(Y)\big)$$
$$\leq \mathscr{I}(E) \ltimes_{\mathscr{B}} \mathscr{I}(F) \ltimes_{\mathscr{B}} D_v(x, y).$$

The conclusion follows from Theorem 11.2. □

An immediate application of the above theorem is the following result.

Corollary 11.1 *If* ξ *is a fixed point of* (11.17), *then*

$$D_v\big(X(k), \xi\big) \leq \big[\mathscr{I}(F)\big]^{(k)} \ltimes_{\mathscr{B}} D_v\big(x(0), \xi\big). \tag{11.25}$$

Particularly, if

$$\mathrm{Col}_\lambda\big(\big[\mathscr{I}(F)\big]^{(k)}\big) = 0, \quad \lambda \in \Lambda := \{j_1, \ldots, j_s\} \subset \{1, 2, \ldots, n\},$$

and

$$x_\alpha(0) = \xi_\alpha, \quad \forall \alpha \notin \{j_1, \ldots, j_s\},$$

then $X(t) = \xi, t \geq k$.

Definition 11.4 The system (11.17) is said to be globally stable if it is globally convergent. In other words, it has a fixed point as a global attractor (equivalently, the only attractor).

Example 11.4 Consider the system

$$\begin{cases} x_1(t+1) = f_1(x_2(t), x_3(t)), \\ x_2(t+1) = f_2(x_4(t), \\ x_3(t+1) = c_0, \\ x_4(t+1) = f_4(x_3), \end{cases} \tag{11.26}$$

where f_1, f_2, and f_3 can be any logical functions, and c_0 is a logical constant. The incidence matrix of F is

$$\mathscr{I}(F) = \begin{bmatrix} 0 & 1 & 1 & 0 \\ 0 & 0 & 0 & 1 \\ 0 & 0 & 0 & 0 \\ 0 & 0 & 1 & 0 \end{bmatrix}.$$

It is easy to check that $[\mathscr{I}(F)]^{(4)} = 0$. If we assume 0 to be a fixed point of the system (11.26), then the system globally converges to 0.

Summarizing the above arguments, we have the following.

Proposition 11.5 *Assume that 0 is a fixed point of F and that there exists an integer $k > 0$ such that*

$$[\mathscr{I}(F)]^{(k)} = 0. \tag{11.27}$$

Then, 0 is the global attractor.

Note that if $x_e = (e_1, e_2, \dots, e_n)$ is a fixed point of the system (11.17), then the above method is still useful for testing whether x_e is a global attractor. Consider the coordinate transformation

$$z_i = \begin{cases} x_i, & e_i = 0, \\ \neg x_i, & e_i = 1. \end{cases} \tag{11.28}$$

It is now easy to convert the system (11.17) into a system under z as

$$z(t+1) = \tilde{F}(z(t)). \tag{11.29}$$

If there exists a $k > 0$ such that $[\mathscr{I}(\tilde{F})]^{(k)} = 0$, then x_e is a global attractor of the system (11.17).

It is easy to prove the following result.

Proposition 11.6 *For a Boolean matrix $H \in \mathcal{B}_{n \times n}$ the following are equivalent:*

(i) *There exists a $k > 0$ such that $H^{(k)} = 0$.*
(ii) *There exists a permutation matrix P such that $P^T \ltimes_{\mathcal{B}} H \ltimes_{\mathcal{B}} P$ is a strictly lower triangular (equivalently, upper triangular) matrix.*

In fact, when $H = \mathscr{I}(F)$ is an incidence matrix, P brings about a reordering variables.

Unfortunately, this method is sufficient but not necessary, as demonstrated by the following example.

Example 11.5 Consider the system

$$\begin{cases} x_1(t+1) = x_1(t) \wedge x_2(t), \\ x_2(t+1) = x_1(t) \wedge (\neg x_2(t)). \end{cases} \tag{11.30}$$

It is easy to check that 0 is its global attractor. However, its incidence matrix is

$$\mathscr{I}(F) = \begin{bmatrix} 1 & 1 \\ 1 & 1 \end{bmatrix},$$

and

$$\left[\mathscr{I}(F)\right]^{(k)} = \mathscr{I}(F) \neq 0, \quad k \geq 1.$$

So, what is the necessary and sufficient condition for a Boolean network to be globally convergent? We have the following result.

Theorem 11.4 *The Boolean network (11.17) is globally convergent if and only if there exists a $k > 0$ such that*

$$\mathscr{I}(F^k) = 0. \tag{11.31}$$

Proof (Necessity) If the system is globally convergent, then after T_t steps (where T_t is the transient period) all the states converge to the global attractor ξ. Therefore, when $k \geq T_t$, (11.31) is true.

(Sufficiency) Now, assume (11.31) to be true. Then, for any X we have that $F^k(X)$ is constant, say

$$F^k(X) = \xi, \quad \forall X \in \mathscr{D}^n.$$

Then, for any number $t \geq k$,

$$F^t(X) = F^k\left(F^{t-k}(X)\right) = \xi. \qquad \square$$

Remark 11.1

1. Proposition 11.5 and the method immediately following it are practically useful because the size of the incidence matrix is $n \times n$, which is of the order of $O(n)$.

2. In Theorem 11.3, F^k is not directly computable. It can only be calculated by the algebraic form of F, say L_F, which is of size $2^n \times 2^n$, so it is difficult to use if n is not small.

3. According to Theorem 11.3 it is clear that

$$\mathscr{I}\left(F^k\right) \leq \left[\mathscr{I}(F)\right]^{(k)}, \quad k \geq 1, \tag{11.32}$$

but they are not generally equal.

Definition 11.5 Let $F : \mathscr{D}^n \to \mathscr{D}^m$. F is called a constant mapping if there exists a constant $Z_0 \in \mathscr{D}^m$ such that

$$F(X) = Z_0, \quad \forall X \in \mathscr{D}^n. \tag{11.33}$$

It is easy to verify the following results, which follow directly from the definitions .

Proposition 11.7

1. F is a constant mapping if and only if its structure matrix, M_F, satisfies (for a fixed $z_0 \in \Delta_{2^m}$)

$$\mathrm{Col}_i(M_F) = z_0, \quad 1 \leq i \leq 2^n. \tag{11.34}$$

2. $\mathscr{I}(F) = 0$ if and only if F is a constant mapping.

Recall Proposition 11.5. In fact, the condition that 0 is a fixed point is not necessary for stability. Since we consider the topology of the network state space to be discrete, stability means global convergence. Because, from (11.32), the condition (11.27) ensures that F^s is constant for $s \geq k$, say $F^s(X) = \xi$, $\forall X$ and $s \geq k$, it follows that the system globally converges to ξ. We present this as a corollary.

Corollary 11.2 *Consider system* (11.17). *It is globally stable if condition* (11.27) *holds.*

Proposition 11.5 is one of the main tools for stability analysis and stabilizer design, so some further discussion is necessary. First, we would like to point out that the incidence matrix $\mathscr{I}(F)$ of a Boolean network is coordinate-dependent. The following example shows this.

Example 11.6 Consider the following system:

$$\begin{cases} x_1(t+1) = [x_1(t) \wedge (x_2(t) \, \bar{\vee} \, x_3(x))] \vee (\neg x_1(t) \wedge x_3(t)), \\ x_2(t+1) = [x_1(t) \wedge (\neg x_2(t))] \vee (\neg x_1 \wedge x_2), \\ x_3(t+1) = [x_1(t) \wedge (\neg(x_2(t) \wedge x_3(t)))] \vee [\neg x_1(t) \wedge (x_2(t) \vee x_3(t))]. \end{cases} \tag{11.35}$$

We can express it compactly as

$$x(t+1) = F\big(x(t)\big).$$

It is easy to check that 0 is a fixed point of (11.35). The incidence matrix of this system is

$$\mathscr{I}(F) = \begin{bmatrix} 1 & 1 & 1 \\ 1 & 1 & 0 \\ 1 & 1 & 1 \end{bmatrix}.$$

There is no way to convert this into a strictly lower triangular form by reordering the variables. In algebraic form it is easy to calculate that system (11.35) can be expressed as

$$x(t+1) = Lx(t), \tag{11.36}$$

where $x(t) = x_1(t)x_2(t)x_3(t)$ and

$$L = \delta_8[8\,3\,1\,5\,1\,5\,3\,8].$$

We now consider a coordinate transformation:

$$\begin{cases} z_1 = [x_1 \wedge \neg(x_3)] \vee [(\neg x_1) \wedge (x_2 \,\bar{\vee}\, x_3)], \\ z_2 = [x_1 \wedge (x_2 \,\bar{\vee}\, x_3)] \vee [(\neg x_1) \wedge x_3], \\ z_3 = x_2. \end{cases} \tag{11.37}$$

In vector form, we can easily calculate that

$$z = z_1 z_2 z_3 = Tx,$$

where

$$T = \delta_8[7\,1\,6\,4\,5\,3\,2\,8].$$

In coordinate frame z, we then have

$$z(t+1) = TLT^{\mathrm{T}}z(t) := \tilde{L}z(t), \tag{11.38}$$

where $\tilde{L}$ is

$$\tilde{L} = \delta_8[6\,6\,5\,5\,7\,7\,8\,8].$$

Recall that a sequence of 2×2^n matrices, called retrievers, were defined in Chap. 7 as

$$S_k^n = \delta_2[\underbrace{1\cdots1}_{2^{n-k}}\underbrace{2\cdots2}_{2^{n-k}}\cdots\underbrace{1\cdots1}_{2^{n-k}}\underbrace{2\cdots2}_{2^{n-k}}], \quad k = 1,\ldots,n. \tag{11.39}$$

Using these, a procedure was proposed in Chap. 7 to recover a system from the transition matrix of its algebraic form (11.38). Next, we recover the system from $\tilde{L}$.

Using retrievers S_i^3, $i = 1, 2, 3$, we have

$$\begin{cases} z_1(t+1) = M_1 z(t), \\ z_2(t+1) = M_2 z(t), \\ z_3(t+1) = M_3 z(t), \end{cases} \tag{11.40}$$

where

$$M_1 = S_1^3 \tilde{L} = \delta_2[2\,2\,2\,2\,2\,2\,2\,2],$$

$$M_2 = S_2^3 \tilde{L} = \delta_2[1\,1\,1\,1\,2\,2\,2\,2],$$

$$M_3 = S_3^3 \tilde{L} = \delta_2[2\,2\,1\,1\,1\,1\,2\,2].$$

It is easy to convert the componentwise algebraic form (11.40) back to logical form, denoted by $z(t+1) = \tilde{F}(z(t))$, as

$$\begin{cases} z_1(t+1) = 0, \\ z_2(t+1) = z_1(t), \\ z_3(t+1) = z_1(t) \,\bar{\vee}\, z_2(t). \end{cases} \tag{11.41}$$

Now, consider the system (11.41) [i.e., the system (11.35) under the coordinates z]. Its incidence matrix is

$$\mathscr{I}(\tilde{F}) = \begin{bmatrix} 0 & 0 & 0 \\ 1 & 0 & 0 \\ 1 & 1 & 0 \end{bmatrix},$$

which is strictly lower triangular. Since $x_1 = x_2 = x_3 = 0$ is a fixed point of (11.35), we conclude that the system (11.35) globally converges to zero.

Example 11.6 shows that in some cases a coordinate transformation can help to find a nice incidence matrix to ensure global convergence.

A question which now naturally arises is: If a network is globally stable, can we always find a coordinate transformation such that under the new coordinate frame the system has a strict triangular form? Unfortunately, the answer is "no". Let us return to Example 11.5. Since $n = 2$ there are $2^2! = 24$ coordinate transformations. We list them in increasing order as

$$T_1 = I_2, \qquad T_2 = \delta_4[1\,2\,4\,3], \qquad T_3 = \delta_4[1\,3\,2\,4], \dots, \qquad T_{24} = \delta_4[4\,3\,2\,1].$$

It follows that under the new coordinate frames we have

$$T_2 : \begin{cases} z_1 = x_1, \\ z_2 = x_1 \leftrightarrow x_2 \end{cases} \implies \begin{cases} z_1(t+1) = z_1(t) \wedge z_2(t), \\ z_2(t+1) = \neg z_1(t). \end{cases}$$

$$T_3 : \begin{cases} z_1 = x_2, \\ z_2 = x_1 \leftrightarrow x_2 \end{cases} \implies \begin{cases} z_1(t+1) = z_1(t) \wedge z_2(t), \\ z_2(t+1) = (\neg z_1(t)) \wedge z_2(t). \end{cases}$$

$$\vdots$$

$$T_{24} : \begin{cases} z_1 = \neg x_1, \\ z_2 = x_2 \end{cases} \implies \begin{cases} z_1(t+1) = z_1(t) \wedge z_2(t), \\ z_2(t+1) = z_2(t) \rightarrow z_1(t). \end{cases}$$

We have a variety of forms, but unfortunately none has an incidence matrix in strictly triangular form. Hence, when the condition of Proposition 11.5 fails, even under all possible coordinate transformations, we have to invoke Theorem 11.4.

11.3 Stabilization of Boolean Control Networks

Consider the Boolean control network (11.18). As before, we use the notation $x(t) = \ltimes_{i=1}^{n} x_i(t)$ and $u(t) = \ltimes_{i=1}^{m} u_i(t)$.

Definition 11.6 The global stabilization problem for the system (11.18) is to find, if possible, $u(t)$ such that the system becomes globally convergent. If $u(t) = Wx(t)$ consists of a set of logical functions, then the control is called the state feedback control.

Proposition 11.5 and the arguments thereafter are the main tools used in this section.

We first give a simple example.

Example 11.7 Consider the following system:

$$\begin{cases} x_1(t+1) = x_3(t) \vee u(t), \\ x_2(t+1) = \neg x_1(t), \\ x_3(t+1) = x_1(t) \leftrightarrow x_2(t). \end{cases} \tag{11.42}$$

It is obvious that as long as we can delete $x_3(t)$ by using control u, the system is globally stable because the incidence matrix becomes strictly lower triangular. This is easily done. We may choose either an open-loop control $u(t) = 1$ or a closed-loop control $u(t) = \neg x_3(t)$.

To obtain a general design method, we first recall the expression of logical state variables. Let $x_1, \ldots, x_n$ be n logical state variables. In scalar form we have $x_i \in \mathcal{D}$, $i = 1, \ldots, n$, and in vector form we write $x = \ltimes_{i=1}^{n} x_i \in \Delta_{2^n}$.

Define a set of vectors:

$$s_k^n = [\underbrace{1 \cdots 1}_{2^{n-k}} \underbrace{0 \cdots 0}_{2^{n-k}} \cdots \underbrace{1 \cdots 1}_{2^{n-k}} \underbrace{0 \cdots 0}_{2^{n-k}}], \quad k = 1, \ldots, n. \tag{11.43}$$

Remark 11.2

1. $s_k^n \in \mathcal{B}_{1 \times 2^n}$, $k = 1, \ldots, n$, so the logical operators are applicable to them.

2. Comparing (11.43) with (11.39), we see that s_k^n can be obtained from the first row of S_k^n by replacing 2 by 0.

We then define a matrix as

$$\mathscr{S}^n = \begin{bmatrix} s_1^n \\ s_2^n \\ \vdots \\ s_n^n \end{bmatrix} \in \mathscr{B}_{n \times 2^n}.$$

The following proposition is easily verifiable.

Proposition 11.8

1. *Converting from scalar form to vector form, we have*

$$x = \left[(x_1 \leftrightarrow s_1^n) \wedge (x_2 \leftrightarrow s_2^n) \wedge \cdots \wedge (x_n \leftrightarrow s_n^n) \right]^{\mathrm{T}}, \quad \forall x_i \in \mathscr{D}. \tag{11.44}$$

2. *Converting from vector form to scalar form, we have*

$$X = \mathscr{S}^n x. \tag{11.45}$$

Example 11.8 Let $X = (1, 0, 1, 0)^{\mathrm{T}}$. In vector form, we then have

$$\begin{aligned} x &= \left[1 \leftrightarrow (1111111100000000) \right] \wedge \left[0 \leftrightarrow (1111000011110000) \right] \\ &\quad \wedge \left[1 \leftrightarrow (1100110011001100) \right] \wedge \left[0 \leftrightarrow (1010101010101010) \right] \\ &= (1111111100000000) \wedge (0000111100001111) \\ &\quad \wedge (1100110011001100) \wedge (0101010101010101) \\ &= (0000010000000000) = \delta_{16}^6. \end{aligned}$$

If $x = \delta_{16}^9$, then

$$X = \mathscr{S}^4 x = (0, 1, 1, 1).$$

Next, we give a systematic analysis of the stabilizer design for Boolean control networks. First, we define a mapping $\pi : \mathscr{B}_{2^n \times 2^n} \to \mathscr{B}_{n \times n}$ as

$$\begin{aligned} \pi(L) &= \left[\left[(\mathscr{S}^n L) \,\bar{\vee}\, (\mathscr{S}^n L M_n) \right] \ltimes_{\mathscr{B}} \mathbf{1}_{2^n}, \left[(\mathscr{S}^n L) \,\bar{\vee}\, (\mathscr{S}^n L)(I_2 \otimes M_n) \right] \ltimes_{\mathscr{B}} \mathbf{1}_{2^n}, \right. \\ &\quad \left. \ldots, \left[(\mathscr{S}^n L) \,\bar{\vee}\, (\mathscr{S}^n L)(I_{2^{n-1}} \otimes M_n) \right] \ltimes_{\mathscr{B}} \mathbf{1}_{2^n} \right], \quad L \in \mathscr{B}_{2^n \times 2^n}. \tag{11.46} \end{aligned}$$

Note that M_n is the structure matrix of negation.

We then have the following result concerning how to build the incidence matrix from L.

Theorem 11.5 *Consider the Boolean network (11.17) [equivalently, (11.21)], with its algebraic form*

$$x(t+1) = Lx(t), \quad x(t) \in \Delta_{2^n}, \qquad (11.47)$$

where $L \in \mathcal{L}_{2^n \times 2^n}$. The incidence matrix of F can be obtained from L by the following formula:

$$\mathcal{I}(F) = \pi(L). \qquad (11.48)$$

Proof Applying (11.45) to (11.47) yields

$$X(t+1) = \begin{bmatrix} x_1(t+1) \\ x_2(t+1) \\ \vdots \\ x_n(t+1) \end{bmatrix}$$

$$= \mathcal{S}^n L x_1(t) \cdots x_n(t)$$

$$= \mathcal{S}^n L M_n \big(\neg x_1(t)\big) x_2(t) \cdots x_n(t).$$

It is now clear that $x_i(t+1)$ is independent of $x_1(t)$ if and only if

$$\mathrm{Row}_i\big(\mathcal{S}^n L\big) = \mathrm{Row}_i\big(\mathcal{S}^n L M_n\big),$$

if and only if

$$\mathrm{Row}_i\big(\mathcal{S}^n L\big) \, \bar{\vee} \, \mathrm{Row}_i\big(\mathcal{S}^n L M_n\big) = 0,$$

if and only if

$$\mathrm{Row}_i\big[\big(\mathcal{S}^n L\big) \bar{\vee} \big(\mathcal{S}^n L M_n\big) \ltimes_{\mathcal{B}} 1_{2^n}\big] = 0.$$

Hence, the first column of $\mathcal{I}(F)$ is

$$\mathrm{Col}_1\big(\mathcal{I}(F)\big) = \big[\big(\mathcal{S}^n L\big) \bar{\vee} \big(\mathcal{S}^n L M_n\big)\big] \ltimes_{\mathcal{B}} 1_{2^n}.$$

Similarly,

$$X(t+1) = \mathcal{S}^n L x_1(t) \cdots x_j(t) \cdots x_n(t)$$

$$= \mathcal{S}^n L x_1(t) \cdots M_n\big(\neg x_j(t)\big) \cdots x_n(t)$$

$$= \mathcal{S}^n (I_{2^{j-1}} \otimes M_n) L x_1(t) \cdots \big(\neg x_j(t)\big) \cdots x_n(t).$$

A similar argument shows that $x_i(t+1)$ is independent of $x_j(t)$ if and only if

$$\mathrm{Row}_i\big[\big(\mathcal{S}^n L\big) \bar{\vee} \big(\mathcal{S}^n L M_n (I_{2^{j-1}} \otimes M_n)\big) \ltimes_{\mathcal{B}} 1_{2^n}\big] = 0.$$

Hence,

$$\mathrm{Col}_j\big(\mathcal{I}(F)\big) = \big[\big(\mathcal{S}^n L\big) \bar{\vee} \big(\mathcal{S}^n L (I_{2^{j-1}} \otimes M_n)\big)\big] \ltimes_{\mathcal{B}} 1_{2^n}, \quad j = 2, \ldots, n.$$

Equation (11.48) follows. $\square$

We give an example to depict it.

Example 11.9 Assume that the system (11.17) has as its network transition matrix

$$L = \delta_{16}[1\ 9\ 9\ 13\ 4\ 12\ 12\ 16\ 2\ 10\ 10\ 14\ 1\ 9\ 9\ 13]. \tag{11.49}$$

$\mathscr{S}^4$ can then be calculated as

$$\mathscr{S}^4 = \begin{bmatrix} 1 & 1 & 1 & 1 & 1 & 1 & 1 & 1 & 0 & 0 & 0 & 0 & 0 & 0 & 0 & 0 \\ 1 & 1 & 1 & 1 & 0 & 0 & 0 & 0 & 1 & 1 & 1 & 1 & 0 & 0 & 0 & 0 \\ 1 & 1 & 0 & 0 & 1 & 1 & 0 & 0 & 1 & 1 & 0 & 0 & 1 & 1 & 0 & 0 \\ 1 & 0 & 1 & 0 & 1 & 0 & 1 & 0 & 1 & 0 & 1 & 0 & 1 & 0 & 1 & 0 \end{bmatrix} \in \mathscr{B}_{4 \times 16}.$$

Using (11.46), we can calculate that

$$\left[(\mathscr{S}^4 L) \bar{\vee} (\mathscr{S}^4 L M_n) \right] \ltimes_{\mathscr{B}} \mathbf{1}_{2^4} = [0\ \ 0\ \ 1\ \ 1]^{\mathrm{T}},$$

$$\left[(\mathscr{S}^4 L) \bar{\vee} (\mathscr{S}^4 L (I_2 \otimes M_n)) \right] \ltimes_{\mathscr{B}} \mathbf{1}_{2^4} = [0\ \ 0\ \ 1\ \ 1]^{\mathrm{T}},$$

$$\left[(\mathscr{S}^4 L) \bar{\vee} (\mathscr{S}^4 L (I_{2^2} \otimes M_n)) \right] \ltimes_{\mathscr{B}} \mathbf{1}_{2^4} = [1\ \ 1\ \ 0\ \ 0]^{\mathrm{T}},$$

$$\left[(\mathscr{S}^4 L) \bar{\vee} (\mathscr{S}^4 L (I_{2^3} \otimes M_n)) \right] \ltimes_{\mathscr{B}} \mathbf{1}_{2^4} = [1\ \ 1\ \ 0\ \ 0]^{\mathrm{T}},$$

that is,

$$\mathscr{I}(F) = \pi(L) = \begin{bmatrix} 0 & 0 & 1 & 1 \\ 0 & 0 & 1 & 1 \\ 1 & 1 & 0 & 0 \\ 1 & 1 & 0 & 0 \end{bmatrix}.$$

In fact, using the standard procedure, we can uniquely recover the system from L as

$$\begin{cases} x_1(t+1) = x_3(t) \wedge x_4(t), \\ x_2(t+1) = x_3(t) \vee x_4(t), \\ x_3(t+1) = x_1(t) \to x_2(t), \\ x_4(t+1) = x_1(t) \leftrightarrow x_2(t). \end{cases} \tag{11.50}$$

This verifies the validity of the $\mathscr{I}(F)$ obtained .

Next, we consider the stabilization problem. Consider the system (11.18), with its algebraic form

$$x(t+1) = Lu(t)x(t), \quad x(t) \in \Delta_{2^n}, u(t) \in \Delta_{2^m}, \tag{11.51}$$

where $L \in \mathscr{L}_{2^n \times 2^{n+m}}$.

Using Propositions 11.5 and 11.6, a sufficient condition for stabilization with open-loop control is given by the following lemma.

Lemma 11.1 *The system (11.18) is stabilizable by a constant control u if $\pi(L(u))$ has a strictly lower (or upper) triangular form.*

Note that since the incidence matrix is coordinate-dependent, coordinate transformations have to be taken into consideration.

Using the formula (8.18), we have the following result.

Theorem 11.6 *The system* (11.18), *with its algebraic form* (11.51), *is stabilizable by a constant control u if there is a coordinate transformation $z = Tx$ such that $\pi(TL(I_{2^m} \otimes T^T)u)$ has a strictly lower (or upper) triangular form.*

Note that both the possible values of u and the number of possible coordinate transforms are finite (2^m and $2^n!$, respectively), so, theoretically, both Lemma 11.1 and Theorem 11.6 are verifiable.

Example 11.10 Consider the following system:

$$\begin{cases} x_1(t+1) = \neg x_2(t), \\ x_2(t+1) = \neg x_4(t) \leftrightarrow ((x_4(t) \wedge (x_2(t) \bar{\vee} x_3(t))) \vee u(t)), \\ x_3(t+1) = \neg((x_4(t) \wedge (x_2(t) \bar{\vee} x_3(t))) \vee u(t)), \\ x_4(t+1) = (x_4(t) \vee (x_2(t) \bar{\vee} x_3(t))) \wedge u(t). \end{cases} \quad (11.52)$$

After a "trial and error" approach to simplifying the system, we use the following coordinate transformation:

$$\begin{cases} z_1 = x_4, \\ z_2 = x_2 \bar{\vee} x_3, \\ z_3 = \neg x_3, \\ z_4 = \neg x_1. \end{cases} \quad (11.53)$$

Its inverse can be easily calculated as

$$\begin{cases} x_1 = \neg z_4, \\ x_2 = z_2 \leftrightarrow z_3, \\ x_3 = \neg z_3, \\ x_4 = z_1. \end{cases} \quad (11.54)$$

The system then becomes

$$\begin{cases} z_1(t+1) = (z_1(t) \vee z_2(t)) \wedge u(t), \\ z_2(t+1) = \neg z_1(t), \\ z_3(t+1) = (z_1(t) \wedge z_2(t)) \vee u(t), \\ z_4(t+1) = z_2(t) \leftrightarrow z_3(t). \end{cases} \quad (11.55)$$

It is now clear that if we choose

$$u(t) = 0, \quad (11.56)$$

then the system becomes

$$\begin{cases} z_1(t+1) = 0, \\ z_2(t+1) = \neg z_1(t), \\ z_3(t+1) = (z_1(t) \wedge z_2(t)), \\ z_4(t+1) = z_2(t) \leftrightarrow z_3(t). \end{cases} \tag{11.57}$$

It is obvious that the incidence matrix of the system (11.57) is

$$\mathscr{I}(F) = \begin{bmatrix} 0 & 0 & 0 & 0 \\ 1 & 0 & 0 & 0 \\ 1 & 1 & 0 & 0 \\ 0 & 1 & 1 & 0 \end{bmatrix},$$

which is in strictly lower triangular form. We conclude that the constant control $u(t) = 0$ stabilizes the system (11.52).

Remark 11.3 Let

$$TL(I_{2^m} \otimes T^{\mathrm{T}}) = [B_1 \ B_2 \ \cdots \ B_{2^n}],$$

where $B_i = \mathrm{Blk}_i(TL(I_{2^m} \otimes T^{\mathrm{T}})) \in \mathscr{L}_{2^n \times 2^n}$, $i = 1, \dots, 2^n$. Theorem 11.6 then becomes the statement that there exists an i such that $\pi(B_i)$ is strictly lower (upper) triangular.

Next, we consider the closed-loop control. Let $u(t)$ be a set of logical functions of $x(t)$. We can then always express it as

$$u(t) = Gx(t), \tag{11.58}$$

where $G \in \mathscr{L}_{2^m \times 2^n}$. Plugging this into (11.51) yields

$$x(t+1) = LGx^2(t) = LG\Phi_n x(t), \tag{11.59}$$

where Φ_n is defined in Chap. 4 [equation (4.6)] as

$$\Phi_n = \prod_{i=1}^{n} I_{2^{n-1}} \otimes \left[(I_2 \otimes W_{[2, 2^{n-i}]}) M_r \right], \tag{11.60}$$

with $M_r = \delta_4[1, 4]$.

The following result is now obvious.

Theorem 11.7 *The system* (11.18) *is stabilizable by a closed-loop control* $u = Gx$ *if* $\pi(LG\Phi_n)$ *has a strictly lower (or upper) triangular form. Moreover, if there exists a coordinate transformation* $z = Tx$ *such that* $\pi(TLG\Phi_n T^{\mathrm{T}})$ *has a strictly lower (or upper) triangular form, then the control also stabilizes the system.*

Example 11.11 Consider the system

$$\begin{cases} x_1(t+1) = \neg x_2(t), \\ x_2(t+1) = \neg x_4(t) \leftrightarrow ((x_4(t) \wedge (x_2(t) \bar{\vee} x_3(t))) \vee u(t)), \\ x_3(t+1) = \neg((x_4(t) \wedge (x_2(t) \bar{\vee} x_3(t))) \vee u(t)), \\ x_4(t+1) = (x_4(t) \vee (x_2(t) \bar{\vee} x_3(t))) \vee u(t). \end{cases} \quad (11.61)$$

In fact, it is obtained from (11.52) by changing the nature of the inputs. Using the same coordinate transformation as in Example 11.10, we have

$$\begin{cases} z_1(t+1) = (z_1(t) \vee z_2(t)) \vee u(t), \\ z_2(t+1) = \neg z_1(t), \\ z_3(t+1) = (z_1(t) \wedge z_2(t)) \wedge u(t), \\ z_4(t+1) = z_2(t) \leftrightarrow z_3(t). \end{cases} \quad (11.62)$$

One can check that constant (open-loop) controls cannot stabilize the system. If we use a closed-loop control

$$u(t) = \neg z_1(t) \wedge \neg z_2(t),$$

then the system (11.62) becomes

$$\begin{cases} z_1(t+1) = 1, \\ z_2(t+1) = \neg z_1(t), \\ z_3(t+1) = 0, \\ z_4(t+1) = z_2(t) \leftrightarrow z_3(t). \end{cases} \quad (11.63)$$

Obviously, this is globally stable. Converting the control back to the original coordinate frame, we conclude that

$$u(t) = \neg x_4(t) \wedge \neg (x_2(t) \bar{\vee} x_3(t))$$

stabilizes the system (11.61).

The advantage of using metric-based analysis is that the matrix involved is of a small size. The disadvantage is that the condition is only a sufficient one. Next, we search for necessary and sufficient condition.

As discussed above, the global stability of a free Boolean network is equivalent to the existence of a $k > 0$ such that $F^k =$ constant. In vector form it is equivalent to L^k having equal columns, which is the global attractor. Now, consider the stabilization by a constant control u. The control-dependent transition matrix is then Lu. Using the properties of the semi-tensor product, it is easy to calculate that

$$(Lu)^k = L[(I_{2^m} \otimes L)\Phi_m]^{k-1} u. \quad (11.64)$$

Since k should be less than or equal to the transient time, i.e., $k \leq T_t \leq 2^n$, we can get an easily verifiable necessary and sufficient condition as follows. Note that $L[(I_{2^m} \otimes L)\Phi_m]^{k-1}$ is a $2^n \times 2^{n+m}$ matrix. We split it into 2^m square blocks as

$$L[(I_{2^m} \otimes L)\Phi_m]^{k-1} := [L_1^k \; L_2^k \; \cdots \; L_{2^m}^k]. \tag{11.65}$$

Using this notation and in accordance with the above argument, we have the following necessary and sufficient condition.

Theorem 11.8 *The system* (11.18) *is stabilizable by a constant control u if and only if there exists a matrix of constant mapping*

$$L_j^k, \quad 1 \leq k \leq 2^n, 1 \leq j \leq 2^m.$$

Moreover, corresponding to each matrix of constant mapping L_j^k, the stabilizing control is $u = \delta_{2^m}^j$.

The following example illustrates this result.

Example 11.12 Consider the following system:

$$\begin{cases} x_1(t+1) = (x_1(t) \vee x_2(t)) \wedge u, \\ x_2(t+1) = (x_2(t) \wedge u) \rightarrow x_1. \end{cases} \tag{11.66}$$

It is easy to calculate that

$$L = \delta_4[1\,1\,2\,3\,3\,3\,3\,3].$$

Since $\Phi_1 = M_r$, according to Theorem 11.8, we have to calculate

$$L[(I_2 \otimes L)M_r]^k, \quad k \geq 1,$$

to see if we can find a constant mapping block. In fact when $k = 2$ we have

$$L[(I_2 \otimes L)M_r]^2 = \delta_4[1\,1\,1\,1\,3\,3\,3\,3].$$

We conclude that if we use control $u = 1$, then the system is stabilized at $x = \delta_4^1$ (i.e., $x_1 = 1$ and $x_2 = 1$), and if we use $u = 0$, the system is stabilized at $x = \delta_4^3$ (i.e., $x_1 = 0, x_2 = 1$).

Consider the state feedback control as in (11.58). Using the expression (11.59) and the above argument, the following result is obvious.

Theorem 11.9 *The system* (11.18) *is stabilizable by a closed-loop control $u = Gx$ if and only if there exists a $2^m \times 2^n$ logical matrix G and an integer $1 \leq k \leq 2^n$ such that $(LG\Phi_n)^k$ is a matrix of constant mapping.*

We give an example.

Example 11.13 Consider the following system:

$$\begin{cases} x_1(t+1) = [x_2(t) \vee (\neg x_2(t) \wedge (x_3(t) \vee x_4(t)))] \wedge u(t), \\ x_2(t+1) = (x_2(t) \wedge (x_3(t) \vee x_4(t))) \vee [x_1(t) \wedge (\neg x_2(t) \\ \qquad\qquad \wedge \neg(x_3(t) \vee x_4(t)))], \\ x_3(t+1) = (x_1(t) \wedge (x_3(t) \leftrightarrow x_4(t))) \vee [\neg x_1(t) \wedge ((x_2(t) \\ \qquad\qquad \wedge (x_3(t) \leftrightarrow x_4(t))) \wedge (\neg x_2(t) \wedge (x_3(t) \wedge x_4(t))))], \\ x_4(t+1) = x_1(t) \wedge \neg x_4(t) \vee [\neg x_1(t) \wedge ((x_2(t) \wedge \neg x_4(t)) \\ \qquad\qquad \vee (\neg x_2(t) \wedge \neg(x_3(t) \rightarrow x_4(t))))]. \end{cases} \qquad (11.67)$$

It is easy to verify that if we choose

$$G = \begin{bmatrix} 1 & 1 & 0 & 0 \\ 0 & 0 & 1 & 1 \end{bmatrix}^3,$$

then

$$(LG\Phi_4)^{14} = \delta_{16}[\underbrace{16 \cdots 16}_{16}].$$

Note that

$$u(t) = Gx(t) = x_1(t),$$

which globally stabilizes the system (11.67) to $X = (0, 0, 0, 0)$.

Next, we briefly discuss the case where the system is required to converge to a particular state x_0. In this case the problem is slightly simpler. In addition to the above stability requirements, we need to ensure that x_0 is a fixed point of the control system. We now give this as a corollary.

Corollary 11.3

1. *The system (11.18) is globally stabilized to x_0 by a constant control u if and only if u satisfies*

$$Lux_0 = x_0, \qquad (11.68)$$

   and there exists an integer $k > 0$ such that $(Lu)^k$ is a constant mapping.
2. *The system (11.18) is globally stabilized to x_0 by a closed-loop control $u = Gx$ if and only if G satisfies*

$$LG\Phi_n x_0 = x_0, \qquad (11.69)$$

   and there exists an integer $k > 0$ such that $(LG\Phi_n)^k$ is a constant mapping.

Finally, we consider the stabilization by an open-loop control, $u(t) = \ltimes_{i=1}^{m} u_i(t)$, $t = 1, 2, \ldots$. We assume that we want to stabilize it to x_0.

First, it is obvious that a necessary condition is that there exists a control $u_e \in \Delta_{2^m}$ such that

$$Lu_e x_0 = x_0. \tag{11.70}$$

Second, note that

$$x(t_0 + k + 1) = Lu(t_0 + k)Lu(t_0 + k - 1) \cdots Lu(t_0 + 1)x(t_0).$$

To make all trajectories converge to x_0, there must be a $k > 0$ such that

$$Lu(k)Lu(k - 1) \cdots Lu(1)x \equiv x_0, \quad \forall x \in \Delta_{2^n}.$$

This is equivalent to

$$\text{Col}\big(Lu(k)Lu(k - 1) \cdots Lu(1)\big) = \{x_0\}. \tag{11.71}$$

Observe that

$$Lu(k)Lu(k - 1) \cdots Lu(1)$$
$$= L(I_{2^m} \otimes L)(I_{2^{2m}} \otimes L) \cdots (I_{2^{(k-1)m}} \otimes L) \ltimes_{i=k}^{1} u(i)$$
$$:= \big[L_1^k, L_2^k, \ldots, L_{2^{km}}^k\big] \ltimes_{i=k}^{1} u(i). \tag{11.72}$$

It is now clear that if there exists a $1 \leq j \leq 2^{km}$ such that L_j^k corresponds to the constant mapping $\psi(x) \equiv x_0$, then we can choose the control

$$\ltimes_{i=k}^{1} u(i) = \delta_{2^{mn}}^{j}$$

such that (11.71) holds.

Summarizing the above arguments, we have the following theorem.

Theorem 11.10 *The system* (11.18) *is globally stabilized to x_0 by an open-loop control $u(t), t = 1, 2, \ldots$, if and only if:*

(i) *There exist an integer $k > 0$ and an L_j^k, $1 \leq j \leq 2^{km}$, such that*

$$\text{Col}(L_j^k) = \{x_0\}.$$

(ii) *There exists a $u_e \in \Delta_{2^m}$ such that* (11.70) *holds.*

We give an example for this.

Example 11.14 Consider the following system:

$$\begin{cases} x_1(t+1) = x_1(t) \vee u_1(t), \\ x_2(t+1) = (x_2(t) \vee x_3(t)) \leftrightarrow u_1(t), \\ x_3(t+1) = (u_1(t) \rightarrow x_2(t)) \vee x_3(t), \\ x_4(t+1) = (x_3(t) \wedge u_2(t)) \rightarrow x_4(t). \end{cases} \quad (11.73)$$

Set $x(t) = \ltimes_{i=1}^4 x_i(t)$ and $u(t) = \ltimes_{i=1}^2 u_i(t)$. Using vector form, (11.73) can be expressed as

$$x(t+1) = Lu(t)x(t), \quad (11.74)$$

where

$$\begin{aligned} L = \delta_{16}[&1\ 2\ 1\ 1\ 1\ 2\ 7\ 7\ 1\ \ 2\ \ 1\ \ 1\ \ 1\ \ 2\ \ 7\ 7 \\ &1\ 1\ 1\ 1\ 1\ 1\ 7\ 7\ 1\ \ 1\ \ 1\ \ 1\ \ 1\ \ 1\ \ 7\ 7 \\ &5\ 6\ 5\ 5\ 5\ 6\ 1\ 1\ 13\ 14\ 13\ 13\ 13\ 14\ 9\ 9 \\ &5\ 5\ 5\ 5\ 5\ 5\ 1\ 1\ 13\ 13\ 13\ 13\ 13\ 13\ 9\ 9]. \end{aligned}$$

According to (11.72), we calculate

$$L(I_{2^2} \otimes L)(I_{2^{2 \times 2}} \otimes L) \cdots (I_{2^{2(k-1)}} \otimes L)$$

to see whether we can find a constant mapping block. In fact when $k = 2$ we have

$$M = L(I_{2^2} \otimes L)$$

$$:= [M_1, M_2, \ldots, M_{16}],$$

where

$$\begin{aligned} M = \delta_{16}[&1\ 2\ 1\ 1\ 1\ 2\ 7\ 7\ 1\ \ 2\ \ 1\ \ 1\ \ 1\ \ 2\ \ 7\ \ 7 \\ &1\ 1\ 1\ 1\ 1\ 1\ 7\ 7\ 1\ \ 1\ \ 1\ \ 1\ \ 1\ \ 1\ \ 7\ \ 7 \\ &1\ 2\ 1\ 1\ 1\ 2\ 1\ 1\ 1\ \ 2\ \ 1\ \ 1\ \ 1\ \ 2\ \ 1\ \ 1 \\ &1\ 1\ 1\ 1\ 1\ 1\ 1\ 1\ 1\ \ 1\ \ 1\ \ 1\ \ 1\ \ 1\ \ 1\ \ 1 \\ &1\ 1\ 1\ 1\ 1\ 1\ 7\ 7\ 1\ \ 1\ \ 1\ \ 1\ \ 1\ \ 1\ \ 7\ \ 7 \\ &1\ 1\ 1\ 1\ 1\ 1\ 7\ 7\ 1\ \ 1\ \ 1\ \ 1\ \ 1\ \ 1\ \ 7\ \ 7 \\ &1\ 1\ 1\ 1\ 1\ 1\ 1\ 1\ 1\ \ 1\ \ 1\ \ 1\ \ 1\ \ 1\ \ 1\ \ 1 \\ &1\ 1\ 1\ 1\ 1\ 1\ 1\ 1\ 1\ \ 1\ \ 1\ \ 1\ \ 1\ \ 1\ \ 1\ \ 1 \\ &5\ 6\ 5\ 5\ 5\ 6\ 1\ 1\ 5\ \ 6\ \ 5\ \ 5\ \ 5\ \ 6\ \ 1\ \ 1 \\ &5\ 5\ 5\ 5\ 5\ 5\ 1\ 1\ 5\ \ 5\ \ 5\ \ 5\ \ 5\ \ 5\ \ 1\ \ 1 \\ &5\ 6\ 5\ 5\ 5\ 6\ 5\ 5\ 13\ 14\ 13\ 13\ 13\ 14\ 13\ 13 \\ &5\ 5\ 5\ 5\ 5\ 5\ 5\ 5\ 13\ 13\ 13\ 13\ 13\ 13\ 13\ 13 \\ &5\ 5\ 5\ 5\ 5\ 5\ 1\ 1\ 5\ \ 5\ \ 5\ \ 5\ \ 5\ \ 5\ \ 1\ \ 1 \\ &5\ 5\ 5\ 5\ 5\ 5\ 1\ 1\ 5\ \ 5\ \ 5\ \ 5\ \ 5\ \ 5\ \ 1\ \ 1 \\ &5\ 5\ 5\ 5\ 5\ 5\ 5\ 5\ 13\ 13\ 13\ 13\ 13\ 13\ 13\ 13 \\ &5\ 5\ 5\ 5\ 5\ 5\ 5\ 5\ 13\ 13\ 13\ 13\ 13\ 13\ 13\ 13], \end{aligned} \quad (11.75)$$

and $\mathrm{Blk}_i(M) \in \mathscr{L}_{16 \times 16}$, $i = 1, \ldots, 16$. From (11.75), we know that

$$\mathrm{Blk}_4(M) = \mathrm{Blk}_7(M) = \mathrm{Blk}_8(M) = \delta_{16}[1\ 1\ 1\ 1\ 1\ 1\ 1\ 1\ 1\ 1\ 1\ 1\ 1\ 1\ 1\ 1].$$

Equivalently,

$$\mathrm{Col}\big(\mathrm{Blk}_4(M)\big) = \mathrm{Col}\big(\mathrm{Blk}_7(M)\big) = \mathrm{Col}\big(\mathrm{Blk}_8(M)\big) = \{\delta_{16}^1\}.$$

On the other hand, choosing $u_e = \delta_4^1 \sim (1, 1)$ [or $u_e = \delta_4^2 \sim (1, 0)$], we have

$$L u_e \delta_{16}^1 = \delta_{16}^1.$$

From Theorem 11.10, the system (11.73) is globally stabilized to $x_0 = \delta_{16}^1 \sim$ (1, 1, 1, 1) by an open-loop control $u(t)$ [or $\bar{u}(t)$ and $\tilde{u}(t)$], where

$$u(t) = \begin{cases} \delta_4^4 \sim (0, 0), & t = 1, \\ \delta_4^1 \sim (1, 1), & t = 2, \\ u_e, & t \geq 3, \end{cases}$$

$$\bar{u}(t) = \begin{cases} \delta_4^3 \sim (0, 1), & t = 1, \\ \delta_4^2 \sim (1, 0), & t = 2, \\ u_e, & t \geq 3, \end{cases}$$

$$\tilde{u}(t) = \begin{cases} \delta_4^4 \sim (0, 0), & t = 1, \\ \delta_4^2 \sim (1, 0), & t = 2, \\ u_e, & t \geq 3. \end{cases}$$

In the following we would like to discuss further the conditions in Theorem 11.10. Condition (i) says that all the trajectories can reach the preassigned fixed point x_0. One may doubt whether condition (ii) (which means that x_0 is a fixed point under a certain control) is necessary. The following example shows that condition (ii) is indeed necessary.

Example 11.15 Consider the following system:

$$\begin{cases} x_1(t + 1) = \neg(x_1(t) \wedge u(t)), \\ x_2(t + 1) = (u(t) \wedge (x_2(t) \to x_1(t))) \vee (\neg u(t) \wedge (x_1(t) \leftrightarrow x_2(t))). \end{cases} \quad (11.76)$$

Set $x(t) = \ltimes_{i=1}^2 x_i(t)$. In vector form, (11.76) can be expressed as

$$x(t + 1) = L u(t) x(t)$$
$$= \delta_4[3\ 3\ 2\ 1\ 1\ 2\ 2\ 1] u(t) x(t). \quad (11.77)$$

For any initial state $\xi \in \Delta_4$, if we choose $u(1) = \delta_2^2$, then

$$x(2) = L u(1) \xi$$
$$= \delta_4[1\ 2\ 2\ 1] \xi.$$

Next, if we choose $u(2) = \delta_2^1$, then

$$
\begin{aligned}
x(3) &= Lu(2)x(2) \\
&= \left(\delta_4[3\ 3\ 2\ 1]\right)\left(\delta_4[1\ 2\ 2\ 1]\right)\xi \\
&= \delta_4[3\ 3\ 3\ 3]\xi \\
&= \delta_4^3, \quad \forall \xi \in \Delta_4.
\end{aligned}
$$

From (11.77), it is obvious that there does not exist an u_e such that

$$
Lu_e\delta_4^3 = \delta_4^3.
$$

In step 3, regardless of the value of $u(3)$ [$u(3) = \delta_2^1$ or $u(3) = \delta_2^2$], the dynamics of the Boolean network will leave the state $x_0 = \delta_4^3$ fixed. Therefore, the system (11.76) cannot be globally stabilized to x_0 by an open-loop control $u(t)$.

References

1. Cheng, D., Qi, H., Li, Z., Liu, J.B.: Stability and stabilization of Boolean networks. Int. J. Robust Nonlinear Control (2010). doi:10.1002/rnc.1581 (to appear)
2. Robert, F.: Discrete Iterations: A Metric Study. Springer, Berlin (1986). Translated by J. Rolne

Chapter 12
Disturbance Decoupling

12.1 Problem Formulation

Assume that in a Boolean control network there are some disturbance inputs. We
then have a disturbed Boolean control network. In general, its dynamics is described
as

$$
\begin{cases}
x_1(t+1) = f_1(x_1(t), \ldots, x_n(t), u_1(t), \ldots, u_m(t), \xi_1(t), \ldots, \xi_q(t)), \\
\quad\vdots \\
x_n(t+1) = f_n(x_1(t), \ldots, x_n(t), u_1(t), \ldots, u_m(t), \xi_1(t), \ldots, \xi_q(t)), \\
\quad y_j(t) = h_j(x(t)), \quad j = 1, \ldots, p,
\end{cases}
\tag{12.1}
$$

where $\xi_i(t)$, $i = 1, \ldots, q$, are disturbances. Let $x(t) = \ltimes_{i=1}^n x_i(t)$, $u(t) = \ltimes_{i=1}^m u_i(t)$,
$\xi(t) = \ltimes_{i=1}^q \xi_i(t)$, and $y(t) = \ltimes_{i=1}^p y_i(t)$. The algebraic form of (12.1) is then expressed as

$$
\begin{aligned}
x(t+1) &= Lu(t)\xi(t)x(t), \\
y(t) &= Hx(t),
\end{aligned}
\tag{12.2}
$$

where $L \in \mathscr{L}_{2^n \times 2^{n+m+q}}$, $H \in \mathscr{L}_{2^p \times 2^n}$.

We consider the following example.

Example 12.1 A disturbed Boolean control network is defined by the following
equation:

$$
\begin{cases}
A(t+1) = B(t) \wedge \xi(t), \\
B(t+1) = C(t) \vee u_1(t), \\
C(t+1) = D(t) \wedge [(B(t) \rightarrow \xi(t)) \vee u_1(t)], \\
D(t+1) = \neg C(t) \vee [\xi(t) \wedge u_2(t)], \\
y(t) = C(t) \wedge D(t).
\end{cases}
\tag{12.3}
$$

D. Cheng et al., *Analysis and Control of Boolean Networks*,
Communications and Control Engineering,
DOI 10.1007/978-0-85729-097-7_12, © Springer-Verlag London Limited 2011

Roughly speaking, the disturbance decoupling problem involves finding suitable controls such that, for the closed-loop system, the outputs are not affected by the disturbances. Consider the system (12.3). If we choose controllers

$$u_1(t) = B(t), \qquad u_2(t) = 0,$$

then the closed-loop system becomes

$$
\begin{cases}
A(t+1) = B(t) \wedge \xi(t), \\
B(t+1) = C(t) \vee B(t), \\
C(t+1) = D(t), \\
D(t+1) = \neg C(t), \\
y(t) = C(t) \wedge D(t).
\end{cases}
\tag{12.4}
$$

It is obvious that the disturbance will not affect the output.

We now give a rigorous definition.

Definition 12.1 Consider the system (12.1). The DDP is solvable if we can find a feedback control

$$u(t) = \phi\big(x(t)\big) \tag{12.5}$$

and a coordinate transformation $z = T(x)$ such that under the z coordinate frame, the closed-loop system becomes

$$
\begin{cases}
z^1(t+1) = F^1(z(t), \phi(x(t)), \xi(t)), \\
z^2(t+1) = F^2(z^2(t)), \\
y(t) = G\big(z^2(t)\big).
\end{cases}
\tag{12.6}
$$

From Definition 12.1 one sees that to solve the DDP problem there are two key issues: (i) finding a regular coordinate subspace z^2 which contains outputs, and (ii) designing a control such that the complement coordinate sub-basis z^1 and the disturbances ξ can be deleted from the dynamics of z^2. These will be investigated in the following two sections.

12.2 Y-friendly Subspace

Definition 12.2 Let $\mathcal{X} = \mathcal{F}_\ell\{x_1, \ldots, x_n\}$ be the state space and $Y = \{y_1, \ldots, y_p\} \subset \mathcal{X}$. A regular subspace $\mathcal{Z} \subset \mathcal{X}$ is called a Y-friendly subspace if $y_i \in \mathcal{Z}$, $i = 1, \ldots, p$. A Y-friendly subspace of minimum dimension is called a minimum Y-friendly subspace.

This section is devoted to finding the output-friendly subspace. First, we consider one variable y. Since $y \in \mathcal{X}$, we its algebraic expression is

$$y = \delta_2[i_1, i_2, \ldots, i_{2^n}]x := hx. \tag{12.7}$$

Let

$$n_j = \left|\{k \mid i_k = j, 1 \le k \le 2^n\}\right|, \quad j = 1, 2,$$

where $|\cdot|$ is the cardinal number of the set. We then have the following result.

Lemma 12.1 *Assume that $Y = \mathscr{F}_\ell\{y\}$ has algebraic form (12.7). There is a Y-friendly subspace of dimension r if and only if n_1 and n_2 have a common factor 2^{n-r}.*

Proof (Necessity) Assume that there is a Y-friendly subspace $\mathscr{Z} = \mathscr{F}_\ell\{z_1, \ldots, z_r\}$ with $\{z_1, \ldots, z_r\}$ as its regular sub-basis, and let $z = \ltimes_{i=1}^r z_i$. Then,

$$z = T_0 x = (t_{i,j}) x,$$

where $T_0 \in \mathscr{L}_{2^r \times 2^n}$. Since $y \in \mathscr{Z}$, we have

$$y = Gz = GT_0 x,$$

where $G \in \mathscr{L}_{2 \times 2^r}$. G can then be expressed as

$$G = \delta_2[j_1, \ldots, j_{2^r}].$$

Hence,

$$h = \delta_2[i_1, i_2, \ldots, i_{2^n}] = \delta_2[j_1, \ldots, j_{2^r}]T_0.$$

Let $m_s = |\{k \mid j_k = s, 1 \le k \le 2^r\}|$, $s = 1, 2$. Using Corollary 8.2, a straightforward computation shows that h has $2^{n-r}m_1$ columns which are equal to δ_2^1 and $2^{n-r}m_2$ columns which are equal to δ_2^2. That is, $n_1 = 2^{n-r}m_1$ and $n_2 = 2^{n-r}m_2$. The conclusion follows.

(Sufficiency) Let $y = hx$ be as in (12.7), where $n_1 = 2^{n-r}m_1$ columns of h equal δ_2^1 and $n_2 = 2^{n-r}m_2$ columns equal δ_2^2. It suffices to construct a Y-friendly subspace of dimension r. We construct a logical matrix $T_0 \in \mathscr{L}_{2^r \times 2^n}$, as follows. Let $J_1 = \{k \mid h_k = \delta_2^1\}$ and $J_2 = \{k \mid h_k = \delta_2^2\}$, where $h_k = \text{Col}_k(h)$. Simply letting $I_1 = \{1, \ldots, m_1\}$ and $I_2 = \{m_1 + 1, \ldots, 2^r\}$, we can split T_0 into 2×2 minors as follows: $T_0^{i,j} = \{t_{r,s} \mid r \in I_i \text{ and } s \in J_j\}$, $i, j = 1, 2$. We set these to be

$$T_0^{1,1} = \left.\begin{bmatrix} \mathbf{1}_{2^{n-r}}^T & \cdots & 0 \\ & \ddots & \\ 0 & \cdots & \mathbf{1}_{2^{n-r}}^T \end{bmatrix}\right\} m_1, \qquad T_0^{1,2} = 0,$$

$$T_0^{2,2} = \left.\begin{bmatrix} \mathbf{1}_{2^{n-r}}^T & \cdots & 0 \\ & \ddots & \\ 0 & \cdots & \mathbf{1}_{2^{n-r}}^T \end{bmatrix}\right\} m_2, \qquad T_0^{2,1} = 0.$$

We are now ready to verify that T_0, constructed in this way, satisfies (8.33). According to Theorem 8.2, $z = T_0 x$ forms a regular sub-basis.

Next, we define G as

$$G = \delta_2[\underbrace{1, \ldots, 1}_{m_1}, \underbrace{2, \ldots, 2}_{m_2}].$$

A straightforward computation shows that $GT_0 = h$, which means that

$$GT_0 x = hx = y. \qquad \qquad \square$$

For ease of statement, we call a factor of the form 2^s a 2-type factor. In sub-basis construction, only 2-type factors are counted.

From the proof of Lemma 12.1 the following result is obvious.

Corollary 12.1 *Assume that 2^{n-r} is the largest common 2-type factor of n_1 and n_2. The minimum Y-friendly subspace is then of dimension r.*

Next, we consider the multi-output case. Let $Y = \{y_1, \ldots, y_p\} \subset \mathscr{X}$ be p logical functions, and let $y = \ltimes_{i=1}^{p} y_i$. Then, y can be expressed in its algebraic form as

$$y = \delta_{2^p}[i_1, i_2, \ldots, i_{2^n}]x := Hx. \qquad (12.8)$$

Let

$$n_j = \left| \{ k \mid i_k = j, 1 \leq k \leq 2^n \} \right|, \quad j = 1, \ldots, 2^p.$$

Using the same argument as for the single output case, it is easy to prove the following result. (In fact, the following Algorithm 12.1 could be considered as a constructive proof.)

Theorem 12.1 *Assume that $y = \ltimes_{i=1}^{p} y_i$ has algebraic form (12.8).*

1. *There is a Y-friendly subspace of dimension r if and only if n_j, $j = 1, \ldots, 2^p$, have a common factor 2^{n-r}.*
2. *Assume that 2^{n-r} is the largest common 2-type factor of n_j, $j = 1, \ldots, 2^p$. The minimum Y-friendly subspace is then of dimension r.*

We give an algorithm for constructing a Y-friendly subspace. Assume that 2^{n-r} is a common factor of n_i, writing $n_i = m_i \cdot 2^{n-r}$, $i = 1, \ldots, 2^p$. We split the set of $\mathrm{Col}(H)$ into 2^p subsets as J_i, $i = 1, \ldots, 2^p$. $k \in J_i$ if and only if the kth column of H satisfies $\mathrm{Col}_k(H) = \delta_{2^p}^{i}$. Constructing the required Y-friendly subspace is equivalent to constructing a logical matrix $T_0 \in \mathscr{L}_{2^r \times 2^n}$ such that we can find a logical matrix $G \in \mathscr{L}_{2^p \times 2^r}$ satisfying

$$GT_0 = H.$$

Algorithm 12.1

Step 1. Split the rows of T_0 into 2^p blocks as follows: I_1 consists of the first m_1 rows, I_2 consists of the following m_2 rows, and so on, until I_{2^p} consists of the last m_{2^p} rows. (Note that $\sum_{i=1}^{2^p} m_i = 2^r$.) Partition T_0 into $2^p \times 2^p$ minors as

$$T_0^{i,j} = \{t_{r,s} \mid r \in I_i, s \in J_j\}, \quad i, j = 1, \ldots, 2^p.$$

Step 2. Note that $T_0^{i,j}$ is an $m_i \times (m_j 2^{n-r})$ minor. Set it as

$$T_0^{i,j} = \begin{cases} I_{m_i} \otimes 1_{2^{n-r}}^{\mathrm{T}}, & i = j, \\ 0, & \text{otherwise.} \end{cases} \tag{12.9}$$

Step 3. Set

$$z = \ltimes_{i=1}^r z_i := T_0 x.$$

Recover $z_i, i = 1, \ldots, r$, from T_0. (We refer to Chap. 7 for the recovery technique.)

Proposition 12.1 *Assume that 2^{n-r} is a common factor of n_i. The $z_i, i = 1, \ldots, r$, obtained from Algorithm 12.1 then form a regular sub-basis of an r-dimensional Y-friendly subspace.*

Proof Define a block diagonal matrix

$$G = \begin{bmatrix} 1_{m_1}^{\mathrm{T}} & 0 & \cdots & 0 \\ 0 & 1_{m_2}^{\mathrm{T}} & \cdots & 0 \\ \vdots & & & \\ 0 & 0 & \cdots & 1_{m_{2^p}}^{\mathrm{T}} \end{bmatrix}. \tag{12.10}$$

By the construction of T_0, it is easy to check that

$$y = GT_0 x = Gz. \qquad \square$$

We are particularly interested in constructing the minimum Y-friendly subspace. We give an example to show how to construct it.

Example 12.2 Let $\mathcal{X} = \mathcal{F}_\ell\{x_1, x_2, x_3, x_4\}$,

$$\begin{aligned} y_1 &= f_1(x_1, x_2, x_3, x_4) = (x_1 \leftrightarrow x_3) \wedge (x_2 \,\bar{\vee}\, x_4), \\ y_2 &= f_2(x_1, x_2, x_3, x_4) = x_1 \wedge x_3. \end{aligned} \tag{12.11}$$

We look for the minimum Y-friendly subspace. Setting $y = y_1 y_2$ and $x = x_1 x_2 x_3 x_4$, it is easy to calculate that

$$\begin{aligned} y_1 &= M_c M_e x_1 x_3 M_p x_2 x_4 \\ &= M_c M_e (I_4 \otimes M_p) x_1 x_3 x_2 x_4 \end{aligned}$$

$$= M_c M_e (I_4 \otimes M_p)(I_2 \otimes W_{[2]}) x_1 x_2 x_3 x_4$$

$$:= M_1 x,$$

where M_1 is the structure matrix of f_1, which can be easily calculated as

$$M_1 = M_c M_e (I_4 \otimes M_p)(I_2 \otimes W_{[2]})$$

$$= \delta_2[2\,1\,2\,2\,1\,2\,2\,2\,2\,2\,1\,2\,2\,1\,2].$$

Similarly, $y_2 = M_2 x$ with

$$M_2 = \delta_2[1\,1\,2\,2\,1\,1\,2\,2\,2\,2\,2\,2\,2\,2].$$

Finally, we have $y = Mx$, where

$$M = \delta_4[3\,1\,4\,4\,1\,3\,4\,4\,4\,4\,4\,2\,4\,4\,2\,4].$$

From M one easily sees that $n_1 = n_2 = n_3 = 2$ and $n_4 = 10$. Since the only common 2-type factor is $2 = 2^{n-r}$, we can have the minimum Y-friendly subspace of dimension $r = 3$. To construct T_0 we have

$$J_1 = \{2, \underline{5}\}, \qquad J_2 = \{12, 15\}, \qquad J_3 = \{1, 6\},$$

$$J_4 = \{3, 4, 7, 8, 9, 10, 11, 13, 14, 16\}.$$

Now, since $m_1 = m_2 = m_3 = 1$ and $m_4 = 5$, we have $I_1 = \{1\}$, $I_2 = \{2\}$, $I_3 = \{3\}$, and $I_4 = \{4, 5, 6, 7, 8\}$. Setting $B^{1,1}$ equal to $\mathbf{1}_2^{\mathrm{T}}$ yields that the 2nd and 5th columns of T_0 are equal to δ_8^1. Similarly, the 12th and 15th columns are equal to δ_8^2, etc. Finally, T_0 is obtained as

$$T_0 = \delta_8[3\,1\,4\,4\,1\,3\,5\,5\,6\,6\,7\,2\,7\,8\,2\,8]. \tag{12.12}$$

Correspondingly, we can construct G by formula (12.10) as

$$G = \delta_4[1\,2\,3\,4\,4\,4\,4\,4]. \tag{12.13}$$

Finally, we construct the minimum Y-friendly subspace which has sub-basis, say, $\{z_1, z_2, z_3\}$. Setting $z = z_1 z_2 z_3$, we have

$$z = T_0 x.$$

Define $z_i := E_i x$, $i = 1, 2, 3$. The structure matrices E_i can then be uniquely calculated from T_0 as

$$E_1 = \delta_2[1\,1\,1\,1\,1\,1\,2\,2\,2\,2\,2\,1\,2\,2\,1\,2],$$

$$E_2 = \delta_2[2\,1\,2\,2\,1\,2\,1\,1\,1\,1\,2\,1\,2\,2\,1\,2], \tag{12.14}$$

$$E_3 = \delta_2[1\,1\,2\,2\,1\,1\,1\,1\,2\,2\,1\,2\,1\,2\,2\,2].$$

We can then use Proposition 7.2 to find the logical expression of z_i from its structure matrix E_i. It is easy to calculate that

$$z_1 = \{x_1 \wedge [x_2 \vee (\neg x_2 \wedge x_3)]\} \vee \{\neg x_1 \wedge ([x_2 \wedge \neg(x_3 \vee x_4)]$$
$$\vee [\neg x_2 \wedge (\neg x_3 \wedge x_4)])\},$$
$$z_2 = \{x_1 \wedge [(x_2 \wedge (x_3 \wedge \neg x_4)) \vee (\neg x_2 \wedge (x_3 \rightarrow x_4))]\}$$
$$\vee \{\neg x_1 \wedge [(x_2 \wedge (x_3 \vee (\neg x_3 \wedge \neg x_4))) \vee (\neg x_2 \wedge (\neg x_3 \wedge x_4))]\},$$
$$z_3 = \{x_1 \wedge [(x_2 \wedge x_3) \vee \neg x_2]\} \vee \{\neg x_1 \wedge [(x_2 \wedge (\neg x_3 \wedge x_4))$$
$$\vee (\neg x_2 \wedge (x_3 \wedge x_4))]\}.$$

$$(12.15)$$

Similarly, from (12.13) we can easily calculate that

$$y_1 = \delta_2[1\,1\,2\,2\,2\,2\,2\,2]z,$$
$$y_2 = \delta_2[1\,2\,1\,2\,2\,2\,2\,2]z.$$

It follows that

$$y_1 = z_1 \wedge z_2,$$
$$y_2 = z_1 \wedge z_3.$$

$$(12.16)$$

A question which now naturally arises is whether the minimum Y-friendly subspace is unique. First, we consider the number of bases of the subspace.

Proposition 12.2 *Assume the algebraic form of $\{y_1, \ldots, y_p\}$ is $y = Hx$ and that the numbers $n_j = |\{s \mid H_s = \delta_{2^p}^j\}|$, $j = 1, \ldots, 2^p$, have common factor 2^{n-r}. There are then*

$$N_r = \prod_{i=1}^{2^p} \frac{(m_i \cdot 2^{n-r})!}{[(2^{n-r})!]^{m_i}}$$

$$(12.17)$$

different choices of sub-basis which form the r-dimensional Y-friendly subspaces.

Proof The question is equivalent to asking how many different T_0 there are. First, we fix the row assignment, that is, the assignment of $I_1, I_2, \ldots, I_{2^p}$. Note that I_i has m_i rows, and J_i has $m_i \cdot 2^{n-r}$ columns. It is obvious that to get the same GT_0 we can, for each row in I_i, choose any 2^{n-r} columns and assign them to be 1. Hence, we have N^i different choices for values in J_i columns, where

$$N^i = \binom{m_i \cdot 2^{n-r}}{2^{n-r}} \binom{(m_i - 1) \cdot 2^{n-r}}{2^{n-r}} \cdots \binom{2^{n-r}}{2^{n-r}} = \frac{(m_i \cdot 2^{n-r})!}{[(2^{n-r})!]^{m_i}}.$$

Since $N_r = \prod_{i=1}^{2^p} N^i$, (12.17) follows immediately.

The only thing remaining to be clarified is that the row assignment for I_i, $i = 1, \ldots, 2^p$, is fixed in advance. In fact, we can choose any m_1 rows for I_1, then m_2 rows from remaining rows for I_2, and so on. For this purpose, we can introduce a row permutation. Let $P \in \mathcal{M}_{2^r \times 2^r}$ be a permutation matrix, so a new T_0 can be obtained by $\tilde{T}_0 = PT_0$. We then have a new regular sub-basis, $\tilde{z} = Pz = PT_0 z$, which is just a coordinate transformation of z, so they generate the same regular subspace and will be considered as the same. It is worth noting that we have

$$y = \tilde{G}\tilde{z}, \quad \text{where } \tilde{G} = GP^{\mathrm{T}}. \qquad \qquad \square$$

In fact, N_r is a huge number when n is large. Fortunately, different sub-bases may determine the same subspace. We give a simple example to illustrate this.

Example 12.3 Let $y = x_1 \wedge x_3 \in \mathscr{F}_\ell\{x_1, x_2, x_3\}$. Using Theorem 12.1, it is easy to verify that a minimum Y-friendly subspace is of dimension 2. Moreover, we can easily show $\{x_1, x_3\}$ to be a basis, that is, $y \in \mathscr{Z}_1 = \mathscr{F}_\ell\{x_1, x_3\}$. It is also easy to verify that $\mathscr{Z}_2 = \mathscr{F}_\ell\{x_1, x_1 \bar{\vee} x_3\}$, $\mathscr{Z}_3 = \mathscr{F}_\ell\{x_3, x_1 \bar{\vee} x_3\}$, $\mathscr{Z}_4 = \mathscr{F}_\ell\{x_3, x_1 \leftrightarrow x_3\}$, etc. are also minimum Y-friendly subspaces. Fortunately, it is easy to check that they are all the same.

Now, let $\mathscr{X} = \mathscr{F}_\ell\{x_1, \ldots, x_n\}$ be the state space and

$$\mathscr{Z}^i = \mathscr{F}_\ell\{z_1^i, \ldots, z_k^i\} \subset \mathscr{X}, \quad i = 1, 2,$$

where $\{z_1^i, \ldots, z_k^i\}$, $i = 1, 2$, are regular sub-bases with

$$z_j^i = f_j^i(x_1, \ldots, x_n), \quad i = 1, 2, j = 1, \ldots, k. \qquad (12.18)$$

We wish to know when $\mathscr{Z}_1 = \mathscr{Z}_2$.

Let $z^i = \ltimes_{j=1}^k z_j^i$, $i = 1, 2$. The logical equations (12.18) have algebraic forms

$$z^i = T_i x, \quad i = 1, 2. \qquad (12.19)$$

Assume that $\mathscr{Z}_1 = \mathscr{Z}_2$. There then exists a nonsingular $P \in \mathscr{L}_{2^k \times 2^k}$ such that

$$T_1 x = P T_2 x$$

and it follows that $T_1 = P T_2$. Hence,

$$P = T_1 \big(T_2^{\mathrm{T}} (T_2 T_2^{\mathrm{T}})^{-1} \big). \qquad (12.20)$$

Plugging this into $T_1 = P T_2$ yields

$$T_1 = T_1 T_2^{\mathrm{T}} (T_2 T_2^{\mathrm{T}})^{-1} T_2. \qquad (12.21)$$

Conversely, if (12.21) holds and the P defined in (12.20) is a coordinate transformation matrix, then we have $T_1 x = P T_2 x$, i.e., $z^1 = P z^2$. According to Theorem 8.1, $\mathscr{Z}_1 = \mathscr{Z}_2$. Summarizing the above argument yields the following theorem.

Theorem 12.2 *Using the above notation, two regular subspaces of equal dimension $\mathscr{L}_1 = \mathscr{L}_2$ if and only if (i) their structure matrices T_1 and T_2 satisfy (12.21), and (ii)*

$$T_1 T_2^{\mathrm{T}} \left(T_2 T_2^{\mathrm{T}} \right)^{-1} \in \mathscr{L}_{2^k \times 2^k}$$

is nonsingular.

Using Theorem 12.2, we can easily show that the minimum Y-friendly subspace is, in general, not unique. This is shown by the following example.

Example 12.4 Consider Example 12.2 again. Using formula (12.17), there are $10!/32$ different choices of sub-basis. These come from setting the entries of the $T_0^{4,4}$, which corresponds to $I_4 \times J_4$.

We may construct some other subspaces by using other structure matrices.

1. Choosing

$$T_0' = \delta_8[3\ 1\ 5\ 5\ 1\ 3\ 7\ 7\ 8\ 8\ 4\ 2\ 4\ 6\ 2\ 6],$$

we can generate another Y-friendly minimum subspace. Note that this choice is legal because we only put two 4s, two 5s, two 6s, two 7s, and two 8s into the slots of J_4. It is easy to check that if we set $T_2 := T_0$ and $T_1 := T_0'$, then (12.21) is satisfied. Moreover,

$$T_1 T_2^{\mathrm{T}} \left(T_2 T_2^{\mathrm{T}} \right)^{-1} = \delta_8[1\ 2\ 3\ 5\ 7\ 8\ 4\ 6] \in \mathscr{L}_{2^k \times 2^k}$$

is nonsingular, so $z = T_0 x$ and $z' = T_0' x$ generate the same Y-friendly subspace.

2. We may consider another legal choice:

$$T_0'' = \delta_8[3\ 1\ 5\ 6\ 1\ 3\ 7\ 8\ 7\ 8\ 4\ 2\ 4\ 5\ 2\ 6].$$

If we set $T_2 := T_0$ and $T_1 := T_0''$, then (12.21) fails to be satisfied. Thus, $z = T_0 x$ and $z'' = T_0'' x$ generate two different Y-friendly minimum regular subspaces.

The second example shows that the Y-friendly minimum regular subspace is not generally unique.

12.3 Control Design

In the previous section, the problem of finding a Y-friendly subspace was investigated. Assume that a Y-friendly subspace is obtained as z^2. We can then find z^1 such that $z = \{z^1, z^2\}$ forms a new coordinate frame. Under this z, the system (12.1) can be expressed as

$$\begin{cases} z^1(t+1) = F^1(z(t), u(t), \xi(t)), \\ z^2(t+1) = F^2(z(t), u(t), \xi(t)), \\ \quad y(t) = G\left(z^2(t) \right). \end{cases} \tag{12.22}$$

Equation (12.22) is called the output-friendly form. Comparing it with (12.6), one sees that solving the DDP reduces to finding $u(t) = u(z(t))$ such that

$$F^2\big(z(t), u(z(t)), \xi(t)\big) = \tilde{F}^2\big(z^2(t)\big). \tag{12.23}$$

Assume $z^2 = (z_1^2, \ldots, z_k^2)$ is of dimension k. We define a set of functions,

$$e_1(z^2) = z_1^2 \wedge z_2^2 \wedge \cdots \wedge z_k^2; \qquad e_2(z^2) = z_1^2 \wedge z_2^2 \wedge \cdots \wedge \neg z_k^2;$$

$$e_3(z^2) = z_1^2 \wedge \cdots \wedge \neg z_{k-1}^2 \wedge z_k^2; \qquad e_4(z^2) = z_1^2 \wedge \cdots \wedge \neg z_{k-1}^2 \wedge \neg z_k^2; \tag{12.24}$$

$$\cdots$$

$$e_{2^k}(z^2) = \neg z_1^2 \wedge \neg z_2^2 \wedge \cdots \wedge \neg z_k^2.$$

It is easy to check that

$$\mathscr{L}^2 := \mathscr{F}_\ell\{z^2\} = \mathscr{F}_\ell\{e_i \mid 1 \le i \le 2^k\}.$$

We call $\{e_i \mid 1 \le i \le 2^k\}$, defined in (12.24), a conjunctive basis of $\mathscr{L}^2$ (or z^2). Using Proposition 7.2, each equation of F^2, denoted by F_i^2, can be expressed as

$$F_j^2(z(t), u(t), \xi(t)) = \bigvee_{i=1}^{2^k} \big[e_i(z^2(t)) \wedge Q_j^i(z^1(t), u(t), \xi(t))\big], \quad j = 1, \ldots, k. \tag{12.25}$$

Proposition 12.3 $F^2(z(t), u(t), \xi(t)) = F^2(z^2(t))$ *if and only if, in the expression* (12.25),

$$Q_j^i\big(z^1(t), u(t), \xi(t)\big) = \text{const.}, \quad j = 1, \ldots, k, i = 1, \ldots, 2^p. \tag{12.26}$$

Proof Sufficiency is trivial. For necessity, assume that for a special pair i, j, the Q_i^j is not constant. Consider the corresponding e_i. If its z_s^2 factor is z_s^2, set $z_s^2 = 1$. Otherwise, if this factor is $\neg z_s^2$, set $z_s^2 = 0$, $s = 1, \ldots, k$. We then have

$$e_i(z^2) = 1, \quad e_j(z^2) = 0, \quad j \ne i.$$

Now, since Q_i^j is not constant, when $Q_i^j = 1$, we have $F_i^2 = 1$, and when $Q_i^j = 0$, we have $F_i^2 = 0$. So, for fixed z^2, F_i^2 can have different values, which means that F_i^2 is not a function with of z^2 alone. $\quad\square$

We are now ready to give the condition for the solvability of the DDP. Taking into consideration the above argument, the following result is obvious.

Theorem 12.3 *Consider the system* (12.1). *The DDP is solvable if and only if:*

(i) *There exists an output-friendly coordinate sub-basis such that, using this sub-basis, the system is expressed as (12.22).*
(ii) *In (12.22), when F^2 is expressed as in (12.25), there exists a feedback control $u(t) = u(z(t))$ such that (12.26) is satisfied.*

Remark 12.1 An output-friendly coordinate sub-basis, say z^2, is obtained. A complement set of logical variables, z^1, must be chosen to form a new coordinate frame $z = \{z^1, z^2\}$. It is easy to check that the choice of z^1 does not affect the solvability of the DDP.

Now, assume that we have (12.22) with z^2 being the minimum output-friendly subspace. We can then search for the feedback control. Set the feedback as

$$u(z^1(t)) = Uz^1(t), \tag{12.27}$$

where $U \in \mathcal{L}_{2^m \times 2^{n-k}}$.

Note that there are only finitely many U. If there is a control u such that all the functions in (12.26) are constant, then we are done. Otherwise, there is no u which deletes ξ from all functions in (12.26), which means that the DDP is not solvable. If there is a u which does delete ξ from all functions in (12.26), but there are some functions of z^1, say $\{\eta_1, \ldots, \eta_s\} \subset \mathcal{F}_\ell\{z^1\}$, which remain undeleted, then we add $\{\eta_i \mid i = 1, \ldots, s\}$ to $\{y_1, \ldots, y_p\}$ and find a minimum $(y_i, \eta_j \mid i = 1, \ldots, p; j = 1, \ldots, s)$-friendly subspace, say V. We then turn the closed-loop system into the form of (12.22) with $\mathcal{F}_\ell\{z^2\} = V$ to see whether it has the form of (12.6).

To see that a possible solution with an output-friendly subspace can be obtained by starting from a minimum subspace, we have to show that an output-friendly subspace contains a minimum output-friendly subspace. We have the following result.

Proposition 12.4 *Let V be a $Y = \{y_1, \ldots, y_p\}$-friendly subspace. There then exists a minimum Y-friendly subspace, $W \subset V$.*

Proof Let

$$y = \ltimes_{i=1}^p y_i = Hx,$$

and let n_i, $i = 1, \ldots, 2^p$, denote the numbers of columns of H, which equal $\delta_{2^p}^i$. Let 2^s be the largest common 2-type factor of $\{n_i\}$. The minimum Y-friendly subspace then has dimension $n - s$. Let $\{v_1, \ldots, v_t\}$ be a basis of V. If $t = n - s$, then we are done. Therefore, we assume that $t > n - s$. Since V is a Y-friendly subspace, by writing $v = \ltimes_{i=1}^t v_i$, we can express

$$y = Gv, \quad \text{where } G \in \mathcal{L}_{2^p \times 2^t}. \tag{12.28}$$

Denote by r_i, $i = 1, \ldots, 2^p$, the numbers of columns of G which are equal to $\delta_{2^p}^i$. Let 2^j be the largest 2-type common factor of r_i, and write $r_i = m_i 2^j$. Since V is a regular subspace, we let $v = Ux$, where $U \in \mathscr{L}_{2^t \times 2^n}$. Since v is a regular sub-basis, (8.33) holds. Note that since

$$y = Gv = GUx,$$

we have to calculate GU. Using the construction of G and the property (8.33) of U, it is easy to verify that each column of $\delta_{2^p}^i$ yields 2^{n-t} columns of $\delta_{2^p}^i$ in GU. Hence, we have

$$r_i \cdot 2^{n-t} = m_i \cdot 2^{n-t+j}$$

instances of $\delta_{2^p}^i$ in GU, $i = 1, \ldots, 2^p$. This means that the largest common 2-type factor of $\{n_i\}$ is 2^{n-t+j}. It follows that $n - t + j = s$. Equivalently,

$$j = t - (n - s).$$

Going back to (12.28), since the dimension of V is t, and the r_i have largest common 2-type factors 2^j, we can find a minimum Y-friendly subspace of V of dimension $t - j = t - [t - (n - s)] = n - s$. It follows from the dimension that this minimum Y-friendly subspace of V is also a minimum Y-friendly subspace of $\mathscr{X} = \mathscr{F}_\ell\{x_1, \ldots, x_n\}$. $\qquad\square$

We give an example to describe this.

Example 12.5 Consider the following system:

$$\begin{cases} x_1(t+1) = x_4(t) \,\bar{\vee}\, u_1(t), \\ x_2(t+1) = (x_2(t) \,\bar{\vee}\, x_3(t)) \wedge \neg\xi(t), \\ x_3(t+1) = [(x_2(t) \leftrightarrow x_3(t)) \vee \xi(t)] \,\bar{\vee}\, [(x_1 \leftrightarrow x_5) \vee u_2(t)], \\ x_4(t+1) = [u_1(t) \rightarrow (\neg x_2(t) \vee \xi(t))] \wedge (x_2(t) \leftrightarrow x_3(t)), \\ x_5(t+1) = (x_4(t) \,\bar{\vee}\, u_1(t)) \leftrightarrow [(u_2(t) \wedge \neg x_2(t)) \vee x_4(t)], \\ y(t) = x_4(t) \wedge (x_1(t) \leftrightarrow x_5(t)), \end{cases} \tag{12.29}$$

where $u_1(t), u_2(t)$ are controls, $\xi(t)$ is a disturbance, and $y(t)$ is the output.

Setting $x(t) = \ltimes_{i=1}^5 x_i(t)$, $u = u_1(t)u_2(t)$, we express (12.29) in algebraic form as

$$\begin{aligned} x(t+1) &= Lu(t)\xi(t)x(t), \\ y(t) &= h(t), \end{aligned} \tag{12.30}$$

where

$$L = \delta_{32}[30\ 30\ 14\ 14\ 32\ 32\ 16\ 16\ 32\ 32\ 15\ 15\ 30\ 30\ 13\ 13$$
$$30\ 30\ 14\ 14\ 32\ 32\ 16\ 16\ 32\ 32\ 15\ 15\ 30\ 30\ 13\ 13$$
$$32\ 32\ 16\ 16\ 20\ 20\ 4\ \ 4\ \ 20\ 20\ 3\ \ 3\ \ 30\ 30\ 13\ 13$$
$$32\ 32\ 16\ 16\ 20\ 20\ 4\ \ 4\ \ 20\ 20\ 3\ \ 3\ \ 30\ 30\ 13\ 13$$
$$30\ 26\ 14\ 10\ 32\ 28\ 16\ 12\ 32\ 28\ 16\ 12\ 30\ 26\ 14\ 10$$
$$26\ 30\ 10\ 14\ 28\ 32\ 12\ 16\ 28\ 32\ 12\ 16\ 26\ 30\ 10\ 14$$
$$32\ 28\ 16\ 12\ 20\ 24\ 4\ \ 8\ \ 20\ 24\ 4\ \ 8\ \ 30\ 26\ 14\ 10$$
$$28\ 32\ 12\ 16\ 24\ 20\ 8\ \ 4\ \ 24\ 20\ 8\ \ 4\ \ 26\ 30\ 10\ 14$$
$$13\ 13\ 29\ 29\ 15\ 15\ 31\ 31\ 15\ 15\ 32\ 32\ 13\ 13\ 30\ 30$$
$$13\ 13\ 29\ 29\ 15\ 15\ 31\ 31\ 15\ 15\ 32\ 32\ 13\ 13\ 30\ 30$$
$$13\ 13\ 29\ 29\ 3\ \ 3\ \ 19\ 19\ 3\ \ 3\ \ 20\ 20\ 13\ 13\ 30\ 30$$
$$13\ 13\ 29\ 29\ 3\ \ 3\ \ 19\ 19\ 3\ \ 3\ \ 20\ 20\ 13\ 13\ 30\ 30$$
$$13\ 9\ \ 29\ 25\ 15\ 11\ 31\ 27\ 15\ 11\ 31\ 27\ 13\ 9\ \ 29\ 25$$
$$9\ \ 13\ 25\ 29\ 11\ 15\ 27\ 31\ 11\ 15\ 27\ 31\ 9\ \ 13\ 25\ 29$$
$$13\ 9\ \ 29\ 25\ 3\ \ 7\ \ 19\ 23\ 3\ \ 7\ \ 19\ 23\ 13\ 9\ \ 29\ 25$$
$$9\ \ 13\ 25\ 29\ 7\ \ 3\ \ 23\ 19\ 7\ \ 3\ \ 23\ 19\ 9\ \ 13\ 25\ 29],$$

$$h = \delta_2[1\ 2\ 2\ 2\ 1\ 2\ 2\ 2\ 1\ 2\ 2\ 2\ 1\ 2\ 2\ 2\ 2\ 1\ 2\ 2\ 2\ 1\ 2\ 2\ 2\ 1\ 2\ 2\ 2\ 1\ 2\ 2].$$

First, we have to find the minimum output-friendly subspace. Observing h, we have $n_1 = 8$ and $n_2 = 24$. We then have the largest common 2-type factor $2^s = 2^3$, and $m_1 = 1$, $m_2 = 3$. Hence, we know that the minimum output-friendly subspace is of dimension $n - s = 5 - 3 = 2$. Using Algorithm 12.1, we may choose

$$T_0 = \delta_4[1\ 2\ 3\ 4\ 1\ 2\ 3\ 4\ 1\ 2\ 3\ 4\ 1\ 2\ 3\ 4\ 2\ 1\ 4\ 3\ 2\ 1\ 4\ 3\ 2\ 1\ 4\ 3\ 2\ 1\ 4\ 3]$$

and

$$G = \delta_2[1\ 2\ 2\ 2].$$

From T_0 we can find the output-friendly sub-basis, denoting it by $\{z_4, z_5\}$, with $z_4 = M_4 x$ and $z_5 = M_5 x$. We can then easily calculate M_4 and M_5 from T_0. In fact, for the two-factor case, we have the following simple rule: For M_1 each column of δ_4^1 or δ_4^2 of T_0 yields a column δ_2^1 in the corresponding column of M_1, otherwise we have δ_2^2; for M_2 each column of δ_4^1 or δ_4^3 of T_0 yields a δ_2^1 in the corresponding column of M_2, otherwise we have δ_2^2. Hence, we have

$$M_4 = \delta_2[1\ 1\ 2\ 2\ 1\ 1\ 2\ 2\ 1\ 1\ 2\ 2\ 1\ 1\ 2\ 2\ 1\ 1\ 2\ 2\ 1\ 1\ 2\ 2\ 1\ 1\ 2\ 2\ 1\ 1\ 2\ 2],$$
$$M_5 = \delta_2[1\ 2\ 1\ 2\ 1\ 2\ 1\ 2\ 1\ 2\ 1\ 2\ 1\ 2\ 1\ 2\ 2\ 1\ 2\ 1\ 2\ 1\ 2\ 1\ 2\ 1\ 2\ 1\ 2\ 1\ 2\ 1].$$

Using Corollary 8.2, we simply set $z_i = M_i x$, $i = 1, 2, 3$, where M_i are chosen as follows:

$$M_1 = \delta_2[1\,1\,1\,1\,1\,1\,1\,1\,1\,1\,1\,1\,1\,1\,1\,1\,2\,2\,2\,2\,2\,2\,2\,2\,2\,2\,2\,2\,2\,2\,2\,2],$$

$$M_2 = \delta_2[2\,2\,2\,2\,2\,2\,2\,2\,1\,1\,1\,1\,1\,1\,1\,1\,2\,2\,2\,2\,2\,2\,2\,2\,1\,1\,1\,1\,1\,1\,1\,1],$$

$$M_3 = \delta_2[1\,1\,1\,1\,2\,2\,2\,2\,2\,2\,2\,2\,1\,1\,1\,1\,1\,1\,1\,1\,2\,2\,2\,2\,2\,2\,2\,2\,1\,1\,1\,1].$$

It is easy to check that the Boolean matrix B_z of $\{z_1, z_2, z_3, z_4, z_5\}$ has no equal columns. Therefore, it is a coordinate transformation. From M_i, the z_i can be calculated as

$$\begin{cases} z_1 = x_1, \\ z_2 = \neg x_2, \\ z_3 = x_2 \leftrightarrow x_3, \\ z_4 = x_4, \\ z_5 = x_1 \leftrightarrow x_5. \end{cases} \tag{12.31}$$

Setting $z = \ltimes_{i=1}^{5} z_i$ and $x = \ltimes_{i=1}^{5} x_i$, the algebraic form of (12.31) is $z = Tx$, with

$$T = \delta_{32}[9 \quad 10\,11\,12\,13\,14\,15\,16\,5\,6\,7\,8\,1\,2\,3\,4\,26\,25$$
$$28\,27\,30\,29\,32\,31\,22\,21\,24\,23\,18\,17\,20\,19].$$

Conversely, we have $x = T^T z$, with

$$T^T = [13\,14\,15\,16\,9\,10\,11\,12\,1\,2\,3\,4\,5\,6\,7\,8\,30\,29$$
$$32\,31\,26\,25\,28\,27\,18\,17\,20\,19\,22\,21\,24\,23].$$

The inverse mapping of the coordinate transformation (12.31) becomes

$$\begin{cases} x_1 = z_1, \\ x_2 = \neg z_2, \\ x_3 = z_2 \bar{\vee} z_3, \\ x_4 = z_4, \\ x_5 = z_1 \leftrightarrow z_5. \end{cases}$$

Under the coordinate frame z, equation (12.30) now becomes

$$z(t+1) = Tx(t+1) = TLu(t)\xi(t)x(t) = TLu(t)\xi(t)T^T z(t)$$
$$= TL(I_8 \otimes T^T)u(t)\xi(t)z(t) := \tilde{L}u(t)\xi(t)z(t)$$

and

$$y(t) = hx(t) = hT^T z(t) := \tilde{h}z(t),$$

where

$$\tilde{L} = \delta_{32}[17\ 17\ 1\quad 1\quad 19\ 19\ 3\quad 3\quad 17\ 17\ 2\quad 2\quad 19\ 19\ 4\quad 4$$
$$17\ 17\ 1\quad 1\quad 19\ 19\ 3\quad 3\quad 17\ 17\ 2\quad 2\quad 19\ 19\ 4\quad 4$$
$$17\ 17\ 1\quad 1\quad 27\ 27\ 11\ 11\ 19\ 19\ 4\quad 4\quad 27\ 27\ 12\ 12$$
$$17\ 17\ 1\quad 1\quad 27\ 27\ 11\ 11\ 19\ 19\ 4\quad 4\quad 27\ 27\ 12\ 12$$
$$17\ 21\ 2\quad 6\quad 19\ 23\ 4\quad 8\quad 17\ 21\ 2\quad 6\quad 19\ 23\ 4\quad 8$$
$$17\ 21\ 2\quad 6\quad 19\ 23\ 4\quad 8\quad 17\ 21\ 2\quad 6\quad 19\ 23\ 4\quad 8$$
$$17\ 21\ 2\quad 6\quad 27\ 31\ 12\ 16\ 19\ 23\ 4\quad 8\quad 27\ 31\ 12\ 16$$
$$17\ 21\ 2\quad 6\quad 27\ 31\ 12\ 16\ 19\ 23\ 4\quad 8\quad 27\ 31\ 12\ 16$$
$$1\quad 1\quad 17\ 17\ 3\quad 3\quad 19\ 19\ 1\quad 1\quad 18\ 18\ 3\quad 3\quad 20\ 20$$
$$1\quad 1\quad 17\ 17\ 3\quad 3\quad 19\ 19\ 1\quad 1\quad 18\ 18\ 3\quad 3\quad 20\ 20$$
$$1\quad 1\quad 17\ 17\ 11\ 11\ 27\ 27\ 1\quad 1\quad 18\ 18\ 11\ 11\ 28\ 28$$
$$1\quad 1\quad 17\ 17\ 11\ 11\ 27\ 27\ 1\quad 1\quad 18\ 18\ 11\ 11\ 28\ 28$$
$$1\quad 5\quad 18\ 22\ 3\quad 7\quad 20\ 24\ 1\quad 5\quad 18\ 22\ 3\quad 7\quad 20\ 24$$
$$1\quad 5\quad 18\ 22\ 3\quad 7\quad 20\ 24\ 1\quad 5\quad 18\ 22\ 3\quad 7\quad 20\ 24$$
$$1\quad 5\quad 18\ 22\ 11\ 15\ 28\ 32\ 1\quad 5\quad 18\ 22\ 11\ 15\ 28\ 32$$
$$1\quad 5\quad 18\ 22\ 11\ 15\ 28\ 32\ 1\quad 5\quad 18\ 22\ 11\ 15\ 28\ 32],$$

$$\tilde{h} = \delta_2[1\ 2\ 2\ 2\ 1\ 2\ 2\ 2\ 1\ 2\ 2\ 2\ 1\ 2\ 2\ 2\ 1\ 2\ 2\ 2\ 1\ 2\ 2\ 2\ 1\ 2\ 2\ 2\ 1\ 2\ 2\ 2].$$

A mechanical procedure can then convert the original system into a Y-friendly co-ordinate frame z as

$$\begin{cases} z_1(t+1) = z_4(t)\ \bar{\vee}\ u_1(t), \\ z_2(t+1) = z_3(t) \vee \xi(t), \\ z_3(t+1) = z_5(t) \vee u_2(t), \\ z_4(t+1) = [u_1(t) \to (z_2(t) \vee \xi(t))] \wedge z_3(t), \\ z_5(t+1) = (u_2(t) \wedge z_2(t)) \vee z_4(t), \\ y = z_4 \wedge z_5. \end{cases} \tag{12.32}$$

In the output-friendly subspace (z_4, z_5), we may now choose

$$u_1(t) = z_2(t) = \neg x_2(t), \qquad u_2(t) = 0.$$

Now the only unlimited variable, which is outside of this space, is z_3. Enlarging the output-friendly subspace to include z_3, one sees that the closed-loop system is in such a form that the DDP is solved. Since, in system (12.32), the controls which solve the DDP are obvious, we do not need to use the general formula.

12.4 Canalizing Boolean Mapping

In this section we consider the canalizing Boolean mapping. It will be used to solve the DDP via constant controls.

Definition 12.3 A Boolean function $y = f(x_1, \ldots, x_n)$ is called a canalizing (or forcing) Boolean function (CBF) if there exist an $i \in \{1, \ldots, n\}$ and $u, v \in \mathcal{D}$ such that

$$f(x_1, \ldots, x_{i-1}, u, x_{i+1}, \ldots, x_n) = v, \quad \forall x_j, j \neq i. \tag{12.33}$$

If (12.33) holds, x_i is called the canalizing variable with canalizing value u and canalized value v, and f is said to be a (u, v)-type canalizing Boolean function.

It was pointed out by Kauffman [4] that canalizing Boolean functions allow us to deduce large-scale order in the underlying ontogeny of genetic regulatory systems.

For our purposes, we define a generalized version of a CBF.

Definition 12.4 A mapping $F : \mathcal{D}^n \to \mathcal{D}^p$, determined by

$$y_j = f_j(x_1, \ldots, x_n), \quad j = 1, \ldots, p,$$

is called a multi-input multi-output canalizing (or forcing) Boolean mapping [or, more briefly, a canalizing Boolean mapping (CBM)] if there exist a proper subset $\Lambda = \{\lambda_1, \ldots, \lambda_k\} \subset \{1, \ldots, n\}$ and $u_1, \ldots, u_k, v_1, \ldots, v_p \in \mathcal{D}$ such that

$$f_j(x_1, \ldots, x_n)|_{x_{\lambda_i} = u_i, i = 1, \ldots, k} = v_j, \quad j = 1, \ldots, p. \tag{12.34}$$

If (12.34) holds, then $x_\lambda, \lambda \in \Lambda$ are called the canalizing variables with canalizing values $u = (u_1, \ldots, u_k)$ and canalized values $v = (v_1, \ldots, v_p)$, and F is said to be a (u, v)-type canalizing Boolean mapping.

In the following we look for a necessary and sufficient condition for a given mapping to be a CBM. Of course, the results obtained are also applicable to CBFs.

Recall that $M \in \mathcal{L}_{2^p \times 2^n}$ is called a constant mapping matrix (CMM) if it is a structure matrix of a constant mapping, that is, if there exists an s, $1 \leq s \leq 2^p$, such that

$$\mathrm{Col}_i(M) = \delta_{2^p}^s, \quad \forall i.$$

First, we assume that $\Lambda = \{1, 2, \ldots, k\}$. We then have the following result.

Theorem 12.4 *Let* $F : \mathcal{D}^n \to \mathcal{D}^p$ *be defined by*

$$y_j = f_j(x_1, \ldots, x_n), \quad j = 1, \ldots, p. \tag{12.35}$$

Its algebraic form is $y = M_F x$*, with* $x = \ltimes_{i=1}^n x_i$ *and* $y = \ltimes_{i=1}^p y_i$*. Split* M_F *into* 2^k *equal-sized blocks as*

$$M_F = \begin{bmatrix} M^1 & M^2 & \cdots & M^{2^k} \end{bmatrix}.$$

F is then a CBM with canalizing variables $x_1, \ldots, x_k$ *if and only if there exits an* s*,* $1 \leq s \leq 2^k$*, such that* M^s *is a CMM.*

Proof First, let $z_1, \ldots, z_s$ be a set of logical variables and $z = \ltimes_{i=1}^s z_i$. Then, $z \in \Delta_{2^s}$ and $\{z_i \mid i = 1, \ldots, s\}$ can be uniquely calculated from z, so we have only to consider y and x. Let $x^1 = \ltimes_{i=1}^k x_i$ and $x^2 = \ltimes_{i=k+1}^n x_i$. Then,

$$y = M_F x^1 x^2 = (M_F x^1) x^2 := M_2 x^2.$$

Now, assume that M^t is a CMM, where $1 \le t \le 2^k$. We can then choose $x_0^1 = \delta_{2^k}^t$. By the definition of the semi-tensor product, we have $M_2 = M_F x_0^1 = M^t$, which is a CMM. That is, y is a CBM with x^1 as canalizing variables.

Conversely, if all M^t are not CMMs, then for any $x_0^1 \in \Delta_{2^k}$ the $M_2 = M_F x_0^1$ is not a CMM. Hence, y is not a CBM with x^1 as canalizing variables. □

Next, we consider the general case. Let

$$\Lambda = \{i_1, \ldots, i_k\} \subset \{1, 2, \ldots, n\}.$$

Without loss of generality, we assume $i_1 < i_2 < \cdots < i_k$. Using the aforementioned notation, we define

$$\tilde{M}_F = M_F \prod_{j=1}^k W_{[2, 2^{i_j + k - j}]}.$$

Splitting this as

$$\tilde{M}_F = \begin{bmatrix} \tilde{M}^1 & \tilde{M}^2 & \cdots & \tilde{M}^{2^k} \end{bmatrix},$$

we then have the following conclusion.

Corollary 12.2 *F is a CBM with canalizing variables x_λ, $\lambda \in \Lambda = \{i_1, \ldots, i_k\}$, if and only if there is a $1 \le s \le 2^k$ such that $\tilde{M}^s$ is a CBM.*

Proof We use a swap matrix to rearrange the order of products. A straightforward computation then shows that

$$x = \ltimes_{i=1}^n x_i = \prod_{j=1}^k W_{[2, 2^{i_j + k - j}]} x_{i_1} \cdots x_{i_k} \ltimes_{\substack{i=1 \\ i \notin \Lambda}}^n x_i,$$

and the conclusion follows. □

As an immediate consequence, we have the following result.

Corollary 12.3 *Let $y = f(x_1, \ldots, x_n)$ be a logical function with structure matrix M_f. Let*

$$M_f W_{[2, 2^{i-1}]} = \begin{bmatrix} M^1 & M^2 \end{bmatrix},$$

Table 12.1 Type of f

i	1	1	2	2
Columns	δ_2^1	δ_2^2	δ_2^1	δ_2^2
Type	$(1,1)$	$(1,0)$	$(0,1)$	$(0,0)$

where $M^i \in \mathscr{L}_{2 \times 2^{n-1}}$, $i = 1, 2$. Then f is a canalizing Boolean function with canal-izing variable x_i if and only if at least one of M^1 or M^2 is a CMM. Moreover, if M^i ($i = 1$ or $i = 2$) is a CMM with its columns equal to δ_2^1 (or δ_2^2), then the type of f is shown in Table 12.1.

12.5 Solving DDPs via Constant Controls

It was proposed in the above that the DDP can be solved in two steps: First, convert (12.1) into an output-friendly form (12.22); then, in (12.22), try to design a control such that the dynamics of the output-related part, x^2, will be independent of x^1 and ξ_i, $i = 1, \ldots, q$.

Converting (12.1) into the output-friendly form (12.22) was discussed in Sect. 12.2. Now, assume that (12.22) is obtained. Write the second part of its state equations as

$$x^2(t+1) = F^2\big(x(t), u(t), \xi(t)\big). \tag{12.36}$$

The DDP is then solvable if we can find controls such that the F^2 in (12.36) is independent of x^1 and ξ_i, $i = 1, \ldots, q$, that is, if we can find state feedback $u(t) = u(x(t))$ such that

$$F^2\big(x(t), u(x(t)), \xi(t)\big) = F^2(x^2). \tag{12.37}$$

Using the algorithm developed in Sect. 12.3, F^2 can be expressed as

$$F^2\big(x(t), u(t), \xi(t)\big) = \bigvee_{i=1}^{2^k} e_i(x^2) \wedge Q^i\big(x^1(t), u(t), \xi(t)\big). \tag{12.38}$$

Note that here

$$Q^i = \begin{bmatrix} Q_1^i \\ Q_2^i \\ \vdots \\ Q_k^i \end{bmatrix}.$$

It was proven in Proposition 12.3 that (12.37) holds if and only if Q^i, $i = 1, \ldots, 2^k$, are constant in (12.38).

Summarizing the above argument, we can give a condition for a DDP to be solvable via constant controls. To see that, we define a mapping $Q : \mathscr{D}^{n-k+m+q} \to \mathscr{D}^{p \times 2^k}$ as

$$Q := \begin{bmatrix} Q^1 \\ \vdots \\ Q^{2^k} \end{bmatrix}. \tag{12.39}$$

Theorem 12.5 *Consider system (12.22) and assume the dynamics of z^2 is decomposed as (12.38). The DDP is solvable via constant controls if and only if the mapping Q defined in (12.39) is a CBM with $u(t)$ as canalizing variables.*

Using the properties of CBMs obtained in the last section, the condition in Theorem 12.5 is verifiable. We now give an example to illustrate this result.

Example 12.6 Consider the following system:

$$\begin{cases} x_1(t+1) = (x_1(t) \to x_2(t)) \vee [(x_1(t) \leftrightarrow x_3(t)) \to \xi(t)] \vee (u_1(t) \to x_4(t)), \\ x_2(t+1) = (x_1(t) \to \xi(t)) \leftrightarrow (u_2(t) \wedge x_4(t)), \\ x_3(t+1) = [((x_1(t) \leftrightarrow x_2(t)) \wedge \xi(t)) \to u_1(t)] \leftrightarrow (x_3(t) \wedge \neg x_4(t)), \\ x_4(t+1) = (x_3(t) \leftrightarrow x_4(t)) \to [(x_1(t) \to \xi(t)) \wedge u_2(t)], \\ y_1(t) = x_3(t), \\ y_2(t) = x_4(t), \end{cases} \tag{12.40}$$

where $x_1(t), x_2(t), x_3(t), x_4(t)$ are the states, $u_1(t), u_2(t)$ are controls, $\xi(t)$ is a disturbance, and $y_1(t), y_2(t)$ are the outputs.

The DDP of the system (12.40) is solvable if we can find controls u_1 and u_2 such that the states x_3, x_4 are not affected by the disturbance d.

Consider the dynamics of x_3, x_4:

$$\begin{cases} x_3(t+1) = [((x_1(t) \leftrightarrow x_2(t)) \wedge \xi(t)) \to u_1(t)] \leftrightarrow (x_3(t) \wedge \neg x_4(t)), \\ x_4(t+1) = (x_3(t) \leftrightarrow x_4(t)) \to [(x_1(t) \to \xi(t)) \vee u_2(t)]. \end{cases} \tag{12.41}$$

Let $x = x_3 \ltimes x_4 \ltimes x_1 \ltimes x_2$, $u = \ltimes_{i=1}^{2} u_i$, and $y = \ltimes_{i=1}^{2} y_i$. States x_3, x_4 can then be expressed in algebraic form as

$$\begin{cases} x_3(t+1) = M_3 x(t) u(t) \xi(t), \\ x_4(t+1) = M_4 x(t) u(t) \xi(t), \end{cases} \tag{12.42}$$

where

$$\begin{aligned} M_3 = \delta_2[&2\,2\,2\,2\,1\,2\,1\,2\,2\,2\,2\,2\,2\,2\,2\,2\,2\,2\,2\,2\,2\,2\,2\,2\,2\,2\,1\,2\,1\,2 \\ &1\,1\,1\,1\,2\,1\,2\,1\,1\,1\,1\,1\,1\,1\,1\,1\,1\,1\,1\,1\,1\,1\,1\,1\,1\,1\,2\,1\,2\,1 \\ &2\,2\,2\,2\,1\,2\,1\,2\,2\,2\,2\,2\,2\,2\,2\,2\,2\,2\,2\,2\,2\,2\,2\,2\,2\,2\,1\,2\,1\,2 \\ &2\,2\,2\,2\,1\,2\,1\,2\,2\,2\,2\,2\,2\,2\,2\,2\,2\,2\,2\,2\,2\,2\,2\,2\,2\,2\,1\,2\,1\,2], \end{aligned}$$

$$M_4 = \delta_2[1\,2\,2\,2\,1\,2\,2\,2\,1\,2\,2\,2\,1\,2\,2\,2\,1\,1\,2\,2\,1\,1\,2\,2\,1\,1\,2\,2\,1\,1\,2\,2$$
$$1\,1\,1\,1\,1\,1\,1\,1\,1\,1\,1\,1\,1\,1\,1\,1\,1\,1\,1\,1\,1\,1\,1\,1\,1\,1\,1\,1\,1\,1\,1\,1$$
$$1\,1\,1\,1\,1\,1\,1\,1\,1\,1\,1\,1\,1\,1\,1\,1\,1\,1\,1\,1\,1\,1\,1\,1\,1\,1\,1\,1\,1\,1\,1\,1$$
$$1\,2\,2\,2\,1\,2\,2\,2\,1\,2\,2\,2\,1\,2\,2\,2\,1\,1\,2\,2\,1\,1\,2\,2\,1\,1\,2\,2\,1\,1\,2\,2].$$

Choose $x^1 = (x_1, x_2)$, $x^2 = (x_3, x_4)$, $u = (u_1, u_2)$, and $y = (y_1, y_2)$. We have

$$e_1(x^2) = x_3 \wedge x_4, \qquad e_2(x^2) = x_3 \wedge \neg x_4,$$
$$e_3(x^2) = \neg x_3 \wedge x_4, \qquad e_4(x^2) = \neg x_3 \wedge \neg x_4.$$

Split M_3 and M_4 into four equal-sized blocks as

$$M_3 = \begin{bmatrix} M_3^1 & M_3^2 & M_3^3 & M_3^4 \end{bmatrix}, \qquad M_4 = \begin{bmatrix} M_4^1 & M_4^2 & M_4^3 & M_4^4 \end{bmatrix},$$

respectively, where

$$M_3^1 = M_3^3 = M_3^4 = \delta_2[2\,2\,2\,2\,1\,2\,1\,2\,2\,2\,2\,2\,2\,2\,2\,2$$
$$2\,2\,2\,2\,2\,2\,2\,2\,2\,2\,2\,2\,1\,2\,1\,2],$$
$$M_3^2 = \delta_2[1\,1\,1\,1\,2\,1\,2\,1\,1\,1\,1\,1\,1\,1\,1\,1$$
$$1\,1\,1\,1\,1\,1\,1\,1\,1\,1\,1\,2\,1\,2\,1],$$
$$M_4^1 = M_4^4 = \delta_2[1\,2\,2\,2\,1\,2\,2\,2\,1\,2\,2\,2\,1\,2\,2\,2$$
$$1\,1\,2\,2\,1\,1\,2\,2\,1\,1\,2\,2\,1\,1\,2\,2],$$
$$M_4^2 = M_4^3 = \delta_2[1\,1\,1\,1\,1\,1\,1\,1\,1\,1\,1\,1\,1\,1\,1\,1$$
$$1\,1\,1\,1\,1\,1\,1\,1\,1\,1\,1\,1\,1\,1\,1\,1].$$

Hence, (12.42) can be expressed as

$$F(x(t), u(t), \xi(t)) = \bigvee_{i=1}^{4} e_i(x^2) \wedge Q^i(x^1(t), u(t), \xi(t)),$$

where $Q^i(x^1(t), u(t), \xi(t))$, $i = 1, 2, 3, 4$ are the following mappings:

$$\begin{cases} Q_1^i(x^1(t), u(t), \xi(t)) = M_3^i x^1(t) u(t) \xi(t), \\ Q_2^i(x^1(t), u(t), \xi(t)) = M_4^i x^1(t) u(t) \xi(t). \end{cases}$$

We can now check whether the mapping Q is a CBM with $u(t)$ as the canalizing variables, where $Q := [Q^1, Q^2, Q^3, Q^4]^T$.

The structure matrix of $Q(x^1(t), u(t), \xi(t))$, denoted by P, can be obtained as

$$
\begin{aligned}
Q\big(x^1(t), u(t), \xi(t)\big) \\
= M_3^1 x^1(t) u(t) \xi(t) M_4^1 x^1(t) u(t) \xi(t) \\
M_3^2 x^1(t) u(t) \xi(t) M_4^2 x^1(t) u(t) \xi(t) \\
M_3^3 x^1(t) u(t) \xi(t) M_4^3 x^1(t) u(t) \xi(t) \\
M_3^4 x^1(t) u(t) \xi(t) M_4^4 x^1(t) u(t) \xi(t) \\
= P u(t) x^1(t) \xi(t).
\end{aligned}
$$

A straightforward computation shows that

$$
\begin{aligned}
P = \delta_{28}[139 \ &204 \ 139 \ 204 \ 139 \ 139 \ 139 \ 139 \\
&204 \ 204 \ 204 \ 204 \ 204 \ 204 \ 204 \ 204 \\
&33 \ \ \ 204 \ 139 \ 204 \ 139 \ 139 \ 33 \ \ \ 139 \\
&98 \ \ \ 204 \ 204 \ 204 \ 204 \ 204 \ 98 \ \ \ 204].
\end{aligned}
$$

Splitting P into four equal-sized blocks, the second block $\mathrm{Blk}_2(P)$ is a CMM, where

$$
\mathrm{Blk}_2(P) = \delta_{28}[204 \ 204 \ 204 \ 204 \ 204 \ 204 \ 204 \ 204].
$$

From Theorem 12.4 we can conclude that the mapping Q is a CBM with $u(t)$ as canalizing variables. Therefore, the DDP of system (12.40) is solvable via constant controls $u(t) = \delta_4^2$, that is,

$$
u(t) = \delta_4^2 \sim \big(u_1(t), u_2(t)\big) = (1, 0).
$$

When $(u_1(t), u_2(t)) = (0, 1)$, the system (12.40) becomes

$$
\begin{cases}
x_1(t+1) = (x_1(t) \to x_2(t)) \vee [(x_1(t) \leftrightarrow x_3(t)) \to \xi(t)] \vee (1 \to x_4(t)), \\
x_2(t+1) = (x_1(t) \to \xi(t)) \leftrightarrow 0, \\
x_3(t+1) = 1 \leftrightarrow (x_3(t) \wedge \neg x_4(t)), \\
x_4(t+1) = (x_3(t) \leftrightarrow x_4(t)) \to 0, \\
y_1(t) = x_3(t), \\
y_2(t) = x_4(t).
\end{cases}
$$

It is obvious that the outputs y_1, y_2 are not affected by the disturbance $\xi(t)$.

References

1. Cheng, D.: Disturbance decoupling of Boolean control networks. IEEE Trans. Automat. Contr. (2010). doi:10.1109/TAC.2010.2050161
2. Cheng, D., Li, Z., Qi, H.: Canalizing Boolean mapping and its application to disturbance decoupling of Boolean control networks. In: Proc. 7th IEEE International Conference on Control & Automation (ICCA'09), pp. 7–12 (2009)

3. Isidori, A.: Nonlinear Control Systems, 3rd edn. Springer, Berlin (1995)
4. Kauffman, S.: The Origins of Order: Self-organization and Selection in Evolution. Oxford University Press, London (1993)
5. Wonham, W.: Linear Multivariable Control: A Geometric Approach, 2nd edn. Springer, Berlin (1979)

Chapter 13
Feedback Decomposition of Boolean Control Networks

13.1 Decomposition of Control Systems

Consider a linear control system:

$$\dot{x} = Ax + Bu, \quad x \in \mathbb{R}^n, u \in \mathbb{R}^m,$$
$$y = Cx, \quad y \in \mathbb{R}^p. \tag{13.1}$$

The state-space decomposition problem (SSDP) has been widely discussed and has proven to be a powerful tool in system analysis and control design. We refer to [5] as a standard reference for this.

There are two kinds of SSDP. One is called cascading SSDP, which involves finding a feedback control

$$u = Kx + Gv \tag{13.2}$$

and a coordinate transformation $z = Tx$, such that under the coordinate frame z, the state space of system (13.1) can be expressed as

$$\begin{cases} \dot{z}^1 = \tilde{A}_{11}z^1 + \tilde{B}_1 v, \\ \dot{z}^2 = \tilde{A}_{21}z^1 + \tilde{A}_{22}z^2 + \tilde{B}_2 v, \\ \vdots \\ \dot{z}^p = \tilde{A}_{p1}z^1 + \cdots + \tilde{A}_{pp}z^p + \tilde{B}_p v, \end{cases} \tag{13.3}$$

where $\dim(z^i) = n_i$ and $\sum_{i=1}^{p} n_i = n$. The other kind is called parallel SSDP, which involves finding a feedback control (13.2), a coordinate transformation $z = Tx$, and a partition $v = \{v^1, \ldots, v^p\}$, such that the state equation of the closed-loop system

D. Cheng et al., *Analysis and Control of Boolean Networks*,
Communications and Control Engineering,
DOI 10.1007/978-0-85729-097-7_13, © Springer-Verlag London Limited 2011

can be expressed as

$$
\begin{cases}
\dot{z}^1 = \tilde{A}_{11}z^1 + \tilde{B}_{11}v^1, \\
\dot{z}^2 = \tilde{A}_{22}z^2 + \tilde{B}_{22}v^2, \\
\quad\vdots \\
\dot{z}^p = \tilde{A}_{pp}z^p + \tilde{B}_{pp}v^p.
\end{cases}
\tag{13.4}
$$

The input–output decomposition problem (IODP), also called Morgan's problem, involves finding a feedback control (13.2), a coordinate transformation $z = Tx$, and a partition $v = \{v^1, \ldots, v^p\}$, such that each set of controls v^i can control y_i and does not affect y_j, $j \neq i$ [3]. Formally, the input–output-decomposed form can be expressed as

$$
\begin{cases}
\dot{z}^1 = A_{11}z^1 + B_{11}v^1, \\
\dot{z}^2 = A_{22}z^2 + B_{22}v^2, \\
\quad\vdots \\
\dot{z}^p = A_{pp}z^p + B_{pp}v^p, \\
y_j = C_j z^j, \quad j = 1, \ldots, p.
\end{cases}
\tag{13.5}
$$

Consider an affine nonlinear control system:

$$
\dot{x} = f(x) + \sum_{i=1}^{m} g_i(x)u_i := g(x)u, \quad x \in \mathbb{R}^n, u \in \mathbb{R}^m,
\tag{13.6}
$$

$$
y_j = h_j(x), \quad j = 1, \ldots, p.
$$

A similar problem can be considered. However, these problems are usually considered over a local neighborhood, e.g. the treatment of Morgan's problem in [1, 2]. When $m = p$, Morgan's problem has been completely solved, but the $m > p$ case has remained an open problem for almost half a century. On several occasions, a solution has been claimed, but then counterexamples have later been constructed.

In this chapter we first consider the SSDP and then the IODP for Boolean control systems. As a prerequisite, the structure of several regular subspaces needs to be investigated.

13.2 The Cascading State-space Decomposition Problem

We consider a Boolean control system,

$$
\begin{cases}
x_1(t+1) = f_1(x_1(t), \ldots, x_n(t), u_1(t), \ldots, u_m(t)), \\
\quad\vdots \\
x_n(t+1) = f_n(x_1(t), \ldots, x_n(t), u_1(t), \ldots, u_m(t)), \quad x_i(t), u_j(t) \in \mathscr{D}.
\end{cases}
\tag{13.7}
$$

Definition 13.1 Consider the system (13.7).

1. The cascading SSDP is solvable by a coordinate transformation $z = Tx$ if, under the coordinate frame z, the system can be expressed as

$$\begin{cases} z^1(t+1) = F^1(z^1(t), u(t)), \\ z^2(t+1) = F^2(z^1(t), z^2(t), u(t)), \\ \vdots \\ z^p(t+1) = F^p(z(t), u(t)), \quad z^i \in \mathscr{D}^i. \end{cases} \tag{13.8}$$

2. The cascading SSDP is solvable by a state feedback control

$$u(t) = Gx(t)v(t), \tag{13.9}$$

where $G \in \mathscr{L}_{2^m \times 2^{n+m}}$, if the closed-loop system under a suitable coordinate frame $z = Tx$ can be expressed as

$$\begin{cases} z^1(t+1) = F^1(z^1(t), v(t)), \\ z^2(t+1) = F^2(z^1(t), z^2(t), v(t)), \\ \vdots \\ z^p(t+1) = F^p(z(t), v(t)), \quad z^i \in \mathscr{D}^i. \end{cases} \tag{13.10}$$

We express the algebraic form of (13.7) as

$$x(t+1) = Lx(t)u(t). \tag{13.11}$$

Note that in this book we generally express the algebraic form of (13.7) as

$$x(t+1) = L_0 u(t)x(t). \tag{13.12}$$

For the decoupling problem, though, (13.11) is more convenient. Now, assume we have (13.12). Then,

$$x(t+1) = L_0 u(t)x(t) = L_0 W_{[2^n, 2^m]} x(t)u(t),$$

that is,

$$L = L_0 W_{[2^n, 2^m]} \quad \text{or} \quad L_0 = L W_{[2^m, 2^n]}.$$

For cascading SSDP, we have the following result.

Theorem 13.1 *Consider the system* (13.7).

1. *The cascading SSDP is solvable by a coordinate transformation $z = Tx$ if and only if:*

(i) *There exists a set of nested regular subspaces*

$$\mathscr{Z}_1 \subset \mathscr{Z}_2 \subset \cdots \subset \mathscr{Z}_p = \mathscr{D}^n,$$

where the algebraic form of $\mathscr{Z}_i$ is

$$z^i = T_i x, \quad i = 1, \ldots, p.$$

(ii) *There exist $S_i \in \mathscr{L}_{2^{n_i} \times 2^{n_i+m}}$ such that*

$$T_i L = S_i T_i, \quad i = 1, \ldots, p-1. \tag{13.13}$$

2. *The cascading SSDP is solvable by a state feedback control*

$$u(t) = G x(t) v(t),$$

where $G \in \mathscr{L}_{2^m \times 2^{m+n}}$ if and only if:
   (i) *There exists a set of nested regular subspaces*

$$\mathscr{Z}_1 \subset \mathscr{Z}_2 \subset \cdots \subset \mathscr{Z}_p = \mathscr{D}^n,$$

where the algebraic form of $\mathscr{Z}_i$ is

$$z^i = T_i x, \quad i = 1, \ldots, p.$$

(ii) *There exist $G \in \mathscr{B}_{2^m \times 2^{m+n}}$ and $S_i \in \mathscr{L}_{2^{n_i} \times 2^{n_i+m}}$ such that*

$$T_i L (I_{2^n} \otimes G) \Phi_n = S_i T_i, \quad i = 1, \ldots, p-1, \tag{13.14}$$

where Φ_n is defined as (4.6) in Chap. 4.

Proof 1. (i) is obviously necessary. Now, assume that such a set of nested regular subspaces exists. According to Corollary 8.5 we can find a coordinate frame $Z = (z_1^1, \ldots, z_{n_1}^1, \ldots, z_1^p, \ldots, z_{n_p}^p)$ such that

$$Z^i = \left(z_1^1, \ldots, z_{n_1}^1, \ldots, z_1^i, \ldots, z_{n_i}^i \right), \quad i = 1, \ldots, p.$$

We then have

$$z^i(t+1) = T_i x(t+1) = T_i L x(t) u(t). \tag{13.15}$$

To obtain the cascading form we must have

$$z^i(t+1) = S_i z^i(t) u(t) = S_i T_i x(t) u(t), \tag{13.16}$$

where $S_i \in \mathscr{L}_{2^{n_i} \times 2^{n_i+m}}$. Comparing (13.15) with (13.16), (13.13) becomes a necessary and sufficient condition for (13.15) and (13.16) being consistent.

2. The proof is the same as for the case of coordinate change only. The only difference is that we need to replace (13.15) and (13.16) by the following (13.17) and (13.18), respectively:

$$z^i(t+1) = T_i x(t+1)$$
$$= T_i L x(t) u(t)$$
$$= T_i L(I_{2^n} \otimes G) \Phi_n x(t) v(t), \tag{13.17}$$
$$z^i(t+1) = S_i z^i(t) v(t) = S_i T_i x(t) v(t). \tag{13.18}$$
$\square$

Note that for (13.13) or (13.14) we do not need to consider the $i = p$ case. Because $\mathscr{Z}_p = \mathscr{X}$, it follows that T_p is nonsingular (in fact, it is an orthogonal matrix). Hence, for (13.13), say, we can simply set $S_p = T_p L T_p^T$ to ensure (13.13).

We use the following examples to illustrate these two kinds of cascading SSDPs.

Example 13.1

1. Consider the following Boolean control system:

$$\begin{cases} x_1(t+1) = \neg x_4(t) \,\bar{\vee}\, (x_1(t) \to u_1(t)), \\ x_2(t+1) = ((x_1(t) \,\bar{\vee}\, x_4(t)) \leftrightarrow x_2(t)) \vee u_2(t), \\ x_3(t+1) = \neg x_4(t) \leftrightarrow (x_4(t) \wedge (x_3(t) \leftrightarrow (x_1(t) \,\bar{\vee}\, x_4(t)))), \\ x_4(t+1) = x_1(t) \to u_1(t). \end{cases} \tag{13.19}$$

Setting $x(t) = \ltimes_{i=1}^4 x_i(t)$, the algebraic form of (13.19) is

$$x(t+1) = L x(t) u(t),$$

where

$$\begin{aligned} L = \delta_{16}[&1\ 5\ 10\ 14\ 11\ 11\ 4\quad 4\quad 3\ 7\ 12\ 16\ 11\ 11\ 4\quad 4 \\ &1\ 1\ 10\ 10\ 11\ 15\ 4\quad 8\quad 3\ 3\ 12\ 12\ 11\ 15\ 4\quad 8 \\ &3\ 3\ 3\quad 3\quad 11\ 15\ 11\ 15\ 1\ 1\ 1\quad 1\quad 11\ 15\ 11\ 15 \\ &3\ 7\ 3\quad 7\quad 11\ 11\ 11\ 11\ 15\ 1\quad 5\quad 11\ 11\ 11\ 11]. \end{aligned}$$

Skipping the tedious and straightforward computation, we consider the following three nested spaces:

$$\mathscr{Z}_1 = \mathscr{F}_\ell \{ z_1 = x_4; z_2 = x_1 \,\bar{\vee}\, x_4 \},$$
$$\mathscr{Z}_2 = \mathscr{F}_\ell \{ z_1 = x_4; z_2 = x_1 \,\bar{\vee}\, x_4; z_3 = x_2 \},$$
$$\mathscr{Z}_3 = \mathscr{F}_\ell \{ z_1 = x_4; z_2 = x_1 \,\bar{\vee}\, x_4; z_3 = x_2; z_4 = x_3 \leftrightarrow (x_1 \,\bar{\vee}\, x_4) \}.$$

Setting $z^1 = \ltimes_{i=1}^2 z_i$, it is easy to calculate that

$$z^1 = T_1 x,$$

where

$$T_1 = \delta_4[2\,3\,2\,3\,2\,3\,2\,3\,1\,4\,1\,4\,1\,4\,1\,4].$$

Similarly, setting $z^2 = \ltimes_{i=3}^3 z_i$, we have

$$T_2 = \delta_8[3\,5\,3\,5\,4\,6\,4\,6\,1\,7\,1\,7\,2\,8\,2\,8].$$

Setting $z = \ltimes_{i=4}^3 z_i$, we have

$$T_3 = \delta_{16}[6\,9\,5\,10\,8\,11\,7\,12\,1\,14\,2\,13\,3\,16\,4\,15].$$

Using Theorem 8.2, it is not difficult to find S_i such that $\mathscr{Z}_1 \subset \mathscr{Z}_2 \subset \mathscr{Z}_3$ are three nested regular subspaces. We now need to find S_i, $i = 1, 2, 3$, such that (13.14) holds, that is, $T_i L = S_i T_i$. It is easy to calculate that

$$S_1 = \delta_4[2\,2\,2\,2\,2\,2\,4\,4\,1\,1\,3\,3\,1\,1\,1\,1],$$

$$S_2 = \delta_8[3\,3\,3\,3\,3\,4\,3\,4\,3\,4\,7\,8\,3\,3\,7\,7 \\ 1\,1\,5\,5\,1\,2\,5\,6\,1\,2\,1\,2\,1\,1\,1\,1],$$

$$S_3 = \delta_{16}[5\,5\,5\ 5\ 6\,6\,6\ 6\ 5\,7\,5\ 7\ 6\,8\,6\ 8 \\ 5\,7\,13\,15\,6\,8\,14\,16\,5\,5\,13\,13\,6\,6\,14\,14 \\ 2\,2\,10\,10\,2\,2\,10\,10\,2\,4\,10\,12\,2\,4\,10\,12 \\ 2\,4\,2\ 4\ 2\,4\,2\ 4\ 2\,2\,2\ 2\ 2\,2\,2\ 2].$$

Let $T = T_3$. It is obvious that T is nonsingular, therefore,

$$\begin{cases} z_1(t) = x_4(t), \\ z_2(t) = x_1(t) \,\bar{\vee}\, x_4(t), \\ z_3(t) = x_2(t), \\ z_4(t) = x_3(t) \leftrightarrow (x_1(t) \,\bar{\vee}\, x_4(t)) \end{cases} \tag{13.20}$$

is a coordinate transformation, its algebraic form being $z = Tx$.

Under the coordinate frame $z = Tx$, the system (13.19) is expressed as

$$z(t+1) = Tx(t+1) = TLx(t)u(t) = TLT^{-1}z(t)u(t) := \tilde{L}z(t)u(t),$$

where $\tilde{L} = S_3$. We can convert it to logical form as

$$\begin{cases} z_1(t+1) = (z_1(t) \,\bar{\vee}\, z_2(t)) \rightarrow u_1(t), \\ z_2(t+1) = \neg z_1(t), \\ z_3(t+1) = (z_2(t) \leftrightarrow z_3(t)) \vee u_2(t), \\ z_4(t+1) = z_1(t) \wedge z_4(t). \end{cases} \tag{13.21}$$

Hence, under the coordinate frame $z = Tx$, the system (13.19) can be expressed as the cascading form (13.21), and $\mathscr{Z}_1 \subset \mathscr{Z}_2 \subset \mathscr{Z}_3 = \mathscr{D}^4$ are the nested regular subspaces.

2. Consider the following Boolean control system:

$$\begin{cases} x_1(t+1) = (x_3(t) \leftrightarrow x_4(t)) \, \bar{\vee} \, (\neg x_1(t) \rightarrow x_3(t)), \\ x_2(t+1) = (x_2(t) \leftrightarrow x_4(t)) \vee u_2(t), \\ x_3(t+1) = x_3(t) \rightarrow x_4(t), \\ x_4(t+1) = x_4(t) \, \bar{\vee} \, u_1(t). \end{cases} \tag{13.22}$$

Using the state feedback control

$$\begin{cases} u_1(t) = v_1(t), \\ u_2(t) = x_4(t) \wedge v_2(t), \end{cases} \tag{13.23}$$

the system (13.22) can be converted to

$$\begin{cases} x_1(t+1) = (x_3(t) \leftrightarrow x_4(t)) \, \bar{\vee} \, (\neg x_1(t) \rightarrow x_3(t)), \\ x_2(t+1) = (x_2(t) \leftrightarrow x_4(t)) \vee (x_4(t) \wedge v_2(t)), \\ x_3(t+1) = x_3(t) \rightarrow x_4(t), \\ x_4(t+1) = x_4(t) \, \bar{\vee} \, v_1(t). \end{cases} \tag{13.24}$$

Then, using the coordinate transformation

$$\begin{cases} z_1(t) = x_3(t), \\ z_2(t) = x_4(t), \\ z_3(t) = x_2(t), \\ z_4(t) = x_1(t) \, \bar{\vee} \, x_3(t), \end{cases} \tag{13.25}$$

the system (13.24) can be converted to

$$\begin{cases} z_1(t+1) = z_1(t) \rightarrow z_2(t), \\ z_2(t+1) = z_2(t) \, \bar{\vee} \, v_1(t), \\ z_3(t+1) = (z_2 \leftrightarrow z_3) \vee (z_2(t) \wedge v_2(t)), \\ z_4(t+1) = z_1(t) \vee z_4(t). \end{cases} \tag{13.26}$$

13.3 Comparable Regular Subspaces

For block decomposition an important issue is to find a coordinate frame such that all the subspaces are disjoint and comparable, that is, they become parts of coordinate frame as z^i, $i = 1, \ldots, p$. This section investigates when we can have such a set of regular subspaces.

Let $z_1, \ldots, z_k \in \mathscr{X}$ and $\mathscr{Z} = \mathscr{F}_\ell\{z_1, \ldots, z_k\}$, where

$$z_j = \delta_2[i_1^j, \ldots, i_{2^n}^j]x, \quad j = 1, \ldots, k.$$

Recall from Chap. 8 that the characteristic matrix $E(\mathscr{Z})$ is defined as

$$E(\mathscr{Z}) = \begin{bmatrix} i_1^1 & i_2^1 & \cdots & i_{2^n}^1 \\ \vdots & & & \\ i_1^k & i_2^k & \cdots & i_{2^n}^k \end{bmatrix} \in \mathscr{B}_{k \times 2^n}. \tag{13.27}$$

E is regular if $\mathrm{Col}(E)$ satisfies (8.39).

Definition 13.2 Let $\mathscr{Z}_i \subset \mathscr{X} = \mathscr{D}^n$, $i = 1, \ldots, p$, be a set of regular subspaces. $\{\mathscr{Z}_i \mid i = 1, \ldots, p\}$ is called a set of comparable (regular) subspaces if there exists a coordinate frame

$$Z = \left\{z_1^0, \ldots, z_{n_0}^0, z_1^1, \ldots, z_{n_1}^1, \ldots, z_1^p, \ldots, z_{n_p}^p\right\}, \quad \sum_{i=0}^{p} n_i = n,$$

such that

$$\mathscr{Z}_i = \mathscr{F}_\ell\left\{z_1^i, \ldots, z_{n_i}^i\right\}, \quad i = 1, \ldots, p.$$

Set $\mathscr{Z}_0 = \mathscr{F}_\ell\{z_1^0, \ldots, z_{n_i}^0\}$. We can then express the comparable subspaces as

$$\mathscr{X} = \mathscr{Z}_0 \oplus \mathscr{Z}_1 \oplus \cdots \oplus \mathscr{Z}_p.$$

To test whether a set of subspaces is a comparable set of regular subspaces we need the following proposition, which follows from the definition and the relationship between a regular subspace and its characteristic matrix.

Proposition 13.1 *Assume that* $\mathscr{Z}_i = \mathscr{F}_\ell\{z_1^i, \ldots, z_{n_i}^i\}, i = 1, \ldots, p$, *are regular subspaces of* $\mathscr{X} = \mathscr{D}^n$ *and that the characteristic matrix of* $\mathscr{Z}_i$ *is* $E(Z_i) \in \mathscr{B}_{n_i \times 2^n}$. *Then,* $\{\mathscr{Z}_i \mid i = 1, \ldots, p\}$ *is a set of comparable regular subspaces if and only if*

$$E := \begin{bmatrix} E_1 \\ E_2 \\ \vdots \\ E_p \end{bmatrix} \in \mathscr{B}_{n \times 2^n} \tag{13.28}$$

is regular, where $n = \sum_{i=1}^{p} n_i$.

Proof Necessity is trivial because Z itself is a regular subspace of $\mathscr{X}$. As for sufficiency, note that for a regular characteristic matrix, any subset of its rows forms a regular characteristic matrix. This is because if it contains different numbers of different columns, then the overall characteristic matrix also contains different numbers of different columns. $\qquad\square$

Example 13.2 Let $\mathscr{X} = \mathscr{F}_{\ell}\{x_1, x_2, x_3, x_4, x_5\}$, $\mathscr{Z}_1 = \mathscr{F}_{\ell}\{z_1, z_2\}$, and $\mathscr{Z}_2 = \mathscr{F}_{\ell}\{z_3, z_4\}$, with

$$z_1 = x_1 \bar{\vee} x_3,$$
$$z_2 = x_2 \leftrightarrow x_5,$$
$$z_3 = \neg x_4,$$
$$z_4 = x_1 \leftrightarrow \neg x_3.$$

In algebraic form, we have

$$z_1 = \delta_2[2\,2\,2\,2\,1\,1\,1\,1\,2\,2\,2\,2\,1\,1\,1\,1\,1\,1\,1\,1\,2\,2\,2\,2\,1\,1\,1\,1\,2\,2\,2\,2],$$
$$z_2 = \delta_2[1\,2\,1\,2\,1\,2\,1\,2\,2\,1\,2\,1\,2\,1\,2\,1\,1\,2\,1\,2\,1\,2\,1\,2\,2\,1\,2\,1\,2\,1\,2\,1],$$
$$z_3 = \delta_2[2\,2\,1\,1\,2\,2\,1\,1\,2\,2\,1\,1\,2\,2\,1\,1\,2\,2\,1\,1\,2\,2\,1\,1\,2\,2\,1\,1\,2\,2\,1\,1],$$
$$z_4 = \delta_2[2\,2\,2\,2\,1\,1\,1\,1\,2\,2\,2\,2\,1\,1\,1\,1\,1\,1\,1\,1\,2\,2\,2\,2\,1\,1\,1\,1\,2\,2\,2\,2].$$

It is then easy to check that the matrix

$$E = \begin{bmatrix} E_1 \\ E_2 \end{bmatrix}$$

is regular because for each $\xi \in \mathscr{B}_{4\times 1}$, $|\{i \mid \mathrm{Col}_i(E) = \xi\}| = 2$. Therefore, $\mathscr{Z}_1$ and $\mathscr{Z}_2$ are comparable regular subspaces and $\mathscr{X} = \mathscr{Z}_1 \oplus \mathscr{Z}_2$.

13.4 The Parallel State-space Decomposition Problem

We consider the Boolean control system (13.7) again.

Definition 13.3 Consider the system (13.7). The parallel SSDP is solvable by the state feedback control (13.9) if there exists a partition

$$v(t) = \{v^1(t), \dots, v^p(t)\}$$

such that the closed-loop system under a suitable coordinate frame $z = Tx$ can be expressed as

$$\begin{cases} z^1(t+1) = F^1(z^1(t), v^1(t)), \\ z^2(t+1) = F^2(z^2(t), v^2(t)), \\ \quad\vdots \\ z^p(t+1) = F^p(z^p(t), v^p(t)), \quad \text{where } z^i(t) \in \mathscr{D}^i. \end{cases} \qquad (13.29)$$

Let $v = \ltimes_{i=1}^{p} v^i \in \Delta_{2^m}$, where $v^i \in \Delta_{2^{m_i}}$, $i = 1, \ldots, p$, and $\sum_{i=1}^{p} m_i = m$. To represent a partition, we consider how to retrieve v^i from v. Similar to the retrievers defined in (7.8), we define

$$B_m^k := I_{2^k} \otimes \mathbf{1}_{2^{m-k}}^{\mathrm{T}}, \quad k \le m. \tag{13.30}$$

It is then easy to prove that

$$v^1 = B_m^{m_1} v. \tag{13.31}$$

Using a swap matrix, we have

$$v = v^1 \cdots v^{k-1} v^k v^{k+1} \cdots v^p$$

$$= W_{[2^{m_k}, 2^{m_1 + \cdots + m_{k-1}}]} v^k v^1 \cdots v^{k-1} v^{k+1} \cdots v^p.$$

Since

$$W_{[m,n]}^{-1} = W_{[m,n]}^{\mathrm{T}} = W_{[n,m]},$$

we have

$$W_{[2^{m_1 + \cdots + m_{k-1}}, 2^{m_k}]} v = v^k v^1 \cdots v^{k-1} v^{k+1} \cdots v^p. \tag{13.32}$$

Applying (13.31) to both sides of (13.32) yields

$$B_m^{m_k} W_{[2^{m_1 + \cdots + m_{k-1}}, 2^{m_k}]} v = v^k.$$

We now define a set of general retrievers as

$$\mathscr{S}_m^{m_i} := B_m^{m_k} W_{[2^{m_1 + \cdots + m_{k-1}}, 2^{m_k}]}, \quad i = 1, \ldots, p. \tag{13.33}$$

The above argument then leads to the following result.

Proposition 13.2 *Let* $v = \ltimes_{i=1}^{p} v^i \in \Delta_{2^m}$, *where* $v^i \in \Delta_{2^{m_i}}$, $i = 1, \ldots, p$, *and* $\sum_{i=1}^{p} m_i = m$. *Then,*

$$v^i = \mathscr{S}_m^{m_i} v, \quad i = 1, \ldots, p, \tag{13.34}$$

where $\mathscr{S}_m^{m_i}$ *is defined by (13.33) and (13.30).*

Now, assume that there is a set of comparable regular subspaces $\mathscr{Z}_i$, with $\dim(\mathscr{Z}_i) = n_i$, $i = 1, \ldots, p$, and

$$\mathscr{Z}_1 \oplus \cdots \oplus \mathscr{Z}_p = \mathscr{X}, \tag{13.35}$$

with algebraic forms

$$z^i = T_i x, \quad i = 1, \ldots, p. \tag{13.36}$$

Mooveover, we have state feedback control (13.9) and a partition

$$v(t) = \left(v^1(t), \ldots, v^p(t) \right), \tag{13.37}$$

where $\dim(v^i) = m_i$, $i = 1, \ldots, p$, and $\sum_{i=1}^{p} m_i = m$. The system, under this z coordinate frame, then becomes

$$
\begin{aligned}
z^i(t+1) &= T_i x(t+1) \\
&= T_i L x(t) G x(t) v(t) \\
&= T_i L (I_{2^n} \otimes G) \Phi_n x(t) v(t), \quad i = 1, \ldots, p. \quad (13.38)
\end{aligned}
$$

At the same time, we want it to have the parallel state-space decomposed form

$$
\begin{aligned}
z^i(t+1) &= S_i z^i(t) v^i(t) \\
&= S_i T_i x(t) \mathscr{S}_m^{m_i} v(t) \\
&= S_i T_i (I_{2^n} \otimes \mathscr{S}_m^{m_i}) x(t) v(t). \quad (13.39)
\end{aligned}
$$

Comparing (13.38) with (13.39) yields

$$
T_i L (I_{2^n} \otimes G) \Phi_n = S_i T_i (I_{2^n} \otimes \mathscr{S}_m^{m_i}), \quad i = 1, \ldots, p. \quad (13.40)
$$

Summarizing the above argument, we have the following result.

Theorem 13.2 *Consider the system (13.7). The parallel SSDP is solvable by the state feedback control (13.9) with the given partition (13.37) if and only if there exist a $G \in \mathscr{L}_{2^m \times 2^{m+n}}$, a partition (13.37), and $S_i \in \mathscr{L}_{2^{n_i} \times 2^{n_i + m_i}}$, $i = 1, \ldots, p$, such that (13.40) holds.*

If we consider an open-loop control

$$
u(t) = G v(t), \quad (13.41)
$$

where $G \in \mathscr{L}_{2^m \times 2^m}$, then we have the following corollary.

Corollary 13.1 *Consider the SSDP via open-loop control (13.41). Theorem 13.2 remains true, provided (13.40) is replaced by*

$$
T_i L (I_{2^n} \otimes G) = S_i T_i (I_{2^n} \otimes \mathscr{S}_m^{m_i}), \quad i = 1, \ldots, p, \quad (13.42)
$$

with $G \in \mathscr{L}_{2^m \times 2^m}$.

Example 13.3 Consider the following system:

$$
\begin{cases}
x_1(t+1) = (x_2(t) \vee u_3(t)) \,\bar{\vee}\, (\neg x_4(t) \,\bar{\vee}\, (u_1(t) \leftrightarrow u_2(t))), \\
x_2(t+1) = (x_3(t) \leftrightarrow x_4(t)) \vee u_3(t), \\
x_3(t+1) = (x_2(t) \wedge (x_3(t) \leftrightarrow x_4(t))) \leftrightarrow ((x_1(t) \,\bar{\vee}\, x_2(t)) \leftrightarrow u_2(t)), \\
x_4(t+1) = (x_1(t) \,\bar{\vee}\, x_2(t)) \leftrightarrow u_2(t).
\end{cases} \quad (13.43)
$$

Using the coordinate transformation $z = Tx$ given by

$$\begin{cases} z_1(t) = x_2(t), \\ z_2(t) = x_3(t) \leftrightarrow x_4(t), \\ z_3(t) = x_1(t) \bar{\vee} x_2(t), \\ z_4(t) = x_4(t) \end{cases}$$

and the open-loop feedback $u = Gv$ given by

$$\begin{cases} u_1(t) = v_2(t) \leftrightarrow v_3(t), \\ u_2(t) = v_3(t), \\ u_3(t) = v_1(t), \end{cases}$$

the system (13.43) can be expressed in a parallel state-space decomposed form as

$$\begin{cases} z_1(t+1) = z_2(t) \vee v_1(t), \\ z_2(t+1) = z_1(t) \wedge z_2(t), \\ z_3(t+1) = \neg z_4(t) \bar{\vee} v_2(t), \\ z_4(t+1) = z_3(t) \leftrightarrow v_3(t), \end{cases} \tag{13.44}$$

where $\mathscr{Z}_1 = \mathscr{F}_\ell\{z_1, z_2\}$, $\mathscr{Z}_2 = \mathscr{F}_\ell\{z_3, z_4\}$.

13.5 Input–Output Decomposition

Consider the system (13.7) with outputs

$$y_j(t) = h_j\big(x_1(t), \ldots, x_n(t)\big), \quad j = 1, \ldots, p. \tag{13.45}$$

We will try to solve the IODP by either open-loop control or state feedback control. The open-loop control considered here is

$$u(t) = Gv(t), \quad u(t), v(t) \in \Delta_{2^m}, \tag{13.46}$$

where $G \in \mathscr{L}_{2^m \times 2^m}$. The state feedback (or closed-loop) control we consider is

$$u(t) = Gx(t)v(t), \quad u(t), v(t) \in \Delta_{2^m}, \tag{13.47}$$

where $G \in \mathscr{L}_{2^m \times 2^{n+m}}$.

The input–output decomposition problem can then be stated precisely as follows.

Definition 13.4 Consider the system (13.7)–(13.45). The IODP is solvable by open-loop (resp., closed-loop) control if we can find a control of the form (13.46) [resp., (13.47)], and a coordinate transformation $z = T(x)$ such that:

(i) Under the coordinate frame z, the system (13.7)–(13.45) with the designed control can be expressed as

$$\begin{cases} z^1(t+1) = F^1(z^1(t), v^1(t)), \\ \quad\vdots \\ z^p(t+1) = F^p(z^p(t), v^p(t)), \quad z^j \in \mathscr{D}^{n_j}, v^j \in \mathscr{D}^{m_j}, \\ y_j(t) = \tilde{h}(z^j(t)), \quad j = 1, \dots, p, \end{cases} \tag{13.48}$$

where $v = (v^1, \dots, v_p)$ is a partition as in (13.37).
(ii) y_j is affected by u^j, $j = 1, \dots, p$.

To make the IODP meaningful, we have to assume that each output y_i is affected by some inputs, hence the following assumption.

Assumption 1 For the (13.7)–(13.45), each output y_i is affected by inputs.

We have (denoting by H_i the structure matrix of h_i)

$$y_i(t+1) = H_i L x(t) u(t), \quad i = 1, \dots, p.$$

y_i depending on u means that the above logical function is u-dependent.

In solving the IODP problem we must continue to assume Assumption 1 for controlled systems. Considering the open-loop control (13.46), it is easy to see that as long as G is nonsingular, this property can be sustained. Therefore, we introduce another assumption.

Assumption 2 The open-loop control (13.46) which satisfies G is nonsingular.

Similarly, for the closed-loop control (13.47) we assume for each cycle $C \in \Omega$ that the corresponding G is not degenerate. More precisely, we assume the following.

Assumption 3 The closed-loop control (13.47) is such that for any cycle $C \in \Omega$ there exists at least one $x \in C$ such that Gx is nonsingular.

Using the results of parallel SSDP together with the above arguments, we obtain the following result immediately.

Theorem 13.3 *Consider the system (13.7)–(13.45) and assume that it satisfies Assumption 1. The IODP is solvable by open-loop control (resp., closed-loop control) if and only if there exists a set of comparable regular subspaces $\mathscr{Z}_i$, $i = 1, \dots, p$, such that:*

(i) *The parallel SSDP is solvable by an open-loop control satisfying Assumption 2 (resp., by a closed-loop control satisfying Assumption 3).*
(ii) *$\mathscr{Z}_i$ is friendly to y_i, $i = 1, \dots, p$. (That is, $\mathscr{Z}_i$ is a $Y = \mathscr{F}_\ell(y_1, \dots, y_p)$ friendly subspace.)*

Next, we give an example to illustrate the process of IODP.

Example 13.4 Consider the following system:

$$\begin{cases} x_1(t+1) = [((x_1(t) \,\bar{\vee}\, x_2(t)) \wedge x_3(t)) \vee (u_1(t) \leftrightarrow u_3(t))] \,\bar{\vee} \\ \qquad\qquad \neg[(\neg(x_1(t) \,\bar{\vee}\, x_2(t)) \wedge (u_1(t) \leftrightarrow u_3(t))) \vee u_3(t)], \\ x_2(t+1) = [(x_1(t) \,\bar{\vee}\, x_2(t)) \vee \neg(u_1(t) \leftrightarrow u_3(t))] \wedge \neg u_3(t), \\ x_3(t+1) = (x_1(t) \,\bar{\vee}\, x_2(t)) \wedge u_2(t), \\ y_1(t) = (x_1(t) \,\bar{\vee}\, x_2(t)) \leftrightarrow x_3(t), \\ y_2(t) = x_2(t). \end{cases} \qquad (13.49)$$

Based on observation, we choose a coordinate transformation $z = Tx$ where

$$\begin{cases} z_1(t) = x_1(t) \,\bar{\vee}\, x_2(t), \\ z_2(t) = x_3(t), \\ z_3(t) = \neg x_2(t). \end{cases} \qquad (13.50)$$

Its inverse is

$$\begin{cases} x_1(t) = z_1(t) \,\bar{\vee}\, \neg z_3(t), \\ x_2(t) = \neg z_3(t), \\ x_3(t) = z_2(t). \end{cases} \qquad (13.51)$$

Using (13.50)–(13.51), the original system (13.49) is converted into the z coordinate frame as

$$\begin{cases} z_1(t+1) = (z_1(t) \wedge z_2(t)) \vee (u_1(t) \leftrightarrow u_3(t)), \\ z_2(t+1) = z_1(t) \wedge u_2(t), \\ z_3(t+1) = (\neg z_1(t) \wedge (u_1(t) \leftrightarrow u_3(t))) \vee u_3(t), \\ y_1(t) = z_1(t) \leftrightarrow z_2(t), \\ y_2(t) = \neg z_3(t). \end{cases} \qquad (13.52)$$

Consider the control transformation $u = Tv$ given by

$$\begin{cases} u_1(t) = v_1(t) \leftrightarrow v_3(t), \\ u_2(t) = v_2(t), \\ u_3(t) = v_3(t), \end{cases} \qquad (13.53)$$

with its inverse given by

$$\begin{cases} v_1(t) = u_1(t) \leftrightarrow u_3(t), \\ v_2(t) = u_2(t), \\ v_3(t) = u_3(t). \end{cases} \qquad (13.54)$$

The system then becomes

$$\begin{cases} z_1(t+1) = (z_1(t) \wedge z_2(t)) \vee v_1(t), \\ z_2(t+1) = z_1(t) \wedge v_2(t), \\ z_3(t+1) = (\neg z_1(t) \wedge v_1(t)) \vee v_3(t), \\ y_1(t) = z_1(t) \leftrightarrow z_2(t), \\ y_2(t) = \neg z_3(t). \end{cases} \tag{13.55}$$

Finally, we construct an additional state feedback control:

$$\begin{cases} v_1(t) = z_1(t) \wedge w_1(t), \\ v_2(t) = w_2(t), \\ v_3(t) = w_3(t). \end{cases} \tag{13.56}$$

The system (13.55) then becomes

$$\begin{cases} z_1(t+1) = (z_1(t) \wedge z_2(t)) \vee (z_1(t) \wedge w_1(t)), \\ z_2(t+1) = z_1(t) \wedge w_2(t), \\ z_3(t+1) = w_3(t), \\ y_1(t) = z_1(t) \leftrightarrow z_2(t), \\ y_2(t) = \neg z_3(t). \end{cases} \tag{13.57}$$

It is easy to check that (13.57) is an input–output-decomposed form with $w^1(t) = \{w_1(t), w_2(t)\}$ and $w^2(t) = \{w_2(t)\}$.

Note that the overall coordinate transformation is (13.50) [equivalently, (13.51)], and the overall state feedback control is

$$\begin{cases} u_1(t) = (z_1(t) \wedge w_1(t)) \leftrightarrow w_3(t), \\ u_2(t) = w_2(t), \\ u_3(t) = w_3(t). \end{cases} \tag{13.58}$$

References

1. Glumineau, A., Moog, C.: Nonlinear Morgan's problem: case of $(p+1)$ inputs and p outputs. IEEE Trans. Automat. Contr. **37**(7), 1067–1072 (1992)
2. Herrera, A., Lafay, J.: New results about Morgan's problem. IEEE Trans. Automat. Contr. **38**(12), 1834–1838 (1993)
3. Morgan, B.: The synthesis of linear multivariable systems by state variable feedback. In: Proceedings of the Joint Automatic Control Conference, pp. 468–472 (1964)
4. Qi, H., Feng, G.: On decomposition of Boolean networks (2010, submitted)
5. Wonham, W.: Linear Multivariable Control: A Geometric Approach, 2nd edn. Springer, Berlin (1979)

Chapter 14
k-valued Networks

14.1 A Review of k-valued Logic

In this section we briefly review the matrix expression of k-valued logic which was introduced in Chaps. 1 and 3.

Let

$$\mathscr{D}_k = \left\{ 1 = T, \frac{k-2}{k-1}, \ldots, \frac{1}{k-1}, 0 = F \right\},$$

which is the set from which k-valued logical variables take their values.

To use the matrix approach, we identify a scalar logic value with a vector as

$$\frac{i}{k-1} \sim \delta_k^{k-i}, \quad i = 0, 1, \ldots, k-1.$$

Δ_k is also used for vector expression as

$$\Delta_k = \left\{ \delta_k^i \mid i = 1, 2, \ldots, k \right\}.$$

We now summarize some of the main results of k-valued logic and its matrix expression.

The basic operators and their structure matrices are listed as follows (the operators are defined in terms of their scalar values):

- Negation is defined as

$$\neg P := 1 - P, \tag{14.1}$$

and its structure matrix is

$$M_{n,k} = \delta_k[k \; k-1 \; \cdots \; 1]. \tag{14.2}$$

- The rotator $\oslash_k$ is defined as

$$\oslash_k(P) := \begin{cases} P - \frac{1}{k-1}, & P \neq 0, \\ 1, & P = 0, \end{cases} \tag{14.3}$$

D. Cheng et al., *Analysis and Control of Boolean Networks*,
Communications and Control Engineering,
DOI 10.1007/978-0-85729-097-7_14, © Springer-Verlag London Limited 2011

and its structure matrix, $M_{o,k}$, is

$$M_{o,k} = \delta_k[2\ 3\ \cdots\ k\ 1].\tag{14.4}$$

For instance, we have

$$M_{o,3} = \delta_3[2\ 3\ 1], \qquad M_{o,4} = \delta_4[2\ 3\ 4\ 1].\tag{14.5}$$

• The i-confirmor, $\nabla_{i,k}$, $i = 1, \ldots, k$, is defined as

$$\nabla_{i,k}(P) = \begin{cases} 1, & P = \frac{k-i}{k-1} \text{ (equivalently, } P = \delta_k^i), \\ 0, & \text{otherwise.} \end{cases}\tag{14.6}$$

Its structure matrix (using the same notation) is

$$\nabla_{i,k} = \delta_k[\underbrace{k \cdots k}_{i-1}\ 1\ \underbrace{k \cdots k}_{k-i}], \quad i = 1, 2, \ldots, k.\tag{14.7}$$

For instance, we have

$$\nabla_{2,3} = \delta_3[3\ 1\ 3], \qquad \nabla_{2,4} = \delta_4[4\ 1\ 4\ 4], \qquad \nabla_{3,4} = \delta_4[4\ 4\ 1\ 4].\tag{14.8}$$

• Conjunction is defined as

$$P \wedge Q := \min\{P, Q\},\tag{14.9}$$

and its structure matrix is (to save space, we let $n = 3$)

$$M_{c,3} = \delta_3[1\ 2\ 3\ 2\ 2\ 3\ 3\ 3\ 3].\tag{14.10}$$

• Disjunction is defined as

$$P \vee Q := \max\{P, Q\},\tag{14.11}$$

and its structure matrix is ($n = 3$)

$$M_{d,3} = \delta_3[1\ 1\ 1\ 1\ 2\ 2\ 1\ 2\ 3].\tag{14.12}$$

• The conditional is defined as

$$P \to Q := (\neg P) \vee Q,\tag{14.13}$$

and its structure matrix is ($n = 3$)

$$M_{i,3} = \delta_3[1\ 2\ 3\ 1\ 2\ 2\ 1\ 1\ 1].\tag{14.14}$$

• The biconditional is defined as

$$P \leftrightarrow Q := (P \to Q) \wedge (Q \to P),\tag{14.15}$$

and its structure matrix is $(n = 3)$

$$M_{e,3} = \delta_3[1\,2\,3\,2\,2\,2\,3\,2\,1].\tag{14.16}$$

Remark 14.1 In general, there are k^{k^2} binary k-valued logical operators, including constant operators and unary operators as special cases. In the above, we give only a few of them which are commonly used. Moreover, [4] proved that $\{\oslash, \wedge, \vee\}$ form an adequate set, and they are enough to express any k-valued logical operator. In other words, all other k-valued logical operators can be expressed as certain combinations of $\{\oslash, \wedge, \vee\}$.

Some fundamental properties are collected the following:

- If $P \in \Delta_k$, then

$$P^2 = M_{r,k} P,\tag{14.17}$$

where $M_{r,k}$ is the base-k power-reducing matrix.

$$M_{r,k} = \begin{bmatrix} \delta_k^1 & 0_k & \cdots & 0_k \\ 0_k & \delta_k^2 & \cdots & 0_k \\ \vdots & & & \\ 0_k & 0_k & \cdots & \delta_k^k \end{bmatrix},\tag{14.18}$$

where $0_k \in \mathbb{R}_k$ is a zero vector. When $k = 3$,

$$M_{r,3} = \delta_9[1\,5\,9].\tag{14.19}$$

When $k = 4$,

$$M_{r,4} = \delta_{16}[1\,6\,11\,16].\tag{14.20}$$

- Let $f(p_1, p_2, \ldots, p_r)$ be a k-valued logical function. There then exists a structure matrix of f, denoted by M_f, such that

$$f(p_1, p_2, \ldots, p_r) = M_f \ltimes_{i=1}^r p_i.\tag{14.21}$$

- For any $P, Q \in \Delta_k$, we have

$$E_{d,k} P Q = Q, \quad P, Q \in \Delta_k,\tag{14.22}$$

where E_d is the base-k dummy operator defined as

$$E_{d,k} := [\underbrace{I_k\ I_k\ \cdots\ I_k}_{k}].\tag{14.23}$$

In the symbols of the above operators and their corresponding structure matrices there is a second index k for the logical type. In the sequel, when k is fixed and there is no possible confusion, this index k can be omitted.

Table 14.1 Structure matrices of logical operators ($k = 3$)

Operator	Structure matrix	Operator	Structure matrix
¬	$M_n = \delta_3[3\,2\,1]$	∨	$M_d = \delta_3[1\,1\,1\,1\,2\,2\,1\,2\,3]$
⊘	$M_o = \delta_3[3\,1\,2]$	∧	$M_c = \delta_3[1\,2\,3\,2\,2\,3\,3\,3\,3]$
∇_1	$M_{\nabla_1} = \delta_3[1\,1\,1]$	→	$M_i = \delta_3[1\,2\,3\,1\,2\,2\,1\,1\,1]$
∇_2	$M_{\nabla_2} = \delta_3[2\,2\,2]$	↔	$M_e = \delta_3[1\,2\,3\,2\,2\,2\,3\,2\,1]$
∇_3	$M_{\nabla_3} = \delta_3[3\,3\,3]$		

As in the Boolean case, using mod k algebra is sometimes convenient. Since the values in $\mathscr{D}_k$ are not integers, we need to multiply each argument by $k - 1$ to convert them into integers. Then two operators, namely, $+(\text{mod } k)$ and $\times(\text{mod } k)$, are used for calculation. Finally, the results are converted back to fraction by dividing them by $k - 1$.

Definition 14.1

- The binary operator $+_k : \mathscr{D}_k^2 \to \mathscr{D}_k$, called mod k addition, is defined as

$$P +_k Q := \frac{[(k-1) * (P + Q)](\text{mod } k)}{k - 1}. \tag{14.24}$$

- The binary operator $\times_k : \mathscr{D}_k^2 \to \mathscr{D}_k$, called mod k multiplication, is defined as

$$P \times_k Q := \frac{[(k-1)^2 * (P \times Q)](\text{mod } k)}{k - 1}. \tag{14.25}$$

Their structure matrices can be easily computed as (for $k = 3$)

$$M_{+_3} = M_{p,3} = \delta_3[2\,3\,1\,3\,1\,2\,1\,2\,3], \tag{14.26}$$

$$M_{\times_3} = M_{t,3} = \delta_3[2\,1\,3\,1\,2\,3\,3\,3\,3]. \tag{14.27}$$

Assuming $k = 3$, the structure matrices of the previous logical operators are collected in Table 14.1.

14.2 Dynamics of k-valued Networks

A k-valued network consists of a set of nodes $V = \{x_1, x_2, \ldots, x_n\}$ and a list of k-valued logical functions $F = \{f_1, f_2, \ldots, f_n\}$. Both the nodes and the functions take values from $\mathscr{D}_k$. The dynamics of a k-valued network is then described as

$$\begin{cases} x_1(t+1) = f_1(x_1(t), x_2(t), \ldots, x_n(t)), \\ x_2(t+1) = f_2(x_1(t), x_2(t), \ldots, x_n(t)), \\ \vdots \\ x_n(t+1) = f_n(x_1(t), x_2(t), \ldots, x_n(t)). \end{cases} \tag{14.28}$$

It is clear that when $k = 2$, the k-valued network becomes a Boolean network.

As with Boolean networks, we can use the semi-tensor product to convert (14.28) into algebraic form. We briefly describe this process. Define

$$x(t) = x_1(t)x_2(t)\cdots x_n(t), \quad x_i \in \Delta_k.$$

Using Theorem 3.2, we can find the structure matrices $M_i = M_{f_i}, i = 1,\ldots, n$, such that

$$x_i(t+1) = M_i x(t), \quad i = 1, 2, \ldots, n. \tag{14.29}$$

Equations (14.29) are called the componentwise algebraic form of (14.28).

Note that by using the base-k dummy matrix (14.23) we can formally introduce any logical variable into a logical expression without changing its real meaning. Similarly to the Boolean case, we can prove the following result.

Lemma 14.1 *Assume that $P_\ell = A_1 A_2 \cdots A_\ell$, where $A_i \in \Delta_k, i = 1, 2, \ldots, \ell$. Then,*

$$P_\ell^2 = \Phi_{\ell,k} P_\ell, \tag{14.30}$$

where

$$\Phi_{\ell,k} = \prod_{i=1}^{\ell} \left(I_{k^{i-1}} \otimes [I_k \otimes W_{[k,k^{\ell-i}]} M_{r,k}] \right). \tag{14.31}$$

Proof We prove this by induction. When $\ell = 1$, using (14.23), we have

$$P_1^2 = A_1^2 = M_{r,k} A_1.$$

In formula (14.30),

$$\Phi_{1,k} = (I_k \otimes W_{[k,1]}) M_{r,k}.$$

Note that because $W_{[k,1]} = I_k$, it follows that $\Phi_{1,k} = M_{r,k}$. Hence, (14.30) is true for $\ell = 1$. Assume that (14.30) is true for $\ell = s$. Then, for $\ell = s + 1$ we have

$$\begin{aligned}
P_{s+1}^2 &= A_1 A_2 \cdots A_{s+1} A_1 A_2 \cdots A_{s+1} \\
&= A_1 W_{[k,k^s]} A_1 [A_2 \cdots A_{s+1}]^2 \\
&= (I_k \otimes W_{[k,k^s]}) A_1^2 [A_2 \cdots A_{s+1}]^2 \\
&= \left[(I_k \otimes W_{[k,k^s]}) M_{r,k} \right] A_1 [A_2 \cdots A_{s+1}]^2.
\end{aligned}$$

Applying the induction assumption to the last factor of the above expression, we have

$$P_{s+1}^2 = (I_k \otimes W_{[k,k^s]}) M_{r,k} A_1 \left(\prod_{i=1}^{s} I_{k^{i-1}} \otimes \left[(I_k \otimes W_{[k,k^{s-i}]}) M_{r,k} \right] \right)$$

$$A_2 A_3 \cdots A_{s+1}$$

$$= \left[(I_k \otimes W_{[k,k^s]})M_{r,k}\right]\left(\prod_{i=1}^{s} I_{k^i} \otimes \left[(I_k \otimes W_{[k,k^{s-i}]})M_{r,k}\right]\right)P_{s+1}$$

$$= \prod_{i=1}^{s+1} \left(I_{k^{i-1}} \otimes [I_k \otimes W_{[k,k^{s+1-i}]}M_{r,k}]\right)P_{s+1}.$$

$\square$

Using Lemma 14.1, a straightforward computation leads to the following result.

Proposition 14.1 *Equation* (14.28) *can be expressed in algebraic form as*

$$x(t+1) = Lx(t), \quad x \in \Delta_{k^n}, \tag{14.32}$$

where the system transition matrix L is obtained as

$$L = M_1 \prod_{j=2}^{n} \left[(I_k \otimes M_j)\Phi_{n,k}\right] \in \mathscr{L}_{k^n \times k^n},$$

M_i *being defined in* (14.29).

Proof Define $x(t) = \ltimes_{i=1}^{n} x_i(x)$. According to Lemma 14.1 we have

$$x(t)^2 = \Phi_{n,k} x(t).$$

Now, multiplying equations in (14.29) together yields

$$
\begin{aligned}
x(t+1) &= M_1 x(t) M_2 x(t) \cdots M_n x(t) \\
&= M_1 (I_k \otimes M_2) x(t)^2 M_3 x(t) \cdots M_n x(t) \\
&= M_1 (I_k \otimes M_2) \Phi_{n,k} x(t) M_3 x(t) \cdots M_n x(t) \\
&= \cdots \\
&= M_1 (I_k \otimes M_2) \Phi_{n,k} (I_k \otimes M_3) \Phi_{n,k} \cdots (I_k \otimes M_n) \Phi_{n,k} x(t),
\end{aligned}
$$

so

$$
\begin{aligned}
L &= M_1 (I_k \otimes M_2) \Phi_{n,k} (I_k \otimes M_3) \Phi_{n,k} \cdots (I_k \otimes M_n) \Phi_{n,k} \\
&= M_1 \prod_{j=2}^{n} \left[(I_k \otimes M_j)\Phi_{n,k}\right].
\end{aligned}
$$

$\square$

For a particular system, we may obtain the system matrix by direct computation, using properties of the semi-tensor product. We give an example to illustrate the process of computing L.

Example 14.1 Consider the following k-valued network:

$$\begin{cases} A(t+1) = A(t), \\ B(t+1) = A(t) \to C(t), \\ C(t+1) = B(t) \vee D(t), \\ D(t+1) = \neg B(t), \\ E(t+1) = \neg C(t). \end{cases} \quad (14.33)$$

Defining $x(t) = A(t)B(t)C(t)D(t)E(t)$, we then have

$$x(t+1) = A(t)M_{i,k}A(t)C(t)M_{d,k}B(t)D(t)M_{n,k}B(t)M_{n,k}C(t).$$

Since there is no $E(t)$ on the left-hand side, we have to introduce it by using the dummy matrix:

$$x(t+1) = A(t)M_{i,k}A(t)C(t)M_{d,k}B(t)D(t)M_{n,k}B(t)M_{n,k}C(t)E_{d,k}E(t)C(t). \quad (14.34)$$

Using the pseudo-commutative property of the semi-tensor product, we can move $A(t), B(t), \ldots, E(t)$ to the last part of the product in the right-hand side of (14.34). Then, using the base-k power-reducing matrix to reduce the powers of $A(t), B(t), \ldots, E(t)$ to 1, we finally obtain the algebraic form of (14.33) as

$$x(t+1) = Lx(t),$$

where

$$L = (I_k \otimes M_i)R_k\big(I_k \otimes \big(I_k \otimes M_d\big(I_k \otimes \big(I_k \otimes M_n(I_k \otimes M_n E_d)\big)\big)\big)\big)$$
$$(I_k \otimes W_{[k]})(I_{k^3} \otimes W_{[k]})(I_{k^2} \otimes W_{[k]})(I_{k^5} \otimes W_{[k]})$$
$$(I_{k^4} \otimes W_{[k]})\big(I_k \otimes R_k(I_k \otimes R_k)\big).$$

When $k = 3$, we can calculate the network transition matrix L. It is the following 243×243 matrix:

$$
\begin{aligned}
L = \delta_{243}[&9 \quad 9 \quad 9 \quad 9 \quad 9 \quad 9 \quad 9 \quad 9 \quad 9 \quad 35 \quad 35 \quad 35 \quad 35 \quad 35 \quad 35 \quad 35 \\
&35 \quad 35 \quad 61 \quad 61 \quad 61 \quad 61 \quad 61 \quad 61 \quad 61 \quad 61 \quad 61 \quad 6 \quad 6 \quad 6 \quad 15 \quad 15 \\
&15 \quad 15 \quad 15 \quad 15 \quad 32 \quad 32 \quad 32 \quad 41 \quad 41 \quad 41 \quad 41 \quad 41 \quad 41 \quad 58 \quad 58 \quad 58 \\
&67 \quad 67 \quad 67 \quad 67 \quad 67 \quad 67 \quad 3 \quad 3 \quad 3 \quad 12 \quad 12 \quad 12 \quad 21 \quad 21 \quad 21 \quad 29 \\
&29 \quad 29 \quad 38 \quad 38 \quad 38 \quad 47 \quad 47 \quad 47 \quad 55 \quad 55 \quad 55 \quad 64 \quad 64 \quad 64 \quad 73 \quad 73 \\
&73 \quad 90 \quad 90 \quad 90 \quad 90 \quad 90 \quad 90 \quad 90 \quad 90 \quad 90 \quad 116 \, 116 \, 116 \, 116 \, 116 \, 116 \\
&116 \, 116 \, 116 \, 115 \, 115 \, 115 \, 115 \, 115 \, 115 \, 115 \, 115 \, 115 \, 87 \quad 87 \quad 87 \quad 96 \\
&96 \quad 96 \quad 96 \quad 96 \quad 96 \quad 113 \, 113 \, 113 \, 122 \, 122 \, 122 \, 122 \, 122 \, 122 \, 112 \, 112 \\
&112 \, 121 \, 121 \, 121 \, 121 \, 121 \, 121 \, 84 \quad 84 \quad 84 \quad 93 \quad 93 \quad 93 \quad 102 \, 102 \, 171 \\
&171 \, 171 \, 171 \, 171 \, 171 \, 171 \, 102 \, 110 \, 110 \, 110 \, 119 \, 119 \, 119 \, 128 \, 128 \, 128 \\
&109 \, 109 \, 109 \, 118 \, 118 \, 118 \, 127 \, 127 \, 127 \, 171 \, 171 \, 170 \, 170 \, 170 \, 170 \, 170 \\
&170 \, 170 \, 170 \, 170 \, 169 \, 169 \, 169 \, 169 \, 169 \, 169 \, 169 \, 169 \, 169 \, 168 \, 168 \, 168 \\
&177 \, 177 \, 177 \, 177 \, 177 \, 177 \, 167 \, 167 \, 167 \, 176 \, 176 \, 176 \, 176 \, 176 \, 176 \, 166 \\
&166 \, 166 \, 175 \, 175 \, 175 \, 175 \, 175 \, 175 \, 165 \, 165 \, 165 \, 174 \, 174 \, 174 \, 183 \, 183 \\
&183 \, 164 \, 164 \, 164 \, 173 \, 173 \, 173 \, 182 \, 182 \, 182 \, 163 \, 163 \, 163 \, 172 \, 172 \, 172 \\
&181 \, 181 \, 181].
\end{aligned}
$$

14.3 State Space and Coordinate Transformations

As with Boolean (control) networks, to use the state-space approach, the state space and its subspaces have to be defined carefully. In the following definition, they are only defined as a set and subsets, but they can be considered as a topological space and subspaces equipped with the discrete topology.

Let $x_1, \dots, x_n \in \mathscr{D}_k = \{0, \frac{1}{k-1}, \dots, 1\}$ be a set of logical variables. Denote by $\mathscr{F}_\ell(x_1, \dots, x_n)$ the set of logical functions of $\{x_1, \dots, x_n\}$. It is obvious that $\mathscr{F}_\ell$ is a finite set with cardinality k^{k^n}.

Definition 14.2 Consider the k-valued logical network (14.28).

(1) The state space of (14.28) is defined as

$$\mathscr{X} = \mathscr{F}_\ell(x_1, \dots, x_n). \tag{14.35}$$

(2) If $y_1, \dots, y_s \in \mathscr{X}$, then

$$\mathscr{Y} = \mathscr{F}_\ell(y_1, \dots, y_s) \subset \mathscr{X} \tag{14.36}$$

is called a subspace of $\mathscr{X}$.
(3) If $\{x_{i_1}, \dots, x_{i_s}\} \subset \{x_1, \dots, x_n\}$, then

$$\mathscr{L} = \mathscr{F}_\ell(x_{i_1}, \dots, x_{i_s}) \tag{14.37}$$

is called an s-dimensional natural subspace of $\mathscr{X}$.

Let $F : \mathscr{D}_k^n \to \mathscr{D}_k^m$ be defined by

$$z_i = f_i(x_1, \dots, x_n), \quad i = 1, \dots, m. \tag{14.38}$$

In vector form, we have $x_i, z_j \in \Delta_k$. Setting $x = \ltimes_{i=1}^n x_i$, $z = \ltimes_{i=1}^m z_i$, we have the following result.

Theorem 14.1 *Given a logical mapping* $F : \mathscr{D}_k^n \to \mathscr{D}_k^m$, *as described by* (14.38), *there is a unique matrix,* $M_F \in \mathscr{L}_{k^m \times k^n}$, *called the structure matrix of* F, *such that*

$$z = M_F x. \tag{14.39}$$

Note that when $m = 1$ the mapping becomes a logical function and M_F is called the structure matrix of the function.

Definition 14.3 Let $\mathscr{X} = \mathscr{F}_\ell(x_1, \dots, x_n)$ be the state space of (14.28). Assume that there exist $z_1, \dots, z_n \in \mathscr{X}$ such that

$$\mathscr{X} = \mathscr{F}_\ell(z_1, \dots, z_n).$$

The logical mapping $F : (x_1, \dots, x_n) \mapsto (z_1, \dots, z_n)$ is then called a coordinate transformation of the state space.

The following proposition is obvious.

Proposition 14.2 *A mapping $T : \mathscr{D}_k^n \to \mathscr{D}_k^n$ is a coordinate transformation if and only if T is one-to-one and onto (i.e., bijective).*

It is easy to prove the following result.

Theorem 14.2 *A mapping $T : \mathscr{D}_k^n \to \mathscr{D}_k^n$ is a coordinate transformation if and only if its structure matrix $M_T \in \mathscr{L}_{k^n \times k^n}$ is nonsingular.*

We give an example to illustrate the above theorem for a 3-valued mapping.

Example 14.2 Consider the following mapping T:

$$\begin{cases} z_1 = \oslash(x_3), \\ z_2 = (\nabla_1(x_1) \wedge x_3) \vee [\nabla_2(x_1) \wedge (\oslash^2(\neg(\nabla_2(x_3))))] \\ \qquad \vee (\nabla_3(x_1) \wedge \nabla_3(x_3)), \\ z_3 = \oslash^2(x_2), \end{cases} \tag{14.40}$$

where $\nabla_{i,3}, i = 1, 2, 3$, and $\oslash_3$ are defined as in Sect. 14.1. The logical variables $x_i \in \mathscr{D}_3$, $i = 1, 2, 3$. Define $x = \ltimes_{i=1}^3 x_i$ and $z = \ltimes_{i=1}^3 z_i$. Based on Theorem 14.1, there exists a unique matrix $M_T \in \mathscr{L}_{3^3 \times 3^3}$ such that

$$z(t) = M_T x(t), \tag{14.41}$$

where

$$M_T = \delta_{81}[20\ 5\ 17\ 21\ 6\ 18\ 19\ 4\ 16\ 23\ 2\ 14\ 24\ 3\ 15\ 22\ 1\ 13\ 26\ 8\ 11\ 27\ 9\ 12\ 25\ 7\ 10],$$

which is nonsingular. Thus, T is a coordinate transformation in the 3-valued network.

Definition 14.4 A subspace $\mathscr{Z} \subset \mathscr{X}$ is called a regular subspace if there is a coordinate frame $\{z\}$ such that under $\{z\}$ the subspace $\mathscr{Z}$ is a natural subspace.

Let $\mathscr{Z} = \mathscr{F}_\ell\{z_1, \ldots, z_r\}$, where $z_i \in \mathscr{X}$, $i = 1, \ldots, k$. Then, z_i are logical functions of $\{x_j\}$, which are expressed as

$$z_j = g_j(x_1, \ldots, x_n), \quad j = 1, \ldots, r. \tag{14.42}$$

Express (14.42) in vector form as

$$z_j = M_j x, \quad j = 1, \ldots, r, \tag{14.43}$$

where

$$M_j = \delta_k[\xi_1^j, \ldots, \xi_{k^n}^j], \quad j = 1, \ldots, r.$$

Equation (14.42) can be further be expressed in one equation as

$$z = Gx = \begin{bmatrix} \ell_{1,1} & \cdots & \ell_{1,k^n} \\ \vdots & & \\ \ell_{r,1} & \cdots & \ell_{r,k^n} \end{bmatrix} x. \tag{14.44}$$

Similarly to Theorem 8.2, we can prove the following theorem.

Theorem 14.3 $\mathscr{L}$ *is a regular subspace if and only if*

$$\sum_{i=1}^{k^n} \ell_{j,i} = k^{n-r}, \quad j = 1, \ldots, r. \tag{14.45}$$

Define the characteristic matrix as

$$E(\mathscr{L}) = \begin{bmatrix} \xi_1^1 & \cdots & \xi_{k^n}^1 \\ \vdots & & \\ \xi_1^r & \cdots & \xi_{k^n}^r \end{bmatrix}.$$

Similarly to Corollary 8.2, we can prove the following result.

Proposition 14.3 $\mathscr{L}$ *is a regular subspace if and only if* $E(\mathscr{L})$ *has equal distinct columns. Precisely,* $\mathscr{L}$ *is a regular subspace if and only if for each vector* $\alpha := (\alpha_1, \ldots, \alpha_r)^\mathrm{T} \in \mathbb{R}^r$ *with* $\alpha_j \in \{1, 2, \ldots, k\}$, $j = 1, \ldots, r$, *we have*

$$\left| \{i \mid \mathrm{Col}_i\big(E(\mathscr{L})\big) = \alpha\} \right| = k^{n-r}, \quad \forall \alpha.$$

Invariant subspaces have been defined for Boolean networks. Here, we generalize them to k-valued logical networks. In addition, we give a geometric description of them.

Definition 14.5 Let $\mathscr{X}$, $\mathscr{L}$ be defined as in (14.35) and (14.37), and $\mathscr{L} \subset \mathscr{X}$.

(1) A mapping $P : \mathscr{D}_k^n \to \mathscr{D}_k^s$, defined from (the domain of) $\mathscr{X}$ to (the domain of) $\mathscr{L}$ as

$$P : (x_1, \ldots, x_n) \mapsto (x_{i_1}, \ldots, x_{i_s}),$$

is called a natural projection from $\mathscr{X}$ to $\mathscr{L}$.
(2) Given $F : \mathscr{D}_k^n \to \mathscr{D}_k^n$, $\mathscr{L}$ is called an invariant subspace (with respect to F) if there exists a mapping $\bar{F}$ such that the following graph (Fig. 14.1) is commutative.

Let $X := (x_1, \ldots, x_n)^\mathrm{T} \in \mathscr{D}_k^n$. We can then compactly express the system (14.28) as

$$X(t+1) = F\big(X(t)\big), \quad X \in \mathscr{D}_k^n. \tag{14.46}$$

Fig. 14.1 Invariant subspace

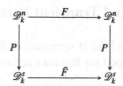

Definition 14.6 Consider the system (14.28) [equivalently, (14.46)]. $\mathscr{L}$ is an invariant subspace if it is invariant with respect to F.

Consider a logical mapping $G : \mathscr{D}_k^n \to \mathscr{D}_k^s$. It can be expressed as

$$z_i = g_1(x_1, \ldots, x_n), \quad i = 1, \ldots, s. \tag{14.47}$$

Setting $z = \ltimes_{i=1}^s z_i$ and $x = \ltimes_{i=1}^n x_i$, the algebraic form of the mapping G is

$$z = M_G x := \begin{bmatrix} g_{11} & \cdots & g_{1,k^n} \\ \vdots & & \\ g_{k^s,1} & \cdots & g_{k^s,k^n} \end{bmatrix} x. \tag{14.48}$$

The algebraic form of multivalued system (14.28) is

$$x(t+1) = Lx(t),$$

where $L \in \mathscr{L}_{k^n \times k^n}$, $x = \ltimes_{i=1}^n x_i$, $x_i \in \mathscr{D}_k$, $i = 1, \ldots, n$.
Using the above notation we have the following theorem.

Theorem 14.4 $\mathscr{L}_0$ *is an invariant subspace with respect to the multivalued system* (14.28) *if and only if one of the following two equivalent conditions is satisfied:*

(i)

$$\text{Row}(M_G L) \subset \text{Span}_\mathscr{B} \text{Row}(M_G). \tag{14.49}$$

(ii) *There exists an* $H \in \mathscr{L}_{k^s \times k^s}$ *such that*

$$M_G L = H M_G. \tag{14.50}$$

Theorem 14.3 is similar to Theorem 8.2, and Theorem 14.4 is similar to Theorem 8.4 and Corollary 8.6 for the Boolean case. The corresponding proofs are effectively the same.

14.4 Cycles and Transient Period

Consider the topological structure of k-valued logical networks. Using the same technique developed for Boolean networks, we can obtain the following results for cycles.

Theorem 14.5 *Consider the k-valued logical network* (14.28).

1. $\delta_{k^n}^i$ *is a fixed point if and only if, in its algebraic form* (14.32), *the diagonal element* ℓ_{ii} *of the network matrix L equals* 1. *It follows that the number of fixed points of system* (14.32), *denoted by* N_e, *equals the number of i for which* $\ell_{ii} = 1$. *Equivalently,*

$$N_e = \operatorname{tr}(L). \tag{14.51}$$

2. *The number of length-s cycles,* N_s, *is inductively given by*

$$
\begin{cases}
N_1 = N_e, \\
N_s = \dfrac{\operatorname{tr}(L^s) - \sum_{t \in \mathscr{P}(s)} t N_t}{s}, & 2 \le s \le k^n.
\end{cases} \tag{14.52}
$$

3. *The set of elements on cycles of length s, denoted by* $\mathscr{C}_s$, *is*

$$\mathscr{C}_s = \mathscr{D}_a(L^s) \setminus \bigcup_{t \in \mathscr{P}(s)} \mathscr{D}_a(L^t), \tag{14.53}$$

where $\mathscr{D}_a(L)$ *is the set of diagonal nonzero columns of L.*

Example 14.3 Recall Example 14.1. A straightforward computation shows that

$$\operatorname{tr}(L^t) = 5, \quad t = 1, 3, \ldots,$$

and

$$\operatorname{tr}(L^t) = 11, \quad t = 2, 4, \ldots.$$

Using Theorem 14.5 we conclude that there are five fixed points and three cycles of length 2. Moreover, we can find the fixed points and the cycles of length 2 as follows.

To find the fixed points, we consider the network matrix L. It is easily shown that the 9th, 41st, 90th, 122nd, and 171st columns of L are diagonal nonzero columns. Therefore the five fixed points are δ_{35}^9, δ_{35}^{41}, δ_{35}^{90}, δ_{35}^{122}, and δ_{35}^{171}. Using conversion formula (4.58), we can convert the fixed points back to standard form as

$$E_1 = \delta_{35}^9 \sim (1, 1, 1, 0, 0),$$

$$E_2 = \delta_{35}^{41} \sim (1, 0.5, 0.5, 0.5, 0.5),$$

$$E_3 = \delta_{35}^{90} \sim (0.5, 1, 1, 0, 0),$$

$$E_4 = \delta_{35}^{122} \sim (0.5, 0.5, 0.5, 0.5, 0.5),$$

$$E_5 = \delta_{35}^{171} \sim (0, 1, 1, 0, 0).$$

For the cycles, we consider L^2. Searching for diagonal nonzero columns of L^2 yields three cycles of length 2:

$$(1, 1, 0.5, 0.5, 0) \to (1, 0.5, 1, 0, 0.5) \to (1, 1, 0.5, 0.5, 0),$$
$$(1, 1, 0, 1, 0) \to (1, 0, 1, 0, 1) \to (1, 1, 0, 1, 0),$$
$$(0.5, 1, 0.5, 0.5, 0) \to (0.5, 0.5, 1, 0, 0.5) \to (0.5, 1, 0.5, 0.5, 0).$$

There are no other cycles.

For the transient period, we also have the following theorem.

Theorem 14.6 *For the system* (14.28) *the transient period is*

$$T_t = r_0 = \min\left\{ r \mid L^r \in \{L^{r+1}, L^{r+2}, \ldots, L^{k^n}\} \right\}. \tag{14.54}$$

Moreover, let $T > 0$ be the smallest positive number satisfying $L^{r_0} = L^{r_0+T}$. Then, T is the least common multiple of the lengths of all cycles.

Since the proof is similar to that of the Boolean case, we leave it to the reader.

Example 14.4 Recall Example 14.1. It is easy to check that the first repeating power for L^k is $L^4 = L^6$, so $r_0 = 4$. That is, $T_t = 4$, $T = 2$. Therefore the transient period is 4, which means that any initial state will enter an attractor after at most four steps.

14.5 Network Reconstruction

Assume for a k-valued logical system that the network matrix L is given. We have to reconstruct the logical network and its dynamics from the network matrix. As with Boolean networks, we first define a set of retrievers. For notational compactness, we first define a set of column vectors:

$$\Sigma_i = [\underbrace{\delta_k^1, \ldots, \delta_k^1}_{i}, \underbrace{\delta_k^2, \ldots, \delta_k^2}_{i}, \ldots, \underbrace{\delta_k^k, \ldots, \delta_k^k}_{i}].$$

We then define the retrievers:

$$S_{1,k}^n = [\Sigma_{k^{n-1}}],$$

$$S_{2,k}^n = [\underbrace{\Sigma_{k^{n-2}}, \ldots, \Sigma_{k^{n-2}}}_{k}],$$

$$S_{3,k} = [\underbrace{\Sigma_{k^{n-3}}, \ldots, \Sigma_{k^{n-3}}}_{k^2}],$$ (14.55)

$$\vdots$$

$$S_{n,k}^n = [\underbrace{\Sigma_1, \ldots, \Sigma_1}_{k^{n-1}}].$$

Proposition 14.4 *Assume the network matrix L of the system (14.28) is known. The structure matrices of f_i, $i = 1, \ldots, n$, are then*

$$M_i = S_{i,k}^n L, \quad i = 1, 2, \ldots, n.$$ (14.56)

Next, we have to determine which node is connected to i, in order to remove fabricated variables from the ith logical equation. We have the following result.

Proposition 14.5 *Consider the system (14.28). If M_i satisfies*

$$M_i W_{[k,k^{j-1}]}(M_{o,k} - I_k) = 0,$$

$$M_i W_{[k,k^{j-1}]}((M_{o,k})^2 - I_k) = 0,$$ (14.57)

$$\vdots$$

$$M_i W_{[k,k^{j-1}]}((M_{o,k})^{k-1} - I_k) = 0,$$

then node j is not in the neighborhood of node i. In other words, the edge $j \to i$ does not exist. The equation of x_i can then be replaced by

$$x_i(t+1) = M_i' x_1(t) \cdots x_{j-1}(t) x_{j+1}(t) \cdots x_n(t),$$ (14.58)

where

$$M_i' = M_i W_{[k,k^{j-1}]} \delta_k^1.$$

Proof Using the properties of the semi-tensor product, we can rewrite the ith equation of (14.28) as

$$x_i(t+1) = M_i W_{[k,k^{j-1}]} x_j(t) x_1(t) \cdots x_{j-1}(t) x_{j+1}(t) \cdots x_n(t).$$

We now replace $x_j(t)$ by $\oslash(x_j(t))$, $\oslash^2(x_j(t))$, $\ldots$, $\oslash^{k-1}(x_j(t))$, that is, all possible values of $x_j(t)$. If such replacements do not affect the overall structure matrix, it means $x_i(t+1)$ is independent of $x_j(t)$. The invariance of replacement is illustrated by (14.57). As for (14.58), since $x_j(t)$ does not affect $x_i(t+1)$, we can simply set $x_j(t) = \delta_k^1$ to simplify the expression. $\qquad\square$

Example 14.5 Given a Boolean network

$$\begin{cases} A(t+1) = f_1(A(t), B(t), C(t), D(t)), \\ B(t+1) = f_2(A(t), B(t), C(t), D(t)), \\ C(t+1) = f_3(A(t), B(t), C(t), D(t)), \\ D(t+1) = f_4(A(t), B(t), C(t), D(t)), \end{cases} \qquad (14.59)$$

where $A(t), B(t), C(t), D(t) \in \mathscr{D}_3 = \{0, 0.5, 1\}$, assume its network matrix $L \in M_{81 \times 81}$ is

$$\begin{aligned} L = \delta_{81}[\ &3\ 6\ 9\ 29\ 41\ 44\ 55\ 67\ 79\ 3\ 6\ 9\ 29\ 41\ 44\ 28 \\ &40\ 52\ 3\ 6\ 9\ 2\ 14\ 17\ 1\ 13\ 25\ 6\ 6\ 9\ 32\ 41 \\ &44\ 58\ 67\ 79\ 6\ 6\ 9\ 32\ 41\ 44\ 31\ 40\ 52\ 6\ 6\ 9 \\ &5\ 14\ 17\ 4\ 13\ 25\ 9\ 9\ 9\ 35\ 44\ 44\ 61\ 70\ 79\ 9 \\ &9\ 9\ 35\ 44\ 44\ 34\ 43\ 52\ 9\ 9\ 9\ 8\ 17\ 17\ 7\ 16\ 25]. \end{aligned}$$

We reconstruct the system. Using retrievers S_i^3 we have

$$M_i = S_{i,3}L, \quad i = 1, 2, 3, 4,$$

which are

$$\begin{aligned} M_1 = \delta_3[\ &1\ 1\ 1\ 2\ 2\ 2\ 3\ 3\ 3\ 1\ 1\ 1\ 2\ 2\ 2\ 2\ 2\ 2\ 1\ 1\ 1 \\ &1\ 1\ 1\ 1\ 1\ 1\ 1\ 1\ 2\ 2\ 2\ 3\ 3\ 3\ 1\ 1\ 1\ 2\ 2\ 2 \\ &2\ 2\ 2\ 1\ 1\ 1\ 1\ 1\ 1\ 1\ 1\ 1\ 1\ 1\ 2\ 2\ 2\ 3\ 3\ 3 \\ &1\ 1\ 1\ 2\ 2\ 2\ 2\ 2\ 2\ 1\ 1\ 1\ 1\ 1\ 1\ 1\ 1\ 1], \end{aligned}$$

$$\begin{aligned} M_2 = \delta_3[\ &1\ 1\ 1\ 1\ 2\ 2\ 1\ 2\ 3\ 1\ 1\ 1\ 1\ 2\ 2\ 1\ 2\ 3\ 1\ 1\ 1 \\ &1\ 2\ 2\ 1\ 2\ 3\ 1\ 1\ 1\ 1\ 2\ 2\ 1\ 2\ 3\ 1\ 1\ 1\ 1\ 2\ 2 \\ &1\ 2\ 3\ 1\ 1\ 1\ 1\ 2\ 2\ 1\ 2\ 3\ 1\ 1\ 1\ 1\ 2\ 2\ 1\ 2\ 3 \\ &1\ 1\ 1\ 1\ 2\ 2\ 1\ 2\ 3\ 1\ 1\ 1\ 1\ 2\ 2\ 1\ 2\ 3], \end{aligned}$$

$$\begin{aligned} M_3 = \delta_3[\ &1\ 2\ 3\ 1\ 2\ 3\ 1\ 2\ 3\ 1\ 2\ 3\ 1\ 2\ 3\ 1\ 2\ 3\ 1\ 2\ 3 \\ &1\ 2\ 3\ 1\ 2\ 3\ 2\ 2\ 3\ 2\ 2\ 3\ 2\ 2\ 3\ 2\ 2\ 3\ 2\ 2\ 3 \\ &2\ 2\ 3\ 2\ 2\ 3\ 2\ 2\ 3\ 2\ 2\ 3\ 3\ 3\ 3\ 3\ 3\ 3\ 3\ 3\ 3 \\ &3\ 3\ 3\ 3\ 3\ 3\ 3\ 3\ 3\ 3\ 3\ 3\ 3\ 3\ 3\ 3\ 3\ 3], \end{aligned}$$

$$\begin{aligned} M_4 = \delta_3[\ &3\ 3\ 3\ 2\ 2\ 2\ 1\ 1\ 1\ 3\ 3\ 3\ 2\ 2\ 2\ 1\ 1\ 1\ 3\ 3\ 3 \\ &2\ 2\ 2\ 1\ 1\ 1\ 3\ 3\ 3\ 2\ 2\ 2\ 1\ 1\ 1\ 3\ 3\ 3\ 2\ 2\ 2 \\ &1\ 1\ 1\ 3\ 3\ 3\ 2\ 2\ 2\ 1\ 1\ 1\ 3\ 3\ 3\ 2\ 2\ 2\ 1\ 1\ 1 \\ &3\ 3\ 3\ 2\ 2\ 2\ 1\ 1\ 1\ 3\ 3\ 3\ 2\ 2\ 2\ 1\ 1\ 1]. \end{aligned}$$

Next, to remove fabricated variables, it is easy to verify that

$$\begin{cases} M_1 M_{o,3} - M_1 = 0, & M_1 (M_{o,3})^2 - M_1 = 0, \\ M_1 W_{[3]}(M_{o,3} - I_3) \neq 0, & M_1 W_{[3]}((M_{o,3})^2 - I_3) \neq 0, \\ M_1 W_{[3,3^2]}(M_{o,3} - I_3) \neq 0, & M_1 W_{[3,3^2]}((M_{o,3})^2 - I_3) \neq 0, \\ M_1 W_{[3,3^3]}(M_{o,3} - I_3) = 0, & M_1 W_{[3,3^3]}((M_{o,3})^2 - I_3) = 0. \end{cases}$$

Therefore we conclude that $A(t+1)$ depends on $B(t)$ and $C(t)$ only. Using the same procedure, we know that $B(t+1)$ depends only on $C(t)$ and $D(t)$, that $C(t+1)$ depends only on $A(t)$ and $D(t)$, and that $D(t+1)$ depends only on $C(t)$. To remove the fabricated variables $A(t)$ and $D(t)$ from the first equation, we set $A(t) = D(t) = \delta_3^1$ and get

$$\begin{aligned} A(t+1) &= M_1 \delta_3^1 B(t) C(t) \delta_3^1 \\ &= M_1 \delta_3^1 W_{[3,9]} \delta_3^1 B(t) C(t) \\ &= \delta_3 [1\,2\,3\,1\,2\,2\,1\,1\,1] B(t) C(t). \end{aligned} \tag{14.60}$$

In a similar way, we can remove the fabricated variables from the other equations. Finally we get

$$B(t+1) = \delta_3 [1\,1\,1\,1\,2\,2\,1\,2\,3] C(t) D(t),$$

$$C(t+1) = \delta_3 [1\,2\,3\,2\,2\,3\,3\,3\,3] D(t) A(t),$$

$$D(t+1) = \delta_3 [3\,2\,1] C(t).$$

Converting back to logical equations, we have

$$\begin{cases} A(t+1) = B(t) \to C(t), \\ B(t+1) = C(t) \vee D(t), \\ C(t+1) = D(t) \wedge A(t), \\ D(t+1) = \neg C(t). \end{cases} \tag{14.61}$$

In general, converting an algebraic form back to its logical form is not easy, so we now describe a mechanical procedure for doing this.

Proposition 14.6 *Assume a k-valued logical variable L has algebraic expression*

$$L = L(A_1, A_2, \ldots, A_n) = M_L A_1 A_2 \cdots A_n, \tag{14.62}$$

where $M_L \in \mathscr{L}_{k \times k^n}$ is the structure matrix of logical function L. Split this into k equal-sized blocks as

$$M_L = [M_{L_1}, M_{L_2}, \ldots, M_{L_k}],$$

where $M_{L_i} \in \mathscr{L}_{k \times k^{n-1}}$. Then, L can be expressed as

$$L = \left[\nabla_{1,k}(A_1) \wedge L_1(A_2, \ldots, A_n) \right] \vee \left[\nabla_{2,k}(A_1) \wedge L_2(A_2, \ldots, A_n) \right]$$
$$\vee \cdots \vee \left[\nabla_{k,k}(A_1) \wedge L_k(A_2, \ldots, A_n) \right],$$

where L_i has M_{L_i} as its structure matrix, $i = 1, \ldots, k$. That is, in vector form,

$$L_i(A_2, \ldots, A_n) = M_{L_i} A_2 \cdots A_n, \quad i = 1, \ldots, k.$$

Using Proposition 14.6 we can obtain the logical expression of L recursively. We give an example to describe this.

Example 14.6 Let L be a logical variable, and

$$L = M_L ABCD,$$

where $A, B, C, D \in \Delta_3$ and

$$\begin{aligned} M_L = \delta_3[& 1\ 2\ 3\ 2\ 2\ 2\ 3\ 2\ 1\ 2\ 2\ 2\ 2\ 2\ 3\ 2\ 2\ 3\ 2\ 1\ 3\ 2\ 1\ 3\ 2\ 1 \\ & 2\ 2\ 2\ 2\ 2\ 3\ 2\ 2\ 2\ 2\ 2\ 2\ 2\ 2\ 2\ 2\ 3\ 2\ 2\ 2\ 2\ 2\ 2\ 2 \\ & 1\ 1\ 1\ 2\ 2\ 2\ 3\ 3\ 3\ 2\ 2\ 2\ 2\ 2\ 2\ 2\ 3\ 3\ 2\ 1\ 2\ 2\ 2\ 1\ 2\ 3]. \end{aligned} \quad (14.63)$$

Then,

$$M_L = \left[\nabla_1(A) \wedge L_1(B, C, D) \right] \vee \left[\nabla_2(A) \right.$$
$$\left. \wedge L_2(B, C, D) \right] \vee \left[\nabla_3(A) \wedge L_3(B, C, D) \right], \quad (14.64)$$

and

$$\begin{aligned} M_{L_1} = \delta_3[& 1\ 2\ 3\ 2\ 2\ 2\ 3\ 2\ 1\ 2\ 2\ 2\ 2 \\ & 2\ 2\ 3\ 2\ 2\ 3\ 2\ 1\ 3\ 2\ 1\ 3\ 2\ 1], \end{aligned} \quad (14.65)$$

$$\begin{aligned} M_{L_2} = \delta_3[& 2\ 2\ 2\ 2\ 2\ 3\ 2\ 2\ 2\ 2\ 2\ 2 \\ & 2\ 2\ 2\ 2\ 3\ 2\ 2\ 2\ 2\ 2\ 2\ 2\ 2], \end{aligned} \quad (14.66)$$

$$\begin{aligned} M_{L_3} = \delta_3[& 1\ 1\ 1\ 2\ 2\ 2\ 3\ 3\ 3\ 2\ 2\ 2 \\ & 2\ 2\ 2\ 3\ 3\ 2\ 1\ 2\ 2\ 2\ 1\ 2\ 3]. \end{aligned} \quad (14.67)$$

Next, consider L_1:

$$L_1(B, C, D) = M_{L_1} BCD$$
$$= \left[\nabla_1(B) \wedge L_{11}(C, D) \right] \vee \left[\nabla_2(B) \right.$$
$$\left. \wedge L_{12}(C, D) \right] \vee \left[\nabla_3(B) \wedge L_{13}(C, D) \right], \quad (14.68)$$

where

$$M_{L_{11}} = \delta_3[\ 1\ \ 2\ \ 3\ \ 2\ \ 2\ \ 2\ \ 3\ \ 2\ \ 1\ ],$$

$$M_{L_{12}} = \delta_3[2 \quad 2 \quad 2 \quad 2 \quad 2 \quad 2 \quad 3 \quad 2 \quad 2],$$
$$M_{L_{13}} = \delta_3[3 \quad 2 \quad 1 \quad 3 \quad 2 \quad 1 \quad 3 \quad 2 \quad 1].$$

Hence, we have

$$L_{11}(C, D) = C \leftrightarrow D,$$
$$L_{12}(C, D) = M_{L_{12}} C D,$$
$$L_{13}(C, D) = \neg D.$$

In the same way, we have the following expression:

$$L_2(B, C, D) = \left[\nabla_1(B) \wedge L_{21}(C, D)\right] \vee \left[\nabla_2(B)\right.$$
$$\left. \wedge L_{22}(C, D)\right] \vee \left[\nabla_3(B) \wedge L_{23}(C, D)\right], \qquad (14.69)$$
$$L_3(B, C, D) = \left[\nabla_1(B) \wedge L_{31}(C, D)\right] \vee \left[\nabla_2(B)\right.$$
$$\left. \wedge L_{32}(C, D)\right] \vee \left[\nabla_3(B) \wedge L_{33}(C, D)\right]. \qquad (14.70)$$

Putting this all together, we have

$$L = \left[\nabla_1(A) \wedge \left[\left[\nabla_1(B) \wedge L_{11}(C, D)\right]\right.\right.$$
$$\vee \left[\nabla_2(B) \wedge L_{12}(C, D)\right] \vee \left[\nabla_3(B) \wedge L_{13}(C, D)\right]\right]$$
$$\vee \left[\nabla_2(A) \wedge \left[\left[\nabla_1(B) \wedge L_{21}(C, D)\right]\right.\right.$$
$$\vee \left[\nabla_2(B) \wedge L_{22}(C, D)\right] \vee \left[\nabla_3(B) \wedge L_{23}(C, D)\right]\right]$$
$$\vee \left[\nabla_3(A) \wedge \left[\left[\nabla_1(B) \wedge L_{31}(C, D)\right]\right.\right.$$
$$\vee \left[\nabla_2(B) \wedge L_{32}(C, D)\right] \vee \left[\nabla_3(B) \wedge L_{33}(C, D)\right]\right]. \qquad (14.71)$$

Remark 14.2 Note that we can also write down the split form of all binary operators. For instance,

$$L_{12}(C, D) = \delta_3[2\,2\,2\,2\,2\,2\,3\,2\,2] C D$$
$$= \left[\nabla_1(C) \wedge \delta_3[2\,2\,2] D\right] \vee \left[\nabla_2(C) \wedge \delta_3[2\,2\,2] D\right]$$
$$\vee \left[\nabla_3(C) \wedge \delta_3[3\,2\,2] D\right]$$
$$= \left[\nabla_1(C) \wedge \delta_3^2\right] \vee \left[\nabla_2(C) \wedge \delta_2^3\right] \vee \left[\nabla_3(C) \wedge \psi(D)\right], \qquad (14.72)$$

where the structure matrix of the unary logical operator ψ is $\delta_3[3\,2\,2]$.

14.6 k-valued Control Networks

Let u_i, $i = 1, \ldots, m$, be a set of controls. These are also k-valued logical variables. Moreover, let h_i, $i = 1, \ldots, p$, be k-valued output logical functions. We then have a k-valued control network with state dynamics

$$\begin{cases} x_1(t+1) = f_1(x_1(t), x_2(t), \ldots, x_n(t), u_1(t), \ldots, u_m(t)), \\ x_2(t+1) = f_2(x_1(t), x_2(t), \ldots, x_n(t), u_1(t), \ldots, u_m(t)), \\ \vdots \\ x_n(t+1) = f_n(x_1(t), x_2(t), \ldots, x_n(t), u_1(t), \ldots, u_m(t)) \end{cases} \quad (14.73)$$

and outputs

$$y_j = h_j(x_1(t), x_2(t), \ldots, x_n(t)), \quad j = 1, \ldots, p. \quad (14.74)$$

The controls could be a free k-valued logical sequence or it could be generated by a control dynamics as

$$\begin{cases} u_1(t+1) = g_1(u_1(t), u_2(t), \ldots, u_m(t)), \\ u_2(t+1) = g_2(u_1(t), u_2(t), \ldots, u_m(t)), \\ \vdots \\ u_m(t+1) = g_m(u_1(t), u_2(t), \ldots, u_m(t)). \end{cases} \quad (14.75)$$

First, we consider the case where the control is generated by a control dynamics. The system (14.73)–(14.75) can then also be expressed in algebraic form as

$$x(t+1) = Lu(t)x(t) = L(u)x(t), \quad x \in \mathscr{D}_k^n,$$
$$y(t) = Hx(t), \quad y \in \mathscr{D}_k^p, \quad (14.76)$$
$$u(t+1) = Gu(t), \quad u \in \mathscr{D}_k^m.$$

We give an example.

Example 14.7 We reconsider Example 6.1 and assume that the logical variables can now take values from $\mathscr{D}_3 = \{0, 0.5, 1\}$. Set $x(t) = B(t) \ltimes C(t) \ltimes D(t)$. Converting the system (6.5) into its algebraic form, we have

$$x(t+1) = L(u)x(t),$$
$$y(t) = M_{n,3}x(t), \quad (14.77)$$
$$u(t+1) = u(t),$$

where $x(t) \in \Delta_{3^3}$.

$L(u)$ can be easily calculated as

$$L(u) = M_{i,3}u(I_3 \otimes M_{d,3})(I_{27} \otimes M_{n,3})W_{[3]}W_{[3,27]}M_{r,3}.$$

When $u_1 = \delta_3^1$,

$$L(u_1) = \delta_{27}[3 \ \ 3 \ 3 \ 12 \ 12 \ 12 \ 21 \ 21 \ 21 \ 2 \ 5 \ 5 \ 11 \ 14 \ 14 \ 20 \ 23 \ 23 \ 1 \ 4 \ 7 \ 10 \ 13 \ 16 \ 19 \\ 22 \ 25],$$

when $u_2 = \delta_3^2$,

$$L(u_2) = \delta_{27}[3 \quad 3\ 3\ 12\ 12\ 12\ 12\ 12\ 12\ 2\ 5\ 5\ 11\ 14\ 14\ 11\ 14\ 14\ 1\ 4\ 7\ 10\ 13\ 16\ 10$$
$$13\ 16],$$

and when $u_3 = \delta_3^3$,

$$L(u_3) = \delta_{27}[3\ 3\ 3\ 3\ 3\ 3\ 3\ 3\ 3\ 2\ 5\ 5\ 2\ 5\ 5\ 2\ 5\ 5\ 1\ 4\ 7\ 1\ 4\ 7\ 1\ 4\ 7].$$

Now, δ_3^1, δ_3^2, and δ_3^3 are fixed points of the control network. It is easy to deduce that for $u = \delta_3^1$ there are two fixed points, $(1, 1, 0)$ and $(0.5, 0.5, 0.5)$, and two cycles of length 2, which are $(1, 0.5, 0.5) \rightarrow (0.5, 1, 0) \rightarrow (1, 0.5, 0.5)$ and $(1, 0, 1) \rightarrow (0, 1, 0) \rightarrow (1, 0, 1)$. For $u = \delta_3^2$ there are two fixed points, $(1, 1, 0)$ and $(0.5, 0.5, 0.5)$, and one cycle of length 2, which is $(1, 0.5, 0.5) \rightarrow (0.5, 1, 0) \rightarrow (1, 0.5, 0.5)$. When $u = \delta_3^3$ there is only one fixed point, $(1, 1, 0)$.

Definition 14.7 Consider the k-valued logical system (14.73) with control (14.75). Given initial state $x(0) = x_0$ and destination state x_d, the latter is said to be reachable from x_0 (at the sth step) with fixed (designable) input structure (G) if we can find u_0 (and G) such that $x(u, 0) = x_0$ and $x(u, s) = x_d$ (for some $s \geq 1$).

We use $\Theta^G(t, 0)$ to denote the input-state transfer matrix in a k-valued network, which can be calculated as

$$\Theta^G(t, 0) = LG^{t-1}\left(I_{k^m} \otimes LG^{t-2}\right)\left(I_{k^{2m}} \otimes LG^{t-3}\right) \cdots \left(I_{k^{(t-1)m}} \otimes L\right)$$
$$(I_{k^{(t-2)m}} \otimes \Phi_{m,k})(I_{k^{(t-3)m}} \otimes \Phi_{m,k}) \cdots (I_{k^m} \otimes \Phi_{m,k})\Phi_{m,k}, \quad (14.78)$$

where $\Phi_{m,k}$ is defined in Lemma 14.1 as

$$\Phi_{m,k} = \prod_{i=1}^{m} I_{k^{i-1}} \otimes \left[(I_k \otimes W_{[k,k^{m-i}]})M_{r,k}\right].$$

It is then easy to prove that for the system (14.76),

$$x(t) = \Theta^G(t, 0)u(0)x(0).$$

We will now discuss two cases.

Case 1: We have fixed s and fixed G.

From the definition of the transfer matrix, the following result is obvious.

Theorem 14.7 *Consider the system* (14.73) *with control* (14.75), *where G is fixed.* x_d *is s-step reachable from x_0 if and only if*

$$x_d \in \text{Col}\left\{\Theta^G(s, 0)W_{[k^n, k^m]}x_0\right\}. \quad (14.79)$$

We give an example to describe this result.

Example 14.8 Reconsider Example 9.1, but now assume that the logical variables may take three different values, $\{0, 0.5, 1\}$. Convert it to its algebraic form,

$$x(t + 1) = M_e B(t)C(t)M_d C(t)u_1(t)M_c A(t)u_2(t) = Lu(t)x(t),$$

where $L \in \mathcal{L}_{27 \times 243}$ is

$$
\begin{aligned}
L = \delta_{27}[1 \quad &10\ 19\ 10\ 10 \quad 10\ 19\ 10\ 1 \quad 2 \quad 11\ 20\ 11\ 11\ 11\ 20\ 11 \\
&2 \quad 3 \quad 12\ 21\ 12 \quad 12\ 12\ 21\ 12\ 3 \quad 2 \quad 11\ 20\ 11\ 11\ 11\ 20 \\
&11\ 2 \quad 2 \quad 11\ 20\ 11\ 11\ 11\ 20\ 11\ 2 \quad 3 \quad 12\ 21\ 12\ 12\ 12 \\
&21\ 12\ 3 \quad 3 \quad 12\ 21\ 12\ 12\ 12\ 21\ 12\ 3 \quad 3 \quad 12\ 21\ 12\ 12 \\
&12\ 21\ 12\ 3 \quad 3 \quad 12\ 21\ 12\ 12\ 12\ 21\ 12\ 3 \quad 1 \quad 13\ 22\ 10 \\
&13\ 13\ 19\ 13\ 4 \quad 2 \quad 14\ 23\ 11\ 14\ 14\ 20\ 14\ 5 \quad 3 \quad 15\ 24 \\
&12\ 15\ 15\ 21\ 15\ 6 \quad 2 \quad 14\ 23\ 11\ 14\ 14\ 20\ 14\ 5 \quad 2 \quad 14 \\
&23\ 11\ 14\ 14\ 20\ 14\ 5 \quad 3 \quad 15\ 24\ 12\ 15\ 15\ 21\ 15\ 6 \quad 3 \\
&15\ 24\ 12\ 15\ 15\ 21\ 15\ 6 \quad 3 \quad 15\ 24\ 12\ 15\ 15\ 21\ 15\ 6 \\
&3 \quad 15\ 24\ 12\ 15\ 15\ 21\ 15\ 6 \quad 1 \quad 13\ 25\ 10\ 13\ 16\ 19\ 13 \\
&7 \quad 2 \quad 14\ 26\ 11\ 14\ 17\ 20\ 14\ 8 \quad 3 \quad 15\ 27\ 12\ 15\ 18\ 21 \\
&15\ 9 \quad 2 \quad 14\ 26\ 11\ 14\ 17\ 20\ 14\ 8 \quad 2 \quad 14\ 26\ 11\ 14\ 17 \\
&20\ 14\ 8 \quad 3 \quad 15\ 27\ 12\ 15\ 18\ 21\ 15\ 9 \quad 3 \quad 15\ 27\ 12\ 15 \\
&18\ 21\ 15\ 9 \quad 3 \quad 15\ 27\ 12\ 15\ 18\ 21\ 15\ 9 \quad 3 \quad 15\ 27\ 12 \\
&15\ 18\ 21\ 15\ 9].
\end{aligned}
$$

Assume g_1 and g_2 are fixed as

$$
\begin{cases}
g_1(u_1(t), u_2(t)) = \neg u_2(t), \\
g_2(u_1(t), u_2(t)) = u_1(t).
\end{cases}
\tag{14.80}
$$

Choose $A(0) = 0.5$, $B(0) = 0$, $C(0) = 0.5$, and $s = 5$. If we let $u(t) = u_1(t)u_2(t)$, then

$$u(t + 1) = u_1(t + 1)u_2(t + 1) = M_n u_2(t)u_1(t) = M_n W_{[3]}u(t).$$

Hence,

$$G = M_n W_{[3]} = \delta_9[7\ 4\ 1\ 8\ 5\ 2\ 9\ 6\ 3] \in \mathcal{L}_{9 \times 9}.$$

It is easy to calculate $\Phi_{2,3}$ as

$$
\begin{aligned}
\Phi_{2,3} &= (I_3 \otimes W_{[3]})M_{r_3}(I_3 \otimes M_{r_3}) \\
&= \delta_{81}[1\ 11\ 21\ 31\ 41\ 51\ 61\ 71\ 81].
\end{aligned}
$$

Finally, using formula (14.78) yields $\Theta(5, 0) \in \mathcal{L}_{27 \times 243}$ as

$$
\begin{aligned}
\Theta(5, 0) = LG^4 &\left(I_{3^2} \otimes LG^3\right)\left(I_{3^4} \otimes LG^2\right)(I_{3^6} \otimes LG)(I_{3^8} \otimes L)(I_{3^6} \otimes \Phi_{3,3}) \\
&(I_{3^4} \otimes \Phi_{3,3})(I_{3^2} \otimes \Phi_{3,3})(I_3 \otimes \Phi_{3,3})\Phi_{3,3},
\end{aligned}
$$

which is

$$\delta_{27}[21\ 20\ 19\ 20\ 20\ 20\ 19\ 20\ 21\ 21\ 20\ 19\ 20\ 20\ 20\ 19\ 20\ 21$$
$$21\ 20\ 19\ 20\ 20\ 20\ 19\ 20\ 21\ 11\ 11\ 11\ 11\ 11\ 11\ 11\ 11\ 11$$
$$11\ 11\ 11\ 11\ 11\ 11\ 11\ 11\ 12\ 11\ 11\ 11\ 11\ 11\ 11\ 11\ 12$$
$$3\ 3\ 3\ 3\ 3\ 3\ 3\ 3\ 3\ 3\ 3\ 3\ 3\ 3\ 3\ 3\ 3\ 3\ 3\ 3\ 3\ 3\ 3$$
$$3\ 3\ 3\ 14\ 14\ 23\ 14\ 14\ 14\ 23\ 14\ 14\ 14\ 14\ 14\ 14\ 14\ 14\ 14$$
$$14\ 14\ 14\ 14\ 14\ 14\ 14\ 14\ 14\ 14\ 14\ 14\ 14\ 14\ 14\ 14\ 14\ 14$$
$$14\ 14\ 14\ 14\ 14\ 14\ 14\ 14\ 14\ 14\ 14\ 14\ 14\ 14\ 14\ 14\ 14\ 14$$
$$14\ 14\ 15\ 15\ 15\ 15\ 15\ 15\ 6\ 15\ 15\ 15\ 15\ 15\ 15\ 15\ 15\ 6$$
$$15\ 15\ 15\ 15\ 15\ 15\ 15\ 15\ 6\ 15\ 15\ 27\ 27\ 27\ 27\ 27\ 27\ 27$$
$$27\ 27\ 15\ 15\ 15\ 15\ 15\ 15\ 15\ 15\ 15\ 3\ 3\ 3\ 3\ 3\ 3\ 3\ 3$$
$$14\ 14\ 14\ 14\ 14\ 14\ 14\ 14\ 14\ 14\ 14\ 14\ 14\ 14\ 14\ 14\ 14\ 14$$
$$15\ 14\ 14\ 15\ 14\ 14\ 15\ 14\ 14\ 9\ 18\ 27\ 9\ 18\ 27\ 9\ 18\ 27\ 9$$
$$18\ 27\ 9\ 18\ 27\ 9\ 18\ 27\ 9\ 18\ 27\ 9\ 18\ 27\ 9\ 18\ 27].$$

Setting the initial value as $X_0 = (A(0), B(0), C(0)) = (0.5, 1, 1)$, we then have

$$x_0 = A(0)B(0)C(0) = \delta_{27}^{10}.$$

Using Theorem 14.7, we have the reachable set as

$$\Theta(5, 0)W_{[27,9]}x_0 = \delta_{27}\{21, 11, 3, 14, 15, 9\}.$$

Converting them to ternary form, we have

$$R_5(X_0) = \{(0, 1, 0), (0.5, 1, 0.5), (1, 1, 0), (0.5, 0.5, 0.5), (0.5, 0.5, 0), (1, 0, 0)\}.$$

Finally, we have to find the initial control u_0 which drives the trajectory to the assigned x_d. Since

$$x_d = \Theta(5, 0)W_{[27,9]}x_0u_0 = \delta_{27}[21\ 11\ 3\ 14\ 14\ 15\ 15\ 14\ 9]u_0,$$

it is obvious that to reach, say, $\delta_{27}^{21} \sim (0, 1, 0)$, the control should be $u_0 = \delta_9^1$, i.e., $u_1(0) = \delta_3^1 \sim 1$ and $u_2(0) = \delta_3^1 \sim 1$. Similarly, to reach all six points in $R_5(X_0)$ at step 5, the corresponding initial controls $u_i(0)$ are given in the following Table 14.2.

Remark 14.3 The $\Theta^G(s, 0)$ can be calculated inductively, and the algorithm is similar to the one in Chap. 7.

Case 2: We have fixed s and a set of G.

Since there are $m_0 = (k^m)^{k^m}$ possible distinct G's, we may express each G in condensed form and order them in "increasing order". For example, when $m = 2, k = 3$, we have $G_1 = \delta_9[1\ 1\ 1\ 1\ 1\ 1\ 1\ 1\ 1]$, $G_2 = \delta_9[1\ 1\ 1\ 1\ 1\ 1\ 1\ 1\ 2], \ldots,$ $G_{99} = \delta_9[9\ 9\ 9\ 9\ 9\ 9\ 9\ 9\ 9]$. In general, we may consider a subset $\Lambda \subset \{1, 2, \ldots, m_0\}$ and allow G to be chosen from the admissible set: $\{G_\lambda \mid \lambda \in \Lambda\}$.

Table 14.2 The desired states and the corresponding controls, $x(0) = (0.5, 1, 1)$

x_d	$u(0)$	$u_1(0)$	$u_2(0)$
$\delta_{27}^{21} \sim (0, 1, 0)$	$\delta_9^1 \sim (0, 1, 0)$	$\delta_3^1 \sim 1$	$\delta_3^1 \sim 1$
$\delta_{27}^{11} \sim (0.5, 1, 0.5)$	$\delta_9^2 \sim (0, 1, 0)$	$\delta_3^1 \sim 1$	$\delta_3^1 \sim 0.5$
$\delta_{27}^{3} \sim (1, 1, 0)$	$\delta_9^3 \sim (0, 1, 0)$	$\delta_3^1 \sim 1$	$\delta_3^1 \sim 0$
$\delta_{27}^{14} \sim (0.5, 0.5, 0.5)$	$\delta_9^4 \sim (0, 1, 0)$	$\delta_3^2 \sim 0.5$	$\delta_3^1 \sim 1$
	$\delta_9^5 \sim (0, 1, 0)$	$\delta_3^2 \sim 0.5$	$\delta_3^2 \sim 0.5$
	$\delta_9^8 \sim (0, 1, 0)$	$\delta_3^3 \sim 0$	$\delta_3^2 \sim 0.5$
$\delta_{27}^{15} \sim (0.5, 0.5, 0)$	$\delta_9^6 \sim (0, 1, 0)$	$\delta_3^2 \sim 0.5$	$\delta_3^3 \sim 0$
	$\delta_9^7 \sim (0, 1, 0)$	$\delta_3^3 \sim 0$	$\delta_3^1 \sim 1$
$\delta_{27}^{9} \sim (1, 0, 0)$	$\delta_9^9 \sim (0, 0, 0)$	$\delta_3^3 \sim 0$	$\delta_3^3 \sim 0$

Corollary 14.1 *Consider the system* (14.73) *with control* (14.74), *where* $G \in \{G_\lambda \mid \lambda \in \Lambda\}$. *Then,* x_d *is reachable from* x_0 *if and only if*

$$x_d \in \bigcup_{\lambda \in \Lambda} \mathrm{Col}\left\{\Theta^{G_\lambda}(s, 0) W_{[k^n, k^m]} x_0\right\}. \tag{14.81}$$

Example 14.9 Recall Example 9.1, with network dynamics (9.7). We change it to a 3-valued network, but still assume that $X_0 = (1, 0, 1)$ and $s = 5$. Assume that $\Xi = \{G_1, G_2, G_3, G_4\}$, where $G_1 = \delta_9[1\,2\,3\,4\,5\,6\,7\,8\,9]$, $G_2 = \delta_9[1\,5\,8\,9\,7\,4\,6\,3\,2]$, $G_3 = \delta_9[1\,8\,9\,6\,5\,7\,3\,2\,4]$, $G_4 = \delta_9[9\,8\,5\,6\,4\,2\,3\,1\,7]$, and the corresponding $G_5^i(X_0) = \mathrm{Col}\left\{\Theta^i(5, 0) W_{[3^n, 3^m]} x_0\right\}$ are

$$G_5^1(X_0) = \delta_{27}\{2, 11, 21, 14, 14, 15, 14, 14, 9\},$$
$$G_5^2(X_0) = \delta_{27}\{2, 14, 12, 14, 15, 15, 11, 26, 15\},$$
$$G_5^3(X_0) = \delta_{27}\{2, 11, 17, 27, 14, 6, 15, 14, 12\},$$
$$G_5^4(X_0) = \delta_{27}\{23, 17, 11, 11, 15, 15, 15, 21, 15\}.$$

The reachable set is then

$$G_5^1(X_0) = \delta_{27}\{2, 11, 21, 14, 15, 9\},$$
$$G_5^2(X_0) = \delta_{27}\{2, 14, 12, 15, 11, 6\},$$
$$G_5^3(X_0) = \delta_{27}\{2, 11, 17, 27, 14, 6, 15, 12\},$$
$$G_5^4(X_0) = \delta_{27}\{23, 17, 11, 15, 21\}.$$

The reachable set at the fifth step is thus

$$\bigcup_{i=1}^{4} G_5^i(X_0) = \delta_{27}\{2, 6, 9, 11, 12, 14, 15, 17, 21, 23, 26, 27\}.$$

Now, assume that we want to reach $(A(5), B(5), C(5)) = (0.5, 1, 0)$, which is δ_{27}^{12} since the third component of $G_5^2(X_0)$ is 12. (We have some other choices, such as the 9th component of $G_5^3(X_0)$, etc.) We can therefore choose G_2 and $u(0) = u_1(0)u_2(0) = \delta_9^3$ to drive $(0.5, 0, 0.5)$ to $(0.5, 1, 0)$ at the fifth step.

We can reconstruct the control dynamics from the logical matrix, G_2. Converting $G_2 = \delta_9[1\,5\,8\,9\,7\,4\,6\,3\,2]$ back to standard form, we have

$$G_2 = \delta_9[1\,5\,8\,9\,7\,4\,6\,3\,2].$$

From $u_1(0)u_2(0) = \delta_9^3$, we have $u_1(0) = \delta_3^1$ and $u_2(0) = \delta_3^3$. To reconstruct control dynamics, we need retrievers

$$S_{1,3} = \delta_3[1\,1\,1\,2\,2\,2\,3\,3\,3], \qquad S_{2,3} = \delta_3[1\,2\,3\,1\,2\,3\,1\,2\,3].$$

We then have the structure matrices

$$M_1 = S_{1,3}G = \delta_3[1\,2\,3\,3\,3\,2\,2\,1\,1],$$
$$M_2 = S_{2,3}G = \delta_3[1\,2\,2\,3\,1\,1\,3\,3\,2].$$

It follows that

$$u_1(t+1) = \delta_3[1\,2\,3\,3\,3\,2\,2\,1\,1]u_1(t)u_2(t),$$
$$u_2(t+1) = \delta_3[1\,2\,2\,3\,1\,1\,3\,3\,2]u_1(t)u_2(t).$$

We leave the investigation of other cases to the reader. Next, we consider the controllability of a multivalued logical network with control a k-valued sequence. We give the following definition.

Definition 14.8 Consider the k-valued logical system (14.73) and assume that an initial state of the network x_0^i, $i = 1, \ldots, n$, and a destination of the network x_d^i, $i = 1, \ldots, n$, at the sth step are given. The control problem via a free control sequence is then to find a sequence of δ_k^i vectors $u(0), \ldots, u(s-1)$ such that $x_i(0) = x_0^i$, $x_i(s) = x_d^i$, $i = 1, \ldots, n$.

Defining $\tilde{L} = LW_{[k^n, k^m]}$, the second equation in (14.76) can be expressed as

$$x(t+1) = \tilde{L}x(t)u(t). \tag{14.82}$$

Using this repetitively yields

$$x(s) = \tilde{L}^s x(0)u(0)u(1)\cdots u(s-1). \tag{14.83}$$

Therefore the answer to this kind of control problem is obvious.

Fig. 14.2 A 3-valued control network

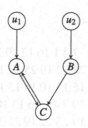

Theorem 14.8 x_d^i *is reachable from* x_0^i, $i = 1, \ldots, n$, *at the sth step by controls of* k-*valued sequences of length s if and only if*

$$x_s \in \text{Col}\{\tilde{L}^s x_0\}, \tag{14.84}$$

where $x_s = \ltimes_{i=1}^n x_d^i$, $x_0 = \ltimes_{i=1}^n x_0^i$.

Remark 14.4 Note that (14.84) means that x_s is equal to a column of $\tilde{L}^s x_0$. For example, if x_s equals the kth column of $\tilde{L}^s x_0$, then the controls should be

$$u(0)u(1) \cdots u(s-1) = \delta_{m^s}^k, \tag{14.85}$$

which uniquely determines all u_i, $i = 0, 1, \ldots, s - 1$.

The following example is taken from [2], but here we allow the values of the nodes in the network to be from $\mathscr{D}_3 = \{0, 0.5, 1\}$.

Example 14.10 Reconsider Fig. 14.2 from Example 9.6. We now consider it as a 3-valued logical control network.

Its system of logical equations is

$$\begin{cases} A(t+1) = C(t) \wedge u_1(t), \\ B(t+1) = \neg u_2(t), \\ C(t+1) = A(t) \vee B(t), \end{cases} \tag{14.86}$$

and its algebraic form is

$$\begin{cases} A(t+1) = M_{c,3}C(t)u_1(t), \\ B(t+1) = M_{n,3}u_2(t), \\ C(t+1) = M_{d,3}A(t)B(t). \end{cases} \tag{14.87}$$

Let $x(t) = A(t)B(t)C(t)$, $u(t) = u_1(t)u_2(t)$. We can then express the system by

$$x(t+1) = \tilde{L}x(t)u(t), \tag{14.88}$$

where $\tilde{L} \in \mathcal{L}_{27 \times 243}$ is

$$\tilde{L} = \delta_{27}[\ 7\ 4\ 1\ 16\ 13\ 10\ 25\ 22\ 19\ 16\ 13\ 10\ 16\ 13\ 10\ 25\ 22\ 19$$
$$25\ 22\ 19\ 25\ 22\ 19\ 25\ 22\ 19\ 7\ 4\ 1\ 16\ 13\ 10\ 25\ 22\ 19$$
$$16\ 13\ 10\ 16\ 13\ 10\ 25\ 22\ 19\ 25\ 22\ 19\ 25\ 22\ 19\ 25\ 22$$
$$19\ 7\ 4\ 1\ 16\ 13\ 10\ 25\ 22\ 19\ 16\ 13\ 10\ 16\ 13\ 10\ 25\ 22$$
$$19\ 25\ 22\ 19\ 25\ 22\ 19\ 25\ 22\ 19\ 7\ 4\ 1\ 16\ 13\ 10\ 25\ 22$$
$$19\ 16\ 13\ 10\ 16\ 13\ 10\ 25\ 22\ 19\ 25\ 22\ 19\ 25\ 22\ 19\ 25$$
$$22\ 19\ 8\ 5\ 2\ 17\ 14\ 11\ 26\ 23\ 20\ 17\ 14\ 11\ 17\ 14\ 11\ 26$$
$$23\ 20\ 26\ 23\ 20\ 26\ 23\ 20\ 26\ 23\ 20\ 8\ 5\ 2\ 17\ 14\ 11\ 26$$
$$23\ 20\ 17\ 14\ 11\ 17\ 14\ 11\ 26\ 23\ 20\ 26\ 23\ 20\ 26\ 23\ 20$$
$$26\ 23\ 20\ 7\ 4\ 1\ 16\ 13\ 10\ 25\ 22\ 19\ 16\ 13\ 10\ 16\ 13\ 10$$
$$25\ 22\ 19\ 25\ 22\ 19\ 25\ 22\ 19\ 25\ 22\ 19\ 8\ 5\ 2\ 17\ 14\ 11$$
$$26\ 23\ 20\ 17\ 14\ 11\ 17\ 14\ 11\ 26\ 23\ 20\ 26\ 23\ 20\ 26\ 23$$
$$20\ 26\ 23\ 20\ 9\ 6\ 3\ 18\ 15\ 12\ 27\ 24\ 21\ 18\ 15\ 12\ 18\ 15$$
$$12\ 27\ 24\ 21\ 27\ 24\ 21\ 27\ 24\ 21\ 27\ 24\ 21].$$

We now assume that $(A(0), B(0), C(0)) = (0, 0, 0)$. We want to know if a designed state can be reached at the sth step. If, for example, $s = 3$, then using Theorem 14.8 we calculate $\tilde{L}^3 x_0 \in M_{3^3 \times 3^6}$ as

$$\tilde{L} = \delta_{27}[27\ 24\ 21\ 27\ 24\ 21\ 27\ 24\ 21\ 26\ 23\ 20\ 26\ 23\ 20\ 26\ 23$$
$$20\ 25\ 22\ 19\ 25\ 22\ 19\ 25\ 22\ 19\ 27\ 24\ 21\ 27\ 24\ 21\ 27$$
$$24\ 21\ 26\ 23\ 20\ 26\ 23\ 20\ 26\ 23\ 20\ 25\ 22\ 19\ 25\ 22\ 19$$
$$25\ 22\ 19\ 27\ 24\ 21\ 27\ 24\ 21\ 27\ 24\ 21\ 26\ 23\ 20\ 26\ 23$$
$$20\ 26\ 23\ 20\ 25\ 22\ 19\ 25\ 22\ 19\ 25\ 22\ 19\ 18\ 15\ 12\ 18$$
$$15\ 12\ 27\ 24\ 21\ 17\ 14\ 11\ 17\ 14\ 11\ 26\ 23\ 20\ 16\ 13\ 10$$
$$16\ 13\ 10\ 25\ 22\ 19\ 18\ 15\ 12\ 18\ 15\ 12\ 27\ 24\ 21\ 17\ 14$$
$$11\ 17\ 14\ 11\ 26\ 23\ 20\ 16\ 13\ 10\ 16\ 13\ 10\ 25\ 22\ 19\ 18$$
$$15\ 12\ 18\ 15\ 12\ 27\ 24\ 21\ 17\ 14\ 11\ 17\ 14\ 11\ 26\ 23\ 20$$
$$16\ 13\ 10\ 16\ 13\ 10\ 25\ 22\ 19\ 9\ 6\ 3\ 18\ 15\ 12\ 27\ 24\ 21$$
$$8\ 5\ 2\ 17\ 14\ 11\ 26\ 23\ 20\ 7\ 4\ 1\ 16\ 13\ 10\ 25\ 22\ 19\ 9$$
$$6\ 3\ 18\ 15\ 12\ 27\ 24\ 21\ 8\ 5\ 2\ 17\ 14\ 11\ 26\ 23\ 20\ 7\ 4$$
$$1\ 16\ 13\ 10\ 25\ 22\ 19\ 9\ 6\ 3\ 18\ 15\ 12\ 27\ 24\ 21\ 8\ 5\ 2$$
$$17\ 14\ 11\ 26\ 23\ 20\ 7\ 4\ 1\ 16\ 13\ 10\ 25\ 22\ 19\ 27\ 24\ 21$$
$$27\ 24\ 21\ 27\ 24\ 21\ 26\ 23\ 20\ 26\ 23\ 20\ 26\ 23\ 20\ 25\ 22$$
$$19\ 25\ 22\ 19\ 25\ 22\ 19\ 27\ 24\ 21\ 27\ 24\ 21\ 27\ 24\ 21\ 26$$
$$23\ 20\ 26\ 23\ 20\ 26\ 23\ 20\ 25\ 22\ 19\ 25\ 22\ 19\ 25\ 22\ 19$$
$$27\ 24\ 21\ 27\ 24\ 21\ 27\ 24\ 21\ 26\ 23\ 20\ 26\ 23\ 20\ 26\ 23$$
$$20\ 25\ 22\ 19\ 25\ 22\ 19\ 25\ 22\ 19\ 18\ 15\ 12\ 18\ 15\ 12\ 27$$
$$24\ 21\ 17\ 14\ 11\ 17\ 14\ 11\ 26\ 23\ 20\ 16\ 13\ 10\ 16\ 13\ 10$$
$$25\ 22\ 19\ 18\ 15\ 12\ 18\ 15\ 12\ 27\ 24\ 21\ 17\ 14\ 11\ 17\ 14$$

11 26 23 20 16 13 10 16 13 10 25 22 19 18 15 12 18
15 12 27 24 21 17 14 11 17 14 11 26 23 20 16 13 10
16 13 10 25 22 19 9 6 3 18 15 12 27 24 21 8 5 2 17
14 11 26 23 20 7 4 1 16 13 10 25 22 19 9 6 3 18 15
12 27 24 21 8 5 2 17 14 11 26 23 20 7 4 1 16 13 10
25 22 19 9 6 3 18 15 12 27 24 21 8 5 2 17 14 11 26
23 20 7 4 1 16 13 10 25 22 19 27 24 21 27 24 21 27
24 21 26 23 20 26 23 20 26 23 20 25 22 19 25 22 19
25 22 19 27 24 21 27 24 21 27 24 21 26 23 20 26 23
20 26 23 20 25 22 19 25 22 19 25 22 19 27 24 21 27
24 21 27 24 21 26 23 20 26 23 20 26 23 20 25 22 19
25 22 19 25 22 19 18 15 12 18 15 12 27 24 21 17 14
11 17 14 11 26 23 20 16 13 10 16 13 10 25 22 19 18
15 12 18 15 12 27 24 21 17 14 11 17 14 11 26 23 20
16 13 10 16 13 10 25 22 19 18 15 12 18 15 12 27 24
21 17 14 11 17 14 11 26 23 20 16 13 10 16 13 10 25
22 19 9 6 3 18 15 12 27 24 21 8 5 2 17 14 11 26 23
20 7 4 1 16 13 10 25 22 19 9 6 3 18 15 12 27 24 21
8 5 2 17 14 11 26 23 20 7 4 1 16 13 10 25 22 19 9
6 3 18 15 12 27 24 21 8 5 2 17 14 11 26 23 20 7 4
1 16 13 10 25 22 19].

A routine from the Toolbox (see Appendix A) shows that at the third step all states can be reached. Choose one state, say $\delta_{27}^{25} \sim (0, 0, 1)$. Note that in the 19th, 22nd, 25th, ... columns of $\tilde{L}^3 x_0$ we have δ_{27}^{25}, which means that controls $\delta_{729}^{19}, \delta_{729}^{22}$, δ_{729}^{25}, or ... can drive the initial state $(0, 0, 0)$ to the destination state $(0, 0, 1)$. We choose, for example,

$$u_1(0)u_2(0)u_1(1)u_2(1)u_1(2)u_2(2) = \delta_{729}^{19}.$$

Converting this to ternary form yields $(1, 1, 1, 0, 1, 1)$, which means that the corresponding controls are

$$u_1(0) = 1, \qquad u_2(0) = 1; \qquad u_1(1) = 1,$$
$$u_2(1) = 0; \qquad u_1(2) = 1, \qquad u_2(2) = 1.$$

It is easy to directly check that this set of controls works. We may check some others. Choosing, say, δ_{729}^{22} and converting it to ternary form as $(1, 1, 1, 0, 0.5, 1)$, we have

$$u_1(0) = 1, \qquad u_2(0) = 1; \qquad u_1(1) = 1,$$
$$u_2(1) = 0; \qquad u_1(2) = 0.5, \qquad u_2(2) = 1.$$

This also works.

In general it is easy to calculate that when $s = 1$, the reachable set from $(0, 0, 0)$ is

$$\{(0, 0, 0), (0, 0.5, 0), (0, 1, 0)\}.$$

When $s = 2$ the reachable set is

$$\{(0, 0, 0), (0, 0.5, 0), (0, 1, 0), (0, 0, 0.5), (0, 0.5, 0.5),$$
$$(0, 1, 0.5), (0, 0.5, 1), (0, 0, 1), (0, 1, 1)\}.$$

In this chapter we considered only the topological structure and the controllability of k-valued logical (control) networks. It is easily seen that the methods developed and the results obtained for Boolean (control) networks can be easily extended to k-valued networks. We are not going to repeat all of the other control problems for the k-valued case, but leave them for the reader to explore the similar results.

14.7 Mix-valued Logic

Consider a set of logical variables $\{x_1, \ldots, x_n\}$. If $x_i \in \mathcal{D}_{k_i}$, then how do we define the logical operators between them? We call such a set of logical variables and operators a mix-valued logic. The problem basically comes from the mix-valued logical dynamical (control) systems. We first introduce them (we also refer to Sect. 16.6 for their properties).

Definition 14.9

1. Consider a logical dynamical system

$$\begin{cases} x_1(t+1) = f_1(x_1(t), \ldots, x_n(t)), \\ \vdots \\ x_n(t+1) = f_n(x_1(t), \ldots, x_n(t)), \end{cases} \tag{14.89}$$

where $x_i \in \mathcal{D}_{k_i}$, $f_i : \prod_{j=1}^{n} \mathcal{D}_{k_j} \to \mathcal{D}_{k_i}$, $i = 1, \ldots, n$, are logical functions. If k_i, $i = 1, \ldots, n$, are not identically equal, then the system (14.89) is called a mix-valued logical system.

2. Consider a logical control system

$$\begin{cases} x_1(t+1) = f_1(x_1(t), \ldots, x_n(t), u_1(t), \ldots, u_m(t)), \\ \vdots \\ x_n(t+1) = f_n(x_1(t), \ldots, x_n(t), u_1(t), \ldots, u_m(t)), \\ y_l(t) = h_l(x_1(t), \ldots, x_n(t)), \quad l = 1, \ldots, p, \end{cases} \tag{14.90}$$

where $x_i \in \mathcal{D}_{k_i}$, $u_j \in \mathcal{D}_{s_j}$, $y_l \in \mathcal{D}_{q_l}$, $f_i : \prod_{i=1}^{n} \mathcal{D}_{k_i} \times \prod_{j=1}^{m} \mathcal{D}_{s_j} \to \mathcal{D}_{k_i}$, $i = 1, \ldots, n$, and $h_l : \prod_{i=1}^{n} \mathcal{D}_{k_i} \to \mathcal{D}_{q_l}$, $l = 1, \ldots, p$, are logical functions. If k_i, s_j, and q_l are not identically equal, then the system (14.90) is called a mix-valued logical control system.

Investigating mix-valued logical systems, we first encounter a problem: how to define logical operators for mix-valued logical variables. Definitions for all of them will be very massy, and they may be of less logical meaning. Thus far, the question has not been considered by logicians. We avoid this and assume that only k-valued logical operators are allowed. We then need to define a projection from $\mathscr{D}_p$ to $\mathscr{D}_q$.

Definition 14.10 The projection $\phi_{[q,p]}: \mathscr{D}_p \to \mathscr{D}_q$ is defined as follows. If $x \in \mathscr{D}_p$, then $\phi_{[q,p]}(x) := \xi$, where $\xi \in \mathscr{D}_q$, satisfying

$$|\xi - x| = \min_{y \in \mathscr{D}_q} |x - y|.$$

If there are two such solutions, $\xi_1 > x$ and $\xi_2 < x$, then $\phi_{[q,p]}(x) = \xi_1$ is called the up-round projection and $\phi_{[q,p]}(x) = \xi_2$ is called the down-round projection.

In the sequel, we assume that the default projection is the up-round projection unless otherwise stated. In vector form, we have $x \in \Delta_p$ and $\phi_{[q,p]}(x) \in \Delta_q$. Hence, there exists a unique $\Phi_{[q,p]} \in \mathscr{L}_{p \times q}$, called the structure matrix of $\phi_{[q,p]}$, such that

$$\phi_{[q,p]}(x) = \Phi_{[q,p]}x. \tag{14.91}$$

We give a simple example to illustrate this.

Example 14.11 Consider $\mathscr{D}_3 = \{0, \frac{1}{2}, 1\}$ and $\mathscr{D}_4 = \{0, \frac{1}{3}, \frac{2}{3}, 1\}$. Then,

1.

$$\phi_{[4,3]}(0) = 0, \qquad \phi_{[4,3]}\left(\frac{1}{2}\right) = \frac{2}{3}, \qquad \phi_{[4,3]}(1) = 1.$$

Hence,

$$\Phi_{[4,3]} = \delta_4[1\ 2\ 4].$$

2.

$$\phi_{[3,4]}(0) = 0, \qquad \phi_{[3,4]}\left(\frac{1}{3}\right) = \frac{1}{2}, \qquad \phi_{[3,4]}\left(\frac{2}{3}\right) = \frac{1}{2}, \qquad \phi_{[3,4]}(1) = 1.$$

Hence,

$$\Phi_{[3,4]} = \delta_3[1\ 2\ 2\ 3].$$

Next, we define the logical operators between mix-valued logical variables.

Definition 14.11

1. Let σ be a unary operator on $\mathscr{D}_k$, and $x \in \mathscr{D}_p$. Then,

$$\sigma(x) := \sigma\left(\phi_{[k,p]}(x)\right) \in \mathscr{D}_k.$$

2. Let σ be a binary operator on $\mathcal{D}_k$, and $x \in \mathcal{D}_p$, $y \in \mathcal{D}_q$. Then,

$$x\sigma y := \big(\phi_{[k,p]}(x)\big)\sigma\big(\phi_{[k,q]}(y)\big) \in \mathcal{D}_k.$$

Example 14.12

1. Let $\oslash = \oslash_3$ be a unary operator on $\mathcal{D}_3$ and $x = \frac{3}{7} \in \mathcal{D}_8$. Then,

$$\oslash_3(x) = \oslash_3\big(\phi_{[3,8]}(x)\big) = \oslash\left(\frac{1}{2}\right) = 0.$$

2. Let $\nabla_{2,4}$ be a unary operator on $\mathcal{D}_4$ and $x = \frac{1}{2} \in \mathcal{D}_3$. Then,

$$\nabla_{2,4}(x) = \nabla_{2,4}\big(\phi_{[4,3]}(x)\big) = \nabla_{2,4}\left(\frac{2}{3}\right) = 1.$$

3. Let $\wedge = \wedge_3$ be a binary operator on $\mathcal{D}_3$, $x = \frac{1}{3} \in \mathcal{D}_4$, and $y = \frac{4}{5} \in \mathcal{D}_6$. Then,

$$x \wedge_3 y = \big(\phi_{[3,4]}(x)\big) \wedge_3 \big(\phi_{[3,6]}(y)\big) = \frac{1}{2} \wedge_3 1 = \frac{1}{2}.$$

4. Let $\vee = \vee_3$ be a binary operator on $\mathcal{D}_3$, and x and y as in part 3. Then,

$$x \vee_3 y = \big(\phi_{[3,4]}(x)\big) \vee_3 \big(\phi_{[3,6]}(y)\big) = \frac{1}{2} \vee_3 1 = 1.$$

We can now consider the logical expression of a mix-valued logical system. Consider either (14.89) or (14.90). Since $x_i \in \mathcal{D}_{k_i}$, $i = 1, \ldots, n$, we can automatically assume the logical operators on the ith equation are all k_i-operators, that is, the operators on $\mathcal{D}_{k_i}$. As for the outputs of (14.90), the type of operators in the kth output equation depends on the type of y_k.

We give an example to illustrate this.

Example 14.13 Consider a mix-valued logical control system:

$$\begin{cases} x_1(t+1) = x_1(t) \vee x_2(t), \\ x_2(t+1) = x_1(t) \wedge (x_2(t) \leftrightarrow u(t)), \\ \quad y(t) = \neg x_1(t), \end{cases} \tag{14.92}$$

where $x_i(t) \in \mathcal{D}_{k_i}$, $u(t) \in \mathcal{D}_s$, and $y(t) \in \mathcal{D}_q$. In componentwise algebraic form, we then have

$$x_1(t+1) = M_{d,k_1}x_1(t)\Phi_{[k_1,k_2]}x_2(t) = M_{d,k_1}(I_{k_1} \otimes \Phi_{[k_1,k_2]})x_1(t)x_2(t),$$

$$x_2(t+1) = M_{c,k_2}\Phi_{[k_2,k_1]}x_1(t)M_{e,k_2}x_2(t)\Phi_{[k_2,s]}u(t)$$

$$= M_{c,k_2}\Phi_{[k_2,k_1]}(I_{k_1} \otimes M_{e,k_2})x_1(t)x_2(t)\Phi_{[k_2,s]}u(t)$$

$$= M_{c,k_2}\Phi_{[k_2,k_1]}(I_{k_1} \otimes M_{e,k_2})W_{[k_2,k_1k_2]}\Phi_{[k_2,s]}u(t)x_1(t)x_2(t),$$

$$y(t) = M_{n,q}\Phi[q, k_1]x_1(t).$$

In particular, assume that $k_1 = 2$, $k_2 = 3$, $s = 2$, and $q = 3$. We can then calculate the algebraic form of (14.92) as follows.

Note that $\Phi_{[2,3]} = \delta_2[1\ 1\ 2]$, $\Phi_{[3,2]} = \delta_3[1\ 3]$, $M_{d,2} = \delta_2[1\ 1\ 1\ 2]$, $M_{c,3} = \delta_3[1\ 2\ 3\ 2\ 2\ 3\ 3\ 3\ 3]$, $M_{e,3} = \delta_3[1\ 2\ 3\ 2\ 2\ 2\ 3\ 2\ 1]$, $M_{n,3} = \delta_3[3\ 2\ 1]$, and $W_{[3,6]} = \delta_3[1\ 4\ 7\ 10\ 13\ 16\ 2\ 5\ 8\ 11\ 14\ 17\ 3\ 6\ 9\ 12\ 15\ 18]$. Therefore we have

$$x_1(t+1) = M_{d,2}(I_2 \otimes \Phi_{[2,3]})x_1(t)x_2(t)$$

$$= \delta_2[1\ 1\ 1\ 1\ 1\ 2]x_1(t)x_2(t)$$

$$= \delta_2[1\ 1\ 1\ 1\ 1\ 2\ 1\ 1\ 1\ 1\ 1\ 2]u(t)x_1(t)x_2(t)$$

$$=: Pu(t)x_1(t)x_2(t),$$

$$x_2(t+1) = M_{c,3}\Phi_{[3,2]}(I_2 \otimes M_{e,3})W_{[3,6]}\Phi_{[3,2]}u(t)x_1(t)x_2(t)$$

$$= \delta_3[1\ 2\ 3\ 3\ 3\ 3\ 3\ 2\ 1\ 3\ 3\ 3]u(t)x_1(t)x_2(t)$$

$$=: Qu(t)x_1(t)x_2(t),$$

$$y(t) = M_{n,3}\Phi[3, 2]x_1(t)$$

$$= \delta_3[3\ 3\ 3\ 1\ 1\ 1]x_1(t)x_2(t).$$

Set $x(t) = x_1(t) \ltimes x_2(t)$. Suppose that the algebraic form of the mix-valued logical control system (14.92) is

$$\begin{cases} x(t+1) = Lu(t)x(t), \\ y(t) = Hx(t), \end{cases}$$

where $L = \delta_6[\ell_1, \dots, \ell_{12}]$, $H = \delta_3[3\ 3\ 3\ 1\ 1\ 1]$. From Proposition 8.1, we have $\text{Col}_i(L) = \text{Col}_i(P) \ltimes \text{Col}_i(Q)$. Hence,

$$L = \delta_6[1\ 2\ 3\ 3\ 3\ 6\ 3\ 2\ 1\ 3\ 3\ 6].$$

Let $k = \prod_{i=1}^{n} k_i$, $s = \prod_{j=1}^{m} s_j$, and $q = \prod_{k=1}^{p} q_i$. We then know that (14.89) has the algebraic form

$$x(t+1) = Lx(t), \tag{14.93}$$

where $L \in \mathscr{L}_{k \times k}$.

Similarly, (14.90) has the algebraic form

$$x(t+1) = Lu(t)x(t),$$
$$y(t) = Hx(t), \tag{14.94}$$

where $L \in \mathscr{L}_{k \times ks}$ and $H \in \mathscr{L}_{q \times k}$.

From Example 14.13 one sees that from the logical form of a mix-valued logical (control) system, to construct its algebraic form is easy. It is now pertinent to ask whether we can always obtain the logical form of a mix-valued logical (control) system from its algebraic form (14.93) or (14.94) as in the Boolean or the *k*-valued case. Unfortunately, we generally cannot.

In the following we consider (14.94) only. (14.93) can be considered as a particular case. From (14.94) we can easily obtain its componentwise algebraic form as

$$x_i(t+1) = M_i u(t)x(t), \quad i = 1, \dots, n,$$
$$y_\alpha(t) = h_\alpha x(t), \quad \alpha = 1, \dots, p, \tag{14.95}$$

where $M_i \in \mathscr{L}_{k_i \times ks}$ and $h_\alpha \in \mathscr{L}_{q_\alpha \times k}$. If we have a logical expression, then a straightforward computation shows that

$$x_i(t+1) = N_i \Phi_{[k_i,s_1]} u_1 \cdots \Phi_{[k_i,s_m]} u_m \Phi_{[k_i,k_1]} x_1 \cdots \Phi_{[k_i,k_n]} x_n$$
$$= N_i \Gamma_i u(t)x(t), \quad i = 1, \dots, n, \tag{14.96}$$

where $N_i \in \mathscr{L}_{k_i \times k_i^{n+m}}$ and, setting $I_{s_0} = 1$ and $k_0 = 0$,

$$\Gamma_i = \ltimes_{\alpha=0}^{m-1} (I_{s_0 + \cdots + s_\alpha} \otimes \Phi_{k_i,s_{\alpha+1}}) \ltimes_{\beta=0}^{n-1} (I_{s+k_0+\cdots+k_\beta} \otimes \Phi_{k_i,k_{\beta+1}}). \tag{14.97}$$

Similarly, assuming that the outputs have mix-valued logical form, we then have $g_\alpha \in \mathscr{L}_{q_\alpha \times q_\alpha^n}$ such that

$$y_\alpha(t) = g_\alpha \Phi_{[q_\alpha,k_1]} x_1(t) \cdots \Phi_{[q_\alpha,k_n]} x_n(t)$$
$$= g_\alpha \Xi_\alpha x(t),$$

where, denoting $I_{k_0} = 1$,

$$\Xi_\alpha = \ltimes_{i=0}^{n-1} (I_{k_0 + \cdots + k_i} \otimes \Phi_{q_\alpha,k_{i+1}}). \tag{14.98}$$

Summarizing the above argument yields the following theorem.

Theorem 14.9 *The algebraic form* (14.95) *has a logical realization if and only if*

$$\begin{cases} N_i \Gamma_i = M_i, & i = 1, \dots, n, \\ g_\alpha \Xi_\alpha = h_\alpha, & \alpha = 1, \dots, p \end{cases} \tag{14.99}$$

has solution $\{N_i, i = 1, \dots, n; g_\alpha, \alpha = 1, \dots, p\}$.

Example 14.14 Consider a mix-valued logical control system, with algebraic form

$$x_1(t+1) = Lu(t)x_1(t)x_2(t), \tag{14.100}$$

where $x_1(t) \in \Delta_2$, $x_2(t) \in \Delta_3$, $u(t) \in \Delta_2$, and with structure matrix L given by

$$L = \delta_2[1\,2\,2\,1\,1\,1\,1\,2\,1\,1\,1\,1].$$

Suppose that the system (14.100) has the logical realization

$$x_1(t+1) = f\big(x_1(t), x_2(t), u(t)\big), \tag{14.101}$$

where $x_1(t) \in \mathscr{D}_2$, $x_2(t) \in \mathscr{D}_3$, $u(t) \in \mathscr{D}_2$. Assume that the system (14.101) can also be expressed as

$$
\begin{aligned}
x_1(t+1) &= Nu(t)x_1(t)\Phi_{[2,3]}x_2(t) \\
&= N\Gamma u(t)x_1(t)x_2(t), \tag{14.102}
\end{aligned}
$$

where $N \in \mathscr{L}_{2\times8}$, $\Phi_{[2,3]} = \delta_2[1\ 1\ 2]$, and

$$
\begin{aligned}
\Gamma &= I_4 \otimes \Phi_{[2,3]} \\
&= \delta_8[1\ 1\ 2\ 3\ 3\ 4\ 5\ 5\ 6\ 7\ 7\ 8].
\end{aligned}
$$

From (14.100) and (14.102), there exists $N \in \mathscr{L}_{2\times8}$ such that

$$N\delta_8[1\ 1\ 2\ 3\ 3\ 4\ 5\ 5\ 6\ 7\ 7\ 8] = \delta_2[1\ 2\ 2\ 1\ 1\ 1\ 1\ 2\ 1\ 1\ 1\ 1]. \tag{14.103}$$

It is obvious that we cannot find the matrix N satisfying (14.103). Thus, the algebraic form (14.100) does not have a logical realization.

References

1. Adamatzky, A.: On dynamically non-trivial three-valued logics: oscillatory and bifurcatory species. Chaos Solitons Fractals **18**, 917–936 (2003)
2. Akutsu, T., Hayashida, M., Ching, W., Ng, M.: Control of Boolean networks: hardness results and algorithms for tree structured networks. J. Theor. Biol. **244**(4), 670–679 (2007)
3. Li, Z., Cheng, D.: Algebraic approach to dynamics of multi-valued networks. Int. J. Bifurc. Chaos **20**(3), 561–582 (2010)
4. Luo, Z.K.: The Theory of Multi-valued Logic and Its Application. Science Press, Beijing (1992) (in Chinese)
5. Volkert, L., Conrad, M.: The role of weak interactions in biological systems: the dual dynamics model. J. Theor. Biol. **193**(2), 287–306 (1998)

Chapter 15
Optimal Control

15.1 Input-State Transfer Graphs

We consider a control network of the form

$$\begin{cases} x_1(t+1) = f_1(x_1(t), \ldots, x_n(t), u_1(t), \ldots, u_m(t)), \\ x_2(t+1) = f_2(x_1(t), \ldots, x_n(t), u_1(t), \ldots, u_m(t)), \\ \vdots \\ x_n(t+1) = f_n(x_1(t), \ldots, x_n(t), u_1(t), \ldots, u_m(t)), \end{cases} \tag{15.1}$$

where $x_i, u_j \in \mathscr{D}_k$, the x_i being state variables, u_j being controls, and f_i being logical functions. (15.1) is compactly expressed as

$$X(t+1) = F\big(X(t), U(t)\big), \tag{15.2}$$

where $X = (x_1, \ldots, x_n)$ and $U = (u_1, \ldots, u_m)$. When $k = 2$, (15.1) becomes a Boolean control network. In this chapter we consider general k. The outputs of the control network are omitted because we are not concerned with outputs in this chapter.

In vector form, we have $x_i, u_i \in \Delta_k$. If we let $x(t) = \ltimes_{i=1}^n x_i(t)$, $u(t) = \ltimes_{i=1}^m u_i(t)$, then (15.1) can be expressed in algebraic form as

$$x(t+1) = Lu(t)x(t), \tag{15.3}$$

where $x(t) \in \Delta_{k^n}$, $u(t) \in \Delta_{k^m}$, $L \in \mathscr{L}_{k^n \times k^{m+n}}$.

The payoff function of the network at time t is denoted by $P(X(t), U(t)) : \mathscr{D}_k^n \times \mathscr{D}_k^m \to \mathbb{R}$ [using vector form, the equivalent mapping is (using the same notation) $P(x(t), u(t)) : \Delta_{k^n} \times \Delta_{k^m} \to \mathbb{R}$]. Set $S(t) = (X(t), U(t))$ or $s(t) = u(t) \ltimes x(t)$. We consider as performance criterion the average payoff or ergodic payoff [3]. From ini-

D. Cheng et al., *Analysis and Control of Boolean Networks*,
Communications and Control Engineering,
DOI 10.1007/978-0-85729-097-7_15, © Springer-Verlag London Limited 2011

tial state x_0, under control $u(t)$, the trajectory of the network is $x(t, x_0, u)$ [or, simply, $x(t)$]. The average payoff of $x(t, x_0, u)$ is defined as

$$J\big(x(t, x_0, u)\big) = J(u) = \overline{\lim_{T \to \infty}} \frac{1}{T} \sum_{t=1}^{T} P\big(x(t), u(t)\big). \tag{15.4}$$

The aim of the optimal control problem is to find the optimal control $u^*(t)$ to maximize the objective function $J(u)$, that is,

$$J\big(u^*\big) = \max_u J(u). \tag{15.5}$$

To solve the optimal control problem, we have to answer the following questions: (i) Does the optimal control $u^*(t)$ exist? (ii) If the optimal control does exist, is it unique? (iii) How do we design it? In what follows, we will answer these questions.

We first define the cycles in $\mathscr{S} = \mathscr{D}_k^n \times \mathscr{D}_k^m$ (in vector form we have the equivalent $\mathscr{S} = \Delta_{k^{m+n}}$). The following definition was first proposed in [2].

Definition 15.1 A directed graph whose nodes are the elements of $\mathscr{S}$ is called the input-state transfer graph (ISTG) of the system (15.1) if its edges are constructed as follows: For any two nodes $S_p = (U_p, X_p) \in \mathscr{S}$ and $S_q = (U_q, X_q) \in \mathscr{S}$, there is a directed edge $\overrightarrow{S_p S_q}$ if and only if

$$X_q = F(U_p, X_p).$$

In vector form, we can also use $s_p = (u_p, x_p) = (\delta_{k^m}^\alpha, \delta_{k^n}^\beta)$ to represent a node, but a more convenient definition is $s_p = u_p x_p = \delta_{k^{m+n}}^\gamma \in \Delta_{k^{m+n}}$. The last expression is reasonable because s_p has a unique decomposition into (u_p, x_p). If we let $\delta_{k^{m+n}}^\gamma = \delta_{k^m}^\alpha \delta_{k^n}^\beta$, then

$$\gamma = (\alpha - 1)k^n + \beta$$

or, equivalently,

$$\alpha = \left[\frac{\gamma}{k^n}\right] + 1, \qquad \beta = \gamma \ (\mathrm{mod}\ k^n).$$

Using the algebraic form, we know that there is a directed edge $\overrightarrow{S_p S_q}$ [or, equivalently, (s_p, s_q)] if and only if

$$x_q = L u_p x_p \quad (\text{or } x_q = L s_p).$$

If we now assume $s_p = \delta_{k^{m+n}}^\xi$ and $s_q = \delta_{k^{m+n}}^\eta$, then the edge (s_p, s_q) can also be expressed as

$$\delta_{k^{m+n}}(\xi, \eta).$$

The topological structure of ISTGs plays a key role in optimal control problems.

Fig. 15.1 Input-state transfer graph of (15.1)

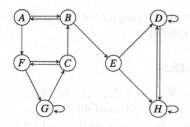

Definition 15.2

1. A state $s' \in \mathscr{S}$ is said to be reachable from $s = (u, x) \in \mathscr{S}$ if there is a path, consisting of directed edges, starting from s and ending at s'. We denote the reachable set of s by $R(s)$.
2. The ISTG is called strongly connected if, for all $s \in \mathscr{S}$,

$$R(s) = \mathscr{S}.$$

We now give an example to illustrate an ISTG.

Example 15.1 Consider the following Boolean control network:

$$\begin{cases} x_1(t+1) = (x_1(t) \wedge u(t)) \vee x_2(t), \\ x_2(t+1) = (u(t) \wedge x_1(t) \wedge \neg x_2(t)) \vee (\neg u(t) \wedge x_1(t)). \end{cases} \quad (15.6)$$

Using vector form, its algebraic form is obtained as

$$x(t+1) = Lu(t)x(t), \quad (15.7)$$

where

$$L = \delta_4[2\ 1\ 2\ 4\ 1\ 3\ 2\ 4].$$

Its input-state transfer graph consists of eight states, which are

$$A = (\delta_2^1, \delta_4^1), \qquad B = (\delta_2^1, \delta_4^2), \qquad C = (\delta_2^1, \delta_4^3), \qquad D = (\delta_2^1, \delta_4^4),$$
$$E = (\delta_2^2, \delta_4^1), \qquad F = (\delta_2^2, \delta_4^2), \qquad G = (\delta_2^1, \delta_4^3), \qquad H = (\delta_2^1, \delta_4^4).$$

Using Definition 15.1, we can easily determine its edges. The ISTG is shown in Fig. 15.1.

From Fig. 15.1, it is easily seen that

$$R(A) = R(B) = R(C) = R(F) = R(G) = \mathscr{S}.$$

On the other hand,

$$R(E) = \{D, H\}, \qquad R(H) = \{H, D\}, \qquad R(D) = \{D, H\},$$

so this ISTG is not strongly connected.

We refer to next chapter for the verification of strong connectedness, where it is called controllability.

Definition 15.3

1. Let $s_i \in \mathscr{S}$, $i = 1, 2, \ldots, \ell$. $(s_1, s_2, \ldots, s_\ell)$ is called a path if (s_i, s_{i+1}), $i = 1, \ldots, \ell - 1$, are edges of the ISTG.
2. A path $(s_1, s_2, \ldots)$ is called a cycle if $s_{i+\ell} = s_i$ for all i, and the smallest ℓ is called the length of the cycle. In particular, if $\ell = 1$, then the cycle is also called a fixed point.
3. Suppose $C = (s_1, s_2, \ldots, s_\ell)$ is a cycle and let $s_i = (u_i, x_i)$, $i = 1, \ldots, \ell$. If $x_i \neq x_j$, $1 \leq i < j \leq \ell$, then the cycle C is called a simple cycle.

Example 15.2 Consider the ISTG of Example 15.1. It is easy to see that we have the following cycles: (D), (H), (G), (H, D), (A, B), (F, C), (F, G, C), (A, F, C, B), (A, F, G, C, B), etc.

Let $C := (s_1, \ldots, s_\ell)$ be a cycle in $\mathscr{S}$. The average payoff is defined by

$$P_a(C) = \frac{P(s_1) + \cdots + P(s_\ell)}{\ell}.$$

We then have the following result [2].

Proposition 15.1

1. *Let* $S := (s_1, \ldots, s_T)$ *be a path of the ISTG of the system* (15.1). *Then,*

$$S = \bigcup_{i=1}^{N} C_i \cup R, \tag{15.8}$$

where C_i *are some cycles,* R *is the remainder, and* $|R| \leq k^{m+n}$.
2. *Let* $S := (s_1, \ldots, s_T)$ *be a path of the ISTG of the system* (15.1) *and* $\mathscr{C}$ *be the (finite) set of cycles. If* $C^* \in \mathscr{C}$ *such that*

$$P_a(C^*) = \max\{P_a(C), \forall C \in \mathscr{C}\},$$

then

$$J(S) \leq P_a(C^*). \tag{15.9}$$

Proof 1. Remove all cycles from S one by one. The remainder then has at most k^{m+n} elements because $|\mathscr{S}| = k^{m+n}$. Note that (15.8) is in the sense of "element set", that is, we do not need to worry about whether the elements of a cycle C_i are adjacent. This completes our proof.

2. Let S be decomposed into cycles C_i with lengths ℓ_i, $i = 1, \ldots, N$, and remainder R. Then,

$$\frac{1}{T} \sum_{t=1}^{T} P(S_t) = \sum_{i=1}^{N} \left[\frac{\ell_1}{T} P_a(C_1) + \cdots + \frac{\ell_N}{T} P_a(C_N) \right] + \frac{P(r_1) + \cdots + P(r_q)}{T}$$

$$\leq \frac{\sum_{i=1}^{N} \ell_i}{T} P_a(C^*) + \frac{P(r_1) + \cdots + P(r_q)}{T}$$

$$= \frac{T-q}{T} P_a(C^*) + \frac{P(r_1) + \cdots + P(r_q)}{T}, \tag{15.10}$$

where $\{r_1, \ldots, r_q\} = R$ and hence $q \leq k^{m+n}$. It is now clear that

$$\lim_{T \to \infty} \frac{T-q}{T} P_a(C^*) + \frac{P(r_1) + \cdots + P(r_q)}{T} = P_a(C^*),$$

and the conclusion follows. □

Next, we define the reachable set of a state x_0 by

$$R(x_0) = \bigcup_{u_0 \in \mathscr{D}_k^m} R(s_0 = (u_0, x_0))$$

and the cycles in this reachable set by

$$\mathscr{C}_{x_0} = \{ C \in \mathscr{C} \mid C \subset R(x_0) \}.$$

The optimal cycle $C_{x_0}^* \in \mathscr{C}_{x_0}$ satisfies

$$P_a(C_{x_0}^*) \geq P_a(C), \quad \forall C \in \mathscr{C}_{x_0}.$$

The following result then follows immediately.

Corollary 15.1 *Consider the optimal control of the network* (15.1) *with performance criterion* (15.4). *The optimal control makes the trajectory converge to $C_{x_0}^*$, and the optimal value of the criterion is $J_{\max} = P_a(C_{x_0}^*)$. If $C^* \subset R(x_0) = \mathscr{S}$, then the optimal value $J_{\max} = P_a(C^*)$. If the ISTG is strongly connected, then the optimal value is $P_a(C^*)$, which is independent of the starting point.*

15.2 Topological Structure of Logical Control Networks

To deal with the optimal control of a logical control network, its topological structure needs to be considered first. In particular, from Corollary 15.1 we know that an optical trajectory could converge to a certain cycle, so calculating cycles becomes a key issue. In the sequel, we suppose the ISTG of the control network to be strongly

Fig. 15.2 Input-state transfer
graph

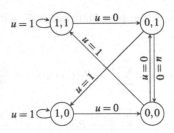

connected. If this is not the case, then we just consider the optimal control within
the reachable set $R(x_0)$ from the initial point x_0.

A k-valued logical control network can be expressed as (15.1) with $x_i, u_i \in \mathscr{D}_k$.
Its algebraic form is (15.3) with $L \in \mathscr{L}_{k^n \times k^{m+n}}$, where $x = \ltimes_{i=1}^n x_i$, $u = \ltimes_{j=1}^m u_j$,
and $x_i, u_i \in \Delta_k$. Therefore we need to investigate the cycles in the control-state
space $\mathscr{S}$. We can prove the following result.

Proposition 15.2 *An edge $\delta_{k^{m+n}}(i, j)$ exists if and only if*

$$\text{Col}_i(L) = \delta_{k^n}^\ell, \quad \text{where } \ell = j \ (\text{mod } k^n). \tag{15.11}$$

Proof By definition, the edge $\delta_{k^{m+n}}(i, j)$ exists if and only if there exists $u(t + 1)$
such

$$u(t + 1) L \delta_{k^{m+n}}^i = \delta_{k^{m+n}}^j. \tag{15.12}$$

It is easy to check that $L \delta_{k^{m+n}}^i = \text{Col}_i(L)$, thus (15.3) yields

$$u(t + 1) \text{Col}_i(L) = \delta_{k^{m+n}}^j. \tag{15.13}$$

Note that $\delta_{k^{m+n}}^j$ can be factorized uniquely into $\delta_{k^m}^\xi \delta_{k^n}^\ell$, where $j = (\xi - 1)k^n + \ell$.
The proposition is thus proved.   □

Example 15.3 Consider the Boolean control network given by

$$x(t + 1) = Lu(t)x(t), \tag{15.14}$$

where $u(t), x(t) \in \Delta$ and

$$L = \delta_2[1\,2\,2\,1].$$

Note that $\delta_4^1 \sim (1, 1)$, $\delta_4^2 \sim (1, 0)$, $\delta_4^3 \sim (0, 1)$, and $\delta_4^4 \sim (0, 0)$, so we can obtain the
graph as follows:

From Fig. 15.2 we can see that $(1, 1)$ and $(1, 0)$ are fixed points and that it has
the following cycles of length less than or equal to 4:

$(0, 1) \to (0, 0)$, $(0, 1) \to (1, 0) \to (0, 0)$,
$(1, 1) \to (0, 1) \to (0, 0)$, $(0, 0) \to (1, 1) \to (0, 1) \to (1, 0)$,
$(1, 1) \to (1, 1) \to (0, 1) \to (0, 0)$, $(1, 0) \to (1, 0) \to (0, 0) \to (0, 1)$.

In a simple case, the fixed points and cycles can be found from the ISTG directly, but when m and n are larger, it is difficult to draw the graph as was done above. Thus, we need to develop formulas to compute all the cycles algebraically.

From (15.3), we have

$$
\begin{aligned}
x(t+d) &= Lu(t+d-1)x(t+d-1) \\
&= Lu(t+d-1)Lu(t+d-2)\cdots Lu(t+1)Lu(t)x(t) \\
&= L(I_{k^m} \otimes L)u(t+d-1)u(t+d-2)Lu(t+d-3) \\
&\quad Lu(t+d-4)\cdots Lu(t)x(t) \\
&:= L_d\big(\ltimes_{\ell=1}^{d} u(t+d-\ell)\big)x(t), \tag{15.15}
\end{aligned}
$$

where

$$
L_d = \prod_{i=1}^{d}(I_{k^{(i-1)m}} \otimes L) \in \mathscr{L}_{k^n \times k^{dm+n}}. \tag{15.16}
$$

Before calculating the cycles, we need some notation.

- For $d \in \mathbb{Z}_+$, $\mathscr{P}(d)$ denotes the set of proper factors of d.
- For $i, k, m \in \mathbb{Z}_+$,

$$
\theta_k^m(d, i) := \big\{(\ell, j) \,\big|\, \exists\, \ell \in \mathscr{P}(d) \text{ and } j \text{ such that } \delta_{k^{dm}}^i = (\delta_{k^{\ell m}}^j)^{\frac{d}{\ell}}\big\}. \tag{15.17}
$$

We now give examples to illustrate the use of this notation.

Example 15.4

1. If $d = 6$, then $\mathscr{P}(d) = \{1, 2, 3\}$.
2. If $m, k, d \in \mathbb{Z}_+$ are given, then, using the obvious formula $\delta_{k^\alpha}^a \delta_{k^\beta}^b = \delta_{k^{\alpha+\beta}}^{(a-1)k^\beta+b}$, there exists at most one j for every $\ell \in \mathscr{P}(d)$ such that $(\ell, j) \in \theta_k^m(i, d)$.

   Suppose $m = k = 2$ and $d = 6$.

   - If $i = 1$, then

     for $\ell = 1$, $\delta_{k^{dm}}^i = \delta_{2^{12}}^1 = (\delta_{2^2}^1)^6 = (\delta_{k^{\ell m}}^j)^{\frac{d}{\ell}}$, so $j = 1$;

     for $\ell = 2$, $\delta_{2^{12}}^1 = (\delta_{2^4}^1)^3$, so $j = 1$;

     for $\ell = 3$, $\delta_{2^{12}}^1 = (\delta_{2^6}^1)^2$, so $j = 1$.

     Hence, $\theta_2^2(6, 1) = \{(1, 1), (2, 1), (3, 1)\}$.

   - If $i = 2$, then for any $\ell \in \mathscr{P}(6)$ and any $1 \le j \le 2^{2\ell}$, $\delta_{2^{12}}^2 \ne (\delta_{2^{2\ell}}^j)^{\frac{d}{\ell}}$, thus there is no solution. That is, $\theta_2^2(6, 2) = \emptyset$.

   - If $i = 2^6 + 2$, then

     for $\ell = 1$ or 2, there is no solution;

     for $\ell = 3$, $\delta_{2^{12}}^{2^6+2} = (\delta_{2^6}^2)^2$, so $j = 2$.

     Therefore $\theta_2^2(6, 2^6 + 2) = \{(3, 2)\}$.

In the following we simply use $\theta(d, i)$ for $\theta_k^m(d, i)$, where the default k and m are assumed to be the type of logic and the number of inputs, respectively. Denote by $\text{Blk}_i(L)$ the ith $n \times n$ block of an $n \times nm$ logical matrix L. We then have the following result.

Theorem 15.1 *The number of cycles of length d in the ISTG of the k-valued logical control network* (15.3) *is inductively determined by*

$$N_d = \frac{1}{d} \sum_{i=1}^{k^{dm}} T\left(\text{Blk}_i(L_d)\right), \tag{15.18}$$

where L_d is defined in (15.16) *and*

$$T\left(\text{Blk}_i(L_d)\right) = \text{tr}\left(\text{Blk}_i(L_d)\right) - \sum_{(\ell, j) \in \theta(d, i)} T\left(\text{Blk}_j(L_\ell)\right). \tag{15.19}$$

Proof Each cycle in $\mathscr{S}$ is a product of cycles in the state space and the control space, so we look for the cycle in the state space first. If $x(t)$ is in a cycle of length d in the state space, then from (15.15) we have

$$x(t) = L_d\left(\ltimes_{\ell=1}^d u(t + d - \ell)\right) x(t).$$

If $u(t + d - 1), \ldots, u(t)$ are fixed, say $\ltimes_{\ell=1}^d u(t + d - \ell) = \delta_{k^{dm}}^i$, then

$$x(t) = \text{Blk}_i(L_d) x(t).$$

If $x(t) = \delta_{k^n}^j$, then the (j, j)-element of $\text{Blk}_i(L_d)$ is 1, so the cycle with length d in the state space under the given controls $u(t + d - 1), \ldots, u(t)$ is

$$\left\{x(t), Lu(t)x(t), L_2 u(t+1)u(t)x(t), \ldots, L_d u(t+d-1) \cdots u(t)x(t)\right\}.$$

Thus, multiplying the cycle and the given u, we obtain a cycle of length d in control-state space. Therefore the number of length-d cycles, including multifold ones, is $\frac{1}{d} \sum_{i=1}^{k^{dm}} \text{tr}(\text{Blk}_i(L_d))$.

It is obvious that if ℓ is a proper factor of d, and $x(t)$ is in the cycle of length ℓ under $\tilde{u}(t + \ell - 1) \cdots \tilde{u}(t) = \delta_{k^{\ell m}}^j$ and the cycle of length d under $u(t + d - 1) \cdots u(t) = \delta_{k^{dm}}^i$, then we can obtain the same cycle in the ISTG if and only if $\delta_{k^{dm}}^i = (\delta_{k^{\ell m}}^j)^{\frac{d}{\ell}}$. Removing these multifold cycles, we obtain (15.18). $\qquad\square$

To see that (15.19) inductively defines all $T(\text{Blk}_i(L_d))$ with respect to d, note that as $d = 1$, we have

$$\theta(i, 1) = \phi, \quad \forall i.$$

Thus, $T(\text{Blk}_i(L_d))$ is well defined for $d = 1$ and hence $d > 1$ can be defined inductively.

For a cycle C of length d, because $s(t) = \delta_{km+n}^{\ell}$ can be decomposed uniquely as $u(t)x(t) = \delta_{km}^{i} \delta_{kn}^{j}$, the cycle can be described as

$$C = \left(\left(\delta_{km}^{i(t)}, \delta_{kn}^{j(t)} \right), \left(\delta_{km}^{i(t+1)}, \delta_{kn}^{j(t+1)} \right), \ldots, \left(\delta_{km}^{i(t+d-1)}, \delta_{kn}^{j(t+d-1)} \right) \right).$$

For compactness, we write this as

$$C = \delta_{km} \times \delta_{kn} \left((i(t), j(t)), (i(t+1), j(t+1)), \ldots, (i(t+d-1), j(t+d-1)) \right). \tag{15.20}$$

We now give an alternative definition of a simple cycle (originally defined in Definition 15.3): A cycle $C = \delta_{km} \times \delta_{kn} ((i(t), j(t)), (i(t+1), j(t+1)), \ldots, (i(t+d-1), j(t+d-1)))$ is called a simple cycle if it satisfies

$$i(\xi) \neq i(\ell), \quad t \leq \xi < \ell \leq t+d-1. \tag{15.21}$$

Example 15.5 Recall Example 15.3. Since

$$L_1 = L = \delta_2[1\,2\,2\,1],$$

we have $\mathrm{tr}(\mathrm{Blk}_1(L_1)) = 2$ and $\mathrm{tr}(\mathrm{Blk}_2(L_1)) = 0$. Hence, δ_2^1 and δ_2^2 are fixed points under the control $u = \delta_2^1$. It follows that $N_1 = 1$ and the fixed points in ISTG are

$$\delta_2 \times \delta_2 ((1, 1)), \qquad \delta_2 \times \delta_2 ((1, 2)),$$

which are simple ones. Next, since

$$L_2 = L(I_2 \otimes L) = \delta_2[1\,2\,2\,1\,2\,1\,1\,2],$$

we have $\mathrm{tr}(\mathrm{Blk}_1(L_2)) = \mathrm{tr}(\mathrm{Blk}_4(L_2)) = 2$, $\mathrm{tr}(\mathrm{Blk}_2(L_2)) = \mathrm{tr}(\mathrm{Blk}_3(L_2)) = 0$, $\delta_4^1 = \delta_2^1 \delta_2^1$, and $\delta_4^4 = \delta_2^2 \delta_2^2$, so

$$T\left(\mathrm{Blk}_1(L_2)\right) = \mathrm{tr}\left(\mathrm{Blk}_1(L_2)\right) - T\left(\mathrm{Blk}_1(L_1)\right) = 0,$$

$$T\left(\mathrm{Blk}_4(L_2)\right) = \mathrm{tr}\left(\mathrm{Blk}_4(L_2)\right) - T\left(\mathrm{Blk}_2(L_1)\right) = 2.$$

$T(\mathrm{Blk}_2(L_2)) = T(\mathrm{Blk}_3(L_2)) = 0$, so $N_2 = 1$. δ_2^1 and δ_2^2 are in cycles of length 2 under $u(t+1)u(t) = \delta_2^2 \delta_2^2$. We can then obtain a cycle of length 2 in ISTG as

$$\delta_2 \times \delta_2 ((2, 1), (2, 2)),$$

which is also simple. Consider

$$L_3 = L(I_2 \otimes L)(I_4 \otimes L) = \delta_2[1\,2\,2\,1\,2\,1\,1\,2\,2\,1\,1\,2\,1\,2\,2\,1].$$

Since $\mathrm{tr}(\mathrm{Blk}_1(L_3)) = \mathrm{tr}(\mathrm{Blk}_4(L_3)) = \mathrm{tr}(\mathrm{Blk}_6(L_3)) = \mathrm{tr}(\mathrm{Blk}_7(L_3)) = 2$, we have $T(\mathrm{Blk}_4(L_3)) = T(\mathrm{Blk}_6(L_3)) = T(\mathrm{Blk}_7(L_3)) = 2$, and $T(\mathrm{Blk}_i(L_3)) = 0$ for $i = 1, 2, 3, 5, 8$. It follows that $N_3 = 2$. δ_2^1 and δ_2^2 are in cycles of length 3 under

$u(t+2)u(t+1)u(t) = \delta_8^4 = \delta_2^1\delta_2^2\delta_2^2$, $\delta_8^6 = \delta_2^2\delta_2^1\delta_2^2$, and $\delta_8^7 = \delta_2^2\delta_2^2\delta_2^1$. We can then obtain the cycles of length 3 in the ISTG as

$$\delta_2 \times \delta_2\big((1,1),(2,1),(2,2)\big), \qquad \delta_2 \times \delta_2\big((2,1),(1,2),(2,2)\big).$$

Finally, since

$$L_4 = L(I_2 \otimes L)(I_4 \otimes L)(I_8 \otimes L)$$

$$= \delta_2[1\,2\,2\,1\,2\,1\,1\,2\,2\,1\,1\,2\,1\,2\,2\,1$$
$$\quad 2\,1\,1\,2\,1\,2\,2\,1\,1\,2\,2\,1\,2\,1\,1\,2],$$

we have $\operatorname{tr}(\mathrm{Blk}_i(L_4)) = 2$, $i = 1, 4, 6, 7, 10, 11, 13, 16$.

Therefore $T(\mathrm{Blk}_i(L_4)) = 2$ for $i = 4$, 6, 7, 10, 11, 13, otherwise $T(\mathrm{Blk}_i(L_4)) = 0$, hence $N_4 = 3$. δ_2^1 and δ_2^2 are in cycles of length 4 under $u(t+3)u(t+2)u(t+1)u(t) = \delta_{16}^4 = \delta_2^1\delta_2^2\delta_2^2\delta_2^2$, $\delta_{16}^6 = \delta_2^1\delta_2^2\delta_2^1\delta_2^2$, $\delta_{16}^7 = \delta_2^1\delta_2^2\delta_2^2\delta_2^1$, $\delta_{16}^{10} = \delta_2^2\delta_2^1\delta_2^1\delta_2^2$, $\delta_{16}^{11} = \delta_2^2\delta_2^1\delta_2^2\delta_2^1$, and $\delta_{16}^{13} = \delta_2^2\delta_2^2\delta_2^1\delta_2^2$. We can then obtain the cycles of length 4 in the ISTG as

$$\delta_2 \times \delta_2\big((1,1),(2,1),(1,2),(2,2)\big),$$

$$\delta_2 \times \delta_2\big((1,2),(1,2),(2,2),(2,1)\big),$$

$$\delta_2 \times \delta_2\big((1,1),(1,1),(2,1),(2,2)\big).$$

This result is the same as what we observed from the graph in Example 15.3.

15.3 Optimal Control of Logical Control Networks

In this section we consider the optimal control and optimal trajectory of logical control networks. Corollary 15.1 generalizes a result in [2] for single-variable Boolean networks to multivariable k-valued logical networks. However, after expression them into graphs, there is no essential difference, so the proofs are similar. Corollary 15.1 ensures that the optimal control can be achieved on a trajectory which converges to a cycle. In the sequel, using the matrix expression of logical functions, we give a method to find the optimal trajectory and obtain a G^*, called the optimal control matrix, such that

$$u^*(t+1) = G^*u^*(t)x^*(t).$$

Proposition 15.3 *The limit*

$$J(u^*) := \lim_{T \to \infty} \frac{1}{T}\sum_{t=1}^{T} P\big(x^*(t), u^*(t)\big) \tag{15.22}$$

always exists.

Proof Consider the system (15.1). According to Corollary 15.1, an optimal trajectory will converge to an attractor. As a limit, $J(u^*)$ is the average over an attractor (fixed point or cycle).   □

Proposition 15.4 *For any cycle C, there exists a simple cycle C_s such that*

$$P_a(C_s) \geq P_a(C). \tag{15.23}$$

Proof Denote by $C = \delta_{k^m} \times \delta_{k^n} ((i(t), j(t)), (i(t+1), j(t+1)), \ldots, (i(t+d-1),$ $j(t+d-1)))$ an arbitrary cycle. If it is a simple cycle, then the result is trivial. Otherwise, assume $\delta_{k^n}^{j(\xi)} = \delta_{k^n}^{j(\ell)}$, $\xi < \ell$, and that $C_1 = \delta_{k^m} \times \delta_{k^n} ((i(\xi), j(\xi)), \ldots,$ $(i(\ell-1), j(\ell-1)))$ is a simple cycle. If $P_a(C_1) \geq P_a(C)$, then we are done. Otherwise, we remove C_1. The remainder then forms a new cycle C_1' because $L\delta_{k^m}^{i(\xi-1)} \delta_{k^n}^{j(\xi-1)} = \delta_{k^n}^{i(\xi)} = \delta_{k^n}^{i(\ell)}$. Now, $P_a(C_1') > P_a(C)$. If C_1' is a simple cycle, then we are done. Otherwise, we can find a simple cycle C_2 such that either it satisfies (15.23) or we can remove it. Continuing this process, we will eventually find a simple cycle C_s such that (15.23) holds.   □

By (15.15), at the dth step, the initial state x_0 can reach

$$R_d(x_0) = \{u(d)L_d \ltimes_{\ell=1}^d u(d-\ell)x_0 \mid \forall u(\ell) \in \Delta_{k^m}, 0 \leq \ell \leq d\},$$

and if $x_0 = \delta_{k^n}^{j(0)}$,

$$R_d(x_0) = \{u(d)\operatorname{Col}_\ell(L_d) \mid \forall u(d) \in \Delta_{k^m}, \ell = j(0) \pmod{k^n}\}.$$

If $\delta_{k^m}^i \delta_{k^n}^j$ is reached from x_0 at the dth step, $d \geq k^n$, then the path from the initial state to $\delta_{k^m}^i \delta_{k^n}^j$ must pass a state at least twice. Similarly to the proof of Proposition 15.4, we can finally conclude that the state $\delta_{k^m}^i \delta_{k^n}^j$ can be reached from x_0 at the d'th step, where $d' < k^n$. Thus,

$$R(x_0) = \bigcup_{d=0}^{k^n-1} R_d(x_0). \tag{15.24}$$

According to the above argument, we can find the optimal cycle C^* from all the simple cycles contained in $R(x_0)$. Denote the shortest path from the initial state to C^* by

$$\left(\delta_{k^m}^{i(0)} \delta_{k^n}^{j(0)}, \delta_{k^m}^{i(1)} \delta_{k^n}^{j(1)}, \ldots, \delta_{k^m}^{i(T_0-1)} \delta_{k^n}^{j(T_0-1)}\right), \tag{15.25}$$

where

$$C^* = \delta_{k^m} \times \delta_{k^n} \left((i(T_0), j(T_0)), \ldots, (i(T_0+d-1), j(T_0+d-1))\right).$$

We call (15.25) the optimal trajectory.

Next, we will prove the existence of the optimal control matrix G^*.

Table 15.1 Payoff bi-matrix

$P_1 \backslash P_2$	L	M	R
L	3, 3	0, 4	9, 2
M	4, 0	4, 4	5, 3
R	2, 9	3, 5	6, 6

Theorem 15.2 *Consider the k-valued logical control network* (15.1) *with objective function* (15.4). *Let the optimal trajectory be* (15.25), *and the optimal control be* $u^*(t)$. *There then exists a logical matrix* $G^* \in \mathscr{L}_{k^n \times k^{n+m}}$, *satisfying*

$$\begin{cases} x^*(t+1) = Lu^*(t)x^*(t), \\ u^*(t+1) = G^*u^*(t)x^*(t). \end{cases} \tag{15.26}$$

Proof According to Proposition 15.4 we can find an optimal cycle from the set of all simple cycles. Because the length of a simple cycle cannot be greater than k^n, we assume the initial state of a trajectory is $\delta_{k^n}^{j(0)}$. We can find all cycles with length less than or equal to k^n which can be reached from the initial state and then determine the optimal trajectory (15.25). It is easy to show that $T_0 + d \leq k^{m+n}$, so we can get $T_0 + d$ columns of the optimal control matrix G^*, which satisfies

$$\mathrm{Col}_s(G^*) = \begin{cases} \delta_{k^m}^{i(\ell+1)}, & s = (i(\ell)-1)k^n + j(\ell), \ell \leq T_0 + d - 2, \\ \delta_{k^m}^{i(T_0)}, & s = (i(T_0+d-1)-1)k^n + j(T_0+d-1), \end{cases} \tag{15.27}$$

and the other columns of G^* [$\mathrm{Col}(G^*) \subset \Delta_{k^m}$] can be arbitrary. Thus, G^* can be constructed. $\qquad\square$

Example 15.6 Recall Example 15.3 and Example 15.5. Set

$$P\big(u(t), x(t)\big) = u^\mathrm{T}(t) \begin{bmatrix} 1 & 2 \\ 3 & 4 \end{bmatrix} x(t)$$

and assume the initial state $x_0 = \delta_2^2$. From the result of Example 15.5 we can see that $C^* = \delta_2 \times \delta_2((2, 1), (2, 2))$ is obviously the optimal cycle. Choosing $u(0) = \delta_2^2$, the optimal cycle and the shortest path from δ_2^2 to the cycle is

$$\delta_2 \times \delta_2\big((2, 2), (2, 1)\big).$$

Therefore $G^* = \delta_2[i \ j \ 2 \ 2]$ and i, j can be either 1 or 2.

Example 15.7 We consider the following infinitely repeated game. Both player 1 and player 2 have three possible actions, $\{L, M, R\}$. The payoff bi-matrix is assumed to be as in Table 15.1.

It is easy to check that (M, M), which means that player 1 chooses M and player 2 also chooses M, is the unique Nash equilibrium of the one-stage game, but it is obvious that (R, R) is more efficient than (M, M). In the infinitely repeated game, assume that player 2's strategy is fixed as follows: he plays R in the first stage, and in the tth stage, if the outcome in the $(t-1)$th stage was (R, R), he plays R, otherwise, he plays M. This is called the "trigger strategy" [1].

Let $L \sim 1$, $M \sim 0.5$, and $R \sim 0$. The above game can be rewritten as

$$x(t+1) = Lu(t)x(t), \tag{15.28}$$

where

$$L = \delta_3[2\,2\,2\,2\,2\,2\,2\,2\,3].$$

$x(t) \in \Delta_3$, as the state, is the action of player 2 at the tth stage; $u(t) \in \Delta_3$, as the control, is the action of player 1 at tth stage.

As we know, the trigger strategy is the Nash equilibrium of an infinitely repeated finite game in which the payoff function is

$$(1-\delta) \sum_{t=1}^{\infty} \delta^{t-1} \pi_t,$$

where π_t is the payoff at the tth stage and δ is the discount factor, when δ is sufficiently close to 1 [1].

Ignoring the discount factor, our respective payoff functions for player 1 and player 2 are

$$J_1 = \varlimsup_{T \to \infty} \frac{1}{T} \sum_{t=1}^{T} P_1(x(t), u(t)),$$

$$J_2 = \varlimsup_{T \to \infty} \frac{1}{T} \sum_{t=1}^{T} P_2(x(t), u(t)),$$

where

$$P_1(x(t), u(t)) = u^T(t) \begin{bmatrix} 3 & 0 & 9 \\ 4 & 4 & 5 \\ 2 & 3 & 6 \end{bmatrix} x(t),$$

$$P_2(x(t), u(t)) = u^T(t) \begin{bmatrix} 3 & 4 & 2 \\ 0 & 4 & 3 \\ 9 & 5 & 6 \end{bmatrix} x(t).$$

A natural question is whether the trigger strategy is still a Nash equilibrium in this game. Player 2 has adopted the trigger strategy, and we want to find the best response for player 1. The question is then converted to one of finding the optimal control of the 3-valued logical control network (15.28) which maximizes J_1.

We now calculate the cycles:

$$L_1 = L = \delta_3[2\,2\,2\,2\,2\,2\,2\,2\,3],$$

thus $\text{tr}(\text{Blk}_1(L_1)) = 1$, $\text{tr}(\text{Blk}_2(L_1)) = 1$, $\text{tr}(\text{Blk}_3(L_1)) = 2$, and $N_1 = 4$. δ_3^2 is a fixed point under $u = \delta_3^i$, $i = 1, 2, 3$, and δ_3^3 is a fixed point under $u = \delta_3^3$, so the fixed points of the system (15.28) are

$$\delta_3 \times \delta_3((1,2)), \qquad \delta_3 \times \delta_3((2,2)), \qquad \delta_3 \times \delta_3((3,2)), \quad \text{and} \quad \delta_3 \times \delta_3((3,3)).$$

$$L_2 = L(I_3 \otimes L) = \delta_3[2\,2\,2\,2\,2\,2\,2\,2\,2\,2\,2\,2\,2\,2\,2\,2\,2\,2\,2\,2\,2\,2\,2\,2\,2\,2\,3],$$

$\text{tr}(\text{Blk}_i(L_2)) = 1$, $i = 1, \ldots, 8$, $\text{tr}(\text{Blk}_9(L_2)) = 2$. For $u(t+1)u(t) = \delta_9^1 = \delta_3^1\delta_3^1$,

$$T(\text{Blk}_1(L_2)) = \text{tr}(\text{Blk}_1(L_2)) - \text{tr}(\text{Blk}_1(L_1)) = 0.$$

Similarly, we obtain $T(\text{Blk}_1(L_2)) = T(\text{Blk}_5(L_2)) = T(\text{Blk}_9(L_2)) = 0$, $T(\text{Blk}_i(L_2)) = 1$, $i = 2, 3, 4, 6, 7, 8$. Thus, $N_2 = 3$. δ_3^2 is in cycles of length 2 with $u(t+1)u(t) = \delta_9^i$, $i = 2, 3, 4, 6, 7, 8$. We can then find the cycles of length 2 as

$$\delta_3 \times \delta_3((1,2),(2,2)), \qquad \delta_3 \times \delta_3((1,2),(3,2)), \qquad \delta_3 \times \delta_3((2,2),(3,2)).$$

$$L_3 = L_x(I_3 \otimes L_x)(I_9 \otimes L_x) = \delta_{81}[\underbrace{2 \cdots 2}_{80}\,3].$$

By (15.19) we have $T(\text{Blk}_i(L_3)) = 1$, $i = 2, \ldots, 13, 15, \ldots, 26$, $T(\text{Blk}_i(L_3)) = 0$, $i = 1, 14, 27$, and $N_3 = 8$. δ_3^2 is in cycles of length 3 with $u(t+2)u(t+1)u(t) = \delta_{27}^i$, $i = 2, \ldots, 13, 15, \ldots, 26$. We can then find the cycles of length 3 as

$$\delta_3 \times \delta_3((1,2),(1,2),(2,2)), \qquad \delta_3 \times \delta_3((1,2),(3,2),(2,2)),$$
$$\delta_3 \times \delta_3((1,2),(1,2),(3,2)), \qquad \delta_3 \times \delta_3((1,2),(3,2),(3,2)),$$
$$\delta_3 \times \delta_3((1,2),(2,2),(2,2)), \qquad \delta_3 \times \delta_3((2,2),(2,2),(3,2)),$$
$$\delta_3 \times \delta_3((1,2),(2,2),(3,2)), \qquad \delta_3 \times \delta_3((2,2),(3,2),(3,2)).$$

There are also many cycles of length greater than or equal to 4, but we have proven that to deal with the optimal control of this game, finding all the cycles of length less than or equal to 3 is sufficient.

As a trigger strategy, from the initial state $x_0 = \delta_3^3$, the reachable set is

$$R(x_0) = \{\delta_3^1\delta_3^2, \delta_3^2\delta_3^2, \delta_3^3\delta_3^2, \delta_3^1\delta_3^3, \delta_3^2\delta_3^3, \delta_3^1\delta_3^3\}.$$

Using the above result, all the simple cycles contained in $R(x_0)$ are $\delta_3 \times \delta_3((1,2))$, $\delta_3 \times \delta_3((2,2))$, $\delta_3 \times \delta_3((3,2))$, and $\delta_3 \times \delta_3((3,3))$, and, among them, $\delta_3 \times \delta_3((3,3))$ is the optimal cycle. Choosing $u^*(0) = \delta_3^3$, we then have

$$G^* = \delta_3[* * * * * * * * 3],$$

where the first eight columns can be arbitrary.

For instance, we can choose

$$G^* = \delta_3[2\,2\,2\,2\,2\,2\,2\,2\,3],$$

which is the trigger strategy. We conclude that the best response for player 1 is to adopt the trigger strategy if player 2 has adopted the trigger strategy. The payoffs are symmetrical, so if player 1 has adopted the trigger strategy, the best response for player 2 is also to adopt the trigger strategy. This means that the trigger strategy is a Nash equilibrium of this game.

15.4 Optimal Control of Higher-Order Logical Control Networks

In general, a μth order logical control network can be expressed as

$$
\begin{cases}
x_1(t+1) = f_1(x_1(t), \ldots, x_n(t), \ldots, x_1(t-\mu+1), \ldots, x_n(t-\mu+1), \\
\qquad\qquad u_1(t), \ldots, u_m(t), \ldots, u_1(t-\mu+1), \ldots, u_m(t-\mu+1)) \\
x_2(t+1) = f_2(x_1(t), \ldots, x_n(t), \ldots, x_1(t-\mu+1), \ldots, x_n(t-\mu+1), \\
\qquad\qquad u_1(t), \ldots, u_m(t), \ldots, u_1(t-\mu+1), \ldots, u_m(t-\mu+1)) \\
\vdots \\
x_n(t+1) = f_n(x_1(t), \ldots, x_n(t), \ldots, x_1(t-\mu+1), \ldots, x_n(t-\mu+1), \\
\qquad\qquad u_1(t), \ldots, u_m(t), \ldots, u_1(t-\mu+1), \ldots, u_m(t-\mu+1)), \\
y_j(t) = h_j(x_1(t), \ldots, x_n(t)), \quad j = 1, \ldots, p.
\end{cases}
\tag{15.29}
$$

To deal with a μth order logical control network, we first consider how to convert it to a first order form. Recall that in last chapter we defined the base-k power-reducing matrix, in (14.18). This can be compactly expressed as

$$
M_{r,k} = \delta_{k^2}[1 \ k+2 \ 2k+3 \ \cdots \ (k-1)k+k].
\tag{15.30}
$$

If we now assume $x = \ltimes_{i=1}^{\mu} x_i \in \Delta_{k^\mu}$, then (14.31) shows that

$$
x^2 = \Phi_{\mu,k} x,
\tag{15.31}
$$

where

$$
\Phi_{\mu,k} := \prod_{i=1}^{\mu} I_{k^{i-1}} \otimes \left[(I_k \otimes W_{[k,k^{\mu-i}]}) M_{r,k} \right] = M_{r,k^\mu},
\tag{15.32}
$$

and we can prove the following retrieval formulas.

Lemma 15.1 *If we assume $x = \ltimes_{i=1}^{n} x_i \in \Delta_{k^n}$, where $x_i \in \Delta_k$, and define*

$$
F_{[m,n],k} = I_{k^m} \otimes \mathbf{1}_{k^{n-m}}^{\mathrm{T}},
$$

$$
E_{[m,n],k} = \mathbf{1}_{k^{n-m}}^{\mathrm{T}} \otimes I_{k^m},
$$

then

$$
F_{[m,n],k} x = \ltimes_{i=1}^{m} x_i, \qquad E_{[m,n],k} x = \ltimes_{i=n-m+1}^{n} x_i.
$$

Proof If $\ltimes_{i=1}^{m} x_i = \delta_{km}^{j}$, then

$$F_{[m,n],k}x = \left(I_{km} \otimes \mathbf{1}_{k^{n-m}}^{\mathrm{T}}\right) \ltimes \delta_{km}^{j} \ltimes_{i=m+1}^{n} x_i = \delta_{km}^{j} \ltimes \mathbf{1}_{k^{n-m}}^{\mathrm{T}} \ltimes_{i=m+1}^{n} x_i = \delta_{km}^{j}.$$

Whatever $\ltimes_{i=1}^{n-m} x_i$ is, we also have

$$E_{[m,n],k}x = E_{[m,n],k} \ltimes_{i=1}^{n-m} x_i \ltimes_{i=n-m+1}^{n} x_i = I_{km} \ltimes_{i=n-m+1}^{n} x_i = \ltimes_{i=n-m+1}^{n} x_i.$$

$\square$

In the following we use the simpler notation M_r, Φ_μ, $F_{[m,n]}$, and $E_{[m,n]}$ for $M_{r,k}$, $\Phi_{\mu,k}$, $F_{[m,n],k}$, and $E_{[m,n],k}$, respectively, where the default k is assumed to be the type of logic. Let $x(t) = \ltimes_{i=1}^{n} x_i(t)$, $u(t) = \ltimes_{i=1}^{m} u_i(t)$. Each equation of the μth order logical control network (15.29) can be written in its componentwise algebraic form as

$$\begin{cases} x_1(t+1) = M_1 u(t-\mu+1) \cdots u(t) x(t-\mu+1) \cdots x(t), \\ x_2(t+1) = M_2 u(t-\mu+1) \cdots u(t) x(t-\mu+1) \cdots x(t), \\ \vdots \\ x_n(t+1) = M_n u(t-\mu+1) \cdots u(t) x(t-\mu+1) \cdots x(t). \end{cases} \tag{15.33}$$

Multiplying the equations in (15.33) together, we obtain

$$x(t+1) = L \ltimes_{i=1}^{\mu} u(t-\mu+i) \ltimes_{i=1}^{\mu} x(t-\mu+i), \tag{15.34}$$

where

$$L = M_1 \prod_{j=2}^{n} \left[(I_{k^{\mu(m+n)}} \otimes M_j) \Phi_{\mu(m+n)} \right].$$

If we let $z(t) = \ltimes_{i=t}^{t+\mu-1} x(i)$, $v(t) = \ltimes_{i=t}^{t+\mu-1} u(i)$, then (15.34) can be converted to

$$x(t+1) = Lv(t-\mu+1)z(t-\mu+1).$$

We then have

$$\begin{aligned} z(t+1) &= \ltimes_{i=t+1}^{t+\mu} x(i) \\ &= \ltimes_{i=t+1}^{t+\mu-1} x(i) L v(t) z(t) \\ &= (I_{k^{(\mu-1)n}} \otimes L) \ltimes_{i=t+1}^{t+\mu-1} x(i) v(t) z(t) \\ &= (I_{k^{(\mu-1)n}} \otimes L) W_{[k^{\mu m+n}, k^{(\mu-1)n}]} v(t) z(t) \ltimes_{i=t+1}^{t+\mu-1} x(i) \\ &= (I_{k^{(\mu-1)n}} \otimes L) W_{[k^{\mu m+n}, k^{(\mu-1)n}]} v(t) x(t) \Phi_{(\mu-1)n} \ltimes_{i=t+1}^{t+\mu-1} x(i) \\ &:= \widetilde{L} v(t) z(t), \tag{15.35} \end{aligned}$$

where

$$\tilde{L} = (I_{k^{(\mu-1)m}} \otimes L) W_{[k^{\mu m+n},k^{(\mu-1)n}]}(I_{k^{\mu m+n}} \otimes \Phi_{(\mu-1)n}).$$

Note that the $v(t), t = 0, 1, \ldots$, here are not completely independent as they should satisfy

$$F_{[(\mu-1)m,\mu m]}v(t+1) = E_{[(\mu-1)m,\mu m]}v(t).$$

Thus, (15.34) can be converted to

$$\begin{cases} z(t+1) = \tilde{L}v(t)z(t), \\ F_{[(\mu-1)m,\mu m]}v(t+1) = E_{[(\mu-1)m,\mu m]}v(t). \end{cases} \tag{15.36}$$

Similarly to (15.15), if $z(t)$ is in a cycle of length d, then we have

$$z(t) = z(t+d) = \tilde{L}_d v(t+d-1)v(t+d-2)\cdots v(t)z(t), \tag{15.37}$$

where

$$\tilde{L}_d = \prod_{i=1}^{d} \left(I_{k^{(i-1)\mu m}} \otimes \tilde{L}\right).$$

$v(t+d-1)v(t+d-2)\cdots v(t)$ can be simplified as

$$\begin{aligned}
v(t+d&-1)\cdots v(t) \\
&= \ltimes_{i=t+d-1}^{t+d+\mu-2} u(i) \ltimes_{i=t+d-2}^{t+d+\mu-3} u(i) \cdots \ltimes_{i=t}^{t+\mu-1} u(i) \\
&= W_{[k^{\mu m}]} \ltimes_{i=t+d-2}^{t+d+\mu-3} u(i) \ltimes_{i=t+d-1}^{t+d+\mu-2} u(i) \ltimes_{i=t+d-3}^{t+d+\mu-4} u(i) \cdots \ltimes_{i=t}^{t+\mu-1} u(i) \\
&= W_{[k^{\mu m}]} \left(u(t+d-2)\Phi_{(\mu-1)m} \ltimes_{i=t+d-1}^{t+d+\mu-3} u(i)u(t+d+\mu-2)\right) \\
&\quad \ltimes_{i=t+d-3}^{t+d+\mu-4} u(i) \cdots \ltimes_{i=t}^{t+\mu-1} u(i) \\
&= W_{[k^{\mu m}]}(I_{k^m} \otimes \Phi_{(\mu-1)m}) \ltimes_{i=t+d-2}^{t+d+\mu-2} u(i) \ltimes_{i=t+d-3}^{t+d+\mu-4} u(i) \cdots \ltimes_{i=t}^{t+\mu-1} u(i) \\
&\;\;\vdots \\
&= \prod_{i=1}^{d-1} \left(W_{[k^{\mu m},k^{(\mu+i-1)m}]}(I_{k^m} \otimes \Phi_{(\mu-1)m})\right) \ltimes_{i=t}^{t+\mu+d-2} u(i) \\
&:= R \ltimes_{i=t}^{t+\mu+d-2} u(i),
\end{aligned}$$

where

$$R = \prod_{i=1}^{d-1} \left(W_{[k^{\mu m},k^{(\mu+i-1)m}]}(I_{k^m} \otimes \Phi_{(\mu-1)m})\right).$$

Moreover, $v(t + d) = v(t)$ must hold, that is,

$$\ltimes_{i=t+d}^{t+d+\mu-1} u(i) = \ltimes_{i=t}^{t+\mu-1} u(i). \tag{15.38}$$

Assuming that $\mu = sd + r$, where $s = [\frac{\mu}{d}]$, $\mu = r \pmod{d}$, the product $v(t+d-1)v(t+d-2)\cdots v(t)$ becomes

$$
\begin{aligned}
&v(t+d-1)\cdots v(t)\\[2pt]
&= R \ltimes_{i=t}^{t+d-1} u(i) \ltimes_{i=t+d}^{t+d+\mu-2} u(i)\\[2pt]
&= R \ltimes_{i=t}^{t+d-1} u(i)\left(\ltimes_{i=t}^{t+d-1} u(i)\right)^{s-1} \ltimes_{i=t}^{t+d+r-2} u(i)\\[2pt]
&= \begin{cases}
R(\Phi_{dm})^{s-1}(I_{k(d-1)m} \otimes W_{[k(d-1)m,k^m]})\Phi_{(d-1)m} \ltimes_{i=t}^{t+d-1} u(i), & r=0,\\[4pt]
R(\Phi_{dm})^s \ltimes_{i=t}^{t+d-1} u(i), & r=1,\\[4pt]
R(\Phi_{dm})^s(I_{k(r-1)m} \otimes W_{[k(r-1)m,k(d-r+1)m]})\\
\quad \times \Phi_{(r-1)m} \ltimes_{i=t}^{t+d-1} u(i), & 2 \le r \le d-1.
\end{cases}
\end{aligned}
$$

Equation (15.37) is then converted to

$$z(t) = \Psi_d \ltimes_{i=t}^{t+d-1} u(i)z(t), \tag{15.39}$$

where

$$
\Psi_d = \begin{cases}
\tilde{L}_d R(\Phi_{dm})^{s-1}(I_{k(d-1)m} \otimes W_{[k(d-1)m,k^m]})\Phi_{(d-1)m}, & r=0,\\[4pt]
\tilde{L}_d R(\Phi_{dm})^s, & r=1,\\[4pt]
\tilde{L}_d R(\Phi_{dm})^s(I_{k(r-1)m} \otimes W_{[k(r-1)m,k(d-r+1)m]})\Phi_{(r-1)m}, & 2 \le r \le d-1.
\end{cases}
$$

Note that in (15.39), $u(i)$, $i = t, t+1, \ldots, t+d-1$ are independent. Referring to the method developed in Sect. 15.2, we can search the cycles of length d of (15.36) by using (15.39) and checking the trace of $\mathrm{Blk}_i(\Psi_d)$: if its (j, j)-element equals 1, then $z(t) = \delta_{k\mu n}^j$ is in a cycle of length d under $u(t)u(t+1)\cdots u(t+d-1) = \delta_{kdm}^i$. Using (15.38) we can get $v(t), \ldots, v(t+d-1)$, and we can then obtain the cycle. Note that when ℓ is a proper factor of d, and $z(t)$ is in the cycles of length ℓ and d simultaneously under $\ltimes_{\xi=t}^{t+\ell-1}\tilde{u}(\xi) = \delta_{k\ell m}^j$ and $\ltimes_{\xi=t}^{t+d-1} u(\xi) = \delta_{kdm}^i$, respectively, we have the same cycle in control-state space if and only if $\delta_{kdm}^i = (\delta_{k\ell m}^j)^{\frac{d}{\ell}}$. To count the number of cycles, we should take out these repeated cycles. Thus, we have the following theorem, similar to Theorem 15.1.

Theorem 15.3 *The number of length-d cycles of the logical control network (15.36) is inductively determined by*

$$N_d = \frac{1}{d}\sum_{i=1}^{kdm} T\left(\mathrm{Blk}_i(\Psi_d)\right), \tag{15.40}$$

where

$$T\big(\mathrm{Blk}_i(\Psi_d)\big) = \mathrm{tr}\big(\mathrm{Blk}_i(\Psi_d)\big) - \sum_{(\ell,j)\in\theta(i,d)} T\big(\mathrm{Blk}_j(\Psi_\ell)\big).$$

Proposition 15.5 *There is a one-to-one correspondence between the cycles of the system* (15.36) *and the cycles of the higher-order logical control network* (15.29).

Proof Since $\delta^i_{k^\mu(m+n)}$ can be decomposed uniquely as $\ltimes^\mu_{\ell=1}\delta^{i(\ell)}_{km}\ltimes^\mu_{\ell=1}\delta^{j(\ell)}_{kn}$, we can construct a function $\pi : \Delta_{k^\mu(m+n)}, \Delta_{k^{m+n}}$ as follows:

$$\pi\big(\delta^i_{k^\mu(m+n)}\big) := F_{[m,\mu m]}(I_{k^{\mu m}} \otimes F_{[n,\mu n]})\delta^i_{k^\mu(m+n)} = \delta^{i(1)}_{km}\delta^{j(1)}_{kn}. \tag{15.41}$$

Denote by Ω_{vz} and Ω_{ux} all the cycles of the system (15.36) and the higher-order logical control network (15.29), respectively. We then define $\psi : \Omega_{vz} \to \Omega_{ux}$ as follows: For any $C = (v(t)z(t),\dots, v(t+d-1)z(t+d-1))$,

$$\psi(C) := \big(\pi\big(v(t)z(t)\big),\dots,\pi\big(v(t+d-1)z(t+d-1)\big)\big). \tag{15.42}$$

Let $u(\xi) = u(\ell)$, $x(\xi) = x(\ell)$, $v(\xi) = v(\ell)$, and $z(\xi) = z(\ell)$, when $\xi = \ell \pmod{d}$. Because

$$L\ltimes^{t+d-1}_{i=t+d-\mu}\pi\big(v(i)z(i)\big) = Lv(t+d-\mu)z(t+d-\mu) = F_{[n,\mu n]}z(t+d),$$

$\psi(C)$ is a cycle in Ω_{ux}. Thus, ψ is well defined. We then prove the following:

(1) ψ *is surjective.* For any cycle $C \in \Omega_{ux}$, $C = (u(t)x(t), u(t+1)\times x(t+1),\dots, u(t+d-1)x(t+d-1))$, let $C_1 = \{v(t)z(t),\dots, v(t+d-1)\times z(t+d-1)\}$, where $v(i) = \ltimes^{i+d-1}_{\xi=i}u(\xi)$, $z(i) = \ltimes^{i+d-1}_{\xi=i}x(\xi)$. We can then easily check that $\psi(C_1) = C$.

(2) ψ *is injective.* If there is another cycle $C_2 = (\tilde{v}(t)\tilde{z}(t),\dots, \tilde{v}(t+d-1)\times \tilde{z}(t+d-1))$ such that $\psi(C_2) = C$, then there exists an $a \le d$ such that $\pi(\tilde{v}(i)\tilde{z}(i)) = u(a+i-1)x(a+i-1)$. That is, the first m factors of $\tilde{v}(i)$ form $u(a+i-1)$, and the first n factors of $\tilde{z}(i)$ form $x(a+i-1)$. By (15.36) we know that the first $(k-1)m$ factors of $\tilde{v}(i+1)$ are equal to the last $(k-1)m$ factors of $\tilde{v}(i)$, while the first $(k-1)n$ factors of $\tilde{z}(i+1)$ are equal to the last $(k-1)n$ factors of $\tilde{z}(i)$. Thus, we obtain

$$\tilde{v}(i) = \ltimes^{a+i+\mu-2}_{\xi=a+i-1}u(\xi) = v(a+i-1),$$

$$\tilde{z}(i) = \ltimes^{a+i+\mu-2}_{\xi=a+i-1}x(\xi) = z(a+i-1).$$

It is then obvious that $C_2 = C_1$. $\qquad\square$

We now consider the optimal control of the μth order logical network. Set

$$\tilde{J}(v) = \lim_{T\to\infty}\frac{1}{T}\sum_{t=1}^{T}\tilde{P}\big(z(t), v(t)\big), \tag{15.43}$$

where

$$\widetilde{P}\big(z(t), v(t)\big) = P\big(F_{[n,\mu n]}z(t), F_{[m,\mu m]}v(t)\big).$$

By Lemma 15.1 it is easy to see that $F_{[n,\mu n]}z(t) = x(t)$ and $F_{[m,\mu m]}v(t) = u(t)$. Maximizing (15.43) is then equivalent to maximizing (15.4).

Proposition 15.4 is no longer true, but it is easy to see that the optimal cycles (in Ω_{vz}) can be found in the cycles with no repeated element. Thus, we can only search from the cycles with lengths less than or equal to the number of elements of the reachable set $R(z_0)$ of the initial state z_0. The following theorem can then be obtained.

Theorem 15.4 *For the μth order logical control network (15.29) with the objective function (15.43), there exists an optimal logical control matrix G^* such that the objective function is maximized and the trajectory of $s^*(t) = u^*(t)x^*(t)$ will become periodic after a certain (finite) time.*

Proof We can use (15.36) and (15.43) to replace (15.29) and (15.4), respectively, to find the optimal control. (15.36) can also be described as a directed graph with finite vertices, so, similarly to Corollary 15.1, we can find the optimal cycle in Ω_{vz}. Using (15.42), the optimal cycle in Ω_{ux} can then be obtained. Denote the shortest path from the initial state $\delta_{kn}^{j(0)}, \ldots, \delta_{kn}^{j(\mu-1)}$ to C^* by

$$\big(\delta_{km}^{i(0)}\delta_{kn}^{j(0)}, \delta_{km}^{i(1)}\delta_{kn}^{j(1)}, \ldots, \delta_{km}^{i(T_0-1)}\delta_{kn}^{j(T_0-1)}, C^*\big),$$

where

$$C^* = \delta_{km} \times \delta_{kn}\big((i(T_0), j(T_0)), (i(T_0+1), j(T_0+1)), \ldots,$$
$$(i(T_0+d-1), j(T_0+d-1))\big).$$

In the following, if $\ell = \xi \pmod{d}$ where $\ell \geq T_0$ and $T_0 \leq \xi \leq T_0 + d - 1$, then we set $i(\ell) = i(\xi)$. Using this convention, we can find G^* satisfying

$$\mathrm{Col}_s(G^*) = \delta_{km}^{i(\ell+1)}, \quad \mu - 1 \leq \ell \leq T_0 + d + \mu - 2, \qquad (15.44)$$

where

$$s = \sum_{\xi=1}^{\mu}(i(\ell - \mu + \xi) - 1)k^{(\mu-\xi)m+\mu n} + \sum_{\zeta=1}^{\mu-1}(j(\ell - \mu + \zeta) - 1)k^{(\mu-\zeta)n} + j(\ell)$$

and the other columns of G^* [$\mathrm{Col}(G^*) \subset \Delta_{km}$] can be arbitrary. The higher-order logical control network (15.29) is then converted to

$$\begin{cases} x^*(t+1) = Lu^*(t-k+1)\cdots u^*(t)x^*(t-k+1)\cdots x^*(t), \\ u^*(t+1) = G^*u^*(t-k+1)\cdots u^*(t)x^*(t-k+1)\cdots x^*(t). \end{cases} \qquad (15.45)$$

$\square$

Table 15.2 Payoff bi-matrix

$P_1 \backslash P_2$	0	1
0	3, 3	0, 5
1	5, 0	1, 1

Example 15.8 We consider the model of the infinite prisoner's dilemma [2].
Player 1 is a machine and player 2 is a person. Their possible actions are

0: the player cooperates with the partner,
1: the player betrays the partner.

The payoff bi-matrix is assumed to be as in Table 15.2.

Assume that the machine strategy, which depends on the μ-memory, is fixed. It is defined as

$$m(t+1) = f_m \big(m(t - \mu + 1), m(t - \mu + 2), \dots, m(t),$$
$$h(t - \mu + 1), h(t - \mu + 2), \dots, h(t) \big), \tag{15.46}$$

where the machine strategy $m(t)$ is considered as the state and f_m is a fixed logical function. The human strategy, $h(t)$, is considered as the control. Denote by $p_h(t) := p_h(m(t), h(t))$ the payoff of the human. Our purpose is to design an optimal control to maximize the average human payoff

$$J = \varlimsup_{T \to \infty} \frac{1}{T} \sum_{t=1}^{T} p_h(t). \tag{15.47}$$

Assuming that the machine uses the strategy "Two Tits For One Tat", it will take the action $m(t+1) = 0$ only under $(h(t-1), h(t), m(t-1), m(t)) = (0, 0, 1, 1)$. Assuming that the initial state and control are $m(0) = m(1) = h(0) = h(1) = 0$, then (15.46) and the human payoff P_h can be rewritten as

$$m(t+1) = Lh(t-1)h(t)m(t-1)m(t), \tag{15.48}$$

where

$$L = \delta^2 [1\,1\,1\,1\,1\,1\,1\,1\,1\,1\,1\,1\,2\,1\,1\,1]$$

and

$$P\big(m(t), h(t)\big) := P_h = h^T(t) \begin{bmatrix} 1 & 5 \\ 0 & 3 \end{bmatrix} m(t).$$

Set $z(t) = m(t)m(t+1)$, $v(t) = h(t)h(t+1)$, and $s(t) = v(t)z(t)$. From (15.35), (15.48) can be converted to

$$z(t+1) = \tilde{L}v(t)z(t), \tag{15.49}$$

where

$$\widetilde{L} = (I_2 \otimes L_m)W_{[8,2]}(I_8 \otimes MR)$$

$$= \delta_4[1\,3\,1\,3\,1\,3\,1\,3\,1\,3\,1\,3\,2\,3\,1\,3].$$

From (15.43),

$$\widetilde{P}(\delta_4^1\delta_4^1) = \widetilde{P}(\delta_4^1\delta_4^2) = \widetilde{P}(\delta_4^2\delta_4^1) = \widetilde{P}(\delta_4^2\delta_4^2) = 1,$$

$$\widetilde{P}(\delta_4^1\delta_4^3) = \widetilde{P}(\delta_4^1\delta_4^4) = \widetilde{P}(\delta_4^2\delta_4^3) = \widetilde{P}(\delta_4^2\delta_4^4) = 5,$$

$$\widetilde{P}(\delta_4^3\delta_4^1) = \widetilde{P}(\delta_4^3\delta_4^2) = \widetilde{P}(\delta_4^4\delta_4^1) = \widetilde{P}(\delta_4^4\delta_4^2) = 0,$$

$$\widetilde{P}(\delta_4^3\delta_4^3) = \widetilde{P}(\delta_4^3\delta_4^4) = \widetilde{P}(\delta_4^4\delta_4^3) = \widetilde{P}(\delta_4^4\delta_4^4) = 3.$$

It is easy to check that the reachable set of the initial state $(x(0), u(0), x(1), u(1)) = (0, 0, 0, 0)$ is

$$R(\delta_4^4\delta_4^4) = \{\delta_4^1\delta_4^1, \delta_4^1\delta_4^3, \delta_4^2\delta_4^1, \delta_4^2\delta_4^3, \delta_4^3\delta_4^1, d_4^3\delta_4^2, \delta_4^3\delta_4^3, \delta_4^4\delta_4^1, \delta_4^4\delta_4^2, \delta_4^4\delta_4^3\},$$

which consists of 10 elements.

By Theorem 15.3, we can obtain the cycles of length less than or equal to 10 with no repeated elements as

$$\begin{cases} C_1 = \delta_4 \times \delta_4\{(1, 1)\}, \\ C_2 = \delta_4 \times \delta_4\{(2, 1), (3, 1)\}, \\ C_3^1 = \delta_4 \times \delta_4\{(1, 1), (2, 1), (3, 1)\}, \\ C_3^2 = \delta_4 \times \delta_4\{(2, 3), (4, 1), (3, 2)\}, \\ C_3^3 = \delta_4 \times \delta_4\{(4, 1), (4, 2), (4, 3)\}, \\ C_4^1 = \delta_4 \times \delta_4\{(1, 3), (2, 1), (4, 1), (3, 2)\}, \\ C_4^2 = \delta_4 \times \delta_4\{(2, 1), (4, 1), (4, 2), (3, 3)\}, \\ C_5^1 = \delta_4 \times \delta_4\{(1, 1), (2, 1), (4, 1), (3, 2), (1, 3)\}, \\ C_5^2 = \delta_4 \times \delta_4\{(2, 1), (4, 1), (3, 2), (2, 3), (3, 1)\}, \\ C_5^3 = \delta_4 \times \delta_4\{(1, 1), (2, 1), (4, 1), (4, 2), (3, 3)\}, \\ C_5^4 = \delta_4 \times \delta_4\{(2, 1), (4, 1), (4, 2), (4, 3), (3, 1)\}, \\ C_6^1 = \delta_4 \times \delta_4\{(1, 1), (2, 1), (4, 1), (3, 2), (2, 3), (3, 1)\}, \\ C_6^2 = \delta_4 \times \delta_4\{(1, 1), (2, 1), (4, 1), (4, 2), (4, 3), (3, 1)\}. \end{cases}$$

A straightforward calculation shows that the optimal cycle is C_3^2, which has average human payoff $\frac{5}{3}$. This result coincides with the one in [2].

The optimal trajectory for the system (15.49) is

$$\delta_4^4 \delta_4^4 \to \delta_4^4 \delta_4^3 \to \delta_4 \times \delta_4 \{(4, 1), (3, 2), (2, 3)\}.$$

Thus, we can find the optimal trajectory for the system (15.48) as

$$\delta_2^2 \delta_2^2 \to \delta_2^2 \times \delta_2^2 \to \delta_2 \times \delta_2 \{(2, 1), (2, 1), (1, 2)\}.$$

Then,

$$G^* = \delta_2[* * * * * * 2 * * 2 1 * * * * 2 2],$$

where $*$ can be chosen arbitrarily from $\{1, 2\}$.

References

1. Gibbons, R.: A Primer in Game Theory. Prentice Hall, New York (1992)
2. Mu, Y., Guo, L.: Optimization and identification in a non-equilibrium dynamic game. In: Proc. CDC-CCC'09, pp. 5750–5755 (2009)
3. Puterman, M.: Markov Decision Processes: Discrete Stochastic Dynamic Programming. Wiley, New York (1994)
4. Zhao, Y., Li, Z., Cheng, D.: Optimal control of logical control networks. IEEE Trans. Automat. Contr. (2010, accepted)

Chapter 16
Input-State Incidence Matrices

16.1 The Input-State Incidence Matrix

Consider a Boolean control network with n network nodes, m input nodes, and p output nodes. Its dynamics is described as

$$
\begin{cases}
x_1(t+1) = f_1(x_1(t), \ldots, x_n(t), u_1(t), \ldots, u_m(t)), \\
\vdots \\
x_n(t+1) = f_n(x_1(t), \ldots, x_n(t), u_1(t), \ldots, u_m(t)), \quad x_i \in \mathcal{D}, \\
\end{cases}
\tag{16.1}
$$
$$
y_j(t) = h_j(x_1(t), \ldots, x_n(t)), \quad j = 1, \ldots, p; \ y_j \in \mathcal{D},
$$

where $f_i : \mathcal{D}^{n+m} \to \mathcal{D}$, $i = 1, \ldots, n$, and $h_i : \mathcal{D}^n \to \mathcal{D}$, $j = 1, \ldots, p$, are logical functions. Its algebraic form is

$$
\begin{aligned}
x(t+1) &= Lu(t)x(t), \\
y(t) &= Hx(t),
\end{aligned}
\tag{16.2}
$$

where $L \in \mathcal{L}_{2^n \times 2^{n+m}}$ and $H \in \mathcal{L}_{2^p \times 2^n}$.

We first ignore the output and consider the input-state transfer graph (ISTG) of the system (16.1). ISTGs were briefly explained in the last chapter. We now give one more example, in order to explain them more thoroughly, as well as to introduce the input-state incidence matrix.

Example 16.1 Consider the Boolean control network Σ, given by

$$
\Sigma : \begin{cases}
x_1(t+1) = (x_1(t) \vee x_2(t)) \wedge u(t), \\
x_2(t+1) = x_1(t) \leftrightarrow u(t).
\end{cases}
\tag{16.3}
$$

Setting $x(t) = x_1(t) \ltimes x_2(t)$, it is easy to calculate that the algebraic form of Σ is

$$
\Sigma : x(t+1) = Lu(t)x(t),
\tag{16.4}
$$

D. Cheng et al., *Analysis and Control of Boolean Networks*,
Communications and Control Engineering,
DOI 10.1007/978-0-85729-097-7_16, © Springer-Verlag London Limited 2011

Fig. 16.1 Input-state transfer
graph

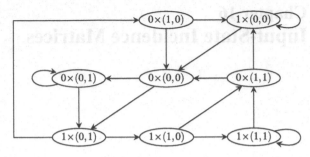

where

$$L = \delta_4[1\ 1\ 2\ 4\ 4\ 4\ 3\ 3].\tag{16.5}$$

According to the dynamic equation (16.3) [equivalently, (16.4)], we can draw the
flow of $(u(t), (x_1(t), x_2(t)))$ on the product space $\mathscr{U} \times \mathscr{X}$, called the input-state
transfer graph, as in Fig. 16.1.

Using vector form, the input-state product space becomes $\Delta \times \Delta_4$. We may define
the points in the input-state product space as $P_1 = \delta_2^1 \times \delta_4^1$, $P_2 = \delta_2^1 \times \delta_4^2, \ldots, P_8 =
\delta_2^2 \times \delta_4^4$. We now construct an 8×8 matrix, $\mathscr{J}(\Sigma)$, in the following way:

$$\mathscr{J}_{ij} = \begin{cases} 1, & \text{there exists an edge from } P_j \text{ to } P_i, \\ 0, & \text{otherwise.} \end{cases}$$

$\mathscr{J}(\Sigma)$, called the input-state incidence matrix of the Boolean control network Σ,
is then

$$\mathscr{J}(\Sigma) = \begin{bmatrix} 1 & 1 & 0 & 0 & 0 & 0 & 0 & 0 \\ 0 & 0 & 1 & 0 & 0 & 0 & 0 & 0 \\ 0 & 0 & 0 & 0 & 0 & 0 & 1 & 1 \\ 0 & 0 & 0 & 1 & 1 & 1 & 0 & 0 \\ 1 & 1 & 0 & 0 & 0 & 0 & 0 & 0 \\ 0 & 0 & 1 & 0 & 0 & 0 & 0 & 0 \\ 0 & 0 & 0 & 0 & 0 & 0 & 1 & 1 \\ 0 & 0 & 0 & 1 & 1 & 1 & 0 & 0 \end{bmatrix}.\tag{16.6}$$

It is easy to see that the incidence matrix of a Boolean control network is indeed
the transpose of the adjacency matrix of the input-state transfer graph (note that it is
different from the adjacency matrix of a Boolean network). However, it is very difficult to find this matrix by drawing the graph since the graph will be very complex
when n and m are not very small.

Next, we explore the structure of the input-state incidence matrix. Comparing
(16.6) with (16.5), it may seem surprising to find that

$$\mathscr{J}(\Sigma) = \begin{bmatrix} L \\ L \end{bmatrix}.$$

In fact, this is also true for the general case. Consider equation (16.2). In fact, the jth column corresponds to the "output" $x(t+1)$ for the P_jth "input" $(u(t), x(t))$ of the dynamical system. If this column $\mathrm{Col}_j(L) = \delta_{2^n}^i$, then the output $x(t+1)$ is exactly the ith element of $x(t) \in \Delta_{2^n}$. Now, since $u(t+1)$ can be arbitrary, it follows that the input-state incidence matrix of the system (16.2) is

$$\mathscr{I} := \mathscr{I}|_{(16.2)} = \left.\begin{bmatrix} L \\ L \\ \vdots \\ L \end{bmatrix}\right\} 2^m \in \mathscr{B}_{2^{m+n} \times 2^{m+n}}, \tag{16.7}$$

where the first block of rows corresponds to $u(t+1) = \delta_{2^m}^1$, the second block corresponds to $u(t+1) = \delta_{2^m}^2$, and so on.

Next, we consider the properties of $\mathscr{I}$. We first introduce a new concept.

Definition 16.1 A matrix $A \in \mathscr{M}_{m \times n}$ is called a row-periodic matrix with period τ if τ is a proper factor of m such that $\mathrm{Row}_{i+\tau}(A) = \mathrm{Row}_i(A)$, $1 \le i \le m - \tau$.

The following property can be verified via a straightforward computation.

Proposition 16.1

1. $A \in \mathscr{M}_{m \times m}$ is a row-periodic matrix with period τ (where $m = \tau k$) if and only if

$$A = \mathbf{1}_\tau A_0,$$

where $A_0 \in \mathscr{M}_{k \times m}$ consists of the first k rows of A, called the basic block of A.
2. If $A \in \mathscr{M}_{m \times m}$ is a row-periodic matrix with period τ (where $m = \tau k$), then so is A^s, $s \in \mathbb{Z}_+$ (where $\mathbb{Z}_+$ is the set of positive integers).

Applying Proposition 16.1 to the incidence matrix, we obtain the following result.

Corollary 16.1 *Consider the system (16.1). Its input-state incidence matrix is*

$$\mathscr{I} = \mathbf{1}_{2^m} \ltimes \mathscr{I}_0, \quad \text{where } \mathscr{I}_0 = L. \tag{16.8}$$

Moreover, the basic block of $\mathscr{I}^s$ is

$$\mathscr{I}_0^s = L \ltimes (\mathbf{1}_{2^m} \ltimes L)^{s-1}. \tag{16.9}$$

Note that since $\mathscr{I}^s$ is a row-periodic matrix, it is easy to see that

$$\mathscr{I}_0^{s+1} = \mathscr{I}_0 \mathscr{I}^s = L \mathscr{I}^s = L\mathbf{1}_{2^m} \ltimes \mathscr{I}_0^s. \tag{16.10}$$

This equation shows that in calculating $\mathscr{I}_0^s$, we do not need to take account of the whole of $\mathscr{I}$ in the calculation. We summarize this in the following proposition.

Proposition 16.2

$$\mathscr{J}_0^{s+1} = M^s L, \qquad (16.11)$$

where

$$M = \sum_{i=1}^{2^m} \mathrm{Blk}_i(L).$$

Proof From (16.10) it is easy to see that $\mathscr{J}_0^{s+1} = M \mathscr{J}_0^s$. Since $\mathscr{J}_0 = L$, (16.11) follows. $\square$

16.2 Controllability

First, we explore the physical meaning of $\mathscr{J}^s$. When $s = 1$ we know that $\mathscr{J}_{ij}$ dictates whether or not there exists a set of controls such that P_i is reachable from P_j in one step (based on whether or not $\mathscr{J}_{ij} = 1$). Is there a similar meaning for $\mathscr{J}^s$? The following result answers this question.

Theorem 16.1 *Consider the system* (16.1) *and assume that the* (i, j)-*element of the* sth *power of its input-state incidence matrix,* $\mathscr{J}_{ij}^s$, *equals* c. *There are then* c *paths from point* P_j *which reach* P_i *at the* sth *step with proper controls.*

Proof We prove this by mathematical induction. When $s = 1$ the conclusion follows from the definition of the input-state incidence matrix. Now, assume that $\mathscr{J}_{ij}^s$ is the number of the paths from P_j to P_i at the sth step. Since a path from P_j to P_i at the $(s + 1)$th step can always be considered as a path from P_j to P_k at the sth step and then from P_k to P_i at the first step, it can be calculated as

$$c = \sum_{k=1}^{2^{m+n}} \mathscr{J}_{ik} \mathscr{J}_{kj}^s,$$

which is exactly $\mathscr{J}_{ij}^{s+1}$. $\square$

From the above theorem the following result is obvious.

Corollary 16.2 *Consider the system* (16.1), *denoting its input-state incidence matrix by* $\mathscr{J}$. P_i *is reachable from* P_j *at the* sth *step if and only if* $\mathscr{J}_{ij}^s > 0$.

The above arguments show that all controllability information is contained in $\{\mathscr{J}^s \mid s = 1, 2, \ldots\}$. By the Cayley–Hamilton theorem of linear algebra, it is easy to see that if $\mathscr{J}_{ij}^s = 0$, $\forall s \le 2^{m+n}$, then $\mathscr{J}^s = 0$, $\forall s$. Next, we consider only

$\{ \mathscr{I}^s \mid s \leq 2^{m+n} \}$. Since they are row-periodic matrices, we need only to consider their basic blocks, $\mathscr{I}_0^s$. We split such a basic block into 2^m blocks as

$$\mathscr{I}_0^s = \begin{bmatrix} \text{Blk}_1(\mathscr{I}_0^s) & \text{Blk}_2(\mathscr{I}_0^s) & \cdots & \text{Blk}_{2^m}(\mathscr{I}_0^s) \end{bmatrix}, \qquad (16.12)$$

where $\text{Blk}_i(\mathscr{I}_0^s) \in \mathscr{M}_{2^n \times 2^n}$, $i = 1, \ldots, 2^m$.

Recall Definition 9.1 for controllability.

Definition 16.2 Consider the system (16.1). Denote its state space by $\mathscr{X} = \mathscr{D}^n$ and let $X_0 \in \mathscr{X}$.

1. $X \in \mathscr{X}$ is said to be reachable from X_0 at time $s > 0$ if we can find a sequence of controls $U(0) = \{u_1(0), \ldots, u_m(0)\}$, $U(1) = \{u_1(1), \ldots, u_m(1)\}$, $\ldots$ such that the trajectory of (16.1) with initial value X_0 and controls $\{U(t)\}$, $t = 0, 1, \ldots$, will reach X at time $t = s$. The reachable set at time s is denoted by $R_s(X_0)$. The overall reachable set is denoted by

$$R(X_0) = \bigcup_{s=1}^{\infty} R_s(X_0).$$

2. The system (16.1) is said to be controllable at X_0 if $R(X_0) = \mathscr{X}$. It is said to be controllable if it is controllable at every $X \in \mathscr{X}$.

From Proposition 16.2, $\text{Blk}_i(\mathscr{I}_0^s) = M^{s-1} \text{Blk}_i(L)$. From the construction, it is clear that $\text{Blk}_i(\mathscr{I}_0^s)$ corresponds to the ith input $u = \delta_{2^m}^i$. Moreover, the jth column of $\text{Blk}_\mu(\mathscr{I}_0^s)$ corresponds to the initial value $x_0 = \delta_{2^n}^j$. The following conclusion is then clear.

Theorem 16.2 *Consider the system (16.1), with input-state incidence matrix $\mathscr{I}$.*

1. $x(s) = \delta_{2^n}^\alpha$ *is reachable from* $x(0) = \delta_{2^n}^j$ *at the* s*th step if and only if*

$$\sum_{i=1}^{2^m} \left(\text{Blk}_i(\mathscr{I}_0^s) \right)_{\alpha j} = \left(M^s \right)_{\alpha j} > 0. \qquad (16.13)$$

2. $x = \delta_{2^n}^\alpha$ *is reachable from* $x(0) = \delta_{2^n}^j$ *if and only if*

$$\sum_{s=1}^{2^{m+n}} \sum_{i=1}^{2^m} \left(\text{Blk}_i(\mathscr{I}_0^s) \right)_{\alpha j} = \sum_{s=1}^{2^{m+n}} \left(M^s \right)_{\alpha j} > 0. \qquad (16.14)$$

3. *The system is controllable at* $x(0) = \delta_{2^n}^j$ *if and only if*

$$\sum_{s=1}^{2^{m+n}} \sum_{i=1}^{2^m} \text{Col}_j \left[\text{Blk}_i(\mathscr{I}_0^s) \right] = \sum_{s=1}^{2^{m+n}} \text{Col}_j(M^s) > 0. \qquad (16.15)$$

4. *The system is controllable if and only if the controllability matrix, $\mathscr{C}$, satisfies*

$$\mathscr{C} = \sum_{s=1}^{2^{m+n}} \sum_{i=1}^{2^m} \mathrm{Blk}_i \left(\mathscr{J}_0^s \right) = \sum_{s=1}^{2^{m+n}} M^s > 0. \tag{16.16}$$

If $A \in \mathscr{M}_{m \times n}$ is a real matrix, then the inequality $A > 0$ is used to mean that all the entries of A are positive, i.e., $a_{i,j} > 0, \forall i, j$.

When controllability is considered, we do not need to consider the number of paths from one state to another. Hence, the real value of each entry of $\mathscr{J}^s$ is less interesting. What we really do need to know is whether it is positive or not. Hence, we can simply use Boolean algebra in the above calculation. We refer to Chap. 11 for more about the following Boolean algebra.

1. If $a, b \in \mathscr{D}$, we can define Boolean addition and the Boolean product, respectively, as

$$a +_{\mathscr{B}} b = a \vee b, \qquad a \times_{\mathscr{B}} b = a \wedge b.$$

$\{\mathscr{D}, +_{\mathscr{B}}, \times_{\mathscr{B}}\}$ forms an algebra, called the Boolean algebra.

2. Let $A = (a_{ij})$, $B = (b_{ij}) \in \mathscr{B}_{m \times n}$. We define

$$A +_{\mathscr{B}} B := (a_{ij} +_{\mathscr{B}} b_{ij}).$$

3. Let $A \in \mathscr{B}_{m \times n}$ and $B \in \mathscr{B}_{n \times p}$. We define $A \times_{\mathscr{B}} B := C \in \mathscr{B}_{m \times p}$, where

$$c_{ij} = \sum_{k=1}^{n} {}_{\mathscr{B}} \, a_{ik} \times_{\mathscr{B}} b_{kj}.$$

In particular, if $A \in \mathscr{B}_{n \times n}$, then

$$A^{(2)} := A \times_{\mathscr{B}} A.$$

We use a simple example to illustrate the Boolean algebra.

Example 16.2 Assume

$$A = \begin{bmatrix} 1 & 0 & 0 \\ 0 & 1 & 0 \\ 1 & 0 & 1 \end{bmatrix}, \qquad B = \begin{bmatrix} 0 & 1 & 0 \\ 1 & 0 & 1 \\ 1 & 0 & 1 \end{bmatrix}.$$

Then,

$$A +_{\mathscr{B}} B = \begin{bmatrix} 1 & 1 & 0 \\ 1 & 1 & 1 \\ 1 & 0 & 1 \end{bmatrix}, \qquad A \ltimes_{\mathscr{B}} B = \begin{bmatrix} 0 & 1 & 0 \\ 1 & 0 & 1 \\ 1 & 1 & 1 \end{bmatrix}.$$

$$A^{(2)} = \begin{bmatrix} 1 & 0 & 0 \\ 0 & 1 & 0 \\ 1 & 0 & 1 \end{bmatrix}, \qquad A^{(s)} = \begin{bmatrix} 1 & 0 & 0 \\ 0 & 1 & 0 \\ 1 & 0 & 1 \end{bmatrix}, \quad s \geq 3.$$

Using Boolean algebra, we have the following alternative condition.

Corollary 16.3 *Results 1, 2, 3, and 4 of Theorem 16.2 remain true if, in the corresponding conditions* (16.13)–(16.16), $\mathscr{J}_0^s$ *is replaced by* $\mathscr{J}_0^{(s)}$, *and* M^s *is replaced by* $M^{(s)}$. *In particular, we call*

$$\mathscr{M}_{\mathscr{C}} := \sum_{s=1}^{2^{m+n}}{}_{\mathscr{B}} \sum_{i=1}^{2^m}{}_{\mathscr{B}} \mathrm{Blk}_i\left(\mathscr{J}_0^{(s)}\right) = \sum_{s=1}^{2^{m+n}}{}_{\mathscr{B}} M^{(s)} \in \mathscr{B}_{2^n \times 2^n} \qquad (16.17)$$

the controllability matrix and write $\mathscr{M}_{\mathscr{C}} = (c_{ij})$. *Then:*

(i) $\delta_{2^n}^i$ *is reachable from* $\delta_{2^n}^j$ *if and only if* $c_{ij} > 0$.
(ii) *The system is controllable at* $\delta_{2^n}^j$ *if and only if* $\mathrm{Col}_j(\mathscr{M}_{\mathscr{C}}) > 0$.
(iii) *The system is controllable if and only if* $\mathscr{M}_{\mathscr{C}} > 0$.

The following example shows how to use Theorem 16.2 or Corollary 16.3.

Example 16.3 Consider the following Boolean control network:

$$\begin{cases} x_1(t+1) = (x_1(t) \leftrightarrow x_2(t)) \vee u_1(t), \\ x_2(t+1) = \neg x_1(t) \wedge u_2(t), \\ y(t) = x_1(t) \vee x_2(t). \end{cases} \qquad (16.18)$$

Setting $x(t) = \ltimes_{i=1}^2 x_i(t)$, $u = \ltimes_{i=1}^2 u_i(t)$, we have

$$\begin{cases} x(t+1) = Lu(t)x(t), \\ y(t) = Hx(t), \end{cases} \qquad (16.19)$$

where

$$L = \delta_4[2\,2\,1\,1\,2\,2\,2\,2\,4\,3\,1\,2\,4\,4\,2],$$

$$H = \delta_2[1\,1\,1\,2].$$

For the system (16.18), the basic block of its input-state incidence matrix $\mathscr{J}_0$ equals L.

1. Is δ_4^1 reachable from $x(0) = \delta_4^2$?
   After a straightforward computation, we have

$$\left(M^{(1)}\right)_{12} = 0, \qquad \left(M^{(2)}\right)_{12} > 0.$$

This means that $x(2) = \delta_4^1$ is reachable from $x(0) = \delta_4^2$ at the second step.
2. Is the system controllable, or controllable at any point?

We check the controllability matrix:

$$\mathscr{M_C} = \sum_{s=1}^{2^4} {}_{\mathscr{B}} M^{(s)} = \begin{bmatrix} 1 & 1 & 1 & 1 \\ 1 & 1 & 1 & 1 \\ 0 & 0 & 1 & 0 \\ 1 & 1 & 1 & 1 \end{bmatrix}.$$

According to Corollary 16.3, we conclude that:

(i) The system is not controllable. However, it is controllable at $x_0 = \delta_4^3 \sim$ (0, 1).

(ii) $x_d = \delta_4^3 \sim (0, 1)$ is not reachable from $x_0 = \delta_4^1 \sim (1, 1)$, $x_0 = \delta_4^2 \sim (1, 0)$, or $x_0 = \delta_4^4 \sim (0, 0)$.

16.3 Trajectory Tracking and Control Design

Assume $x_d \in R(x_0)$. The purpose of this section is to find a control which drives x_0 to x_d. Since the trajectory from x_0 to x_d (driven by a proper sequence of controls) is generally not unique, we only try to find the shortest one. A similar approach can produce all the required trajectories.

Assume that $x_0 = \delta_{2^n}^j$ and $x_d = \delta_{2^n}^i$. Consider the following algorithm.

Algorithm 16.1 Assume that the (i, j)-element of the controllability matrix, $c_{i,j}$, is positive.

• Step 1. Find the smallest s such that in the block-decomposed form (16.12) of $\mathscr{J}_0^{(s)}$, there exists a block, say $\mathrm{Blk}_\alpha(\mathscr{J}_0^{(s)})$, which has as its (i, j)-element

$$\left[\mathrm{Blk}_\alpha\big(\mathscr{J}_0^{(s)}\big)\right]_{ij} > 0. \tag{16.20}$$

Set $u(0) = \delta_{2^m}^\alpha$ and $x(s) = \delta_{2^n}^i$. If $s = 1$, stop; otherwise, go to next step.
• Step 2. Find k, β such that

$$\left[\mathrm{Blk}_\beta(\mathscr{J}_0)\right]_{ik} > 0, \qquad \left[\mathrm{Blk}_\alpha\big(\mathscr{J}_0^{s-1}\big)\right]_{kj} > 0.$$

Set $u(s-1) = \delta_{2^m}^\beta$ and $x(s-1) = \delta_{2^n}^k$.
• Step 3. If $s - 1 = 1$, stop; otherwise, set $s = s - 1$, $i = k$ (that is, replace s by $s - 1$ and replace i by k) and go back to Step 2.

Proposition 16.3 *As long as $x_d \in R(x_0)$, the control sequence $\{u(0), u(1), \ldots, u(s-1)\}$ generated by Algorithm 16.1 can drive the trajectory from x_0 to x_d. Moreover, the corresponding trajectory is $\{x(0) = x_0, x(1), \ldots, x(s) = x_d\}$, which is also produced by the algorithm.*

Proof Since $x_d \in R(x_0)$, by the construction of controllability matrix $\mathcal{M}_{\mathscr{C}}$, there exists a smallest s such that $[\text{Blk}_\alpha(\mathscr{J}_0^{(s)})]_{i,j} > 0$. This means that if $u(0) = \delta_{2m}^\alpha$, $x(0) = \delta_{2n}^j$, then there exists at least one path from $x(0)$ to $x(s) = \delta_{2n}^i$. We then know that x_0 can reach x_d at the sth step if $u(0) = \delta_{2m}^\alpha$. Hence, it is obvious that there must exist k such that x_0 can reach δ_{2n}^k at the $(s-1)$th step with $u(0) = \delta_{2m}^\alpha$, and β such that $u(s-1) = \delta_{2m}^\beta$, which make $L\delta_{2m}^\beta\delta_{2n}^k = \delta_{2n}^i$. Equivalently, we can find k, β such that

$$\left[\text{Blk}_\beta(\mathscr{J}_0)\right]_{ik} > 0, \qquad \left[\text{Blk}_\alpha(\mathscr{J}_0^{(s-1)})\right]_{kj} > 0.$$

In the same way, we can find β' and k' such that $\delta_{2n}^{k'}$ can be reached at $(s-2)$th step and $L\delta_{2m}^{\beta'}\delta_{2n}^{k'} = \delta_{2n}^k$. Continuing this process, the sequence of controls and states from x_0 to x_d can be obtained.   □

Example 16.4 Recall Example 16.3. For $x_0 = \delta_4^2$ and $x_d = \delta_4^1$, we want to find a trajectory from x_0 to x_d. We follow Algorithm 16.1 step by step, as follows:

- Step 1. The smallest s is 2. We can calculate that

$$\left[\text{Blk}_3(\mathscr{J}_0^2)\right]_{12} > 0,$$

so $u(0) = \delta_4^3$, $x(2) = \delta_4^1$.
- Step 2. From a straightforward computation, we have

$$\left[\text{Blk}_1(\mathscr{J}_0)\right]_{14} > 0, \qquad \left[\text{Blk}_3(\mathscr{J}_0^{2-1})\right]_{42} > 0,$$

so $u(1) = \delta_4^1$, $x(1) = \delta_4^4$.
- Step 3. Now $s - 1 = 1$, so we stop the process.

Hence, the control sequence for $x_0 = \delta_4^2 \sim (1, 0)$ and $x_d = \delta_4^1 \sim (1, 1)$ is $\{u(0) = \delta_4^3 \sim (0, 1), u(1) = \delta_4^1 \sim (1, 1)\}$, and the trajectory is $\{x(0) = \delta_4^2 \sim (1, 0), x(1) = \delta_4^4 \sim (0, 0), x(2) = \delta_4^1 \sim (1, 1)\}$. In general, the smallest-step trajectory is not unique. In this example there are four ways drive x_0 to x_d at the second step. In the same way, we can find the other three paths, which are

$$\left\{u(0) = \delta_4^3, u(1) = \delta_4^3\right\}, \left\{x(0) = \delta_4^2, x(1) = \delta_4^4, x(2) = \delta_4^1\right\};$$

$$\left\{u(0) = \delta_4^4, u(1) = \delta_4^1\right\}, \left\{x(0) = \delta_4^2, x(1) = \delta_4^4, x(2) = \delta_4^1\right\};$$

$$\left\{u(0) = \delta_4^4, u(1) = \delta_4^3\right\}, \left\{x(0) = \delta_4^2, x(1) = \delta_4^4, x(2) = \delta_4^1\right\}.$$

16.4 Observability

This section considers the observability of the system (16.1). We denote the outputs as $Y(t) = (y_1(t), \ldots, y_p(t))$ and, alternatively, $y(t) = \ltimes_{k=1}^p y_k(t)$. We adapt Definition 9.7 for observability as follows:

Definition 16.3 Consider the system (16.1).

1. X_1^0 and X_2^0 are said to be distinguishable if there exists a control sequence $\{U(0), U(1), \ldots, U(s)\}$, where $s \geq 0$, such that

$$Y^1(s+1) = y^{s+1}\big(U(s), \ldots, U(0), X_1^0\big) \neq Y^2(s+1)$$

$$= y^{s+1}\big(U(s), \ldots, U(0), X_2^0\big). \tag{16.21}$$

2. The system is said to be observable if any two initial points $X_1^0, X_2^0 \in \Delta_{2^n}$ are distinguishable.

Recall that in Chap. 11 we defined the logical operators for Boolean matrices as follows:

1. Let $A = (a_{ij}) \in \mathscr{B}_{n \times s}$ and σ be a unary operator. $\sigma : \mathscr{B}_{n \times s} \to \mathscr{B}_{n \times s}$ is then defined as

$$\sigma A := (\sigma a_{ij}). \tag{16.22}$$

2. Let $A = (a_{ij})$, $B = (b_{ij}) \in \mathscr{B}_{n \times s}$, and σ be a binary operator. $\sigma : \mathscr{B}_{n \times s} \times \mathscr{B}_{n \times s} \to \mathscr{B}_{n \times s}$ is then defined as

$$A \sigma B := (a_{i,j} \sigma b_{ij}). \tag{16.23}$$

Definition 16.4 Let $A = (a_{ij}) \in \mathscr{B}_{m \times n}$. The weight of A is defined as

$$wt(A) := \sum_{i=1}^{m} \sum_{j=1}^{n} a_{ij}. \tag{16.24}$$

The Boolean weight of A is defined as

$$wb(A) := \begin{cases} 1, & wt(A) > 0, \\ 0, & wt(A) = 0. \end{cases} \tag{16.25}$$

We give a simple example.

Example 16.5 Assume that

$$A = \begin{bmatrix} 1 & 0 & 1 \\ 0 & 0 & 1 \end{bmatrix}, \qquad B = \begin{bmatrix} 1 & 1 & 0 \\ 1 & 0 & 1 \end{bmatrix}.$$

Then,

$$\neg A = \begin{bmatrix} 0 & 1 & 0 \\ 1 & 1 & 0 \end{bmatrix}, \qquad A \bar{\vee} B = \begin{bmatrix} 0 & 1 & 1 \\ 1 & 0 & 0 \end{bmatrix},$$

$$wt(A) = 3, \qquad wt(B) = 4, \qquad wb(A) = wb(B) = 1.$$

Recall (16.12). From the construction of $\mathscr{J}$ and the properties of the semi-tensor product, it is easy to see that Blk_i corresponds to the input $u(0) = \delta_{2^m}^i$. Moreover, each block $\text{Col}_j(\text{Blk}_i)$ corresponds to $x_0 = \delta_{2^n}^j$. To exchange the running order of the indices i and j, we use the swap matrix to define

$$\tilde{\mathscr{J}}_0^{(s)} := \mathscr{J}_0^{(s)} W_{[2^n, 2^m]} \qquad (16.26)$$

and then split it into 2^n blocks as

$$\tilde{\mathscr{J}}_0^{(s)} = [\text{Blk}_1(\tilde{\mathscr{J}}_0^{(s)}) \quad \text{Blk}_2(\tilde{\mathscr{J}}_0^{(s)}) \quad \cdots \quad \text{Blk}_{2^n}(\tilde{\mathscr{J}}_0^{(s)})], \qquad (16.27)$$

where $\text{Blk}_i(\tilde{\mathscr{J}}_0^{(s)}) \in \mathscr{B}_{2^n \times 2^m}$, $i = 1, \ldots, 2^n$.

Each block $\text{Blk}_i(\tilde{\mathscr{J}}_0^{(s)})$ now corresponds to $x_0 = \delta_{2^n}^i$, and in each block, $\text{Col}_j(\text{Blk}_i(\tilde{\mathscr{J}}_0^{(s)}))$ corresponds to $u(0) = \delta_{2^m}^j$.

Using the Boolean algebraic expression, we have the following sufficient condition for observability.

Theorem 16.3 *Consider the system* (16.1) *with algebraic form* (16.2). *If*

$$\bigvee_{s=1}^{2^{m+n}} [(H \ltimes \text{Blk}_i(\tilde{\mathscr{J}}_0^{(s)})) \, \bar{\vee} \, (H \ltimes \text{Blk}_j(\tilde{\mathscr{J}}_0^{(s)}))] \neq 0, \quad 1 \leq i < j \leq 2^n, \quad (16.28)$$

then the system is observable.

Proof According to the construction and the above argument, it is easy to see that (16.28) implies that, at least at first step, the outputs corresponding to $x^0 = \delta_{2^m}^i$ and $x^0 = \delta_{2^m}^j$ are distinct. $\qquad \square$

Theorem 16.3 can be alternatively expressed as follows.

Corollary 16.4 *Consider the system* (16.1) *with algebraic form* (16.2). *Define*

$$O_{ij} := \bigvee_{s=1}^{2^{m+n}} [(H \ltimes \text{Blk}_i(\tilde{\mathscr{J}}_0^{(s)})) \, \bar{\vee} \, (H \ltimes \text{Blk}_j(\tilde{\mathscr{J}}_0^{(s)}))].$$

If

$$\bigwedge_{1 \leq i < j \leq 2^n} wb(O_{ij}) = 1, \qquad (16.29)$$

then the system is observable.

Remark 16.1 Comparing this with the corresponding result in Chap. 9, one of the advantages of this result is that when the step s increases, the corresponding matrices involved in the condition do not increase their dimensions. Thus, it is easily computable. Another advantage is that this condition does not require controllability of the system. The major disadvantage is that this result is not necessary.

We give an example to illustrate this.

Example 16.6 Consider the network (16.18) in Example 16.3.
Let

$$O_{ij} = \bigvee_{s=1}^{2^4} \left[\left(H \ltimes \text{Blk}_i \left(\tilde{\mathscr{J}}_0^{(s)} \right) \right) \bar{\vee} \left(H \ltimes \text{Blk}_j \left(\tilde{\mathscr{J}}_0^{(s)} \right) \right) \right].$$

A straightforward computation yields

$$O_{12} = \begin{bmatrix} 0 & 0 & 1 & 1 \\ 0 & 0 & 1 & 1 \end{bmatrix}, \qquad O_{13} = \begin{bmatrix} 0 & 0 & 0 & 1 \\ 1 & 0 & 0 & 1 \end{bmatrix}, \qquad O_{14} = \begin{bmatrix} 0 & 0 & 0 & 0 \\ 1 & 0 & 1 & 0 \end{bmatrix},$$

$$O_{23} = \begin{bmatrix} 0 & 0 & 1 & 0 \\ 1 & 0 & 1 & 0 \end{bmatrix}, \qquad O_{24} = \begin{bmatrix} 0 & 0 & 1 & 1 \\ 1 & 0 & 1 & 1 \end{bmatrix}, \qquad O_{34} = \begin{bmatrix} 0 & 0 & 0 & 1 \\ 0 & 0 & 1 & 1 \end{bmatrix}.$$

We then have

$$\bigwedge_{1 \le i < j \le 4} wb(O_{ij}) = 1.$$

The system is therefore observable.

16.5 Fixed Points and Cycles

The fixed points and cycles of an input-state transfer graph are very important topological features of a Boolean control network. For instance, the optimal control can always be realized over a fixed point or a cycle. We refer to Chap. 15 or [2]. The input-state incidence matrix can also provide the information about this.

We now adapt Definition 15.3 for fixed points and cycles.

Definition 16.5 Consider the system (16.1). Denote the input-state (product) space by

$$\mathscr{S} = \left\{ (U, X) \,\middle|\, U = (u_1, \dots, u_m) \in \mathscr{D}^p, \; X = (x_1, \dots, x_n) \in \mathscr{D}^n \right\}.$$

Note that $|\mathscr{S}| = 2^{m+n}$.

1. Let $S_i = (U^i, X^i) \in \mathscr{S}$ and $S_j = (U^j, X^j) \in \mathscr{S}$. Let $U^i = (u_1^i, \dots, u_m^i)$, $X_i = (x_1^i, \dots, x_n^i)$, etc. (S_i, S_j) is said to be a directed edge if X^i, U^i, and X^j satisfy (16.1), that is, if

$$x_k^j = f_k(x_1^i, \dots, x_n^i, u_1^i, \dots, u_m^i), \qquad k = 1, \dots, n.$$

The set of edges is denoted by $\mathscr{E} \subset \mathscr{S} \times \mathscr{S}$.

2. The pair $(\mathscr{S}, \mathscr{E})$ forms a directed graph, which is called the input-state transfer graph.
3. $(S_1, S_2, \ldots, S_\ell)$ is called a path if $(S_i, S_{i+1}) \in \mathscr{E}, i = 1, 2, \ldots, \ell - 1$.
4. A path $(S_1, S_2, \ldots)$ is called a cycle if $S_{i+\ell} = S_i$ for all i. The smallest ℓ is called the length of the cycle. In particular, the cycle of length 1 is called a fixed point.

Taking the properties of $\mathscr{J}^s$ into consideration and recalling the argument for the fixed points and cycles of a free Boolean network (without control) in Chap. 5, the following result is obvious.

Theorem 16.4 *Consider the system* (16.1) *with input-state incidence matrix* $\mathscr{J}$.

1. *The number of the fixed points in the input-state dynamic graph is*

$$N_1 = \sum_{i=1}^{2^m} \mathrm{tr}\big(\mathrm{Blk}_i(\mathscr{J}_0)\big) = \mathrm{tr}(M). \tag{16.30}$$

2. *The number of length-s cycles can be calculated inductively as*

$$N_s = \frac{\mathrm{tr}(M^s) - \sum_{k \in \mathscr{P}(s)} k N_k}{s}, \quad 2 \leq s \leq 2^{m+n}. \tag{16.31}$$

We use an example to illustrate this.

Example 16.7 Recall Example 16.1. We can calculate that

$$\mathrm{tr}\, M = 3, \qquad \mathrm{tr}\, M^3 = 6,$$
$$\mathrm{tr}\, M^4 = 15, \qquad \mathrm{tr}\, M^5 = 33,$$
$$\mathrm{tr}\, M^6 = 66, \qquad \mathrm{tr}\, M^7 = 129,$$
$$\mathrm{tr}\, M^8 = 255.$$

Using Theorem 16.4, we conclude that $N_1 = 3$, $N_3 = 1$, $N_4 = 3$, $N_5 = 6$, $N_6 = 10$, $N_7 = 18$, and $N_8 = 30$. It is not easy to convert them from the graph directly.

16.6 Mix-valued Logical Systems

In the multivalued logic case, say in a k-valued logical network, we have $x_i, u_i \in \mathscr{D}_k$ [1]. When the infinitely repeated game is considered, the dynamics of the strategies, depending on one history, may be expressed as in (16.1), but $x_i \in \mathscr{D}_{k_i}$ and $u_\alpha \in \mathscr{D}_{j_\alpha}$. Such a dynamic system is called a mix-valued logical dynamical system. We refer to Sect. 14.7 for a detailed discussion of mix-valued logic.

Set

$$x = \ltimes_{i=1}^{n} x_i \in \prod_{i=1}^{n} \mathscr{D}_{k_i}, \qquad u = \ltimes_{\alpha=1}^{m} u_\alpha \in \prod_{\alpha=1}^{m} \mathscr{D}_{j_\alpha}.$$

In vector form, we then have $x_i \in \Delta_{k_i}$ and $u_\alpha \in \Delta_{j_\alpha}$. Setting $k = \prod_{i=1}^{n} k_i$ and $j = \prod_{\alpha=1}^{m} j_\alpha$, we have

$$x \in \Delta_k, \qquad u \in \Delta_j.$$

In this section, we claim that all the major results obtained in previous sections remain true for mix-valued logical dynamical systems (including multivalued logical control networks as a particular case). We state this as a theorem and omit the proofs since they are identical.

Theorem 16.5 *Consider the system* (16.1) *and assume that it is a mix-valued logical dynamical system, where* $x_i \in \mathscr{D}_{k_i}$, $i = 1, \ldots, n$, $u_\alpha \in \mathscr{D}_{j_\alpha}$, $\alpha = 1, \ldots, m$, *and* $y_\beta \in \mathscr{D}_{\ell_\beta}$, $\beta = 1, \ldots, p$. *That is, each state* x_i, *control* u_α, *and output* y_β *can have different dimensions. We then have the following generalizations:*

1. *Considering the controllability of this mix-valued logical dynamical system, Theorem* 16.2 *and Corollary* 16.3 *remain true.*
2. *Considering the observability of this mix-valued logical dynamical system, Theorem* 16.3 *(equivalently, Corollary* 16.4*) remains true.*
3. *Considering the number of fixed points and the number of cycles of this mixvalued logical dynamical system, Theorem* 16.4 *remains true.*
4. *Considering the trajectories and corresponding controls, Algorithm* 16.1 *remains available.*

To apply the extended results technically, we need to solve the problem of how to calculate $\{x_i\}$ from x and vice versa. Similarly, we also have to calculate $\{u_i\}$ from u and vice versa. We give the following formula.

Proposition 16.4 *Let* $x_i = \delta_{k_i}^{\alpha_i}$, $i = 1, \ldots, n$, *and* $x = \delta_k^\alpha$. *Then:*

1.

$$\alpha = (\alpha_1 - 1) \times \frac{k}{k_1} + (\alpha_2 - 1) \times \frac{k}{k_1 k_2} + \cdots + (\alpha_{n-1} - 1) \times k_n + \alpha_n. \quad (16.32)$$

2.

$$\begin{cases} x_1 = (I_{k/k_1} \otimes 1_{k_1}^{\mathrm{T}})x, \\ x_j = (I_{k/k_j} \otimes 1_{k_j}^{\mathrm{T}})W_{[\prod_{i=1}^{j-1} k_i, k_j]}x. \end{cases} \quad (16.33)$$

Proof Equation (16.32) can be proven via a straightforward computation. The first equality in (16.33) comes from the definition of the semi-tensor product. To prove the second one, we have

$$W_{[\prod_{i=1}^{j-1} k_i, k_j]} x = x_j x_1 \cdots x_{j-1} x_{j+1} \cdots x_n.$$

Applying the first equality to it yields the second equality. □

We use the following example to demonstrate all the extended results in Theorem 16.5.

Example 16.8 Consider the mix-valued dynamical system

$$\begin{cases} x_1(t+1) = f_1(u(t), x_1(t), x_2(t)), \\ x_2(t+1) = f_2(u(t), x_1(t), x_2(t)), \\ y(t) = h(x_1(t), x_2(t)), \end{cases} \tag{16.34}$$

where $x_1(t) \in \mathscr{D}_2$, $x_2(t) \in \mathscr{D}_3$, $u(t) \in \mathscr{D}_2$, $f_1 : \mathscr{D}_2^2 \times \mathscr{D}_3 \to \mathscr{D}_2$, $f_2 : \mathscr{D}_2^2 \times \mathscr{D}_3 \to \mathscr{D}_3$, and $h : \mathscr{D}_2 \times \mathscr{D}_3 \to \mathscr{D}_2$ are mix-valued logical functions.

Using vector form, the system (16.34) can be expressed as

$$\begin{cases} x_1(t+1) = M_1 u(t) x_1(t) x_2(t), \\ x_2(t+1) = M_2 u(t) x_1(t) x_2(t), \quad x_1, u \in \Delta_2, \; x_2 \in \Delta_3, \\ y(t) = H x_1(t) x_2(t), \quad y \in \Delta_2. \end{cases} \tag{16.35}$$

In fact, in the mix-valued case, describing a logical function is not easy. In general it should be described by a truth table. We refer to Sect. 14.7 for a detailed discussion of the logical expression of mix-valued logical systems. In general, we use structure matrices to represent the functions directly. We assume the structure matrices of f_1, f_2, and h are M_1, M_2, and H, respectively, where

$$M_1 = \delta_2[1\,1\,1\,2\,1\,2\,2\,2\,2\,2\,2\,2],$$

$$M_2 = \delta_3[3\,1\,3\,2\,2\,1\,3\,2\,1\,3\,3\,3],$$

$$H = \delta_3[1\,3\,3\,2\,2\,2].$$

Setting $x(t) = x_1(t) x_2(t)$, the algebraic form of (16.34) can be calculated as

$$\begin{cases} x(t+1) = L u(t) x(t), \\ y(t) = H x(t), \end{cases} \tag{16.36}$$

where

$$L = \delta_6[3\,1\,3\,5\,2\,4\,6\,5\,4\,6\,6\,6].$$

1. Consider the controllability of the system. The basic block of the input-state incidence matrix $\mathscr{J}_0$ equals L. From a straightforward computation, the con-

trollability matrix is

$$\mathcal{M_C} = \sum_{s=1}^{12} {}_{\mathcal{B}} M^{(s)} = \begin{bmatrix} 1 & 1 & 1 & 1 & 1 & 1 \\ 1 & 1 & 1 & 1 & 1 & 1 \\ 1 & 1 & 1 & 1 & 1 & 1 \\ 1 & 1 & 1 & 1 & 1 & 1 \\ 1 & 1 & 1 & 1 & 1 & 1 \\ 1 & 1 & 1 & 1 & 1 & 1 \end{bmatrix} > 0.$$

We conclude that the system (16.34) is controllable.

2. Given any two points, say $x_0 = \delta_6^1 \sim (\delta_2^1, d_3^1)$ and $x_d = \delta_6^5 \sim (\delta_2^2, \delta_3^2)$, we want to find a trajectory from x_0 to x_d with proper controls.

- Step 1: The smallest s is 3 for

$$\left[\text{Blk}_1 \ \mathcal{J}_0^3 \right]_{51} > 0,$$

so $u(0) = \delta_2^1$, $x(3) = \delta_6^5$.
- Step 2: We have

$$\left[\text{Blk}_1 \ \mathcal{J}_0 \right]_{54} > 0, \qquad \left[\text{Blk}_1 \ \mathcal{J}_0^{3-1} \right]_{41} > 0,$$

so $u(2) = \delta_2^1$, $x(2) = \delta_6^4$. Then,

$$[\text{Blk}_2 \ \mathcal{J}_0]_{43} > 0, \qquad \left[\text{Blk}_1 \ \mathcal{J}_0^{2-1} \right]_{31} > 0,$$

so $u(1) = \delta_2^2$, $x(1) = \delta_6^3$.
- Step 3: $s - 1 = 1$, and we stop the process.

Hence, the control sequence which drives $x_0 = \delta_6^1$ to $x_d = \delta_6^5$ is $\{u(0) = \delta_2^1, u(1) = \delta_2^2, u(2) = \delta_2^1\}$, and the trajectory is $\{x(0) = \delta_6^1, x(1) = \delta_6^3, x(2) = \delta_6^4, x(3) = \delta_6^5\}$.

3. Next, we calculate the number of fixed points and the numbers of cycles of different lengths. It is easy to calculate that

$$\text{tr } M = 2, \qquad \text{tr } M^2 = 6,$$
$$\text{tr } M^3 = 8, \qquad \text{tr } M^4 = 14,$$
$$\text{tr } M^5 = 37, \qquad \text{tr } M^6 = 60,$$
$$\text{tr } M^7 = 135, \qquad \text{tr } M^8 = 254,$$
$$\text{tr } M^9 = 512, \qquad \text{tr } M^{10} = 1031,$$
$$\text{tr } M^{11} = 2037, \qquad \text{tr } M^{12} = 4112.$$

We conclude that there are $N_1 = 2$ fixed points and N_i cycles of length i, $i = 2, 3, 4, 5, 6, 7, 8, 9, 10, 11$, where $N_2 = 2$, $N_3 = 2$, $N_4 = 2$, $N_5 = 7$, $N_6 = 8$, $N_7 = 19$, $N_8 = 30$, $N_9 = 56$, $N_{10} = 99$, $N_{11} = 185$, and $N_{12} = 337$.

4. Finally, we consider the observability of the system. Let

$$O_{ij} = \bigvee_{s=1}^{2\times 6} [(H \ltimes \mathrm{Blk}_i(\tilde{\mathscr{J}}_0^{(s)})) \; \tilde{\vee} \; (H \ltimes \mathrm{Blk}_j(\tilde{\mathscr{J}}_0^{(s)}))].$$

A straightforward computation yields

$$O_{12} = \begin{bmatrix} 1 & 1 \\ 1 & 0 \\ 1 & 0 \end{bmatrix}, \qquad O_{13} = \begin{bmatrix} 0 & 0 \\ 0 & 1 \\ 0 & 1 \end{bmatrix}, \qquad O_{14} = \begin{bmatrix} 1 & 0 \\ 1 & 0 \\ 1 & 0 \end{bmatrix},$$

$$O_{15} = \begin{bmatrix} 1 & 0 \\ 1 & 0 \\ 1 & 0 \end{bmatrix}, \qquad O_{16} = \begin{bmatrix} 0 & 0 \\ 1 & 0 \\ 1 & 0 \end{bmatrix}, \qquad O_{23} = \begin{bmatrix} 1 & 1 \\ 1 & 1 \\ 1 & 1 \end{bmatrix},$$

$$O_{24} = \begin{bmatrix} 1 & 1 \\ 1 & 0 \\ 0 & 0 \end{bmatrix}, \qquad O_{25} = \begin{bmatrix} 1 & 1 \\ 0 & 0 \\ 1 & 0 \end{bmatrix}, \qquad O_{26} = \begin{bmatrix} 1 & 1 \\ 0 & 0 \\ 1 & 0 \end{bmatrix},$$

$$O_{34} = \begin{bmatrix} 1 & 0 \\ 1 & 1 \\ 1 & 1 \end{bmatrix}, \qquad O_{35} = \begin{bmatrix} 1 & 0 \\ 1 & 1 \\ 1 & 1 \end{bmatrix}, \qquad O_{36} = \begin{bmatrix} 0 & 0 \\ 1 & 1 \\ 1 & 1 \end{bmatrix},$$

$$O_{45} = \begin{bmatrix} 1 & 0 \\ 1 & 0 \\ 1 & 0 \end{bmatrix}, \qquad O_{46} = \begin{bmatrix} 1 & 0 \\ 1 & 0 \\ 1 & 0 \end{bmatrix}, \qquad O_{56} = \begin{bmatrix} 1 & 0 \\ 0 & 0 \\ 0 & 0 \end{bmatrix}.$$

Hence,

$$\bigwedge_{1 \le i < j \le 6} wb(O_{i,j}) = 1.$$

According to Theorem 16.3 or Corollary 16.4, the system is observable.

Finally, we compare the new controllability result with the corresponding result in Chap. 9. The main results in Chap. 9 for free sequences of controls are Theorem 9.3 and Corollary 9.2. Roughly speaking, they claim that the reachable set from x_0 is

$$R(x_0) = \mathrm{Col}\left\{\bigcup_{i=1}^{2^n} \tilde{L}^i x_0\right\}. \tag{16.37}$$

Note that by the properties of semi-tensor product, $\tilde{L}^s \in \mathscr{L}_{2^n \times 2^{n+sm}}$. So, when the step s is not small enough, the size of $\tilde{L}^s$ will be too large to be calculated in a memory-restricted computer. However, the main result in this chapter requires that $\mathscr{J}_0^{(s)}$ is checked. Since $\mathscr{J}_0^{(s)} \in \mathscr{L}_{2^n \times 2^m}$, $\forall s$, it is always easily computable (as long as the first step is computable).

References

1. Li, Z., Cheng, D.: Algebraic approach to dynamics of multi-valued networks. Int. J. Bifurc. Chaos **20**(3), 561–582 (2010)
2. Mu, Y., Guo, L.: Optimization and identification in a non-equilibrium dynamic game. In: Proc. CDC-CCC'09, pp. 5750–5755 (2009)
3. Zhao, Y., Qi, H., Cheng, D.: Input-state incidence matrix of Boolean control networks and its applications. Syst. Control Lett. (2010). doi:10.1016/j.sysconle.2010.09.002

Chapter 17
Identification of Boolean Control Networks

17.1 What Is Identification?

Consider the Boolean control network

$$
\begin{cases}
x_1(t+1) = f_1(x_1(t), \ldots, x_n(t), u_1(t), \ldots, u_m(t)), \\
x_2(t+1) = f_2(x_1(t), \ldots, x_n(t), u_1(t), \ldots, u_m(t)), \\
\vdots \\
x_n(t+1) = f_n(x_1(t), \ldots, x_n(t), u_1(t), \ldots, u_m(t)), \\
y_j(t) = h_j(x_1(t), \ldots, x_n(t)), \quad j = 1, \ldots, p,
\end{cases}
\tag{17.1}
$$

where $x_i(t)$, $u_k(t)$, $y_j(t) \in \mathcal{D}$, $i = 1, \ldots, n$, $k = 1, \ldots, m$, $j = 1, \ldots, p$, are states, inputs (controls), and outputs respectively, and $f_i : \mathcal{D}^{n+m} \to \mathcal{D}^n$, $h_j : \mathcal{D}^n \to \mathcal{D}^p$ are logical functions.

The identification problem is stated as follows.

Definition 17.1 Assume we have a Boolean control network with dynamic structure (17.1). The identification problem involves finding the functions f_i, $i = 1, \ldots, n$, and h_j, $j = 1, \ldots, p$, via certain input–output data $\{U(0), U(1), \ldots\}$, $\{Y(0), Y(1), \ldots\}$. The identification problem is said to be solvable if f_i and h_j can be uniquely determined by using proper inputs $\{U(0), U(1), \ldots\}$.

Note that here we use the following notation: $X(t) := (x_1(t), x_2(t), \ldots, x_n(t))$, $Y(t) := (y_1(t), y_2(t), \ldots, y_p(t))$, and $U(t) := (u_1(t), u_2(t), \ldots, u_m(t))$.

Remark 17.1 From Chap. 10 we know that different models may realize the same input–output mapping, so we may not be able to obtain unique f_i's and h_i's. Most likely, we are only interested in the equivalence classes which realize the same input–output mapping. Therefore the "uniqueness" should be clearly stated.

D. Cheng et al., *Analysis and Control of Boolean Networks*,
Communications and Control Engineering,
DOI 10.1007/978-0-85729-097-7_17, © Springer-Verlag London Limited 2011

By identifying $1 \sim \delta_2^1$ and $0 \sim \delta_2^2$, the algebraic form of (17.1) is obtained as

$$
\begin{cases}
x(t+1) = Lu(t)x(t), \\
y(t) = Hx(t).
\end{cases}
\tag{17.2}
$$

Note that $X = (x_1, \ldots, x_n)$ and $x = \ltimes_{i=1}^n x_i$ are in one-to-one correspondence and can be easily converted from one form to the other. Similarly, Y and y (U and u) are in one-to-one correspondence. Therefore, (17.1) and (17.2) are equivalent, and hence identifying f_i, $i = 1, \ldots, n$, and h_j, $j = 1, \ldots, p$, is equivalent to identifying L and H.

Model construction for a Boolean network was discussed in Chap. 7, where a pure Boolean network without inputs and outputs was investigated. This chapter considers a Boolean network with inputs and outputs.

17.2 Identification via Input-State Data

In this section, we assume that the state is measurable. Alternatively, we may make the following assumption.

Assumption 1 $p = n$ and $y_i(t) = x_i(t)$, $i = 1, \ldots, n$.

Recall the following definition from Chap. 9.

Definition 17.2 The system (17.1) is controllable if, for any initial state $X_0 = (x_1(0), \ldots, x_n(0)) \in \mathscr{D}^n$ and destination state X_d, there is a sequence of controls $U_0, U_1, \ldots$, where $U_t = (u_1(t), \ldots, u_m(t))$, such that the trajectory $X(t, X_0, U)$ satisfies $X(0, X_0, U) = X_0$, and $X(s, X_0, U) = X_d$ for some $s > 0$.

For identifiability from input-state data, we have the following result.

Theorem 17.1 *The system (17.1) is input-state identifiable if it is controllable.*

Proof (Sufficiency) Since the system is controllable, for any $x_d = \delta_{2^n}^i$ we can find a set of controls such that at time s, $x(s) = \delta_{2^n}^i$. Now, to identify $\mathrm{Col}_k(L)$, we can find a unique pair (i, j) such that

$$
\delta_{2^m}^j \delta_{2^n}^i = \delta_{2^{n+m}}^k.
$$

In fact, $i = k \% 2^m$ and $j = \frac{k-i}{2^m} + 1$. Hence, we can first choose control $u_0, u_1, \ldots$ to drive the system to $x(s) = \delta_{2^n}^i$ at a certain time $s > 0$, and then choose $u(s) = \delta_{2^m}^j$. It follows that

$$
\mathrm{Col}_k(L) = x(s+1).
$$

(Necessity) Split $\tilde{L} = LW_{[2^n, 2^m]}$ into 2^n equal-sized blocks as $\tilde{L} = [\tilde{L}_1, \tilde{L}_2, \ldots, \tilde{L}_{2^n}]$. If we now assume that $\delta_{2^n}^i$ is not reachable, then the columns of $\mathrm{Blk}_i(\tilde{L}) = \tilde{L}_i$ can never be shown in the state $x(t)$, $t = 1, 2, \ldots$. Thus, $\mathrm{Blk}_i(\tilde{L})$ is not identifiable. $\qquad\square$

It is now apparent that controllability is the key for identifiability. We refer to Chap. 9 for the necessary and sufficient condition for controllability, and to Chap. 16 for an alternative condition.

Remark 17.2

1. Assume we have enough proper input data $\{U_0, U_1, \ldots, U_T\}$ and the corresponding state data $\{X_0, X_1, \ldots, X_T\}$ such that (in the set-theoretical sense)

$$\{U_0 \times X_0, U_1 \times X_1, \ldots, U_{T-1} \times X_{T-1}\} = \mathscr{D}^{n+m}. \qquad (17.3)$$

L can then be identified in the following way: if, in vector form, $u_i x_i = \delta_{2^{m+n}}^j$, then $\mathrm{Col}_j(L) = x_{i+1}$.
2. Similarly, if we know $\{X_0, X_1, \ldots, X_T\}$ and the corresponding $\{Y_0, Y_1, \ldots, Y_T\}$, such that

$$\{X_0, X_1, \ldots, X_T\} = \mathscr{D}^n, \qquad (17.4)$$

then in vector form $x_i = \delta_{2^n}^j$ implies $\mathrm{Col}_j(H) = y_i$.

Example 17.1 Consider the following Boolean control network:

$$\begin{cases} x_1(t+1) = \neg x_1(t) \vee x_2(t), \\ x_2(t+1) = u(t) \wedge \neg x_1(t) \vee (\neg u(t) \wedge x_1(t) \wedge \neg x_2(t)). \end{cases} \qquad (17.5)$$

Setting $x(t) = \ltimes_{i=1}^2 x_i(t)$, we have

$$x(t+1) = Lu(t)x(t), \qquad (17.6)$$

where

$$L = \delta_4[2\,4\,1\,1\,2\,3\,2\,2].$$

For the system (17.5), the basic block of its input-state incidence matrix $\mathscr{J}_0$ equals L.

Checking the controllability matrix, we have

$$\mathscr{M}_{\mathscr{C}} = \sum_{s=1}^{2^3} M^{(s)} = \begin{bmatrix} 1 & 1 & 1 & 1 \\ 1 & 1 & 1 & 1 \\ 1 & 1 & 1 & 1 \\ 1 & 1 & 1 & 1 \end{bmatrix} > 0.$$

We conclude that the system is identifiable.

Table 17.1 Input-state data

t	0	1	2	3	4	5	6	7	8	9	10	11	12	13	14	15	16	17	18	19
$u(t)$	1	1	1	1	1	2	1	1	1	1	2	1	2	1	1	2	1	2	2	1
$x(t)$	1	2	4	1	2	4	2	4	1	2	4	2	4	2	4	1	2	4	2	3

We can choose a sequence of controls and the initial state randomly, and the sequence of states can be determined. First, we choose 20 controls: the control-state data are given in Table 17.1.

Here, the number i in $u(t)$ [resp., $x(t)$] refers to δ_2^i (resp., δ_4^i). We can obtain L as

$$L = \delta_4[2\ 4 * 1\ 2\ 3 * 2].$$

Some columns of L are not identified because not all input-states $\delta_{2^{m+n}}^j$ are reached by the randomly chosen sequence of control. One way to deal with this problem is to choose a long sequence of input-state data. For example, if we randomly choose a sequence of 100 controls (or even more), then L could be identified.

From the example, we know that any length of input sequence chosen randomly cannot ensure that L can be identified, although the probability is very close to 1 if the sequence is long enough. In fact, we can design an input sequence

$$u(t) = \begin{cases} \delta_{2^m}^i, & \exists s \text{ such that } x(s) = x(t), u(s) = \delta_{2^m}^{i-1}, \forall s < t' < t, x(t') \neq x(t), \\ \delta_{2^m}^1, & \text{otherwise,} \end{cases}$$

$$(17.7)$$

where, when $x(t)$ enters a cycle, we stop the process. We then have the following result.

Theorem 17.2 *If the system* (17.1) *is identifiable, then the logical functions* f_i *can be determined uniquely by the inputs designed in* (17.7).

Proof The state under the input sequence must enter a cycle. Thus, for $\delta_{2^n}^i$ in the cycle, there must exist $t_1 < t_2 < \cdots < t_{2^m}$ such that $x(t_j) = \delta_{2^n}^i$, $u(t_j) = \delta_{2^m}^j$. Hence, none of the states in this cycle can reach the state outside the cycle by changing the control. If the cycle does not contain all the states, then the system is not controllable. $\qquad\square$

Example 17.2 Recall Example 17.1. Using (17.7) we can obtain the input-state data given in Table 17.2.

L can then be identified as

$$L = \delta_4[2\ 4\ 1\ 1\ 2\ 3\ 2\ 2].$$

Table 17.2 Input-state data

t	0	1	2	3	4	5	6	7	8	9	10	11
$u(t)$	1	1	1	2	2	1	1	1	2	2	2	1
$x(t)$	1	2	4	1	2	3	1	2	4	2	3	2

17.3 Identification via Input–Output Data

Definition 17.3 The system (17.1) is observable if there is a sequence of controls $\{U_0, U_1, \ldots\}$, where $U_t = (u_1(t), \ldots, u_m(t))$, such that the initial state X_0 can be determined by the outputs $\{Y_0, Y_1, \ldots\}$.

We introduce the following assumption.

Assumption 2 The system is controllable.

For identification via input–output data we need the observability condition. We refer to Chap. 9 for the necessary and sufficient conditions. In the following we give an alternative condition. This is basically the same as the one in Chap. 9, but it is convenient for identifying x_0.

Split L into 2^m equal-sized blocks as

$$L = \left[\text{Blk}_1(L), \text{Blk}_2(L), \ldots, \text{Blk}_{2^m}(L)\right] := [B_1, B_2, \ldots, B_{2^m}],$$

where $B_i \in \mathscr{L}_{2^n \times 2^n}$, $i = 1, \ldots, 2^m$.

Define a sequence of sets of matrices $\Gamma_i \in \mathscr{L}_2^p \times 2^n$, $i = 0, 1, 2, \ldots$, as follows:

$$
\begin{cases}
\Gamma_0 = \{H\}, \\
\Gamma_1 = \{HB_i \,|\, i = 1, 2, \ldots, 2^m\}, \\
\Gamma_2 = \{HB_i B_j \,\big|\, i, j = 1, 2, \ldots, 2^m\}, \\
\vdots \\
\Gamma_s = \{HB_{i_1} B_{i_2} \cdots B_{i_s} \,|\, i_1, i_2, \ldots, i_s = 1, 2, \ldots, 2^m\}, \\
\vdots
\end{cases}
\tag{17.8}
$$

Note that $\Gamma_s \subset \mathscr{L}_{2^p \times 2^n}$, $\forall s$. It is then easy to prove the following result.

Lemma 17.1

1. *There exists an $s^* > 0$ such that*

$$\Gamma_{s^*+1} \subset \bigcup_{k=1}^{s^*} \Gamma_k. \tag{17.9}$$

2. *Let $s^* > 0$ be the smallest positive integer such that (17.9) holds. Then,*

$$\Gamma_j \subset \bigcup_{k=1}^{s^*} \Gamma_k, \quad \forall j > s^*. \tag{17.10}$$

For notational ease, we also use Γ_s to denote the matrix consisting of its elements arranged in a column. For instance,

$$\Gamma_1 = \begin{bmatrix} HB_1 \\ HB_2 \\ \vdots \\ HB_{2^m} \end{bmatrix}, \qquad \Gamma_2 = \begin{bmatrix} HB_1B_1 \\ HB_1B_2 \\ \vdots \\ HB_{2^m}B_{2^m} \end{bmatrix}, \dots$$

Using these, we construct a matrix, called the observability matrix:

$$\mathcal{M}_{\mathcal{O}} = \begin{bmatrix} \Gamma_0 \\ \Gamma_1 \\ \vdots \\ \Gamma_{s^*} \end{bmatrix}. \tag{17.11}$$

We then have the following theorem.

Theorem 17.3 *Assume that the system (17.1) is controllable. It is then observable if and only if*

$$\text{rank}\,(\mathcal{M}_{\mathcal{O}}) = 2^n. \tag{17.12}$$

Proof Let the initial state be x_0. Since the system is controllable we can find a time sequence $\{t_i^1 \mid i = 1, 2, \dots, n\}$ satisfying

$$t_{i+1}^1 > t_i^1 + 1, \quad i = 0, 1, \dots, 2^m - 1, \ t_0^1 := 0,$$

such that $x(t_i^1) = x_0$. Using $u(t_i^1) = \delta_{2^m}^i, i = 1, \dots, 2^m$, it is easy to see that

$$\begin{bmatrix} y(t_1^1 + 1) \\ y(t_2^1 + 1) \\ \vdots \\ y(t_{2^m}^1 + 1) \end{bmatrix} = \Gamma_1 x_0.$$

In general, assume we have a time sequence $t_{i_1 i_2 \cdots i_s}^s$, $i_k = 1, 2, \dots, 2^m$, $k = 1, \dots, s$, and we convert the multi-index $(i_1\, i_2 \cdots i_s)$ to a single index $\mu(i_1\, i_2 \cdots i_s)$ in "alphabetical order". That is,

$$\mu(1\,1 \cdots 1) = 1, \qquad \mu(1\,1 \cdots 2) = 2, \dots, \qquad \mu\big(2^m\, 2^m \cdots 2^m\big) = 2^{sm}.$$

We assume that this time sequence satisfies

$$t_{i+1}^s > t_i^s + s, \quad i = 0, 1, \dots, 2^{sm} - 1, \ t_0^s := 0.$$

Now, assume (using proper controls) that

$$x\left(t_{i_1 \, i_2 \, \cdots \, i_s}^s\right) = x_0$$

and define a sequence of controls as

$$u\left(t_{\mu(i_1 \, i_2 \, \cdots \, i_s)}^s\right) = \delta_{2^m}^{i_1}, \qquad u\left(t_{\mu(i_1 \, i_2 \, \cdots \, i_s)}^s + 1\right) = \delta_{2^m}^{i_2}, \dots,$$
$$u\left(t_{\mu(i_1 \, i_2 \, \cdots \, i_s)}^s + s - 1\right) = \delta_{2^m}^{i_s}.$$

We then have

$$\begin{bmatrix} y(t_1^s + s) \\ y(t_2^s + s) \\ \vdots \\ y(t_{2^{sm}}^s + s) \end{bmatrix} = \Gamma_s x_0.$$

Note that we assume the sets of time sequences to be sufficiently far separated from each other. Precisely,

$$t_{\underbrace{1 \, 1 \, \cdots \, 1}_{s}}^s > \left(t_{\underbrace{2^m \, 2^m \, \cdots \, 2^m}_{s-1}}^{s-1}\right) + s.$$

Finally, we have

$$\mathcal{M}_{\mathcal{O}} x_0 = \begin{bmatrix} y(0) \\ y(t_1^1 + 1) \\ y(t_2^1 + 1) \\ \vdots \\ y(t_{2^m}^1 + 1) \\ \vdots \\ y(t_1^{s^*} + s^*) \\ y(t_2^{s^*} + s^*) \\ \vdots \\ y(t_{2^{sm}}^{s^*} + s^*) \end{bmatrix} := Y. \tag{17.13}$$

Note that when Y is a set of observed data, x_0 can be uniquely solved as

$$x_0 = \left(\mathcal{M}_{\mathcal{O}}^{\mathrm{T}} \mathcal{M}_{\mathcal{O}}\right)^{-1} \mathcal{M}_{\mathcal{O}}^{\mathrm{T}} Y. \tag{17.14}$$

(Necessity) By the definition of s^* it is easy to see that Y contains all possible outputs. Now, if $\mathrm{rank}(\mathcal{M}_{\mathcal{O}}) < 2^n$, then one sees easily that in addition to x_0, there exists at least one other solution x_0' of (17.13). Then, x_0 and x_0' are not distinguishable.   □

We are now ready to present our main result.

Theorem 17.4 *The system* (17.1) *is identifiable from input–output data with proper controls if and only if the system is controllable and observable.*

Proof (Necessity) In fact, from Theorems 17.1 and 17.3, the necessity in obvious because if the system is not observable, then it is impossible to identify all the states from outputs. If the system is not controllable, then it is impossible to identify L from input-state data.

(Sufficiency) Because the system is controllable, we can assume that we first construct enough input data $\{U_0, U_1, \ldots, U_{T_1}\}$, this generates the corresponding $\{X_0, X_1, \ldots, X_{T_1}\}$, and these collectively satisfy (17.3). Then, by controllability, there exist controls $\{U_t \mid T_1 < t \leq T_2\}$ such that $X_{T_2} = X_0$. Since the system is observable, there exist controls $\{U_t \mid T_2 < t \leq T_3\}$ and $X_{T_2} = X_0$ can be identified by using this control sequence. We can then choose $\{U_t \mid T_3 < t \leq T_4 - 2\}$ and $U_{T_4-1} = U_0$ such that $X_{T_4-1} = X_0$. We know that $X_{T_4} = X_1$ and this can be then be identified by choosing proper controls. Continuing this process, $\{X_0, X_1, \ldots, X_{T_1}\}$ can eventually be identified. Now, using the identified $\{X_i \mid i = 0, 1, \ldots, T_1\}$ and the input data $\{U_i \mid i = 0, 1, \ldots, T_1\}$, we can identify L according to Theorem 17.1. It is easy to see that (17.3) implies (17.4). Then, using $\{X_i \mid i = 0, 1, \ldots, T_1\}$ and $\{Y_i \mid i = 0, 1, \ldots, T_1\}$, we can identify H.   □

17.4 Numerical Solutions

17.4.1 General Algorithm

Consider the system (17.1) again. Assume that we have a coordinate transformation $z = Tx$, where $T \in \mathscr{L}_{2^n \times 2^n}$. Its algebraic form (17.2) then becomes

$$\begin{cases} z(t+1) = \tilde{L}u(t)z(t), \\ y(t) = \tilde{H}z(t), \end{cases} \tag{17.15}$$

where

$$\tilde{L} = TL(I_{2^m} \otimes T^{\mathrm{T}}), \qquad \tilde{H} = HT^{\mathrm{T}}. \tag{17.16}$$

It is obvious that $\{L, H\}$ and $\{\tilde{L}, \tilde{H}\}$ are not distinguishable by any input–output data. Is this a counterexample to Theorem 17.4? In fact, when we state that a system is observable, we implicitly assume that the coordinate x is fixed. Otherwise, it would be impossible to identify x_0 from input–output data. So, precisely speaking, we should say either (i) assume the coordinate frame x is fixed, then Theorem 17.4 holds, or (ii) $\{L, H\}$ are identifiable up to a coordinate transformation.

Keeping this in mind, what we are going to identify is the equivalence class, but not a particular $\{L, H\}$. We then have the following lemma.

Lemma 17.2 *Without loss of generality, we can assume the initial value is fixed, say, $x_0 = \delta_{2^n}^1$.*

Proof Assume that we have a realization $\{L, H\}$ with initial value $x_0 = \delta_{2^n}^i$. Under a coordinate transformation $z = Tx$, with

$$\text{Col}_i(T) = \delta_{2^n}^1,$$

we then have $z_0 = \delta_{2^n}^1$. $\qquad\square$

To evaluate the error, we need a distance.

Definition 17.4 Let $A, B \in \mathscr{B}_{m \times n}$. The distance between A and B, denoted by $d(A, B)$, is then defined as

$$d(A, B) = \frac{1}{2} \sum_{i=1}^{m} \sum_{j=1}^{n} [a_{ij} \,\bar{\vee}\, b_{ij}]. \tag{17.17}$$

Remark 17.3 Let $A, B \in \mathscr{L}_{p \times q}$. It is then easy to see that $d(A, B)$ is the number of different columns of A and B. This is why we introduce the coefficient $\frac{1}{2}$ into the definition.

Assume that we have the input data $\{U_t \mid t = 0, 1, \ldots, T\}$ and the corresponding output data $\{Y_t \mid t = 0, 1, \ldots, T\}$. For each pair $\{L, H\}$ we define the error as follows.

Definition 17.5 Assume the input data $\{U_t \mid t = 0, 1, \ldots, T\}$ and the corresponding output data $\{Y_t \mid t = 0, 1, \ldots, T\}$ are given. For a given (L, H), using initial $x_0 = \delta_{2^n}^1$ and the input data $\{U_t \mid t = 0, 1, \ldots, T\}$, the estimated output data can be calculated as $\{\hat{Y}_t \mid t = 1, \ldots, T\}$. The error is then defined as

$$\varepsilon(L, H) = \sum_{t=1}^{T} d(\hat{Y}_t, Y_t). \tag{17.18}$$

Next, we define a neighborhood of (L, H).

Definition 17.6 Let $r \in \mathbb{Z}_+$. The neighborhood of (L_0, H_0), denoted by $B_r(L_0, H_0)$, is defined as

$$B_r(L_0, H_0) := \left\{ (L, H) \mid d(L, L_0) \le r, d(H, H_0) \le r \right\}. \tag{17.19}$$

We are now ready to present our main algorithm.

Algorithm 17.1 Assume a set of input data $\{U_t \,|\, t = 0, 1, \ldots, T\}$ and the corresponding output data $\{Y_t \,|\, t = 0, 1, \ldots, T\}$ are given.

Step 0. Set $S = \{(L, H) \,|\, L \in \mathscr{L}_{2^n \times 2^{n+m}}; \; H \in \mathscr{L}_{2^p \times 2^n}\}$, $r = r_0$ (default: $r_0 = 1$), and $\varepsilon_{\min} = \infty$.

Step 1. Choose an $(L_0, H_0) \in S$.

   (i) If $\varepsilon(L_0, H_0) = 0$, set $(L^*, H^*) = (L_0, H_0)$ and terminate the algorithm (the solution is obtained).
   (ii) Otherwise, set $\varepsilon_0 := \varepsilon(L_0, H_0)$ and proceed to the next step.

Step 2. Over the neighborhood $B_r(L_0, H_0)$ find a point (L^*, H^*) such that

$$\varepsilon^* = \varepsilon\big(L^*, H^*\big) = \min_{(L,H) \in B_r(L_0, H_0) \cap S} \varepsilon(L, H).$$

   (i) If $\varepsilon^* = 0$, set $(L^*, H^*) = (L_0, H_0)$ and terminate the algorithm (the solution is obtained).
   (ii) Otherwise, if $\varepsilon^* < \varepsilon_0$, replace (L_0, H_0) by (L^*, H^*) and S by $S := S \setminus \{B_r(L_0, H_0)\}$, then return to Step 2.
   (iii) Otherwise, if $\varepsilon_0 < \varepsilon_{\min}$, replace $\varepsilon_{\min}$ by ε_0 and return to Step 1.

Theorem 17.5 *Algorithm* 17.1 *will terminate at a certain step, where a solution* (L^*, H^*) *with* $\varepsilon(L^*, H^*) = 0$ *is provided.*

Proof Since at each iteration the error ε is strictly decreasing, and there are finitely many $\{L, H\}$, the conclusion is obvious.   □

Remark 17.4 If the data contain errors, then we can only obtain an optimal solution (L^*, H^*), satisfying $\varepsilon(L^*, H^*) = \varepsilon_{\min}$.

Example 17.3 Recalling Example 17.1, add the output as

$$y(t) = x_1 \,\bar{\vee}\, x_2.$$

The algebraic form of the Boolean control system is then

$$\begin{cases} x(t + 1) = Lu(t)x(t), \\ y(t) = Hx(t), \end{cases} \tag{17.20}$$

where

$$L = \delta_4[2\,4\,1\,1\,2\,3\,2\,2]$$

and

$$H = \delta_2[2\,1\,1\,2].$$

A straightforward computation shows that the observability matrix is

$$
\mathcal{M}_{\mathcal{O}} = \begin{bmatrix} H \\ HB_1 \\ HB_2 \\ HB_1B_1 \\ \vdots \end{bmatrix} = \begin{bmatrix} 0 & 1 & 1 & 0 \\ 1 & 0 & 0 & 1 \\ 1 & 0 & 0 & 0 \\ 0 & 1 & 1 & 1 \\ 1 & 1 & 1 & 1 \\ 0 & 0 & 0 & 0 \\ 0 & 0 & 1 & 1 \\ 1 & 1 & 0 & 0 \\ \vdots & & & \end{bmatrix}.
$$

From this part of $\mathcal{M}_{\mathcal{O}}$, it is apparent that all of its columns are different, so rank($\mathcal{M}_{\mathcal{O}}$) = 4. Thus, the system is observable and hence identifiable.

Using (17.20) and setting $x_0 = \delta_4^1$, we randomly choose 50 inputs and calculate 50 corresponding outputs as follows:

$$u(t): \quad 1\,2\,1\,2\,1\,2\,1\,2\,2\,2\,1\,2\,2\,2\,2\,2\,1$$
$$1\,1\,2\,2\,1\,2\,2\,1\,2\,2\,1\,2\,2\,2\,2\,2\,2$$
$$1\,2\,1\,1\,2\,2\,1\,1\,2\,1\,1\,2\,1\,1\,1\,2,$$

$$y(t): \quad 2\,1\,1\,2\,1\,2\,1\,2\,1\,1\,1\,2\,1\,1\,1\,1\,1$$
$$2\,2\,1\,1\,1\,2\,1\,1\,2\,1\,1\,2\,1\,1\,1\,1\,1$$
$$1\,2\,1\,2\,2\,1\,1\,2\,1\,1\,2\,1\,1\,2\,1\,2.$$

Using Algorithm 17.1, this terminates at

$$
\begin{cases} \hat{L} = \delta_4[4\,1\,1\,2\,4\,4\,4\,3], \\ \hat{H} = \delta_2[2\,2\,1\,1]. \end{cases}
\tag{17.21}
$$

It seems that it is quite different from the original system (17.20), but note that if we set

$$
x(t) = Tz(t) = \begin{bmatrix} 1 & 0 & 0 & 0 \\ 0 & 0 & 0 & 1 \\ 0 & 0 & 1 & 0 \\ 0 & 1 & 0 & 0 \end{bmatrix} z(t),
$$

which is a coordinate transform, then the system (17.20) can be converted to

$$
\begin{cases} z(t+1) = \tilde{L}u(t)z(t), \\ y(t) = \tilde{H}z(t), \end{cases}
$$

where

$$\tilde{L} = T^{-1}L(I_2 \otimes T) = \delta_4[4\,1\,1\,2\,4\,4\,4\,3] = \hat{L},$$

$$\tilde{H} = HT = \delta_2[2\,2\,1\,1] = \hat{H}.$$

Thus, the system which we identified is equivalent to the original system.

17.4.2 Numerical Solution Based on Network Graph

In practice, for an n-node network, the in-degree of each node is usually much less than n. In this case, the number of candidate structure matrices L can be reduced tremendously. In this subsection we assume that the network graph is known and consider how to identify the system. Note that since the graph is fixed, a coordinate transformation which changes the graph is not allowed. Therefore we have to take the initial state x_0 into consideration.

Example 17.4 Consider the following system:

$$\begin{cases} x_1(t+1) = f_1(x_1(t), x_2(t), u_1(t), u_2(t)), \\ x_2(t+1) = f_2(x_1(t), x_2(t), u_1(t), u_2(t)), \\ y_1(t) = h_1(x_1(t), x_2(t)), \\ y_1(t) = h_2(x_1(t), x_2(t)). \end{cases} \tag{17.22}$$

Assume the observed data are as follows:

$$u(t):\quad \begin{matrix} 4\,1\,3\,2\,4\,4\,2\,1\,4\,2\,3\,4\,4\,3\,1\,2\,4\,4\,2\,4 \\ 1\,2\,4\,1\,1\,1\,1\,3\,2\,1\,1\,3\,2\,4\,2\,2\,4\,3\,1\,3 \\ 4\,1\,3\,2\,4\,3\,3\,2\,2\,1\,1\,3\,2\,3\,1\,3\,2\,4\,4\,3 \\ 2\,4\,4\,3\,4\,3\,2\,2\,2\,3\,3\,2\,4\,3\,2\,3\,3\,2\,3\,3 \\ 4\,4\,3\,4\,1\,4\,2\,2\,4\,3\,1\,1\,4\,1\,2\,3\,2\,2\,1\,4, \end{matrix}$$

$$y(t):\quad \begin{matrix} 2\,3\,4\,4\,1\,3\,3\,3\,4\,3\,3\,4\,3\,3\,4\,2\,1\,3\,3\,3 \\ 3\,4\,1\,3\,4\,2\,2\,2\,4\,1\,4\,2\,4\,1\,3\,3\,3\,3\,4\,2 \\ 4\,3\,4\,4\,1\,3\,4\,4\,1\,3\,4\,2\,4\,1\,4\,2\,4\,1\,3\,3 \\ 4\,1\,3\,3\,4\,3\,4\,1\,3\,3\,4\,4\,1\,3\,4\,1\,4\,4\,1\,4 \\ 4\,3\,3\,4\,3\,4\,3\,3\,3\,4\,2\,2\,3\,4\,1\,4\,1\,3\,4. \end{matrix}$$

If the network graph of the system is known, as in Fig. 17.1, then we can infer that the algebraic form of the system is

$$\begin{cases} x_1(t+1) = L_1 u_1(t) x_2(t), \\ x_2(t+1) = L_2 u_2(t), \\ y_1(t+1) = H_1 x_1(t), \\ y_2(t+1) = H_2 x_2(t), \end{cases}$$

Fig. 17.1 Network graph

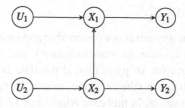

where $L_1 \in \mathscr{L}_{2\times4}$, $L_2, H_1, H_2 \in \mathscr{L}_{2\times2}$. Replacing the neighborhood (17.19) by

$$B_r\left(L_0^i, H_0^j, x_{00}\right)$$

$$= \left\{\left(L^i, H^j, x_0\right) \mid d\left(L^i, L_0^i\right) \le r, \ d\left(H^j, H_0^j\right) \le r, \ d(x_0, x_{00}) \le 1, \right.$$

$$\left. i = 1, 2, \ j = 1, 2\right\},$$

Algorithm 17.1 remains applicable. The system is then identified as

$$\begin{cases} \hat{L}_1 = \delta_2[2\ 1\ 1\ 1], \\ \hat{L}_2 = \delta_2[1\ 2], \\ \hat{H}_1 = \delta_2[2\ 1], \\ \hat{H}_2 = \delta_2[2\ 1], \\ \hat{x}_0 = \delta_4^3. \end{cases} \tag{17.23}$$

In fact, the data are generated from the system

$$\begin{cases} x_1(t+1) = u_1(t) \wedge x_2(t), \\ x_2(t+1) = u_2(t), \\ y_1(t) = x_1(t), \\ y_2(t) = \neg x_2(t), \end{cases} \tag{17.24}$$

with the initial state $x_0 = \delta_4^1$.

Let $x(t) = x_1(t)x_2(t)$, $u(t) = u_1(t)u_2(t)$, $y(t) = y_1(t)y_2(t)$. Its algebraic form is

$$\begin{aligned} x(t+1) &= Lx(t), \\ y(t) &= Hx(t), \end{aligned} \tag{17.25}$$

where

$$L = \delta_4[1\ 3\ 1\ 3\ 2\ 4\ 2\ 4\ 3\ 3\ 3\ 3\ 4\ 4\ 4\ 4],$$
$$H = \delta_4[2\ 1\ 4\ 3].$$

Note that if we set $\tilde{x}_1(t) = \neg x_1(t)$ and $\tilde{x}_2(t) = x_2(t)$, then it is easy to see that (17.23) is equivalent to the original system (17.24) via this coordinate transformation. Moveover, this coordinate transformation does not change the network graph.

Remark 17.5

1. When the system is not observable and/or controllable, there is no unique (L, H) (up to a coordinate transformation), but this does not mean that the data have no realization. In particular, if the data are from a real Boolean network, then one realization (the real one) exists. In fact, it only means there is more than one realization. In this case Algorithm 17.1 remains applicable. Therefore, when applying this algorithm we do not need to worry about whether the system is controllable and/or observable.
2. If the data contain some errors, the algorithm is still useful. In this case we may preassign an acceptable error level such that the algorithm terminates when the true error reaches this.

Example 17.5 Consider the following system:

$$
\begin{cases}
x_1(t+1) = u_1(t) \wedge x_2(t), \\
x_2(t+1) = \neg x_1(t), \\
x_3(t+1) = x_1(t) \leftrightarrow x_2(t), \\
x_4(t+1) = \neg u_2(t), \\
y_1(t) = x_2(t) \, \bar{\vee} \, x_3(t), \\
y_2(t) = x_4(t).
\end{cases}
\tag{17.26}
$$

Let $x(t) = x_1(t)x_2(t)x_3(t)x_4(t)$, $u(t) = u_1(t)u_2(t)$, $y(t) = y_1(t)y_2(t)$. Its algebraic form is then

$$
\begin{aligned}
x(t+1) &= Lx(t), \\
y(t) &= Hx(t),
\end{aligned}
\tag{17.27}
$$

where

$$
\begin{aligned}
L = \delta_{16}[\, &6\ 6\ \ 6\ \ 6\ \ \ 16\ 16\ 16\ 16\ 4\ \ 4\ \ 4\ \ 4\ \ \ 10\ 10\ 10\ 10 \\
&5\ 5\ \ 5\ \ 5\ \ \ 15\ 15\ 15\ 15\ 3\ \ 3\ \ 3\ \ 3\ \ \ 9\ 9\ 9\ 9 \\
&14\ 14\ 14\ 14\ 16\ 16\ 16\ 16\ 11\ 11\ 11\ 11\ 10\ 10\ 10\ 10 \\
&13\ 13\ 13\ 13\ 15\ 15\ 15\ 15\ 11\ 11\ 11\ 11\ 9\ \ 9\ \ 9\ \ 9\,], \\
H = \delta_4[\, &3\ 4\ \ 1\ \ 2\ \ 1\ \ 2\ \ 3\ \ 4\ \ 3\ \ 4\ \ 1\ \ 2\ \ 1\ \ 2\ \ 3\ \ 4\,].
\end{aligned}
$$

It is obvious that the system is not controllable, thus it is not uniquely identifiable. However, Algorithm 17.1 remains applicable.

Let a set of input–output data be generated from (17.26) as

$$
\begin{aligned}
u(t): \quad &1\ 3\ 3\ 2\ 3\ 4\ 2\ 2\ 2\ 2\ 2\ 3\ 3\ 2\ 1\ 2\ 2\ 1\ 3 \\
&3\ 2\ 2\ 3\ 4\ 4\ 2\ 4\ 2\ 4\ 4\ 1\ 2\ 3\ 3\ 3\ 2\ 2\ 1\ 2 \\
&2\ 2\ 4\ 4\ 4\ 3\ 1\ 3\ 4\ 4\ 1\ 3\ 3\ 4\ 3\ 1\ 2\ 1\ 4\ 2 \\
&1\ 4\ 4\ 2\ 2\ 4\ 1\ 4\ 3\ 4\ 1\ 2\ 1\ 4\ 3\ 2\ 2\ 4\ 2\ 1 \\
&1\ 1\ 3\ 1\ 3\ 1\ 4\ 2\ 4\ 1\ 3\ 4\ 2\ 4\ 1\ 1\ 1\ 4\ 1\ 2,
\end{aligned}
$$

$$y(t): \quad 3\ 2\ 4\ 4\ 1\ 2\ 3\ 1\ 1\ 3\ 3\ 1\ 1\ 4\ 4\ 1\ 2\ 3\ 3\ 2$$
$$2\ 4\ 1\ 1\ 4\ 3\ 1\ 1\ 1\ 3\ 1\ 1\ 2\ 1\ 4\ 4\ 1\ 1\ 1\ 4$$
$$3\ 1\ 1\ 3\ 3\ 1\ 1\ 2\ 2\ 3\ 1\ 2\ 2\ 4\ 1\ 1\ 2\ 1\ 4\ 3$$
$$1\ 2\ 3\ 3\ 1\ 1\ 3\ 4\ 1\ 1\ 1\ 2\ 1\ 4\ 3\ 1\ 1\ 1\ 3\ 3$$
$$2\ 2\ 4\ 4\ 2\ 2\ 4\ 1\ 1\ 1\ 4\ 1\ 1\ 1\ 1\ 4\ 2\ 2\ 3\ 4.$$

Let the initial L_0, H_0 be (to avoid computational complexity, we choose the initial structure close enough to the real network to see whether the algorithm works)

$$L_0 = \delta_{16}[\ 7\ 6\ \ 6\ \ 6\ \ 16\ 16\ 16\ 16\ 4\ \ 4\ \ 4\ \ 4\ \ 10\ 10\ 10\ 10$$
$$5\ 5\ \ 5\ \ 5\ \ 15\ 15\ 15\ 15\ 3\ \ 3\ \ 8\ \ 3\ \ 9\ \ 9\ \ 9\ \ 9$$
$$14\ 14\ 14\ 14\ 16\ 16\ 16\ 16\ 11\ 11\ 11\ 11\ 10\ 10\ 10\ 10$$
$$13\ 13\ 13\ 13\ 15\ 15\ 15\ 15\ 11\ 11\ 11\ 11\ 9\ \ 9\ \ 9\ \ 9],$$
$$H_0 = \delta_4[4\ 4\ \ 1\ \ 2\ \ 1\ \ 2\ \ 3\ \ 4\ \ 3\ \ 1\ \ 1\ \ 2\ \ 1\ \ 2\ \ 3\ \ 4].$$

The algorithm then terminates at the second step at

$$\hat{L} = \delta_{16}[\ 6\ 6\ \ 6\ \ 6\ \ 16\ 16\ 16\ 16\ 4\ \ 4\ \ 4\ \ 4\ \ 10\ 10\ 10\ 10$$
$$5\ 5\ \ 5\ \ 5\ \ 15\ 15\ 15\ 15\ 3\ \ 3\ \ 3\ \ 3\ \ 9\ \ 9\ \ 9\ \ 9$$
$$14\ 14\ 14\ 14\ 16\ 16\ 16\ 16\ 11\ 11\ 11\ 11\ 10\ 10\ 10\ 10$$
$$13\ 13\ 13\ 13\ 15\ 15\ 15\ 15\ 11\ 11\ 11\ 11\ 9\ \ 9\ \ 9\ \ 9],$$
$$\hat{H} = \delta_4[4\ 4\ \ 1\ \ 2\ \ 1\ \ 2\ \ 2\ \ 4\ \ 3\ \ 4\ \ 1\ \ 2\ \ 1\ \ 2\ \ 3\ \ 4].$$

Note that $\hat{L} = L$, but $\hat{H}$ is different from H. Since the system is not uniquely identifiable, we here have another realization.

17.4.3 Identification of Higher-Order Systems

A μth order Boolean control network is defined as follows:

$$\begin{cases} x_1(t+1) = f_1(x_1(t), \ldots, x_n(t), u_1(t), \ldots, u_m(t), \ldots, x_1(t-\mu+1), \ldots, \\ \qquad x_n(t-\mu+1), u_1(t-\mu+1), \ldots, u_m(t-\mu+1)), \\ x_2(t+1) = f_2(x_1(t), \ldots, x_n(t), u_1(t), \ldots, u_m(t)), \ldots, x_1(t-\mu+1), \ldots, \\ \qquad x_n(t-\mu+1), u_1(t-\mu+1), \ldots, u_m(t-\mu+1), \\ \vdots \\ x_n(t+1) = f_n(x_1(t), \ldots, x_n(t), u_1(t), \ldots, u_m(t), \ldots, x_1(t-\mu+1), \ldots, \\ \qquad x_n(t-\mu+1), u_1(t-\mu+1), \ldots, u_m(t-\mu+1)), \\ y_j(t) = h_j(x_1(t), \ldots, x_n(t)), \quad j = 1, \ldots, p. \end{cases}$$

$$(17.28)$$

Consider the identification of the system (17.28). Basically, Algorithm 17.1 is still applicable, provided some observations are introduced. These are:

1. Note that

$$L \in \mathscr{L}_{2^n \times 2^{(n+m)\mu}}, \qquad H \in \mathscr{L}_{2^p \times 2^n}.$$

2. Considering the initial values $\{x(0), x(1), \ldots, x(\mu - 1)\}$, we cannot simply assume that all of them are $\delta_{2^n}^1$ because this cannot represent all possible initial values, even under a coordinate transformation. Let

$$[0, \mu - 1] = [\mu_0 = 0, \mu_1) \cup [\mu_1, \mu_2) \cup \cdots \cup [\mu_{2^n-1}, \mu_{2^n} = \mu - 1] \quad (17.29)$$

be a partition with nonincreasing length, where μ_i, $i = 0, 1, \ldots, 2^n$, are nonnegative integers, that is,

$$\mu_i - \mu_{i-1} \geq \mu_{i+1} - \mu_i, \quad i = 1, \ldots, 2^n - 1. \tag{17.30}$$

For each partition (17.29) satisfying (17.30), we assign the initial values as follows:

$$x(i) = \delta_{2^n}^j, \quad \mu_{j-1} \leq i < \mu_j, \ i = 0, 1, \ldots, \mu - 1.$$

It is easy to check that this covers all possible assignments of initial values under a coordinate transformation. For instance, if we assume $\mu = 2$, then we have two different partitions:

$$[0, 1) \cup [1, 2) \cup \emptyset \cup \cdots, \qquad [0, 2) \cup \emptyset \cup \cdots.$$

The corresponding initial-value assignments are

$$\begin{cases} x(0) = \delta_{2^n}^1, \\ x(1) = \delta_{2^n}^2, \end{cases} \quad \text{and} \quad \begin{cases} x(0) = \delta_{2^n}^1, \\ x(1) = \delta_{2^n}^1. \end{cases}$$

17.5 Approximate Identification

Assume that we have a large Boolean network and are particularly interested in a certain function. We may then consider approximating the network with a simpler model. For instance, consider a large Boolean network as in Fig. 17.2. We may

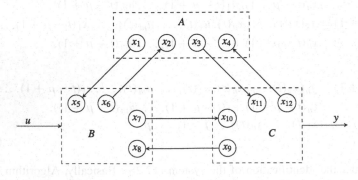

Fig. 17.2 A partitioned network

split it into several parts. Suppose we have three parts, A, B, and C. Inside each part, the nodes are strongly connected and in between parts, the connections are assumed to be weak. We may now ignore the interior nodes and focus solely on the frontier nodes which are related to other parts. Suppose we have frontier nodes $\{x_1, x_2, \ldots, x_{12}\}$. We can then approximate the original network by (17.31) and try to identify these nodes.

$$
\begin{cases}
x_1(t+1) = f_1(x_1(t), x_2(t), x_3(t), x_4(t)), \\
x_2(t+1) = f_2(x_1(t), x_2(t), x_3(t), x_4(t), x_6(t)), \\
x_3(t+1) = f_3(x_1(t), x_2(t), x_3(t), x_4(t)), \\
x_4(t+1) = f_4(x_1(t), x_2(t), x_3(t), x_4(t), x_{12}(t)), \\
x_5(t+1) = f_5(x_1(t), x_5(t), x_6(t), x_7(t), x_8(t), u(t)), \\
x_6(t+1) = f_6(x_5(t), x_6(t), x_7(t), x_8(t), u(t)), \\
x_7(t+1) = f_7(x_5(t), x_6(t), x_7(t), x_8(t), u(t)), \\
x_8(t+1) = f_8(x_1(t), x_5(t), x_6(t), x_7(t), x_8(t), x_9(t), u(t)), \\
x_9(t+1) = f_9(x_9(t), x_{10}(t), x_{11}(t), x_{12}(t)), \\
x_{10}(t+1) = f_{10}(x_7(t), x_9(t), x_{10}(t), x_{11}(t), x_{12}(t)), \\
x_{11}(t+1) = f_{11}(x_3(t), x_9(t), x_{10}(t), x_{11}(t), x_{12}(t)), \\
x_{12}(t+1) = f_{12}(x_9(t), x_{10}(t), x_{11}(t), x_{12}(t)), \\
y(t) = h(x_9(t), x_{10}(t), x_{11}(t), x_{12}(t)).
\end{cases}
\tag{17.31}
$$

We give a numerical example to illustrate this.

Example 17.6 Consider the following system:

$$
\begin{cases}
x_1(t+1) = \neg x_2(t) \wedge x_3(t) \vee u(t), \\
x_2(t+1) = \neg x_3(t) \wedge x_4(t) \vee u(t), \\
\quad \vdots \\
x_{99}(t+1) = \neg x_{100}(t) \wedge x_1(t) \vee u(t), \\
x_{100}(t+1) = \neg x_1(t) \wedge x_2(t) \vee u(t), \\
\\
z_1(t+1) = z_2(t) \wedge \neg z_3(t), \\
z_2(t+1) = z_3(t) \wedge \neg z_4(t), \\
\quad \vdots \\
z_{99}(t+1) = z_{100}(t) \wedge \neg z_1(t), \\
z_{100}(t+1) = x_{100}(t) \wedge \neg z_1(t), \\
y(t) = z_1(t) \wedge \neg z_5(t) \neg z_{100}(t).
\end{cases}
\tag{17.32}
$$

For example, we choose $\xi_1(t) = x_{100}(t)$ and $\xi_2(t) = z_{100}(t)$. We can then construct an approximate system as follows:

$$\begin{cases} \xi_1(t+1) = f_1(\xi_1(t), u(t)), \\ \xi_2(t+1) = f_2(\xi_1(t), \xi_2(t)), \\ y(t) = h(\xi_2(t)). \end{cases} \qquad (17.33)$$

Its algebraic form is

$$\begin{cases} \xi_1(t+1) = L1\xi_1(t)u(t), \\ \xi_2(t+1) = L2\xi_1(t)\xi_2(t), \\ y(t) = H\xi_2(t). \end{cases}$$

Randomly choose a sequence of 100 input data and let $x_i(0) = z_i(0) = \delta_2^2$. Using the dynamics (17.32) yields 100 data as follows:

$u(t)$:   2 2 2 1 1 1 1 1 2 1 1 2 2 1 1 2 1 2 1 2
             2 1 2 2 2 1 2 2 2 1 1 2 1 1 2 1 1 1 2 1
             1 2 1 1 2 2 2 1 2 2 1 1 1 1 2 2 2 2 1 1
             2 1 1 1 2 1 2 1 2 1 2 1 1 1 2 1 1 2 1 1
             2 1 1 2 1 1 2 1 1 2 1 2 2 2 1 2 1 1 1 1,

$y(t)$:   2 2 2 2 1 1 1 1 1 2 1 1 2 2 1 1 2 1 2 1
             2 2 1 2 2 2 1 2 2 2 1 1 2 1 1 2 1 1 1 2
             1 1 2 1 1 2 2 2 1 2 2 1 1 1 1 2 2 2 2 1
             1 2 1 1 1 2 1 2 1 2 1 2 1 1 1 2 1 1 2 1
             1 2 1 1 2 1 1 2 1 1 2 1 2 2 2 1 2 1 1 1.

L_1, L_2, H can then be identified as

$$\begin{cases} L_1 = \delta_2[1\ 2\ 1\ 1], \\ L_2 = \delta_2[2\ 1\ 1\ 1], \\ H = \delta_2[1\ 2], \end{cases}$$

with $\xi_1(0) = \delta_2^1$, $\xi_2(0) = \delta_2^2$, and the smallest error being 29.

We change the input sequence 100 times to see its effect. The errors are as follows:

32 31 36 36 29 27 27 29 29 28 30 34 25 24 25 26 27 30 29 25
26 36 33 24 32 28 32 32 31 28 21 33 33 24 25 24 35 29 28 21
29 26 30 23 23 20 30 30 27 29 26 25 25 27 26 28 20 28 35 23
27 23 27 24 26 25 28 30 27 31 28 27 27 27 28 28 27 36 23 29
28 27 27 32 23 28 29 28 28 26 26 27 27 30 29 25 27 26 27 24.

It is easily seen that the minimum of the 100 times is 20, while the maximum is 36, and the average is 27.76.

References

1. Cheng, D., Zhao, Y.: Identification of Boolean control networks. Automatica (2010, provisionally accepted)

Chapter 18
Applications to Game Theory

18.1 Strategies with Finite Memory

To make the objectives of this chapter clear, we give a rigorous definition for the games which we will consider

Definition 18.1

(1) A static game G consists of three components: (i) n players, named $A_1, \ldots, A_n$, (ii) k_i possible actions for each player A_i, denoted by $x_i \in \mathscr{D}_{k_i}$, $i = 1, \ldots n$, (iii) n payoff functions for the n players, given respectively by

$$c_j(x_1 = i_1, \ldots, x_n = i_n) = c^j_{i_1 i_2 \cdots i_n}, \quad j = 1, \ldots, n. \tag{18.1}$$

(2) A set of actions of all players, denoted by $s = (x_1, \ldots, x_n)$, is called a pure strategy of G, and the set of pure strategies is denoted by S.

Throughout this chapter only pure strategies are considered and so the word "pure" is omitted. Using the above definition, we can define the infinitely repeated version of game G, denoted by G_∞.

Definition 18.2 Consider the game G described in Definition 18.1.

(1) A strategy of the infinitely repeated game G_∞ is

$$s_\infty = \{s_1, \ldots, s_n\} \in S_\infty,$$

where the sequence $s_j = (s_j(0), s_j(1), \ldots)$ is called the strategy of player A_j, in which $s_j(t) = x_j(t)$, called the action of A_j at time t, is a function of the first t historical actions of A_i, $i = 1, \ldots, n$.
(2) A set of logical dynamic functions which determine the strategies of the infinitely repeated game is called a generator of the strategies. More precisely, we

D. Cheng et al., *Analysis and Control of Boolean Networks*, Communications and Control Engineering, DOI 10.1007/978-0-85729-097-7_18, © Springer-Verlag London Limited 2011

have

$$x_j(t+1) = f_j^{t+1}(x_1(t), \ldots, x_n(t), \ldots, x_1(0), \ldots, x_n(0)), \quad j = 1, \ldots, n,$$

(18.2)

where

$$f_j^{t+1} : \underbrace{(\mathscr{D}_{k_1} \times \cdots \times \mathscr{D}_{k_n}) \cdots (\mathscr{D}_{k_1} \times \cdots \times \mathscr{D}_{k_n})}_{t+1} \to \mathscr{D}_{k_j}, \quad j = 1, \ldots, n.$$

For convenience, (18.2) is sometimes also called a strategy.

(3) The payoff functions for G are assumed to be

$$J_j^0 = c_j(x_1, \ldots, x_n), \quad j = 1, \ldots, n.$$

(18.3)

The corresponding payoff functions for G_∞ are the average payoffs defined as [6, 7]

$$J_j = \varlimsup_{T \to \infty} \frac{1}{T} \sum_{t=1}^{T} c_j(x_1(t), \ldots, x_n(t)), \quad j = 1, \ldots, n.$$

(18.4)

(4) A strategy for G_∞ is called a zero-memory strategy if

$$x_j(t+1) = f_j(t+1) = f_j^{t+1}, \quad j = 1, \ldots, n.$$

(18.5)

(5) A strategy for G_∞ is called a μ-memory strategy, with $\mu > 0$, if its generators are

$$x_j(t+1) = f_j(x_1(t), \ldots, x_n(t), \ldots, x_1(t-\mu+1), \ldots, x_n(t-\mu+1)),$$
$$j = 1, \ldots, n,$$

(18.6)

combined with the initial conditions

$$x_j(t_0) = x_{t_0}^j, \quad j = 1, \ldots, n, \ t_0 = 0, 1, \ldots, \mu - 1.$$

Remark 18.1

1. For the zero-memory strategy defined by (18.5), the action of each player could be an arbitrary infinity-time sequence $\{x_j(0), x_j(1), \ldots\}$, $j = 1, \ldots, n$.
2. For a μ-memory strategy, the generating functions f_j are assumed to be time-invariant. If f_j is time-varying, then the value $x_j(t)$ is entirely arbitrary and the strategy degenerates to the zero-memory case.

Throughout this chapter we consider only the μ-memory case where $0 \leq \mu < \infty$.

Using vector form, it is easily seen that, as in Proposition 14.1, we can find a unique $L_j \in \mathscr{L}_{k_j \times k^\mu}$ such that

$$x_j(t+1) = L_j \ltimes_{i=0}^{\mu-1} x(t-i), \quad j = 1, \ldots, n.$$

(18.7)

Equivalently, we can find a unique $L \in \mathscr{L}_{k \times k^\mu}$ such that

$$x(t+1) = L \ltimes_{i=0}^{\mu-1} x(t-i). \tag{18.8}$$

In Chap. 5 (Sect. 5.6) and in [5] it was shown that the topological structure of (18.6) can be obtained by investigating the topological structure of (18.8). We describe a brief constructive process.

Note that $x = \ltimes_{i=1}^{n} x_i$ can uniquely determine the x_i, $i = 1, \ldots, n$. We denote the natural projections by

$$x_i = \pi_i(x), \quad i = 1, \ldots, n.$$

The projections can be determined precisely by Proposition 16.4. That is, if we let $k^i = \prod_{j=1, j \neq i}^{n} k_j = \frac{k}{k_i}$, then we have

$$\begin{cases} x_1 = \pi_1(x) = (I_{k_1} \otimes \mathbf{1}_{k^1})x, \\ x_i = \pi_i(x) = (I_{k_i} \otimes \mathbf{1}_{k^i})W_{[k_1 \times k_2 \times \cdots k_{i-1}, k_i]}x, \quad i > 1. \end{cases} \tag{18.9}$$

We need to show that (18.7) and (18.8) are equivalent, that is, that there is a one-to-one correspondence between L and $\{L_1, \ldots, L_n\}$. We prove this by constructing the conversion formulas, which are themselves very useful.

Using the notation $\kappa_i = \prod_{j=i+1}^{n} k_j$, $i = 1, 2, \ldots, n-1$, we define a set of retrievers as

$$S_1 = I_{k_1} \otimes \mathbf{1}_{\kappa_1}^{\mathrm{T}},$$

$$S_2 = \underbrace{\left[I_{k_2} \otimes \mathbf{1}_{\kappa_2}^{\mathrm{T}}, \ldots, I_{k_2} \otimes \mathbf{1}_{\kappa_2}^{\mathrm{T}} \right]}_{k_1},$$

$$\vdots \tag{18.10}$$

$$S_{n-1} = \underbrace{\left[I_{k_{n-1}} \otimes \mathbf{1}_{\kappa_{n-1}}^{\mathrm{T}}, \ldots, I_{k_{n-1}} \otimes \mathbf{1}_{\kappa_{n-1}}^{\mathrm{T}} \right]}_{k_1 \times k_2 \times \cdots \times k_{n-2}},$$

$$S_n = \underbrace{\left[I_{k_n}, \ldots, I_{k_n} \right]}_{k_1 \times k_2 \times \cdots \times k_{n-1}}.$$

It is then easy to check the following proposition.

Proposition 18.1

1.

$$L = L_1 \prod_{j=2}^{n} [(I_k \otimes L_j)\Phi], \tag{18.11}$$

where

$$\Phi = \prod_{j=1}^{\mu} \left[I_{k^{j-1}} \otimes (I_k \otimes W_{[k,k^{\mu-j}]}) M_{r,k} \right].$$

2.

$$L_i = S_i L, \quad i = 1, 2, \ldots, n. \tag{18.12}$$

The proof of this proposition is similar to the proof of Theorem 7.1. We leave it to the reader. Next, we define the sub-Nash equilibrium for the game.

Definition 18.3 Let $S = \{ s^{\lambda} \mid \lambda \in \Lambda \}$ be a set of strategies of G and $\varepsilon_{\lambda_0} \geq 0$ the smallest nonnegative real number such that

$$c_j\left(x_1^{\lambda_0}, \ldots, x_n^{\lambda_0}\right) + \varepsilon_{\lambda_0} \geq c_j\left(x_1^{\lambda_0}, \ldots, x_j^{\lambda}, \ldots, x_n^{\lambda_0}\right), \quad \forall \lambda, \; j = 1, \ldots, n. \tag{18.13}$$

Then, ε_{λ_0} is called the tolerance of s^{λ_0}.

- If $\varepsilon_{\lambda_0} = 0$, then s^{λ_0} is called a Nash equilibrium.
- If

$$\varepsilon^0 = \min \{ \varepsilon_\lambda \mid \lambda \in \Lambda \} > 0,$$

then there is no Nash equilibrium, and if $\varepsilon_\lambda = \varepsilon^0$, then s^{λ} is called a sub-Nash equilibrium of S.
- For any strategy s^{λ}, if its tolerance is 0, then it is also called a Nash solution of the game; if its tolerance is $\varepsilon > 0$, then it is called an ε-tolerance solution of the game.

We can similarly define the Nash or sub-Nash equilibria and ε-tolerance solutions for G_∞ in this way, except that (18.13) is replaced by

$$J_j\left(s_1^{\lambda_0}, \ldots, s_n^{\lambda_0}\right) + \varepsilon_0 \geq J_j\left(s_1^{\lambda_0}, \ldots, s_j^{\lambda}, \ldots, s_n^{\lambda_0}\right), \quad j = 1, \ldots, n. \tag{18.14}$$

18.2 Cycle Strategy

Definition 18.4 A strategy for G_∞ is called a cycle strategy if it satisfies

$$x_j(t + T_0) = x_j(t), \quad j = 1, \ldots, n, \; \forall t \geq 0. \tag{18.15}$$

Equivalently, (18.15) can be expressed in product form, with $x = \ltimes_{i=1}^{n} x_i$, as

$$x(t + T_0) = x(t), \quad t \geq 0. \tag{18.16}$$

As a convention, in the sequel we always assume that the $T_0 > 0$ in (18.15) or (18.16) is the smallest one, i.e., the period of the strategy. Denote by C the set of cycle strategies. C is then a subset of strategies, i.e., $C \subset S$. Also, denote by C_α the set of cycle strategies with cycle length α, which equals T_0.

If a strategy satisfies (18.15) [or, equivalently, (18.16)], for $t \geq t_0$, with a fixed $t_0 > 0$, the payoff functions J_j, $j = 1, \ldots, n$, have the same values as $t_0 = 0$. We still consider it as a cycle strategy. In general, the possible strategies are infinite in number. We consider the μ-memory $(0 \leq \mu < \infty)$ Nash equilibrium only within C. In the following we will show that if there is a Nash solution, then there is a corresponding cycle strategy as the solution.

Consider a logical control network

$$
\begin{cases}
x_1(t+1) = f_1(x_1(t), \ldots, x_p(t), \ldots, x_1(t - \mu + 1), \ldots, x_p(t - \mu + 1), \\
\qquad\qquad u_1(t), \ldots, u_q(t), \ldots, u_1(t - \mu + 1), \ldots, u_q(t - \mu + 1)), \\
\;\vdots \\
x_p(t+1) = f_p(x_1(t), \ldots, x_p(t), \ldots, x_1(t - \mu + 1), \ldots, x_p(t - \mu + 1), \\
\qquad\qquad u_1(t), \ldots, u_q(t), \ldots, u_1(t - \mu + 1), \ldots, u_q(t - \mu + 1)).
\end{cases}
\tag{18.17}
$$

The performance criterion is

$$
J = \varlimsup_{T \to \infty} \frac{1}{T} \sum_{t=1}^{T} c(x_1(t), \ldots, x_p(t), u_1(t), \ldots, u_q(t)). \tag{18.18}
$$

The design purpose is to find an optimal control $u^*(t) = \ltimes_{i=1}^{q} u_i^*(t)$ such that

$$
J(u^*(t)) = \max_{u(t)} J(u(t)).
$$

It was proven in the last chapter that for any given initial value $\{x_1(0), \ldots, x_p(0), \ldots, x_1(\mu - 1), \ldots, x_p(\mu - 1)\}$, there exists an optimal control as described in the following proposition.

Proposition 18.2 *There exists an optimal control satisfying*

$$
\begin{cases}
u_1(t+1) = g_1(x_1(t), \ldots, x_p(t), \ldots, x_1(t - \mu + 1), \ldots, x_p(t - \mu + 1), \\
\qquad\qquad u_1(t), \ldots, u_q(t), \ldots, u_1(t - \mu + 1), \ldots, u_q(t - \mu + 1)), \\
\;\vdots \\
u_q(t+1) = g_q(x_1(t), \ldots, x_p(t), \ldots, x_1(t - \mu + 1), \ldots, x_p(t - \mu + 1), \\
\qquad\qquad u_1(t), \ldots, u_q(t), \ldots, u_1(t - \mu + 1), \ldots, u_q(t - \mu + 1)).
\end{cases}
\tag{18.19}
$$

Moreover, the optimal value can be obtained on a cycle trajectory starting from the given initial value.

Considering the μ-memory strategy (18.6), we propose an algorithm for solving the game.

Algorithm 18.1

Step 1. Assume that $\{L_1^0, \ldots, L_n^0\}$ are chosen. We consider $x_1(t)$ as a control for the logical control system

$$x_j(t+1) = f_j(x_1(t), \ldots, x_n(t), \ldots, x_1(t-\mu+1), \ldots, x_n(t-\mu+1)),$$

$$j = 2, 3, \ldots, n, \tag{18.20}$$

where f_j, $j = 2, \ldots, n$, are uniquely determined by L_j^0. Solving the optimal control problem with the given initial values $\{x_j^0 \mid j \neq 1\}$ and the performance criterion

$$J = J_1 = \overline{\lim_{T \to \infty}} \frac{1}{T} \sum_{t=1}^{T} c_1(x(t)),$$

the optimal solution $\{L_1^1, x_1^0\}$ can be obtained. (Note that for the same $\{f_j, x_j^0 \mid j \neq 1\}$ there may be more than one solution. We can choose any one of these in the first instance. However, at the next time when the same $\{f_j, x_j^0 \mid j \neq 1\}$ appear, we must choose the same one which was first chosen.) Using this L_1^1, we can uniquely determine a new f_1. We then replace the old f_1 by this new f_1 and take x_1^0 as the initial value.

Step 2. Assume $\{L_1^s, \ldots, L_{j-1}^s, L_{j+1}^{s-1}, \ldots, L_n^{s-1}\}$ with $\{x_i^0 \mid i \neq j\}$ are obtained. We consider the corresponding system

$$x_i(t+1) = f_i(x_1(t), \ldots, x_n(t), \ldots, x_1(t-\mu+1), \ldots, x_n(t-\mu+1), \quad i \neq j. \tag{18.21}$$

Solving the optimal control problem with

$$J = J_j = \overline{\lim_{T \to \infty}} \frac{1}{T} \sum_{t=1}^{T} c_j(x(t)),$$

the L_j^s and x_j^0 are obtained, and then f_j and x_j^0 are updated.

By definition, the following result is then obvious.

Theorem 18.1

1. *In Algorithm* 18.1, *assume that there exists a* k^* *such that*

$$L_i^{k^*} = L_i^{k^*+1}, \quad i = 1, \ldots, n. \tag{18.22}$$

   Then,

$$L_i = L_i^{k^*}, \quad i = 1, \ldots, n,$$

form a set of Nash equilibria. Moreover, the corresponding optimal solution can be chosen as a cycle strategy.

2. *If* $\{L_1^0, \ldots, L_n^0\}$ *with* $\{x_j^0 \mid j = 1, \ldots, n\}$ *is a μ-memory Nash equilibrium, then it is a fixed point of the algorithm.*

18.3 Compounded Games

Definition 18.5 Let G^s, $s = 1, 2$, be two games with $A_1^s, \ldots, A_n^s$ being the two groups of players. Each A_i^s has k_i^s possible actions. Their payoff functions are, respectively,

$$
\begin{aligned}
c_i^1\left(\delta_{k_1^1}^{\alpha_1}, \ldots, \delta_{k_n^1}^{\alpha_n}\right) &= c_{\alpha_1 \cdots \alpha_n}^i, \\
c_i^2\left(\delta_{k_1^2}^{\beta_1}, \ldots, \delta_{k_n^2}^{\beta_n}\right) &= d_{\beta_1 \cdots \beta_n}^i, \quad i = 1, 2, \ldots, n.
\end{aligned}
\tag{18.23}
$$

G is called a compounded game of G^1 and G^2 with weight $(\lambda, 1 - \lambda)$, $0 < \lambda < 1$, denoted by $G = G^1 \circ G^2$, if:

(i) G has n players, A_i, $i = 1, \ldots, n$.
(ii) Each A_i has $k_i^1 \times k_i^2$ possible actions, denoted by $\delta_{k_i^1}^p \times \delta_{k_i^2}^q$, $p = 1, \ldots, k_i^1$ and $q = 1, \ldots, k_i^2$.
(iii) Using (18.23), the payoff functions are

$$
c_i\left(\delta_{k_1^1}^{\alpha_1} \times \delta_{k_1^2}^{\beta_1}, \ldots, \delta_{k_n^1}^{\alpha_n} \times \delta_{k_n^2}^{\beta_n}\right) = \lambda c_{\alpha_1 \cdots \alpha_n}^i + (1 - \lambda) d_{\beta_1 \cdots \beta_n}^i.
\tag{18.24}
$$

In practical terms, a compounded game corresponds to a game between teams of players. We give an example to illustrate this.

Example 18.1 Assume that there are two games G^1 and G^2 with players $\{A, B\}$ and $\{C, D\}$, respectively. Their strategies and payoff functions are described via payoff bi-matrices in Tables 18.1 and 18.2, respectively.

Assume that A and C form a team to play against the team of B and D. The strategies and payoff functions of the compounded game $G = G^1 \circ G^2$ with weight $(0.5, 0.5)$ are then described in Table 18.3.

By definition the following result is obvious.

Table 18.1 Payoff bi-matrix of G^1	$A \backslash B$	1	2	3
	1	1, 1	1, −1	2, 1
	2	0, 1	1, 0	0, 0

Table 18.2 Payoff bi-matrix of G^2

$C \backslash D$	1	2
1	$-1, 1$	$1, 2$
2	$2, 1$	$1, -2$

Table 18.3 Payoff bi-matrix of $G^1 \circ G^2$

$A \circ C \backslash B \circ D$	1×1	1×2	2×1	2×2	3×1	3×2
1×1	$0, 1$	$1, 1.5$	$0, 0$	$1, 0.5$	$0.5, 1$	$1.5, 1.5$
1×2	$1.5, 1$	$1, -0.5$	$1.5, 0$	$1, -1.5$	$2, 1$	$1.5, -0.5$
2×1	$-0.5, 1$	$-0.5, 1$	$0, 0.5$	$1, 1$	$-0.5, 0.5$	$0.5, 1$
2×2	$1, 1$	$0.5, -0.5$	$1.5, 0.5$	$1, -1$	$1, 0.5$	$0.5, -1$

Proposition 18.3 *Let* $G = G^1 \circ G^2$. $\{\xi_1^*, \ldots, \xi_n^*\}$ *and* $\{\eta_1^*, \ldots, \eta_n^*\}$ *are Nash equilibria of* G^1 *and* G^2, *respectively, if and only if* $\{\xi_1^* \times \eta_1^*, \ldots, \xi_n^* \times \eta_n^*\}$ *is a Nash equilibrium of* G.

Proof (Necessity) Assume that $\{\xi_1^*, \ldots, \xi_n^*\}$ and $\{\eta_1^*, \ldots, \eta_n^*\}$ are Nash equilibria of G^1 and G^2, respectively. For $G = G^1 \circ G^2$, we then have the payoffs

$$
\begin{aligned}
& c_j\big(\xi_1^* \times \eta_1^*, \ldots, \xi_n^* \times \eta_n^*\big) \\
&= \lambda c_j^1\big(\xi_1^*, \ldots, \xi_n^*\big) + (1 - \lambda)c_j^2\big(\eta_1^*, \ldots, \eta_n^*\big) \\
&\geq \lambda c_j^1\big(\xi_1^*, \ldots, \xi_j, \ldots, \xi_n^*\big) + (1 - \lambda)c_j^2\big(\eta_1^*, \ldots, \eta_j, \ldots, \eta_n^*\big) \\
&= c_j\big(\xi_1^* \times \eta_1^*, \ldots, \xi_j \times \eta_j, \ldots, \xi_n^* \times \eta_n^*\big), \quad 1 \leq j \leq n.
\end{aligned}
$$

(Sufficiency) Assume that $\{\xi_1^* \times \eta_1^*, \ldots, \xi_n^* \times \eta_n^*\}$ is a Nash equilibrium of G. Then,

$$
c_j\big(\xi_1^* \times \eta_1^*, \ldots, \xi_j^* \times \eta_j^*, \ldots, \xi_n^* \times \eta_n^*\big) \geq c_j\big(\xi_1^* \times \eta_1^*, \ldots, \xi_j \times \eta_j^*, \ldots, \xi_n^* \times \eta_n^*\big),
$$

where $1 \leq j \leq n$. That is,

$$
\begin{aligned}
& \lambda c_j^1\big(\xi_1^*, \ldots, \xi_j^*, \ldots, \xi_n^*\big) + (1 - \lambda)c_j^2\big(\eta_1^*, \ldots, \eta_j^*, \ldots, \eta_n^*\big) \\
&\geq \lambda c_j^1\big(\xi_1^*, \ldots, \xi_j, \ldots, \xi_n^*\big) + (1 - \lambda)c_j^2\big(\eta_1^*, \ldots, \eta_j^*, \ldots, \eta_n^*\big).
\end{aligned}
$$

Hence,

$$
c_j^1\big(\xi_1^*, \ldots, \xi_j^*, \ldots, \xi_n^*\big) \geq c_j^1\big(\xi_1^*, \ldots, \xi_j, \ldots, \xi_n^*\big), \quad 1 \leq j \leq n.
$$

Similarly, we have

$$
c_j^2\big(\eta_1^*, \ldots, \eta_j^*, \ldots, \eta_n^*\big) \geq c_j^1\big(\eta_1^*, \ldots, \eta_j, \ldots, \eta_n^*\big), \quad 1 \leq j \leq n. \qquad \square
$$

An immediate consequence of this proposition is the following.

Corollary 18.1 *G has no Nash equilibrium if and only if any finitely repeated game* $\underbrace{G \circ G \circ \cdots \circ G}_{s}$ *has no Nash equilibrium.*

18.4 Sub-Nash Solution for Zero-Memory Strategies

Consider an infinitely repeated game G_∞ of G. If G has a Nash equilibrium, then according to Corollary 18.1, it, after being compounded finitely many times, is a Nash equilibrium for the corresponding finite-time repeated game, and there is no new zero-memory Nash equilibrium. So, for zero-memory strategies, this case is less interesting. For example, consider the prisoner's dilemma and assume its payoff bi-matrix is as in Table 18.4.

It is clear that $(2, 2)$ is a Nash equilibrium. Now, consider the infinitely repeated prisoner's dilemma: If it is formulated as a game between a machine and a human [6] and if only zero-memory strategies are considered, then the machine can simply choose the strategy of the Nash equilibrium, that is, choose action 2 forever. The human then has no option but simply to choose action 2. Note that this case is not true for the $\mu > 0$ memory case, where new a Nash equilibrium may appear, because in this case, the actions of different steps are no longer independent.

Therefore, for the zero-memory case, when considering the Nash solution, the interesting case is when the original payoff bi-matrix has no Nash equilibrium. However, we know that in this case we have no Nash equilibrium for a finitely repeated game, so we consider the sub-Nash solution.

We give a simple example to illustrate the sub-Nash solution.

Example 18.2 Consider a game G with two players, A and B. The payoff bi-matrix is given in Table 18.5.

It is obvious that there is no Nash equilibrium. It is easy to calculate that the tolerance of $(1, 1)$ and $(1, 2)$ is 2, and that the tolerance of $(2, 1)$ and $(2, 2)$ is 1. Hence, $(2, 1)$ and $(2, 2)$ are sub-Nash equilibria with tolerance 1 and corresponding

Table 18.4 Payoff bi-matrix

$P_1 \backslash P_2$	1	2
1	3, 3	0, 5
2	5, 0	1, 1

Table 18.5 Payoff bi-matrix of G

$A \backslash B$	1	2
1	2, 0	0, 2
2	1, 2	2, 1

Table 18.6 Payoff bi-matrix of $G \circ G$

$A \backslash B$	1×1	1×2	2×1	2×2
1×1	2, 0	1, 1	1, 1	0, 2
1×2	1.5, 1	2, 0.5	0.5, 2	1, 1.5
2×1	1.5, 1	0.5, 2	2, 0.5	1, 1.5
2×2	1, 2	1.5, 1.5	1.5, 1.5	2, 1

payoffs $\{p_A, p_B\} = \{1, 2\}$ and $\{2, 1\}$, respectively. It is very likely that A may not be satisfied with $(2, 1)$ and B may not be satisfied with $(2, 2)$.

Next, we may consider length-2 cycle zero-memory strategies. That is, we consider $G \circ G$, with $\lambda = 0.5$. The payoff bi-matrix is then as shown in Table 18.6. Now, $(2 \times 2, 1 \times 2)$ and $(2 \times 2, 2 \times 1)$ are the two sub-Nash equilibria with tolerance 0.5 and the same corresponding payoffs, $\{1.5, 1.5\}$. This payoff may be acceptable by both A and B. We argue this as follows: A may choose 1×2 to get a better payoff, but as A chooses 1×2, B can choose 2×1 to get a better payoff. Similarly, B may choose 1×1 to get a better payoff, but then A can choose 1×1 to get a better payoff. Hence, to avoid uncertainty, this sub-Nash strategy may be accepted by both A and B.

Continuing this process, we consider cycle strategies of length 3. The tolerance for the sub-Nash equilibria is 0.6667. The sub-Nash equilibria

$$(122, 111), (122, 222), (212, 111),$$
$$(212, 222), (221, 111), (221, 222)$$

have payoffs $\{1.3333, 1.3333\}$, the sub-Nash equilibria

$$(222, 112), (222, 121), (222, 211)$$

have payoffs $\{1.3333, 1.6667\}$, and the sub-Nash equilibria

$$(222, 122), (222, 211), (222, 221)$$

have payoffs $\{1.6667, 1.3333\}$.

For cycle strategies of length 4, the tolerance is 0.5. Sub-Nash equilibria are

$$(2222, 1122), (2222, 1212), (2222, 1221),$$
$$(2222, 2112), (2222, 2121), (2222, 2211),$$

which have the same payoffs, $\{1.5, 1.5\}$.

Note that the longest length of a cycle is 4. No more calculation is necessary. Summarizing the above argument, one sees that the best sub-Nash equilibrium for cycle strategies is of tolerance 0.5. The corresponding sub-Nash solutions are $(2 \times 2, 1 \times 2)$ and $(2 \times 2, 2 \times 1)$.

In principle, for a game with n players, each player A_i with k_i strategies, the longest cycle strategy has length $k = \prod_{j=1}^{n} k_j$. For each length, we can then find the sub-Nash equilibrium (equilibria). Finally, we can find the best sub-Nash equilibrium (equilibria), which has (have) the minimum tolerance. This (these) can be taken as the sub-Nash solution(s) of the zero-memory cycle strategies for the game.

Denote by ε_α the minimum tolerance of cycle strategies. Assume that $\ell \pmod{\alpha} = \beta$, i.e.,

$$\ell = \gamma\alpha + \beta.$$

By choosing an optimal cycle of length α for γ times and an optimal cycle of length β, it is easily seen that

$$\varepsilon_\ell \leq \frac{\gamma\varepsilon_\alpha + \varepsilon_\beta}{\ell}. \tag{18.25}$$

If (18.25) becomes an equality, then the "optimal" cycle of length ℓ is meaningless. Otherwise, we have an improved sub-Nash solution.

18.5 Nash Equilibrium for μ-Memory Strategies

Consider a μ-memory strategy. By Definition 18.2 each such strategy can be determined by

$$x_j(t+1) = f_j\big(x(t), \ldots, x(t-\mu+1)\big), \quad j = 1, \ldots, n, \tag{18.26}$$

with a set of initial values $\{x_j(t) \mid j = 1, \ldots, n; \ t = 0, \ldots, \mu - 1\}$. Now, (18.26) can be expressed equivalently in its algebraic form as

$$x_j(t+1) = L_j \ltimes_{i=1}^{\mu} x(t-\mu+i), \quad j = 1, \ldots, n, \tag{18.27}$$

where $L_j \in \mathscr{L}_{k_j \times k^\mu}$. Hence, a strategy can be determined by a set of logical matrices

$$\{L_j \in \mathscr{L}_{k_j \times k^\mu} \mid j = 1, \ldots, n\}$$

with a set of initial values.

We give an example to illustrate this.

Example 18.3 Consider the infinitely repeated prisoner's dilemma with $\mu = 1$. Both players then have 16 strategies, denoted by

$$s_1 = \delta_2[1111], \quad s_2 = \delta_2[1112], \quad s_3 = \delta_2[1121], \quad s_4 = \delta_2[1122],$$

$$s_5 = \delta_2[1211], \quad s_6 = \delta_2[1212], \quad s_7 = \delta_2[1221], \quad s_8 = \delta_2[1222],$$

$$s_9 = \delta_2[2111], \quad s_{10} = \delta_2[2112], \quad s_{11} = \delta_2[2121], \quad s_{12} = \delta_2[2122],$$

$$s_{13} = \delta_2[3211], \quad s_{14} = \delta_2[2212], \quad s_{15} = \delta_2[2221], \quad s_{16} = \delta_2[2222].$$

The payoff bi-matrices for different initial values are as follows:

- $x_1(0) = \delta_2^1$, $x_2(0) = \delta_2^1$, as in Table 18.7.
- $x_1(0) = \delta_2^1$, $x_2(0) = \delta_2^2$, as in Table 18.8.

Table 18.7 Payoff bi-matrix

A/B	1	2	3	4	5	6	7	8
1	3, 3	3, 3	3, 3	3, 3	3, 3	3, 3	3, 3	3, 3
2	3, 3	3, 3	3, 3	3, 3	3, 3	3, 3	3, 3	3, 3
3	3, 3	3, 3	3, 3	3, 3	3, 3	3, 3	3, 3	3, 3
4	3, 3	3, 3	3, 3	3, 3	3, 3	3, 3	3, 3	3, 3
5	3, 3	3, 3	3, 3	3, 3	3, 3	3, 3	3, 3	3, 3
6	3, 3	3, 3	3, 3	3, 3	3, 3	3, 3	3, 3	3, 3
7	3, 3	3, 3	3, 3	3, 3	3, 3	3, 3	3, 3	3, 3
8	3, 3	3, 3	3, 3	3, 3	3, 3	3, 3	3, 3	3, 3
9	4, 1.5	4, 1.5	2.67, 2.67	2.67, 2.67	4, 1.5	4, 1.5	0, 5	0, 5
10	4, 1.5	4, 1.5	2.67, 2.67	2.67, 2.67	4, 1.5	4, 1.5	0, 5	0, 5
11	5, 0	5, 0	3, 1.33	2.25, 2.25	5, 0	5, 0	3, 1.33	0, 5
12	5, 0	5, 0	3, 0.5	1, 1	5, 0	5, 0	3, 0.5	1, 1
13	4, 1.5	4, 1.5	2.5, 2.5	2.5, 2.5	4, 1.5	4, 1.5	2.25, 2.25	0.5, 3
14	4, 1.5	4, 1.5	2.5, 2.5	2.5, 2.5	4, 1.5	4, 1.5	2, 2	1, 1
15	5, 0	5, 0	3, 1.33	2, 2	5, 0	5, 0	3, 1.33	0.5, 3
16	5, 0	5, 0	3, 0.5	1, 1	5, 0	5, 0	3, 0.5	1, 1

A/B	9	10	11	12	13	14	15	16
1	1.5, 4	1.5, 4	1.5, 4	1.5, 4	0, 5	0, 5	0, 5	0, 5
2	1.5, 4	1.5, 4	1.5, 4	1.5, 4	0, 5	0, 5	0, 5	0, 5
3	1.5, 4	1.5, 4	1.5, 4	1.5, 4	0, 5	0, 5	0, 5	0, 5
4	1.5, 4	1.5, 4	1.5, 4	1.5, 4	0, 5	0, 5	0, 5	0, 5
5	2.67, 2.67	2.67, 2.67	2.5, 2.5	2.5, 2.5	1.33, 3	0.5, 3	1.33, 3	0.5, 3
6	2.67, 2.67	2.67, 2.67	2.5, 2.5	2.5, 2.5	2.25, 2.25	1, 1	2, 2	1, 1
7	5, 0	5, 0	2.25, 2.25	2, 2	1.33, 3	0.5, 3	1.33, 3	0.5, 3
8	5, 0	5, 0	3, 0.5	1, 1	5, 0	1, 1	3, 0.5	1, 1
9	2, 2	1.33, 3	2, 2	1.33, 3	2, 2	0, 5	2, 2	0, 5
10	3, 1.33	1, 1	2.25, 2.25	1, 1	3, 1.33	1, 1	0, 5	1, 1
11	2, 2	1.33, 3	2, 2	1.33, 3	2, 2	0, 5	2, 2	0, 5
12	5, 0	1, 1	3, 0.5	1, 1	5, 0	1, 1	3, 0.5	1, 1
13	2, 2	2.25, 2.25	2, 2	2.5, 2.5	2, 2	0.5, 3	2, 2	0.5, 3
14	3, 1.33	1, 1	2.5, 2.5	1, 1	3, 1.33	1, 1	2, 2	1, 1
15	2, 2	5, 0	2, 2	2, 2	2, 2	0.5, 3	2, 2	0.5, 3
16	5, 0	1, 1	3, 0.5	1, 1	5, 0	1, 1	3, 0.5	1, 1

- $x_1(0) = \delta_2^2$, $x_2(0) = \delta_2^1$, as in Table 18.9.
- $x_1(0) = \delta_2^2$, $x_2(0) = \delta_2^2$, as in Table 18.10.

We can check that there are 53 1-memory Nash equilibria, as in Table 18.11.

In Table 18.11 the double pairs $(n\ a)$, $(m\ b)$ indicate that player A takes a strategy with initial value n and uses the logical matrix s_a, while player B takes a strategy with initial value m and uses the logical matrix s_b.

Table 18.8 Payoff bi-matrix

A/B	1	2	3	4	5	6	7	8
1	3, 3	3, 3	3, 3	3, 3	0, 5	0, 5	0, 5	0, 5
2	3, 3	3, 3	3, 3	3, 3	0, 5	0, 5	0, 5	0, 5
3	3, 3	3, 3	3, 3	3, 3	0, 5	0, 5	0, 5	0, 5
4	3, 3	3, 3	3, 3	3, 3	0, 5	0, 5	0, 5	0, 5
5	3, 3	3, 3	2.5, 2.5	2.5, 2.5	3, 3	0.5, 3	3, 3	0.5, 3
6	3, 3	3, 3	2.5, 2.5	2.5, 2.5	3, 3	1, 1	2, 2	1, 1
7	5, 0	5, 0	3, 3	2, 2	3, 3	0.5, 3	3, 3	0.5, 3
8	5, 0	5, 0	3, 0.5	1, 1	5, 0	1, 1	3, 0.5	1, 1
9	4, 1.5	4, 1.5	2.67, 2.67	2.67, 2.67	0, 5	0, 5	0, 5	0, 5
10	4, 1.5	4, 1.5	2.67, 2.67	2.67, 2.67	0, 5	0, 5	0, 5	0, 5
11	5, 0	5, 0	3, 1.33	2.25, 2.25	0, 5	0, 5	0, 5	0, 5
12	5, 0	5, 0	3, 0.5	1, 1	0, 5	0, 5	0, 5	0, 5
13	4, 1.5	4, 1.5	2.5, 2.5	2.5, 2.5	4, 1.5	0.5, 3	2.25, 2.25	0.5, 3
14	4, 1.5	4, 1.5	2.5, 2.5	2.5, 2.5	4, 1.5	1, 1	2, 2	1, 1
15	5, 0	5, 0	3, 1.33	2, 2	5, 0	0.5, 3	3, 1.33	0.5, 3
16	5, 0	5, 0	3, 0.5	1, 1	5, 0	1, 1	3, 0.5	1, 1

A/B	9	10	11	12	13	14	15	16
1	1.5, 4	1.5, 4	1.5, 4	1.5, 4	0, 5	0, 5	0, 5	0, 5
2	1.5, 4	1.5, 4	1.5, 4	1.5, 4	0, 5	0, 5	0, 5	0, 5
3	1.5, 4	1.5, 4	1.5, 4	1.5, 4	0, 5	0, 5	0, 5	0, 5
4	1.5, 4	1.5, 4	1.5, 4	1.5, 4	0, 5	0, 5	0, 5	0, 5
5	2.67, 2.67	2.67, 2.67	2.5, 2.5	2.5, 2.5	1.33, 3	0.5, 3	1.33, 3	0.5, 3
6	2.67, 2.67	2.67, 2.67	2.5, 2.5	2.5, 2.5	2.25, 2.25	1, 1	2, 2	1, 1
7	5, 0	5, 0	2.25, 2.25	2, 2	1.33, 3	0.5, 3	1.33, 3	0.5, 3
8	5, 0	5, 0	3, 0.5	1, 1	5, 0	1, 1	3, 0.5	1, 1
9	2, 2	1.33, 3	2, 2	1.33, 3	0, 5	0, 5	0, 5	0, 5
10	3, 1.33	1, 1	2.25, 2.25	1, 1	0, 5	0, 5	0, 5	0, 5
11	2, 2	1.33, 3	2, 2	1.33, 3	0, 5	0, 5	0, 5	0, 5
12	5, 0	1, 1	3, 0.5	1, 1	0, 5	0, 5	0, 5	0, 5
13	2, 2	2.25, 2.25	2.5, 2.5	2.5, 2.5	2, 2	0.5, 3	2, 2	0.5, 3
14	3, 1.33	1, 1	2.5, 2.5	2.5, 2.5	3, 1.33	1, 1	2, 2	1, 1
15	5, 0	5, 0	2, 2	2, 2	2, 2	0.5, 3	2, 2	0.5, 3
16	5, 0	5, 0	3, 0.5	1, 1	5, 0	1, 1	3, 0.5	1, 1

18.6 Common Nash (Sub-Nash) Solutions for μ-Memory Strategies

In the previous section we investigated how to find a Nash solution for μ-memory strategies. Since there may be many Nash solutions, we are particularly interested in

Table 18.9 Payoff bi-matrix

A/B	1	2	3	4	5	6	7	8
1	3, 3	3, 3	3, 3	3, 3	3, 3	3, 3	0, 5	0, 5
2	3, 3	3, 3	3, 3	3, 3	3, 3	3, 3	0, 5	0, 5
3	5, 0	5, 0	3, 3	3, 3	5, 0	5, 0	3, 3	0, 5
4	5, 0	5, 0	3, 0.5	1, 1	5, 0	5, 0	3, 0.5	1, 1
5	3, 3	3, 3	2.5, 2.5	2.5, 2.5	3, 3	3, 3	3, 3	0.5, 3
6	3, 3	3, 3	2.5, 2.5	2.5, 2.5	3, 3	3, 3	2, 2	1, 1
7	5, 0	5, 0	3, 3	2, 2	5, 0	5, 0	3, 3	0.5, 3
8	5, 0	5, 0	3, 0.5	1, 1	5, 0	5, 0	3, 0.5	1, 1
9	4, 1.5	4, 1.5	2.67, 2.67	2.67, 2.67	4, 1.5	4, 1.5	0, 5	0, 5
10	4, 1.5	4, 1.5	2.67, 2.67	2.67, 2.67	4, 1.5	4, 1.5	0, 5	0, 5
11	5, 0	5, 0	3, 1.33	2.25, 2.25	5, 0	5, 0	3, 1.33	0, 5
12	5, 0	5, 0	3, 0.5	1, 1	5, 0	5, 0	3, 0.5	1, 1
13	4, 1.5	4, 1.5	2.5, 2.5	2.5, 2.5	4, 1.5	4, 1.5	2.25, 2.25	0.5, 3
14	4, 1.5	4, 1.5	2.5, 2.5	2.5, 2.5	4, 1.5	4, 1.5	2, 2	1, 1
15	5, 0	5, 0	3, 1.33	2, 2	5, 0	5, 0	3, 1.33	0.5, 3
16	5, 0	5, 0	3, 0.5	1, 1	5, 0	5, 0	3, 0.5	1, 1

A/B	9	10	11	12	13	14	15	16
1	1.5, 4	1.5, 4	1.5, 4	1.5, 4	0, 5	0, 5	0, 5	0, 5
2	1.5, 4	1.5, 4	1.5, 4	1.5, 4	0, 5	0, 5	0, 5	0, 5
3	5, 0	5, 0	1.5, 4	1.5, 4	5, 0	5, 0	0, 5	0, 5
4	5, 0	5, 0	3, 0.5	1, 1	5, 0	5, 0	3, 0.5	1, 1
5	2.67, 2.67	2.67, 2.67	2.5, 2.5	2.5, 2.5	1.33, 3	0.5, 3	1.33, 3	0.5, 3
6	2.67, 2.67	2.67, 2.67	2.5, 2.5	2.5, 2.5	2.25, 2.25	1, 1	2, 2	1, 1
7	5, 0	5, 0	2.25, 2.25	2, 2	5, 0	5, 0	1.33, 3	0.5, 3
8	5, 0	5, 0	3, 0.5	1, 1	5, 0	5, 0	3, 0.5	1, 1
9	2, 2	1.33, 3	2, 2	1.33, 3	2, 2	0, 5	0, 5	0, 5
10	3, 1.33	1, 1	2.25, 2.25	1, 1	3, 1.33	1, 1	0, 5	0, 5
11	5, 0	5, 0	2, 2	1.33, 3	5, 0	5, 0	2, 2	0, 5
12	5, 0	5, 0	3, 0.5	1, 1	5, 0	5, 0	3, 0.5	1, 1
13	2, 2	2.25, 2.25	2.5, 2.5	2.5, 2.5	2, 2	0.5, 3	2, 2	0.5, 3
14	3, 1.33	1, 1	2.5, 2.5	2.5, 2.5	3, 1.33	1, 1	2, 2	1, 1
15	5, 0	5, 0	2, 2	2, 2	5, 0	5, 0	2, 2	0.5, 3
16	5, 0	5, 0	3, 0.5	1, 1	5, 0	5, 0	3, 0.5	1, 1

the common Nash equilibrium, which is independent of the initial values. If such a common Nash equilibrium exists, it could be considered as the best solution to G_∞.

If such a common Nash equilibrium does not exist, finding a reasonable solution to the game G_∞ becomes a challenging problem. In this section we consider the common Nash or sub-Nash solution for μ-memory strategies, which is independent of the initial value.

Table 18.10 Payoff bi-matrix

A/B	1	2	3	4	5	6	7	8
1	3, 3	3, 3	3, 3	3, 3	3, 3	0, 5	3, 3	0, 5
2	3, 3	1, 1	3, 3	1, 1	3, 3	1, 1	0, 5	1, 1
3	3, 3	3, 3	3, 3	3, 3	3, 3	0, 5	3, 3	0, 5
4	5, 0	1, 1	3, 0.5	1, 1	5, 0	1, 1	3, 0.5	1, 1
5	3, 3	3, 3	3, 3	2.5, 2.5	3, 3	0.5, 3	3, 3	0.5, 3
6	3, 3	1, 1	2.5, 2.5	1, 1	3, 3	1, 1	2, 2	1, 1
7	3, 3	5, 0	3, 3	2, 2	3, 3	0.5, 3	3, 3	0.5, 3
8	5, 0	1, 1	3, 0.5	1, 1	5, 0	1, 1	3, 0.5	1, 1
9	4, 1.5	4, 1.5	2.67, 2.67	2.67, 2.67	4, 1.5	0, 5	0, 5	0, 5
10	4, 1.5	1, 1	2.67, 2.67	1, 1	4, 1.5	1, 1	0, 5	1, 1
11	5, 0	5, 0	3, 1.33	2.25, 2.25	5, 0	0, 5	3, 1.33	0, 5
12	5, 0	1, 1	3, 0.5	1, 1	5, 0	1, 1	3, 0.5	1, 1
13	4, 1.5	4, 1.5	2.5, 2.5	2.5, 2.5	4, 1.5	0.5, 3	2.25, 2.25	0.5, 3
14	4, 1.5	1, 1	2.5, 2.5	1, 1	4, 1.5	1, 1	2, 2	1, 1
15	5, 0	5, 0	3, 1.33	2, 2	5, 0	0.5, 3	3, 1.33	0.5, 3
16	5, 0	1, 1	3, 0.5	1, 1	5, 0	1, 1	3, 0.5	1, 1

A/B	9	10	11	12	13	14	15	16
1	1.5, 4	1.5, 4	1.5, 4	1.5, 4	0, 5	0, 5	0, 5	0, 5
2	1.5, 4	1, 1	1.5, 4	1, 1	0, 5	1, 1	0, 5	1, 1
3	1.5, 4	1.5, 4	1.5, 4	1.5, 4	0, 5	0, 5	0, 5	0, 5
4	5, 0	1, 1	3, 0.5	1, 1	5, 0	1, 1	3, 0.5	1, 1
5	2.67, 2.67	2.67, 2.67	2.5, 2.5	2.5, 2.5	1.33, 3	0.5, 3	1.33, 3	0.5, 3
6	2.67, 2.67	1, 1	2.5, 2.5	1, 1	2.25, 2.25	1, 1	2, 2	1, 1
7	5, 0	5, 0	2.25, 2.25	2, 2	1.33, 3	0.5, 3	1.33, 3	0.5, 3
8	5, 0	1, 1	3, 0.5	1, 1	5, 0	1, 1	3, 0.5	1, 1
9	2, 2	1.33, 3	2, 2	1.33, 3	2, 2	0, 5	2, 2	0, 5
10	3, 1.33	1, 1	2.25, 2.25	1, 1	3, 1.33	1, 1	0, 5	1, 1
11	2, 2	1.33, 3	2, 2	1.33, 3	2, 2	0, 5	2, 2	0, 5
12	5, 0	1, 1	3, 0.5	1, 1	5, 0	1, 1	3, 0.5	1, 1
13	2, 2	2.25, 2.25	2, 2	2.5, 2.5	2, 2	0.5, 3	2, 2	0.5, 3
14	3, 1.33	1, 1	2.5, 2.5	1, 1	3, 1.33	1, 1	2, 2	1, 1
15	2, 2	5, 0	2, 2	2, 2	2, 2	0.5, 3	2, 2	0.5, 3
16	5, 0	1, 1	3, 0.5	1, 1	5, 0	1, 1	3, 0.5	1, 1

Describing a μ-memory strategy by its generating dynamics as

$$L := \{L_1 \in \mathscr{L}_{k_1 \times k^\mu}, \dots, L_n \in \mathscr{L}_{k_n \times k^\mu}\} \tag{18.28}$$

with initial values

$$x^0 = \left(x_1^0(0), \dots, x_n^0(0), \dots, x_1^0(\mu - 1), \dots, x_n^0(\mu - 1)\right) \in \Delta_{k^\mu},$$

we give the following definition.

Table 18.11 1-memory Nash equilibria

(1 5), (1 3)	(1 5), (1 4)	(1 5), (1 7)	(1 5), (1 8)	(1 5), (2 7)	(1 6), (1 3)
(1 6), (1 4)	(1 6), (1 7)	(1 6), (1 8)	(1 7), (1 3)	(1 7), (1 4)	(1 7), (1 7)
(1 7), (1 8)	(1 7), (2 3)	(1 7), (2 7)	(1 8), (1 3)	(1 8), (1 4)	(1 8), (1 7)
(1 8), (1 8)	(1 14), (2 12)	(1 16), (1 16)	(1 16), (2 6)	(1 16), (2 8)	(1 16), (2 14)
(1 16), (2 16)	(2 4), (1 16)	(2 4), (2 6)	(2 4), (2 8)	(2 4), (2 14)	(2 4), (2 16)
(2 5), (1 7)	(2 5), (2 3)	(2 5), (2 7)	(2 7), (1 3)	(2 7), (1 7)	(2 7), (2 3)
(2 7), (2 7)	(2 8), (1 16)	(2 8), (2 6)	(2 8), (2 8)	(2 8), (2 14)	(2 8), (2 16)
(2 12), (1 16)	(2 12), (2 6)	(2 12), (2 8)	(2 12), (2 14)	(2 12), (2 16)	(2 14), (1 12)
(2 16), (1 16)	(2 16), (2 6)	(2 16), (2 8)	(2 16), (2 14)	(2 16), (2 16)	

Definition 18.6 Consider G_∞. A common μ-memory strategy is defined as (18.28).

(1) $L^* = (L_1^*, \ldots, L_n^*)$ is called a common Nash equilibrium if

$$J_j(L_1^*, \ldots, L_n^*; x^0) \geq J_j(L_1^*, \ldots, L_j, \ldots, L_n^*; x^0), \quad \forall x^0 \in \Delta_{k^\mu}, \ j = 1, \ldots, n. \tag{18.29}$$

(2) Let $\varepsilon_L^{x^0}$ be the tolerance of the strategy $\{L\}$ when x^0 is fixed. The common tolerance of L is defined as

$$\varepsilon_L = \max_{x^0 \in \Delta_{k^\mu}} \varepsilon_L^{x^0}. \tag{18.30}$$

(3) A strategy L^0 is called a common ε_0 sub-Nash equilibrium if

$$\varepsilon_0 = \varepsilon_{L_0} \leq \varepsilon_L, \quad \forall L \in \{\mathscr{L}_{k_1 \times k^\mu}, \ldots, \mathscr{L}_{k_n \times k^\mu}\}.$$

(4) A strategy L is also called a common ε_L sub-Nash solution to G_∞.

If a common Nash equilibrium does not exist, then we will naturally look for a common sub-Nash equilibrium as an acceptable solution to G_∞. If it is difficult to find a common sub-Nash equilibrium, then we may have to accept a reasonable sub-Nash solution. It is easy to see that a common (sub-)Nash equilibrium must be a (sub-)Nash equilibrium with respect to any initial states. Thus, it can be thought of as a refinement of a Nash equilibrium.

We discuss this by means of the following examples. The first example concerns the common Nash equilibrium.

Example 18.4 Recall Example 18.3. Because there are too many Nash equilibria, for further refinement we are particularly interested in common Nash equilibria, which are independent of the initial value. It is easy to check that there are three common Nash equilibria:

Table 18.12 Payoff bi-matrix

A\B	1	2	3
1	2.7, 1.5	0.3, 2.4	1.1, 0.8
2	0.9, 2.1	1.8, 1.3	1.6, 1.4

(i) $s_A = s_{16}$ and $s_B = s_{16}$. Since $s_{16} = \delta_2[2, 2, 2, 2]$, regardless of the initial value, both players just take action 2, which constitutes the original Nash equilibrium of G. This is not particularly interesting.

(ii) $s_A = s_8$ and $s_B = s_8$. Since $s_8 = \delta_2[1, 2, 2, 2]$, if the previous values are $(1, 1)$, then action 1 is taken, otherwise action 2 is taken. This is the famous trigger strategy [4].

(iii) $s_A = s_7$ and $s_B = s_7$. Since $s_7 = \delta_2[1, 2, 2, 1]$, we have

$$x_i(t+1) = \begin{cases} 1, & x_1(t) = x_2(t), \\ 2, & x_1(t) \neq x_2(t). \end{cases}$$

In fact, under the payoff bi-matrix in Table 18.4, this strategy is the best one because it will converge to $(1, 1)$, regardless of the initial state. However, this strategy has poor robustness: if the payoff 5 is changed to $5 + \delta$ for arbitrary $\delta > 0$, then it is easy to check that this strategy is no longer the Nash equilibrium.

The second example concerns sub-Nash equilibria.

Example 18.5 The payoff bi-matrix of a game G is given in Table 18.12.

From Table 18.12 it is easily seen that there is no Nash equilibrium and the only sub-Nash equilibrium is $(2, 3)$, with payoffs $[1.6, 1.4]$ and tolerance $\varepsilon = 0.7$. Consider 1-memory strategies for G_∞. A standard routine shows that there are 57 sub-Nash equilibria with payoff $[1.3, 2]$ and tolerance $\varepsilon = 0.3$, which are listed in Table 18.13.

In Table 18.13 the double pairs $(n\ a)$, $(m\ b)$ indicate that player A takes a strategy M_a^A with initial value n, while player B takes a strategy M_b^B. The sets of strategies are ordered as follows:

$$M^A = \left\{ M_1^A, M_2^A, M_3^A, \ldots, M_{26}^A \right\}$$
$$= \left\{ \delta_2[1\ 1\ 1\ 1\ 1\ 1], \delta_2[1\ 1\ 1\ 1\ 1\ 2], \delta_2[1\ 1\ 1\ 1\ 2\ 1], \ldots, \delta_2[2\ 2\ 2\ 2\ 2\ 2] \right\},$$
$$M^B = \left\{ M_1^B, M_2^B, M_3^B, \ldots, M_{36}^B \right\}$$
$$= \left\{ \delta_3[1\ 1\ 1\ 1\ 1\ 1], \delta_3[1\ 1\ 1\ 1\ 1\ 2], \delta_3[1\ 1\ 1\ 1\ 1\ 3], \ldots, \delta_3[3\ 3\ 3\ 3\ 3\ 3] \right\}.$$

Finally, $(17, 279)$ and $(19, 279)$ are common sub-Nash equilibria with $\varepsilon = 0.3$, which are independent of the initial value.

Table 18.13 1-memory sub-Nash equilibria

(1 17), (1 279)	(1 17), (1 306)	(1 17), (2 279)	(1 17), (2 306)	(1 17), (3 279)
(1 18), (1 279)	(1 18), (1 306)	(1 18), (2 279)	(1 18), (2 306)	(1 18), (3 279)
(1 19), (1 279)	(1 19), (1 306)	(1 19), (2 279)	(1 19), (2 306)	(1 19), (3 279)
(1 20), (1 279)	(1 20), (1 306)	(1 20), (2 279)	(1 20), (2 306)	(1 20), (3 279)
(1 25), (1 279)	(1 25), (1 306)	(1 25), (2 279)	(1 25), (2 306)	(1 26), (1 279)
(1 26), (1 306)	(1 26), (2 279)	(1 26), (2 306)	(1 27), (1 279)	(1 27), (1 306)
(1 27), (2 279)	(1 27), (2 306)	(1 28), (1 279)	(1 28), (1 306)	(1 28), (2 279)
(1 28), (2 306)	(2 17), (1 279)	(2 17), (1 306)	(2 17), (2 279)	(2 17), (3 279)
(2 18), (1 279)	(2 18), (1 306)	(2 18), (2 279)	(2 19), (1 279)	(2 19), (1 306)
(2 19), (2 279)	(2 19), (3 279)	(2 20), (1 279)	(2 20), (1 306)	(2 25), (1 279)
(2 25), (1 306)	(2 26), (1 279)	(2 26), (1 306)	(2 27), (1 279)	(2 27), (1 306)
(2 28), (1 279)	(2 28), (1 306)			

Next, we consider 2-memory strategies. Ignoring initial values, we have $2^{36} \times 3^{36}$ strategies. It is almost impossible to consider all of them in the previous way. We propose the following algorithm for obtaining common sub-Nash solutions with $\mu > 1$ memory strategies.

Algorithm 18.2

Step 1. Let the set of 1-memory strategies (without initial values) be

$$S^1 = \left\{ S_1^1, S_2^1, \dots, S_n^1 \right\},$$

where $S_j^1 = \mathcal{L}_{k_j \times k}$, $j = 1, \dots, n$, are the strategies of player A_j. Find the common sub-Nash equilibria with tolerance ε_1.

- If $\varepsilon_1 = 0$, or $\varepsilon_1 > 0$ and the common sub-Nash equilibrium is unique, then take the common sub-Nash equilibrium as a common ε_1 sub-Nash solution and stop. (For ease of statement, a common Nash solution is considered as a particular common sub-Nash solution with zero tolerance. In this case, the solution may not be unique.)
- Otherwise, denote the set of common sub-Nash equilibria by N^1 and consider it as the set of 1-memory common sub-Nash solutions.

Step μ ($\mu > 1$). Assume that

$$N^{\mu-1} = \left\{ N_1^{\mu-1}, N_2^{\mu-1}, \dots, N_n^{\mu-1} \right\}$$

are obtained. Set

$$S^\mu = \left\{ S_1^\mu, S_2^\mu, \dots, S_n^\mu \right\},$$

where

$$S_j^\mu = \{[s_1^{\mu-1} \, s_2^{\mu-1} \, \cdots \, s_k^{\mu-1}] | s_\alpha^{\mu-1} \in N_j^{\mu-1}, \ \alpha = 1, \ldots, k\}, \quad j = 1, \ldots, n.$$

Find the μ-memory common sub-Nash equilibria over S^μ (just compare with strategies in S^μ) with tolerance ε_μ.

- If $\varepsilon_\mu = 0$, or $\varepsilon_\mu = \varepsilon_{\mu-1}$, or $\varepsilon_\mu > 0$ and the common sub-Nash equilibrium is unique, then take the common sub-Nash equilibrium as a common ε_μ sub-Nash solution and stop. (For the case of $\varepsilon_\mu = 0$, or $\varepsilon_\mu = \varepsilon_{\mu-1}$, the solution may not be unique.)
- Otherwise, denote the set of common sub-Nash equilibria in S^μ by N^μ and consider it as the set of μ-memory common sub-Nash solutions. Go to Step $\mu + 1$.

The following proposition shows that the sub-Nash equilibria obtained by the algorithm have monotonically nonincreasing tolerances.

Proposition 18.4 *Let N^μ be the set of μ-memory common sub-Nash equilibria with tolerance ε_μ. Construct*

$$S^{\mu+1} = \{[L_1^{\mu+1}, \ldots, L_n^{\mu+1}]\} \tag{18.31}$$

with

$$L_i^{\mu+1} = [L_{i1}^\mu, \ldots, L_{ik}^\mu], \quad i = 1, \ldots, n, \tag{18.32}$$

and

$$(L_{1\alpha}^\mu, \ldots, L_{n\alpha}^\mu) \in N^\mu, \quad \alpha = 1, \ldots, k. \tag{18.33}$$

If $L^{\mu+1*} = (L_1^{\mu+1*}, \ldots, L_n^{\mu+1*})$ is a strategy of $(\mu+1)$-memory common sub-Nash equilibrium of $S^{\mu+1}$, then the tolerance of $L^{\mu+1*}$, denoted by $\varepsilon_{\mu+1}$, satisfies

$$\varepsilon_{\mu+1} \le \varepsilon_\mu. \tag{18.34}$$

Proof Let

$$L^{\mu*} = \{L_1^{\mu*}, \ldots, L_n^{\mu*}\} \in N^\mu$$

with tolerance ε_μ. We can then construct $L^{\mu+1}$ by

$$L_j^{\mu+1} = [\underbrace{L_j^{\mu*}, \ldots, L_j^{\mu*}}_{k}], \quad j = 1, \ldots, n. \tag{18.35}$$

By the construction we know that

$$L_j^{\mu+1} \in S^{\mu+1}. \tag{18.36}$$

Using (18.27), we then know that

$$x_j(t+1) = L_j^{\mu+1} x(t-\mu)x(t-\mu+1)\cdots x(t)$$

$$= \underbrace{[L_j^{\mu*}, \ldots, L_j^{\mu*}]}_{k} x(t-\mu)x(t-\mu+1)\cdots x(t)$$

$$= L_j^{\mu*} x(t-\mu+1)\cdots x(t), \quad j=1,\ldots,n,$$

which means that the strategy $L^{\mu+1}$ constructed in (18.35) is exactly the same as the strategy $L^{\mu*}$. Hence, the tolerance of $L^{\mu+1}$ is ε_μ. Now, since it is in $S^{\mu+1}$, $\varepsilon_{\mu+1}$, as the smallest tolerance over $S^{\mu+1}$, surely satisfies (18.34).   □

We now continue the discussion of Example 18.5.

Example 18.6 Consider Example 18.5. Since

$$N^1 = \{(17, 279), (19, 279)\}$$

consists of only two strategies, it is easy to see that

$$M_{17}^A = \delta_2[1\,2\,1\,1\,1\,1], \qquad M_{19}^A = \delta_2[1\,2\,1\,1\,2\,1],$$

$$M_{279}^B = \delta_3[2\,1\,2\,1\,3\,3].$$

A straightforward computation shows that the payoffs of the two common 1-memory sub-Nash equilibria with any initial value are the same, so all the strategies in S^2 have the same payoff. Thus, they are all common Nash equilibria in S^2, which implies that $\varepsilon_2 = 0$. The algorithm then terminates.

We now have to answer the following questions:

- What is the real tolerance of the common Nash equilibrium in S^2?
- Are they the best 2-memory strategies?

We use the following nonparametric test to answer these questions: Choose 300 additional 2-memory strategies randomly and add them to S^2 to form an extended set $\tilde{S}^2$, then find the sub-Nash equilibria over $\tilde{S}^2$. We then have the following results:

- The common Nash equilibria in S^2 have tolerance 0.3 over $\tilde{S}^2$. We conclude that the algorithm does not provide better solutions (with 2-memory) than the 1-memory sub-Nash solutions.
- There is no better sub-Nash solution in $\tilde{S}^2$. Using the nonparametric test, it is easily seen that with the confidence limit 99.5%, we can say that the tolerance of the strategies in S^2 is less than or equal to the tolerance of 99.8% of the overall 2-memory strategies. (We refer to any standard statistics textbook, e.g., [3].)

We conclude that the algorithm provides only two "best sub-Nash solutions", which are with 1-memory. They are:

(1) The strategies of A and B are as in the following (18.37) and (18.38), respectively.

$$x_1(t+1) = \begin{cases} 2, & x_1(t) = 1 \text{ and } x_2(t) = 2, \\ 1, & \text{otherwise.} \end{cases} \tag{18.37}$$

$$x_2(t+1) = \begin{cases} 1, & x_1(t) = 1 \text{ and } x_2(t) = 2, \text{ or } x_1(t) = 2 \text{ and } x_2(t) = 1, \\ 2, & x_1(t) = 1 \text{ and } x_2(t) = 2 \text{ or } 3, \\ 3, & \text{otherwise.} \end{cases} \tag{18.38}$$

(2) The strategy of A is as in the following (18.39) and the strategy of B is as in (18.38).

$$x_1(t+1) = \begin{cases} 2, & x_2(t) = 2, \\ 1, & \text{otherwise.} \end{cases} \tag{18.39}$$

References

1. Cheng, D., Zhao, Y., Li, Z.: Nash and sub-Nash solutions to infinitely repeated games. Preprint (2010)
2. Cheng, D., Zhao, Y., Mu, Y.: Strategy optimization with its application to dynamic games. In: Proc. IEEE CDC'2010 (2010, to appear)
3. Daniel, W.: Applied Nonparametric Statistics. PWS-Kent Pub., Boston (1990)
4. Gibbons, R.: A Primer in Game Theory. Prentice Hall, New York (1992)
5. Li, Z., Cheng, D.: Algebraic approach to dynamics of multi-valued networks. Int. J. Bifurc. Chaos **20**(3), 561–582 (2010)
6. Mu, Y., Guo, L.: Optimization and identification in a non-equilibrium dynamic game. In: Proc. CDC-CCC'09, pp. 5750–5755 (2009)
7. Puterman, M.: Markov Decision Processes: Discrete Stochastic Dynamic Programming. Wiley, New York (1994)

Chapter 19
Random Boolean Networks

19.1 Markov Chains

This section provides some simple background on Markov processes. We review some basic concepts, notation, and properties (without proofs), and refer to some standard textbooks, e.g., [1], for details. Readers familiar with stochastic processes can skip this section.

Definition 19.1

(i) Let Ω be a set, $\mathscr{F}$ an algebra generated by a set of subsets of Ω, and P a probabilistic measure on $(\Omega, \mathscr{F})$. Then, $(\Omega, \mathscr{F}, P)$ is called a probabilistic space.

(ii) Let $\mathscr{B}$ be the Borel set on $\mathbb{R}$. We denote by $(\mathbb{R}, \mathscr{B})$ the Borel-measurable space on $\mathbb{R}$.

(iii) Let $T = \mathbb{Z}_+ = \{0, 1, 2, \ldots\}$. A sequence $\xi(t, \omega)$, $t \in T$, is called a discrete-time real stochastic process if for each $t \in T$, $\xi(t, \cdot) : (\Omega, \mathscr{F}) \to (\mathbb{R}^1, \mathscr{B})$ is a measurable function.

We now give an example of a discrete-time real stochastic process.

Example 19.1 (Bernoulli sequence) A bag contains m red balls (denoted by 0) and n white balls (denoted by 1). A person repetitively draws balls from the bag and each time, after drawing, returns the ball back to the bag. It is easy to see that the probability of drawing $a_1, a_2, \ldots, a_s$ ($a_i = 0$ or 1) is

$$p^{a_1 + a_2 + \cdots + a_s} q^{s - (a_1 + a_2 + \cdots + a_s)}, \tag{19.1}$$

where

$$p = \frac{n}{m + n}, \qquad q = 1 - p.$$

To make this a stochastic process, we need a probabilistic space. Let

$$\Omega = \{\omega_1, \omega_2, \ldots, \omega_s, \ldots \mid \omega_i \in \mathscr{D}, \ i \geq 1\}.$$

D. Cheng et al., *Analysis and Control of Boolean Networks*,
Communications and Control Engineering,
DOI 10.1007/978-0-85729-097-7_19, © Springer-Verlag London Limited 2011

It is then easy to see that there exists a unique probabilistic measure P such that

$$P\big(\{\omega = (\omega_1, \omega_2, \ldots, \omega_s, \ldots) \,|\, \omega_i = a_i,\ i = 1, 2, \ldots, s\}\big)$$
$$= p^{a_1 + a_2 + \cdots + a_s} q^{s - (a_1 + a_2 + \cdots + a_s)}. \tag{19.2}$$

$\mathscr{F}$ is a σ-algebra generated by the cylinder set of Ω, i.e.,

$$\mathscr{F} = \big\{\Omega = (\omega_1, \omega_2, \ldots, \omega_s, \ldots) \,\big|\, \omega_j = a_j,\ j > s;\ s \geq 1\big\}_\sigma.$$

Now, on $(\Omega, \mathscr{F}, P)$ we define

$$\xi(t, \omega) = \omega_t, \quad t \geq 0.$$

$\xi = \{\xi(t, \cdot) \,|\, t \in \mathbb{Z}_+\}$ is then a stochastic process.

Definition 19.2 A discrete-time stochastic process $\{\xi(t) \,|\, t \in \mathbb{Z}_+\}$ on a probabilistic space $(\Omega, \mathscr{F}, P)$ with state space $I = \{1, 2, \ldots\}$ is called a Markov process (or Markov chain) if, for any positive integers $j_1 < j_2 < \cdots < j_\ell < m$,

$$P\big(\xi(m + k) = a_{m+k} \,\big|\, \xi(j_1) = a_{j_1}, \xi(j_2) = a_{j_2}, \ldots, \xi(j_\ell) = a_{j_\ell}, \xi(m) = a_m\big)$$
$$= P\big(\xi(m + k) = a_{m+k} \,\big|\, \xi(m) = a_m\big). \tag{19.3}$$

The probability of the process taking the value j at time $m + k$ and taking the value i at time m is called the k-step transition probability at m, denoted by

$$P\big(\xi(m + k) = j \,\big|\, \xi(m) = i\big) := p_{ij}^{(k)}(m). \tag{19.4}$$

It is obvious that $p_{ij}^{(k)}(m) \geq 0$ and

$$\sum_{j \in I} p_{ij}^{(k)}(m) = 1. \tag{19.5}$$

The matrix

$$P^{(k)}(m) = \big(p_{ij}^{(k)}(m)\big)\big|_{i, j \in I}$$

is called the k-step transition probability matrix, which is, in general, an infinite-dimensional matrix. When $k = 1$, we denote it simply by $P(m)$ and call it the transition probability matrix. It is particularly useful when $P(m) = P$, which is independent of m.

We have the following property.

Proposition 19.1 (Kolmogorov–Chapman equation) *For any two positive integers* k, ℓ, *we have*

$$p_{ij}^{(k+\ell)}(m) = \sum_{r \in I} p_{ir}^{(k)}(m) p_{rj}^{(\ell)}(m + k). \tag{19.6}$$

Equivalently, in matrix form, (19.6) can be written as

$$P^{(k+\ell)}(m) = P^{(k)}(m)P^{(\ell)}(m+k). \tag{19.7}$$

The one-step transition matrix is denoted by $P^1(m) := P(m)$.

Definition 19.3 A Markov chain is said to be homogeneous if $P(m) = P$ is independent of m.

Example 19.2 (Random walk) A particle is moving on a straight line according to the following rule: At time t it is at position i, and at the next moment it moves to $i+1$ with probability p or to $i-1$ with probability $q = 1 - p$. It is then easily seen that $\{\xi(t)\}$ forms a homogeneous Markov chain. The state space is $I = \{0, \pm 1, \pm 2, \ldots\}$ and the transition probabilities are

$$\begin{cases} p_{i\,i+1} = p, \\ p_{i\,i-1} = q, \\ p_{i\,j} = 0, \quad |i-j| > 1, \ i, j \in I. \end{cases} \tag{19.8}$$

It is easy to calculate that

$$p_{ij}^{(n)} = \begin{cases} \binom{n}{(n+j-i)/2} p^{(n+j-i)/2} q^{(n-j+i)/2}, & n+j-i \text{ is even}, \\ 0, & \text{otherwise}. \end{cases} \tag{19.9}$$

For states $i, j \in I$, we define the first arrival time from i to j as

$$T_{ij} = \begin{cases} \min\{n \mid \xi(0) = i, \ \xi(n) = j\}, \\ \infty, \quad \text{if } \{n \mid \xi(0) = i, \ \xi(n) = j\} = \emptyset. \end{cases} \tag{19.10}$$

The probability of arriving at j from i via n steps is

$$f_{ij}^{(n)} = P\{\xi(n) = j, \xi(m) \neq j, \ m = 1, 2, \ldots, n-1 \mid \xi(0) = i\}$$

$$= \sum_{i_1 \neq j} \cdots \sum_{i_{n-1} \neq j} p_{ii_1} p_{ii_2} \cdots p_{ii_{n-1}}, \quad n \geq 1. \tag{19.11}$$

The conditional probability of starting from $\xi(0) = i$ and arriving at j after a finite time is then

$$f_{ij} = \sum_{n=1}^{\infty} f_{ij}^{(n)} = \sum_{n=1}^{\infty} P\{T_{ij} = n\} = P\{T_{ij} < \infty\}. \tag{19.12}$$

Proposition 19.2 *If $f_{jj} = 1$, then $\xi(t)$ returns to j infinitely many times with probability 1. If $f_{jj} < 1$, then $\xi(t)$ returns to j only finitely many times with probability 1.*

Observing this, we give the following definition.

Definition 19.4 For a state $i \in I$, if $f_{ii} = 1$, then the state i is called a recurrent state and if $f_{ii} < 1$, then the state i is called a nonrecurrent (or transient) state.

Proposition 19.3 *i is recurrent if and only if*

$$\sum_{n=1}^{\infty} p_{ii}^n = \infty. \tag{19.13}$$

Assume a state $i \in I$ is recurrent. The average return time is then defined as

$$\mu_i = \sum_{n=1}^{\infty} n f_{ii}^{(n)}. \tag{19.14}$$

Definition 19.5 A recurrent state $i \in I$ is said to be positive recurrent if $\mu_i < \infty$, and it is said to be null recurrent if $\mu_i = \infty$.

Definition 19.6

1. A state $j \in I$ is said to have period t if $\{n \mid p_{jj}^{(n)} > 0\}$ have a common factor t. If $t > 1$, then the state j is called periodic. If $t = 1$, then the state j is called aperiodic.
2. If a state j is positive recurrent and aperiodic, it is called ergodic.

Proposition 19.4 *Assume that i is a recurrent state. It is then null recurrent if and only if*

$$\lim_{n \to \infty} p_{ii}^{(n)} = 0. \tag{19.15}$$

Proposition 19.5

1. *If i is an ergodic state, then*

$$\lim_{n \to \infty} p_{ii}^{(n)} = \frac{1}{\mu_i}. \tag{19.16}$$

2. *If i is positive recurrent with period t, then*

$$\lim_{n \to \infty} p_{ii}^{(nt)} = \frac{t}{\mu_i}. \tag{19.17}$$

Definition 19.7 The state j is said to be reachable from i, written as $i \to j$, if there exists some $n \geq 1$ such that $p_{ij}^{(n)} > 0$. If $i \to j$ and $j \to i$, then i and j are said to be connected, written as $i \leftrightarrow j$.

Proposition 19.6

1. $i \to j$ *if and only if* $f_{ij} > 0$.

2. *If $i \leftrightarrow j$, then they are either both nonrecurrent, or both recurrent. Moreover, if they are both recurrent, then they are either both null recurrent or both positive recurrent.*

3. *If $i \leftrightarrow j$, then they are either both aperiodic or both periodic. Moreover, if they are both periodic, then they have the same period.*

Definition 19.8 Let $C \subset I$ be a subset of the state space. C is called a closed set if, for any $i \in C$ and $j \in C^c$, we have $p_{ij} = 0$. A closed set C is said to be irreducible if the states in C are connected. A Markov chain is irreducible if it does not have a proper closed set.

Proposition 19.7 *$C \subset I$ is closed if and only if one of the following two equivalent conditions is satisfied:*

(i) *For any $i \in C$ and $j \in C^c$,*

$$p_{ij}^{(n)} = 0, \quad \forall n = 1, 2, \ldots . \tag{19.18}$$

(ii) *For any $i \in C$,*

$$\sum_{j \in C} p_{ij}^{(n)} = 1, \quad \forall n = 1, 2, \ldots . \tag{19.19}$$

We now a couple of examples.

Example 19.3

1. Consider a Markov chain $\{\xi(n) \mid n = 1, 2, \ldots\}$ with state space $I = \{1, 2, 3\}$. Its state transition graph is depicted in Fig. 19.1(a), and its transition matrix is

$$P = \begin{bmatrix} \frac{1}{3} & \frac{1}{3} & \frac{1}{3} \\ 0 & 0 & 1 \\ 1 & 0 & 0 \end{bmatrix} . \tag{19.20}$$

- From the graph it is easily seen that each state can be reached from another one, so the graph is connected. Hence, the chain is irreducible.

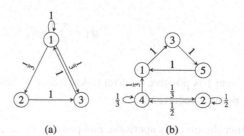

Fig. 19.1 Markov chain   (a)   (b)

- Since $p_{11} = \frac{1}{3}$, we have

$$\{n \mid p_{11}^{(n)} > 0\} = 1.$$

It follows that state 1 is aperiodic. According to Proposition 19.6, all the states are aperiodic.

-

$$f_{11}^{(1)} = \frac{1}{3},$$

$$f_{11}^{(2)} = P\{x(2) = 1, x(1) \neq 1 \mid x(0) = 1\}$$

$$= P\{x(2) = 1, x(1) = 2 \mid x(0) = 1\} + P\{x(2) = 1, x(1) = 3 \mid x(0) = 1\}$$

$$= \frac{1}{3} \cdot 0 + \frac{1}{3} \cdot 1 = \frac{1}{3},$$

$$f_{11}^{(3)} = P\{x(3) = 1, x(2) \neq 1, x(1) \neq 1 \mid x(0) = 1\}$$

$$= P\{x(3) = 1, x(2) = 2, x(1) = 2 \mid x(0) = 1\}$$

$$+ P\{x(3) = 1, x(2) = 2, x(1) = 3 \mid x(0) = 1\}$$

$$+ P\{x(3) = 1, x(2) = 3, x(1) = 2 \mid x(0) = 1\}$$

$$+ P\{x(3) = 1, x(2) = 3, x(1) = 3 \mid x(0) = 1\}$$

$$= 0 + 0 + \frac{1}{3} + 0 = \frac{1}{3},$$

$$f_{11}^{(n)} = 0, \quad n > 3.$$

Since

$$f_{11} = \sum_{n=1}^{\infty} f_{11}^n = 1,$$

state 1 is recurrent. Since

$$\mu_1 = \sum_{n=1}^{\infty} n f_{11}^n = 2 < \infty,$$

state 1 is positive recurrent. According to Proposition 19.6, the whole chain is positive recurrent.

Since the chain is aperiodic and positive recurrent, it is ergodic.

2. Consider a Markov chain $\{\xi(n) \mid n = 1, 2, \ldots\}$ with state space $I = \{1, 2, 3, 4, 5\}$. Its state transition graph is depicted in Fig. 19.1(b), and its transition matrix is

$$P = \begin{bmatrix} 0 & 0 & 1 & 0 & 0 \\ 0 & \frac{1}{2} & 0 & \frac{1}{2} & 0 \\ 0 & 0 & 0 & 0 & 1 \\ \frac{1}{3} & \frac{1}{3} & 0 & \frac{1}{3} & 0 \\ 1 & 0 & 0 & 0 & 0 \end{bmatrix}. \tag{19.21}$$

- Since

$$p_{11}^{(n)} = \begin{cases} 1, & n = 3k, \\ 0, & n \neq 3k, \ k \in \mathbb{Z}_+, \end{cases}$$

we know that state 1 is of period 3, and thus so are states 3 and 5.
- Since

$$f_{11} = \sum_{n=1}^{\infty} f_{11}^{(n)} = 1$$

and

$$\mu_1 = \sum_{n=1}^{\infty} n f_{11}^{(n)} = 3,$$

state 1 is positive recurrent (and thus so are the states 3 and 5).
- It is obvious that $C_1 = \{1, 3, 5\}$ is a closed set. Hence, the chain is not irreducible.
- Since

$$p_{22}^{(1)} = \frac{1}{2},$$

state 2 is aperiodic.
-

$$f_{22}^{(1)} = \frac{1}{2},$$

$$f_{22}^{(k)} = P\{x(k) = 2, x(s) \neq 2, \ k > s \geq 1 \mid x(0) = 2\}$$

$$= 0 + P\{x(k) = 2, x(s) = 4, \ k > s \geq 1 \mid x(0) = 2\}$$

$$= \frac{1}{2}\left(\frac{1}{3}\right)^{k-1}, \quad k = 1, 2, \ldots.$$

Since

$$f_{22} = \sum_{i=1}^{\infty} f_{22}^{(n)} = \frac{1/2}{1 - 1/3} = \frac{3}{4},$$

state 2 is nonrecurrent.

$$f_{44}^{(1)} = \frac{1}{3},$$

$$f_{44}^{(2)} = P\{x(2) = 4, x(1) \neq 4 \mid x(0) = 4\}$$

$$= 0 + P\{x(2) = 4, x(1) = 2 \mid x(0) = 4\} = \frac{1}{6}.$$

Similarly,

$$f_{44}^{(k)} = P\{x(k) = 4, x(s) \neq 4, \ k > s \geq 1 \mid x(0) = 4\}$$

$$= 0 + P\{x(k) = 4, x(s) = 2, \ k > s \geq 1 \mid x(0) = 4\}$$

$$= \frac{1}{3}\left(\frac{1}{2}\right)^{k-1}, \quad k = 1, 2, \dots.$$

Since

$$f_{44} = \sum_{i=1}^{\infty} f_{44}^{(n)} = \frac{1/3}{1 - 1/2} = \frac{2}{3},$$

state 4 is nonrecurrent.

Proposition 19.8 *Assume that $\{\xi(n)\}$ is a finite Markov chain (the state space consists of finitely many elements, that is, $|I| < \infty$). If the chain is irreducible, then all the states are positive recurrent.*

Note that the chain in part 1 of Example 19.3 is irreducible, and hence its states are all positive recurrent.

In practice the most important problem is to investigate the limiting case of the distribution of a Markov chain. We now consider this.

Definition 19.9 Let (p_{ij}) be the transition probabilities of a Markov chain. If there is a nonnegative series $\{\pi_j\}$ such that

$$\begin{cases} \sum_{j=1}^{\infty} \pi_j = 1, \\ \pi_j = \sum_{i=1}^{\infty} \pi_i \cdot p_{ij}, \quad j = 1, 2, \dots, \end{cases} \tag{19.22}$$

then $\{\pi_j\}$ is called the stationary distribution of the Markov chain.

Proposition 19.9 *Suppose we have an irreducible Markov chain and a state j that is aperiodic. Then,*

$$\lim_{n \to \infty} p_{ij}^{(n)} = \frac{1}{\mu_j} \geq 0. \tag{19.23}$$

The $\{\frac{1}{\mu_j}\}$ defined in Proposition 19.9 is called the limiting distribution.

Theorem 19.1 *Assume a Markov chain is irreducible and aperiodic. There then exists a steady-state distribution if and only if the chain is positive recurrent. Moreover, in this case the steady-state distribution is exactly the limiting distribution.*

Note that according to Proposition 19.6, if the state space is finite, i.e., $|I| < \infty$, then the positive recurrence is ensured by irreducibility.

From the above discussion we see that the properties of a Markov chain, particularly those of a homogeneous one, depend completely on its transition matrix. The transition matrix of a homogeneous Markov chain is also called a stochastic matrix. It can also be defined independently (for a finite state space) as follows.

Definition 19.10 $A \in \mathcal{M}_{n \times n}$ is called a stochastic matrix if

(i)
$$a_{ij} \geq 0, \quad \forall i, j = 1, \ldots, n,$$

(ii)
$$\sum_{j=1}^{n} a_{ij} = 1, \quad \forall i = 1, \ldots, n. \tag{19.24}$$

19.2 Vector Form of Random Boolean Variables

We first give a rigorous definition of a random Boolean variable as a variable which can take values from $\mathcal{D}_f = \{r \in \mathbb{R} \mid 0 \leq r \leq 1\}$. Assume $\alpha \in \mathcal{D}_f$. To express α in vector form, we define

$$\Lambda := \left\{ \begin{bmatrix} \alpha \\ 1 - \alpha \end{bmatrix} \middle| \alpha \in \mathcal{D}_f \right\}.$$

We now have a one-to-one correspondence between $\mathcal{D}_f$ and Λ as

$$\alpha \Leftrightarrow \begin{bmatrix} \alpha \\ 1 - \alpha \end{bmatrix}, \quad \forall \alpha \in \mathcal{D}_f.$$

In general, we define

$$\Lambda_n = \left\{ v = (v_1, \ldots, v_n)^{\mathrm{T}} \in \mathbb{R}^n \middle| v_i \geq 0, \ \sum_{i=1}^{n} v_i = 1 \right\}.$$

It is clear that $\Lambda = \Lambda_2$.

Definition 19.11 A matrix $A \in \mathcal{M}_{m \times n}$ is called a random logical matrix if

$$\mathrm{Col}(A) \in \Lambda_m.$$

The set of $m \times n$ random logical matrices is denoted by $\mathcal{L}^r_{m \times n}$.

The following results are fundamental.

Proposition 19.10

1. If $x \in \Lambda_p$ and $y \in \Lambda_q$, then

$$xy := x \ltimes y \in \Lambda_{pq}.$$

2. Let $A \in \mathcal{L}^r_{m \times n}$ and $B \in \mathcal{L}^r_{p \times q}$. If $n = pt$, then $AB := A \ltimes B \in \mathcal{L}^r_{m \times qt}$, and if $nt = p$, then $AB \in \mathcal{L}^r_{mt \times q}$.

Proof We prove item 2. Item 1 can be considered as a particular case of item 2.

First, note that if $A \succ_t B$ (resp., $A \prec_t B$), then $A \ltimes B = A(B \otimes I_t)$ [resp., $A \ltimes B = (A \otimes I_t)B$]. It is easy to see that if L is a random logical matrix, then $L \otimes I_{2^t}$ is also a random logical matrix. Using these two facts, we can assume that $n = p$. The product AB then becomes the conventional matrix product. Let $A = (a_{i,j})$ and $B = (b_{i,j})$. It is obvious that since all the entries of A and B are nonnegative, so are all the entries of AB. Hence, we have only to prove that the sum of the entries of each column of AB is 1. Consider the ith column of AB, which is

$$\mathrm{Col}_i(AB) = \begin{bmatrix} a_{1,1} & a_{1,2} & \cdots & a_{1,2^n} \\ a_{2,1} & a_{2,2} & \cdots & a_{2,2^n} \\ \vdots & & & \\ a_{2^m,1} & a_{2^m,2} & \cdots & a_{2^m,2^n} \end{bmatrix} \begin{bmatrix} b_{1,i} \\ b_{2,i} \\ \vdots \\ b_{2^n,i} \end{bmatrix}$$

$$= \begin{bmatrix} \sum_{k=1}^{2^n} a_{1,k} b_{k,i} \\ \sum_{k=1}^{2^n} a_{2,k} b_{k,i} \\ \vdots \\ \sum_{k=1}^{2^n} a_{2^m,k} b_{k,i} \end{bmatrix}.$$

Since

$$\sum_{j=1}^{2^m} \sum_{k=1}^{2^n} a_{j,k} b_{k,i} = \sum_{k=1}^{2^n} \left(\sum_{j=1}^{2^m} a_{j,k} \right) b_{k,i} = \sum_{k=1}^{2^n} b_{k,i} = 1,$$

the result follows. $\square$

Let $x_i \in \Lambda$, $i = 1, \ldots, n$, and define $x = \ltimes_{i=1}^n x_i$. We then want to know whether the x_i's can be retrieved from x. Using the retrievers S_i^n defined in Chap. 7, we can prove a similar reconstruction result.

Proposition 19.11 *If* $x = \ltimes_{i=1}^{n} x_i$, *where* $x_i \in \Lambda$, $i = 1, \ldots, n$, *then*

$$x_i = S_i^n x, \quad i = 1, 2, \ldots, n. \tag{19.25}$$

Proof Let

$$x_1 = \begin{bmatrix} p \\ 1 - p \end{bmatrix}.$$

Then we have

$$x = x_1 x_2 \cdots x_n = \begin{bmatrix} p x_2 \cdots x_n \\ (1 - p) x_2 \cdots x_n \end{bmatrix}.$$

Write

$$x_2 x_3 \cdots x_n = (\alpha_1, \alpha_2, \ldots, \alpha_{2^{n-1}})^{\mathrm{T}} \in \Lambda_{2^{n-1}}.$$

It is then easy to see that

$$S_1^n x = \begin{bmatrix} p \sum_{i=1}^{2^{n-1}} \alpha_i \\ (1 - p) \sum_{i=1}^{2^{n-1}} \alpha_i \end{bmatrix} = \begin{bmatrix} p \\ 1 - p \end{bmatrix} = x_1. \tag{19.26}$$

Note that

$$W_{[2^{k-1}, 2]} x_1 x_2 \cdots x_n = x_k x_1 \cdots x_{k-1} x_{k+1} \cdots x_n.$$

Using (19.26), we have

$$S_1^n W_{[2^{k-1}, 2]} x = x_k,$$

so it is enough to prove that

$$S_1^n W_{[2^{k-1}, 2]} = S_k^n. \tag{19.27}$$

Recall the factorization formula of a swap matrix,

$$W_{[pq, r]} = (W_{[p, r]} \otimes I_q)(I_p \otimes W_{[q, r]}), \tag{19.28}$$

we have

$$W_{[2^{k-1}, 2]} = (W_{[2]} \otimes I_{2^{k-2}})(I_2 \otimes W_{[2^{k-2}, 2]}). \tag{19.29}$$

Using (19.29), (19.27) can easily be proven by mathematical induction. □

19.3 Matrix Expression of a Random Boolean Network

Recall that a Boolean network with n nodes can be described as

$$\begin{cases} x_1(t+1) = f_1(x_1(t), x_2(t), \ldots, x_n(t)), \\ x_2(t+1) = f_2(x_1(t), x_2(t), \ldots, x_n(t)), \\ \vdots \\ x_n(t+1) = f_n(x_1(t), x_2(t), \ldots, x_n(t)), \end{cases} \tag{19.30}$$

where f_i, $i = 1, 2, \ldots, n$, are logical functions. If M_i is the structure matrix of f_i, $i = 1, 2, \ldots, n$, then (19.30) can be converted into componentwise algebraic form as

$$\begin{cases} x_1(t+1) = M_1 x_1(t) x_2(t) \cdots x_n(t), \\ x_2(t+1) = M_2 x_1(t) x_2(t) \cdots x_n(t), \\ \vdots \\ x_n(t+1) = M_n x_1(t) x_2(t) \cdots x_n(t). \end{cases} \tag{19.31}$$

If we define $x(t) = \ltimes_{i=1}^{n} x_i(t)$, then (19.31) can be converted into algebraic form as

$$x(t+1) = Lx(t). \tag{19.32}$$

The Boolean network (19.30) becomes a random Boolean network if f_i could be chosen from a previously given set of ℓ_i different models [3]. That is,

$$f_i \in \{f_i^1, f_i^2, \ldots, f_i^{\ell_i}\}, \tag{19.33}$$

and the probability of f_i being f_i^j is

$$\Pr\{f_i = f_i^j\} = p_i^j, \quad j = 1, 2, \ldots, \ell_i. \tag{19.34}$$

It is clear that

$$\sum_{j=1}^{\ell_i} p_i^j = 1, \quad i = 1, \ldots, n.$$

Summarizing the above description, we can give a rigorous definition of a random Boolean network.

Definition 19.12 A random Boolean network consists of a finite set of logical functions and probabilities,

$$\{f_i^j, \Pr(f_i = f_i^j) \mid i = 1, \ldots, n, \ j = 1, \ldots, \ell_i\}, \tag{19.35}$$

such that in the Boolean network (19.30), the ith submodel f_i is f_i^j with probability $\Pr(f_i = f_i^j)$.

A matrix K is used to denote the index set of possible models [3]:

$$K = \begin{bmatrix} 1 & 1 & \cdots & 1 & 1 \\ 1 & 1 & \cdots & 1 & 2 \\ \vdots & \vdots & \ddots & \vdots & \vdots \\ 1 & 1 & \cdots & 1 & \ell_n \\ 1 & 1 & \cdots & 2 & 1 \\ 1 & 1 & \cdots & 2 & 2 \\ \vdots & \vdots & \ddots & \vdots & \vdots \\ 1 & 1 & \cdots & 2 & \ell_n \\ \vdots & \vdots & \ddots & \vdots & \vdots \\ \ell_1 & \ell_2 & \cdots & \ell_{n-1} & \ell_n \end{bmatrix}. \tag{19.36}$$

$K \in \mathcal{M}_{N \times n}$ and $N = \prod_{j=1}^{n} \ell_j$.

Each row of K represents a possible network with probability

$$P_i = \Pr\{\text{network } i \text{ is selected}\} = \prod_{j=1}^{n} p_j^{K_{ij}}. \tag{19.37}$$

If we now define

$$x(t) := \ltimes_{i=1}^{n} x_i(t)$$

then, for each network, we have

$$x(t+1) = L_i x(t), \quad i = 1, 2, \ldots, N. \tag{19.38}$$

Hence, the overall expected value of $x(t+1)$ satisfies

$$Ex(t+1) = \sum_{i=1}^{N} P_i L_i Ex(t) := LEx(t). \tag{19.39}$$

It is easy to see that the matrix

$$L := \sum_{i=1}^{N} P_i L_i \in \mathscr{L}_{2^n \times 2^n}^r$$

is a random Boolean matrix. It is called the random network transition matrix.

Since L^T is a probability matrix, we simply say that L is irreducible (resp., aperiodic) if the Markov chain determined by L^T is irreducible (resp., aperiodic).

Using Proposition 19.8 and Theorem 19.1, we have the following.

Proposition 19.12 *If L is irreducible and aperiodic, then there exists a steady-state distribution*

$$\pi_i \geq 0, \quad \sum_{i=1}^{2^n} \pi_i = 1,$$

such that

$$P\left\{x = \ltimes_{i=1}^{n} x_i = \delta_{2^n}^i\right\} = \pi_i, \quad i = 1, \ldots, n.$$

In fact, we have

$$\lim_{t \to \infty} L^t = \begin{bmatrix} \pi_1 & \cdots & \pi_1 \\ \pi_2 & \cdots & \pi_2 \\ \vdots & & \\ \pi_{2^n} & \cdots & \pi_{2^n} \end{bmatrix}.$$

We now give an example.

Example 19.4 Consider the system

$$\begin{cases} A(t+1) = f_1(A(t), B(t), C(t)), \\ B(t+1) = f_2(A(t), B(t), C(t)), \\ C(t+1) = f_3(A(t), B(t), C(t)), \end{cases} \tag{19.40}$$

where

$$\begin{cases} f_1^1 = [A_1(t) \wedge (\neg(A_2(t) \wedge A_3(t)))] \vee [(\neg A_1(t)) \wedge A_2(t)], \\ f_1^2 = [A_1(t) \wedge (\neg(A_3(t) \to A_2(t)))] \vee [(\neg A_1(t)) \wedge (\neg(A_2(t) \leftrightarrow A_3(t)))] \end{cases}$$

with

$$\Pr\left(f_1 = f_1^1\right) = 0.4, \qquad \Pr\left(f_1 = f_1^2\right) = 0.6,$$

$$\begin{cases} f_2^1 = [A_1(t) \wedge (A_2(t) \leftrightarrow A_3(t))] \vee [(\neg A_1(t)) \wedge (\neg(A_2(t)))], \\ f_2^2 = [A_1(t) \wedge A_2(t)] \vee [(\neg A_1(t)) \wedge (A_2(t) \leftrightarrow A_3(t))] \end{cases}$$

with

$$\Pr\left(f_2 = f_2^1\right) = 0.6, \qquad \Pr\left(f_2 = f_2^2\right) = 0.4,$$

and

$$\begin{cases} f_3^1 = A_1(t) \wedge (A_3(t) \to A_2(t)), \\ f_3^2 = [A_1(t) \wedge A_2(t) \wedge A_3(t)] \vee [(\neg A_1(t)) \wedge A_3(t)] \end{cases}$$

with

$$\Pr\left(f_3 = f_3^1\right) = 0.4, \qquad \Pr\left(f_3 = f_3^2\right) = 0.6.$$

The model-index matrix K and the model probabilities are now

$$K = \begin{bmatrix} 1 & 1 & 1 \\ 1 & 1 & 2 \\ 1 & 2 & 1 \\ 1 & 2 & 2 \\ 2 & 1 & 1 \\ 2 & 1 & 2 \\ 2 & 2 & 1 \\ 2 & 2 & 2 \end{bmatrix},$$

$$\begin{aligned} P_1 &= 0.4 \times 0.6 \times 0.4 = 0.096, \\ P_2 &= 0.4 \times 0.6 \times 0.6 = 0.144, \\ P_3 &= 0.4 \times 0.4 \times 0.4 = 0.064, \\ P_4 &= 0.4 \times 0.4 \times 0.6 = 0.096, \\ P_5 &= 0.6 \times 0.6 \times 0.4 = 0.144, \\ P_6 &= 0.6 \times 0.6 \times 0.6 = 0.216, \\ P_7 &= 0.6 \times 0.4 \times 0.4 = 0.096, \\ P_8 &= 0.6 \times 0.4 \times 0.6 = 0.144. \end{aligned}$$

Denote the structure matrix of f_i^j by M_i^j. It is then easy to calculate that

$$M_1^1 = \delta_2[2\,1\,1\,1\,1\,1\,2\,2],$$

$$M_1^2 = \delta_2[2\,2\,1\,2\,2\,1\,1\,2],$$

$$M_2^1 = \delta_2[1\,2\,2\,1\,2\,2\,1\,1],$$

$$M_2^2 = \delta_2[1\,1\,2\,2\,1\,2\,2\,1],$$

$$M_3^1 = \delta_2[1\,1\,2\,1\,2\,2\,2\,2],$$

$$M_3^2 = \delta_2[1\,2\,2\,2\,1\,2\,1\,2].$$

Now, set $x(t) = A(t)B(t)C(t)$. The network matrix of each network can then be calculated using a standard procedure. For example, for the first model we have

$$x(t+1) = M_1^1 x(t) M_2^1 x(t) M_3^1 x(t) := L_1 x(t),$$

where L_1 can be calculated as

$$L_1 = \delta_8[8\,1\,1\,1\,1\,1\,8\,8].$$

Similarly, we can calculate L_i, $i = 2, 3, \ldots, 8$. Finally, the random network matrix of the random Boolean network is found to be

$$L = \sum_{i=1}^{8} P_i L_i$$

$$= \begin{bmatrix} 0 & 0.4 & 0 & 0 & 0 & 0 & 0.6 & 0 \\ 0 & 0 & 0 & 0 & 0.4 & 0 & 0 & 0 \\ 0 & 0 & 0 & 0.4 & 0 & 0 & 0 & 0 \\ 0 & 0 & 1 & 0 & 0 & 1 & 0 & 0 \\ 1 & 0 & 0 & 0 & 0 & 0 & 0 & 1 \\ 0 & 0 & 0 & 0.6 & 0 & 0 & 0 & 0 \\ 0 & 0 & 0 & 0 & 0.6 & 0 & 0 & 0 \\ 0 & 0.6 & 0 & 0 & 0 & 0 & 0.4 & 0 \end{bmatrix}. \tag{19.41}$$

Next, we consider another example.

Example 19.5 [3] The system of equations is as (19.40), where f_1 has two models, f_1^1 and f_1^2. $\Pr(f_1 = f_1^1) = 0.6$, $\Pr(f_1 = f_1^2) = 0.4$, and

$$M_1^1 = \delta_2[1\ 1\ 1\ 2\ 1\ 1\ 1\ 2],$$
$$M_1^2 = \delta_2[1\ 1\ 1\ 2\ 2\ 1\ 1\ 2].$$

f_2 has only one model, and

$$M_2 = \delta_2[1\ 2\ 1\ 1\ 2\ 1\ 1\ 2].$$

f_3 has two models, f_3^1 and f_3^2. $\Pr(f_3 = f_3^1) = 0.5$, $\Pr(f_3 = f_3^2) = 0.5$, and

$$M_3^1 = \delta_2[1\ 1\ 1\ 2\ 1\ 2\ 2\ 2],$$
$$M_3^2 = \delta_2[1\ 2\ 2\ 2\ 2\ 2\ 2\ 2].$$

It is then easy to calculate that

$$L_1 = \delta_8[1\ 3\ 1\ 6\ 3\ 2\ 2\ 8], \qquad P_1 = 0.2,$$
$$L_2 = \delta_8[1\ 4\ 2\ 6\ 4\ 2\ 2\ 8], \qquad P_1 = 0.2,$$
$$L_3 = \delta_8[1\ 3\ 1\ 6\ 7\ 2\ 2\ 8], \qquad P_3 = 0.3,$$
$$L_4 = \delta_8[1\ 4\ 2\ 6\ 8\ 2\ 2\ 8], \qquad P_4 = 0.3.$$

Finally, we have

$$L = \begin{bmatrix} 1 & 0 & 0.5 & 0 & 0 & 0 & 0 & 0 \\ 0 & 0 & 0.5 & 0 & 0 & 1 & 1 & 0 \\ 0 & 0.5 & 0 & 0 & 0.2 & 0 & 0 & 0 \\ 0 & 0.5 & 0 & 0 & 0.2 & 0 & 0 & 0 \\ 0 & 0 & 0 & 1 & 0 & 0 & 0 & 0 \\ 0 & 0 & 0 & 0 & 0 & 0 & 0 & 0 \\ 0 & 0 & 0 & 0 & 0.3 & 0 & 0 & 0 \\ 0 & 0 & 0 & 0 & 0.3 & 0 & 0 & 1 \end{bmatrix}. \tag{19.42}$$

An interesting feature of this system is that there is a "pseudo-steady-state distribution". Define

$$L_s := \lim_{k \to \infty} L^k. \tag{19.43}$$

Such a limit then exists, which is

$$
L_s = \begin{bmatrix}
1 & \frac{5}{8} & \frac{13}{16} & \frac{7}{16} & \frac{7}{16} & \frac{5}{8} & \frac{5}{8} & 0 \\
0 & 0 & 0 & 0 & 0 & 0 & 0 & 0 \\
0 & 0 & 0 & 0 & 0 & 0 & 0 & 0 \\
0 & 0 & 0 & 0 & 0 & 0 & 0 & 0 \\
0 & 0 & 0 & 0 & 0 & 0 & 0 & 0 \\
0 & 0 & 0 & 0 & 0 & 0 & 0 & 0 \\
0 & 0 & 0 & 0 & 0 & 0 & 0 & 0 \\
0 & \frac{3}{8} & \frac{3}{16} & \frac{9}{16} & \frac{9}{16} & \frac{3}{8} & \frac{3}{8} & 1
\end{bmatrix}.
\tag{19.44}
$$

There are two fixed points, $P = \delta_8^1 \sim (1, 1, 1)^\mathrm{T}$ and $Q = \delta_8^8 \sim (0, 0, 0)^\mathrm{T}$. Starting from any initial value, the trajectory will converge to either P or Q with probability 1. However, this is not a genuine steady-state distribution because, starting from different points, the probabilities of convergence to P and Q will vary according to the initial value.

19.4 Some Topological Properties

This section is based on [2]. First, we consider the cycles of a random Boolean network. We consider a fixed point to be a cycle of length 1.

The following result is obvious.

Proposition 19.13 *Consider a random Boolean network Σ. Assume that it has N possible models, Σ_i, with $P_i = P(\Sigma = \Sigma_i) > 0$, $i = 1, \dots, N$. If C is a common cycle of all Σ_i, then C is a cycle of Σ.*

Proposition 19.14 *Consider a random Boolean network Σ. Assume that it has N possible models, Σ_i, with $P_i = P(\Sigma = \Sigma_i) > 0$, $i = 1, \dots, N$. Assume that:*

 (i) *C is a common cycle of all Σ_i,*
(ii) *there is an i^* such that C is the unique attractor of Σ_{i^*}.*

The network then converges to C with probability 1.

Proof Since C is the unique attractor of Σ_{i^*}, there is a transient time T_t such that as $\Sigma = \Sigma_{i^*}$ for a period $[t_1, t_2]$ with $t_2 - t_1 + 1 \geq T_t$, all the trajectories will enter C. Consider the time period $(kT_t, (k+1)T_t]$:

$$
P\{\Sigma(t) = \Sigma_{i^*}(t) \,|\, kT_t < t \leq (k+1)T_t\} = P_{i^*}^{T_t} > 0.
$$

Now, consider the time period $[0, mT_t]$:

$$
P\{\Sigma(t) = \Sigma_{i^*}(t) \,|\, kT_t < t \leq (k+1)T_t; \ 0 \leq k < m\} = 1 - \left(1 - P_{i^*}^{T_t}\right)^m.
$$

As $m \to \infty$ one sees that the probability of Σ_{i*} appearing sequentially over T_t times is 1. Hence, all the trajectories of the network converge to C with probability 1. $\square$

Finally, we consider the random Boolean control network. The system is described as

$$\begin{cases} x_1(t+1) = f_1(x_1(t), x_2(t), \ldots, x_n(t), u_1(t), \ldots, u_m(t)), \\ x_2(t+1) = f_2(x_1(t), x_2(t), \ldots, x_n(t), u_1(t), \ldots, u_m(t)), \\ \vdots \\ x_n(t+1) = f_n(x_1(t), x_2(t), \ldots, x_n(t), u_1(t), \ldots, u_m(t)). \end{cases} \tag{19.45}$$

Now assume f_i can equal one of f_i^j, $j = 1, 2, \ldots, \ell_i$, with probabilities

$$\Pr\{f_i = f_i^j\} = p_i^j > 0, \quad j = 1, 2, \ldots, \ell_i. \tag{19.46}$$

We consider the stabilization problem of (19.45). Using Proposition 19.14, we have the following result.

Corollary 19.1 *Consider the random Boolean control network* (19.45). *Assume that there exists a fixed point x_e and a set of controls*

$$\left(u_1^i, \ldots, u_m^i\right), \quad i = 1, \ldots, N, \tag{19.47}$$

such that for the closed-loop models Σ_i, $i = 1, \ldots, N$:

(i) *x_e is a common fixed point of all Σ_i,*
(ii) *there is an i^* such that x_e is the unique attractor of Σ_{i*}.*

The closed-loop network then converges to x_e with probability 1. In other words, the controls (19.47) *stabilize the network* (19.45).

Example 19.6 Consider the system

$$\begin{cases} A(t+1) = f_1(A(t), B(t), C(t), u(t)), \\ \dot{B}(t+1) = f_2(A(t), B(t), C(t), u(t)), \\ C(t+1) = f_3(A(t), B(t), C(t), u(t)), \end{cases} \tag{19.48}$$

where

$$\begin{cases} f_1^1 = A(t) \wedge C(t), \\ f_1^2 = A(t) \wedge B(t) \end{cases}$$

with

$$\Pr(f_1 = f_1^1) = 0.2, \qquad \Pr(f_1 = f_1^2) = 0.8,$$

$$\begin{cases} f_2^1 = \neg A(t) \vee C(t), \\ f_2^2 = (A(t) \wedge C(t)) \vee u^2(t) \end{cases}$$

with

$$\Pr\left(f_2 = f_2^1\right) = 0.7, \qquad \Pr\left(f_2 = f_2^2\right) = 0.3,$$

and

$$\begin{cases} f_3^1 = (B(t) \leftrightarrow C(t)) \wedge u^1(t), \\ f_3^2 = A(t) \wedge \neg B(t), \end{cases}$$

with

$$\Pr\left(f_3 = f_3^1\right) = 0.4, \qquad \Pr\left(f_3 = f_3^2\right) = 0.6.$$

The model-index matrix K and the model probabilities are

$$K = \begin{bmatrix} 1 & 1 & 1 \\ 1 & 1 & 2 \\ 1 & 2 & 1 \\ 1 & 2 & 2 \\ 2 & 1 & 1 \\ 2 & 1 & 2 \\ 2 & 2 & 1 \\ 2 & 2 & 2 \end{bmatrix}, \qquad \begin{aligned} P_1 &= 0.2 \times 0.7 \times 0.4 = 0.056, \\ P_2 &= 0.2 \times 0.7 \times 0.6 = 0.084, \\ P_3 &= 0.2 \times 0.3 \times 0.4 = 0.024, \\ P_4 &= 0.2 \times 0.3 \times 0.6 = 0.036, \\ P_5 &= 0.8 \times 0.7 \times 0.4 = 0.224, \\ P_6 &= 0.8 \times 0.7 \times 0.6 = 0.336, \\ P_7 &= 0.8 \times 0.3 \times 0.4 = 0.096, \\ P_8 &= 0.8 \times 0.3 \times 0.6 = 0.144. \end{aligned}$$

Using the control

$$\begin{cases} u^1(t) = A(t), \\ u^2(t) = \neg A(t) \wedge \neg C(t), \end{cases} \tag{19.49}$$

we can calculate the network matrices for all the models as follows:

$$L_1 = \delta_8[1\,8\,2\,7\,6\,6\,6\,6],$$

$$L_2 = \delta_8[2\,8\,1\,7\,6\,6\,6\,6],$$

$$L_3 = \delta_8[1\,8\,2\,7\,6\,6\,6\,6],$$

$$L_4 = \delta_8[2\,8\,1\,7\,6\,6\,6\,6],$$

$$L_5 = \delta_8[1\,4\,6\,7\,6\,6\,6\,6],$$

$$L_6 = \delta_8[2\,4\,5\,7\,6\,6\,6\,6],$$

$$L_7 = \delta_8[1\,4\,6\,7\,6\,6\,6\,6],$$

$$L_8 = \delta_8[2\,4\,5\,7\,6\,6\,6\,6].$$

It is easy to show that models 1, 3, 5, and 7 have two fixed points, $(1, 1, 1)^T$ and $(0, 1, 0)^T$, and models 2, 4, 6, and 8 have only one fixed point, $(0, 1, 0)^T$. Hence, $(0, 1, 0)^T$ is the only common fixed point for these models.

We now calculate the network transition matrix of the random Boolean network:

$$
L = \begin{bmatrix}
\frac{2}{5} & 0 & \frac{3}{25} & 0 & 0 & 0 & 0 & 0 \\
\frac{3}{5} & 0 & \frac{2}{25} & 0 & 0 & 0 & 0 & 0 \\
0 & 0 & 0 & 0 & 0 & 0 & 0 & 0 \\
0 & \frac{4}{5} & 0 & 0 & 0 & 0 & 0 & 0 \\
0 & 0 & \frac{12}{25} & 0 & 0 & 0 & 0 & 0 \\
0 & 0 & \frac{8}{25} & 0 & 1 & 1 & 1 & 1 \\
0 & 0 & 0 & 1 & 0 & 0 & 0 & 0 \\
0 & \frac{1}{5} & 0 & 0 & 0 & 0 & 0 & 0
\end{bmatrix}.
$$

We can also calculate that the limit of L is

$$
L_s = \lim_{k \to \infty} L^k = \begin{bmatrix}
0 & 0 & 0 & 0 & 0 & 0 & 0 & 0 \\
0 & 0 & 0 & 0 & 0 & 0 & 0 & 0 \\
0 & 0 & 0 & 0 & 0 & 0 & 0 & 0 \\
0 & 0 & 0 & 0 & 0 & 0 & 0 & 0 \\
0 & 0 & 0 & 0 & 0 & 0 & 0 & 0 \\
1 & 1 & 1 & 1 & 1 & 1 & 1 & 1 \\
0 & 0 & 0 & 0 & 0 & 0 & 0 & 0 \\
0 & 0 & 0 & 0 & 0 & 0 & 0 & 0
\end{bmatrix}.
$$

There is only one fixed point $C = \delta_8^6 \sim (0, 1, 0)^T$. That is, the network converges to C with probability 1.

References

1. Cinlar, E.: Introduction to Stochastic Processes. Prentice Hall, New York (1997)
2. Qi, H., Cheng, D., Hu, X.: Stabilization of random Boolean networks. In: Proc. WCICA'2010, pp. 1968–1973 (2010)
3. Shmulevich, I., Dougherty, E., Kim, S., Zhang, W.: Probabilistic Boolean networks: a rule-based uncertainty model for gene regulatory networks. Bioinformatics **18**(2), 261–274 (2002)

Appendix A
Numerical Algorithms

A.1 Computation of Logical Matrices

In computing a logical matrix L and other related matrices involved in this book, it is easily seen that the dimension grows exponentially with n. To reduce the computational complexity, we present in this section some formulas for the computation of logical matrices, which will be used in the computation of examples in the next section.

A matrix $L \in \mathcal{M}_{m \times n}$ is called a logical matrix if its columns are of the form δ_m^i.

Example A.1

1. The structure matrix of any logical operator is a logical matrix. For instance, M_n, M_d, M_c, M_i, M_e, etc. are all logical matrices.
2. The swap matrix $W_{[m,n]}$ is a logical matrix.
3. The power-reducing matrix M_r (or $M_{r,k}$) is a logical matrix.

Note that from previous examples one may find that for computing system matrix L, only delta matrices are involved.

Now, if $\psi \in \mathcal{L}_{m \times n}$ is a logical matrix, then ψ can be expressed as

$$\psi = \left[\delta_m^{i_1}, \delta_m^{i_2}, \ldots, \delta_m^{i_{2q}} \right].$$

In the text of this book it is denoted as

$$\psi = \delta_m[i_1, \ldots, i_n].$$

In the toolbox it is denoted as

$$\psi = \left([i_1, i_2, \ldots, i_n], m \right).$$

D. Cheng et al., *Analysis and Control of Boolean Networks*,
Communications and Control Engineering,
DOI 10.1007/978-0-85729-097-7, © Springer-Verlag London Limited 2011

We call such an expression the condensed form of a logical matrix. Using this notation we now deduce some formulas which are useful in computations.

Proposition A.1

1. *Assume that* $\psi = ([i_1, i_2, \ldots, i_{2^q}], p)$. *Then*

$$\psi \otimes I_{2^r} = \big([(i_1 - 1)2^r + 1 \quad (i_1 - 1)2^r + 2 \quad \cdots \quad (i_1)2^r$$
$$(i_2 - 1)2^r + 1 \quad (i_2 - 1)2^r + 2 \quad \cdots \quad (i_2)2^r$$

$$\vdots$$

$$(i_{2^q} - 1)2^r + 1 \ (i_{2^q} - 1)2^r + 2 \cdots \ (i_{2^q})2^r], p + r). \qquad (A.1)$$

2. *Assume that* $\psi = ([i_1, i_2, \ldots, i_{2^q}], p)$. *Then*

$$I_{2^r} \otimes \psi = \big([i_1 \qquad\qquad i_2 \qquad\qquad \cdots\ i_{2^q}$$
$$2^p + i_1 \qquad\qquad 2^p + i_2 \qquad\qquad \cdots\ 2^p + i_{2^q}$$
$$2 \times 2^p + i_1 \qquad\qquad 2 \times 2^p + i_2 \qquad\qquad \cdots\ 2 \times 2^p + i_{2^q}$$

$$(2^r - 1) \times 2^p + i_1 \ (2^r - 1) \times 2^p + i_2 \cdots \ (2^r - 1) \times 2^p + i_{2^q}], p + r).$$
$$(A.2)$$

3. *Assume that* $\psi = ([i_1, i_2, \ldots, i_{2^q}], p)$, $\phi = ([j_1, j_2, \ldots, j_{2^r}], q)$. *Then*

$$\psi \phi = \big([i_{j_1}, i_{j_2}, \ldots, i_{j_{2^r}}], p\big). \qquad (A.3)$$

Formulas (A.1)–(A.3) are enough to calculate the transition matrix L of a Boolean (control) network.

Next, let $\mathscr{X} = \mathscr{F}\{x_1, \ldots, x_n\}$, $y_1, \ldots, y_p \in \mathscr{X}$, and $z_1, \ldots, z_q \in \mathscr{X}$. Set $x = \ltimes_{i=1}^{n} x_i$, $y = \ltimes_{i=1}^{p} y_i$, and $z = \ltimes_{i=1}^{q} z_i$. Assume that

$$y = Px, \qquad z = Qx,$$

where $P \in \mathscr{L}_{2^p \times 2^n}$ and $Q \in \mathscr{L}_{2^q \times 2^n}$. We then have the following.

Proposition A.2

$$yz = Wx, \qquad (A.4)$$

where $W \in \mathscr{L}_{2^{p+q} \times 2^n}$ *can be calculated as follows. Denote by* $\text{Col}_i(W)$ *[resp.,* $\text{Col}_i(P)$, $\text{Col}_i(Q)$*] the ith column of W (resp., P, Q). Then*

$$\text{Col}_i(W) = \text{Col}_i(P) \ltimes \text{Col}_i(Q), \quad i = 1, 2, \ldots, 2^n.$$

A.2 Basic Functions

1. Calculate the semi-tensor product of A and B:

```
function c = sp(a,b)
% SP      Semi-Tensor Product of Matrices using Kronecker
    product
%
%    SP(A,B) is to calculate the semi-tensor product of A and
    B.
%    The number of columns of the fisrt matrix must be the
    divisor
%    or multiple of the number of rows of the last matrix.

if ¬(isa(a,'sym') | isa(a,'double'))
    a = double(a);
end
if ¬(isa(b,'sym') | isa(b,'double'))
    b = double(b);
end

if ndims(a) > 2 | ndims(b) > 2
    error('Input arguments must be 2-D.');
end

[m,n] = size(a);
[p,q] = size(b);
if n == p
    c = a*b;
elseif mod(n,p) == 0
    z = n/p;
    c = zeros(m,z*q);
    c = a*kron(b,eye(z));
elseif mod(p,n) == 0
    z = p/n;
    c = zeros(m*z,q);
    c = kron(a,eye(z))*b;
else
    error('dimension error: sp');
end;
```

2. Calculate the semi-tensor product of n (≥ 2) matrices:

```
function r = spn(varargin)

% SPN      Semi-tensor product of matrices with arbitrary number
    of matrices
%
%    SPN(A,B,C,...) calculates the semi-tensor product of
    arbitrary
%    number of matrices which have the proper dimensions.

ni = nargin;
```

```
switch ni
    case 0
        error('No input arguments.')
    case 1
        r = varargin{1};
        return
    case 2
        r = sp(varargin{1},varargin{2});
        return
    otherwise
        r = sp(varargin{1},varargin{2});
        for i = 3:ni
            r = sp(r,varargin{i});
        end
end
```

3. Calculate the swap matrix $W_{[m,n]}$:

```
function w = wij(m,n)

% WIJ      Produces swap matrix
%
%    A = WIJ(N) produces an N^2-by-N^2 swap matrix.
%    A = WIJ(M,N) produces an MN-by-MN swap matrix.

if nargin == 1
    n=m;
end

d = m*n;
w = zeros(d);
for k = 1:d
    j = mod(k,n);
    if j == 0
        j = n;
    end
    i = (k-j)/n+1;
    w((j-1)*m+i,k) = 1;
end;
```

4. Create a semi-tensor product object:

```
function m = stp(a)

% STP/STP semi-tensor product (STP) class constructor
% m = stp(a) creates an STP object from the matrix A
```

5. Create an LM object:

```
function m = lm(varargin)

% LM/LM      logical matrix (LM) class constructor
```

```
%
%  M = LM(A)   creates  an LM object  from  the  matrix A
%  Example: m = lm(eye(3))
%
%  M = LM(V,N)  creates  an LM object  from  a  vector V and a
%     positive  integer  N
%  Example: m = lm([1,2,2,3],4)
```

6. Create the logical matrix for an $n \times n$ identity matrix:

```
function m = leye(n)

% LEYE     Create  an  n-by-n  identity  matrix ,  return  an LM
%     object
%
%   M = LEYE(N)
%
%   Example: m = leye(3),  class(m)

if n < 0
    error('Input  argument  must  be  a  positive  integer')
end

m = lm(1:n,n);
```

7. Create the logical matrix for power-reducing matrix:

```
function Mr = lmr(k,n)

% LMR     Produces  power-reducing  matrix ,  returns  an LM object
%
%   The power-reducing  matrix M satisfies  P^2=MP, where P is a
%     logical  variable.
%
%   M = LMR      for  classical  logic
%   M = LMR(K)   for  k-valued  logic
%
%   Example: m = lmr, m = lmr(2)

if nargin == 0 | isempty(k), k = 2; end;

a = 1:k;
Mr = lm(a+(a-1)*k,k^2);
```

8. Create the logical matrix for negation:

```
function m = lmn(k);

% LMN     Produces  logical  matrix  for  negation ,  returns  an LM
%     object
%
%   M = LMN      for  classical  logic
```

```
%    M = LMN(K)   for k-valued logic
%
%    Example: m = lmn, m = lmn(2)

if nargin == 0 | isempty(k)
    k = 2;
end

m = lm(k:-1:1,k);
```

9. Create the logical matrix for conjunction:

```
function m = lmc(k)

% LMC      Produces logical matrix for conjunction, returns an
    LM object
%
%    M = LMC       for classical logic
%    M = LMC(K)    for k-valued logic
%
%    Example: m = lmc, m = lmc(2)

if nargin == 0 | isempty(k)
    k = 2;
end

m = lm;
m.n = k;

a = 1:k;
p = a(ones(1,k),:);
p = (p(:))';
q = repmat(a,1,k);
b = p≥q;
m.v = p.*b+q.*¬b;
```

10. Create the logical matrix for disjunction:

```
function m = lmd(k)

% LMD      Produces logical matrix for disjunction, returns an
    LM object
%
%    M = LMD       for classical logic
%    M = LMD(K)    for k-valued logic
%
%    Example: m = lmd, m = lmd(2)

if nargin == 0 | isempty(k)
    k = 2;
end

m = lm;
```

```
m.n = k;

a = 1:k;
p = a(ones(1,k),:);
p = (p(:))';
q = repmat(a,1,k);
b = p≤q;
m.v = p.*b+q.*¬b;
```

11. Create the logical matrix for implication:

```
function m = lmi(k)

% LMI     Produces logical matrix for implication, returns an
    LM object
%
%   M = LMI       for classical logic
%   M = LMI(K)    for k-valued logic
%
%   Example: m = lmi, m = lmi(2)

if nargin == 0 | isempty(k)
    k = 2;
end

Md = lmd(k);
Mn = lmn(k);

m = Md*Mn;
```

12. Create the logical matrix for equivalence:

```
function m = lme(k)

% LME     Produces logical matrix for equivalence, returns an
    LM object
%
%   M = LME       for classical logic
%   M = LME(K)    for k-valued logic
%
%   Example: m = lme, m = lme(2)

if nargin == 0 | isempty(k)
    k = 2;
end

Mc = lmc(k);
Mi = lmi(k);
Mr = lmr(k);

m = Mc*Mi*(leye(k^2)+Mi)*(leye(k)+Mr)*(leye(k)+lwij(k))*Mr;
```

13. Create the dummy logical matrix:

```
function m = lmu(k)

% LMU      Produces dummy logical matrix, returns an LM object
%
%     The dummy logical matrix M satisfies MXY = Y, where X, Y
%     are two logical variables
%
%     M = LMU         for classical logic
%     M = LMU(K)      for k-valued logic
%
%     Example: m = lmu, m = lmu(2)

if nargin == 0 | isempty(k)
    k = 2;
end

m = lm(repmat(1:k,1,k),k);
```

A.3 Some Examples

1. Calculate the semi-tensor product:

```
% This example is to show how to perform semi-tensor product

x = [1  2  3  -1];
y = [2  1]';
r1 = sp(x,y)
% r1 = [5,3]

x = [2  1];
y = [1  2  3  -1]';
r2 = sp(x,y)
% r2 = [5;3]

x = [1  2  1  1;
     2  3  1  2;
     3  2  1  0];
y = [1  -2;
     2  -1];
r3 = sp(x,y)
% r3 = [3,4,-3,-5;4,7,-5,-8;5,2,-7,-4]

r4 = spn(x,y,y)
% r4 = [-3,-6,-3,-3;-6,-9,-3,-6;-9,-6,-3,0]
```

2. Examples for semi-tensor product class:

```
% This example is to show the usage of stp class.
```

```
% Many useful methods are overloaded for stp class, thus you
    can use stp object as double.

x = [1 2 1 1;
     2 3 1 2;
     3 2 1 0];
y = [1 -2;
     2 -1];

% Covert x and y to stp class
a = stp(x)
b = stp(y)

% mtimes method is overloaded by semi-tensor product for stp
    class
c0 = spn(x,y,y)
c = a*b*b, class(c)

% Convert an stp object to double
c1 = double(c), class(c1)

% size method for stp class
size(c)

% length method for stp class
length(c)

% subsref method for stp class
c(1,:)

% subsasgn method for stp class
c(1,1) = 3
```

3. Examples for the LM class:

```
% This example is to show the usage of lm class.
% Many methods are overloaded for lm class.

% Consider classical (2-valued) logic here
k = 2;

T = lm(1,k); % True
F = lm(k,k); % False

% Given a logical matrix, and convert it to lm class
A = [1 0 0 0;
     0 1 1 1]
M = lm(A)
% or we can use
% M = lm([1 2 2 2], 2)

% Use m-function to perform semi-tensor product for logical
    matrices
```

```
r1 = lspn(M,T,F)

% Use overloaded mtimes method for lm class to perform semi-
    tensor product
r2 = M*T*F

% Create a 4-by-4 logical matrix randomly
M1 = lmrand(4)
% M1 = randlm(4)

% Convert an lm object to double
double(M1)

% size method for lm class
size(M1)

% diag method for lm class
diag(M1)

% Identity matrix is a special type of logical matrix
I3 = leye(3)

% plus method is overloaded by Kronecher product for lm class
r3 = M1 + I3
% Alternative way to perform Kronecher product of two logical
    matrices
r4 = lkro(M1,I3)

% Create an lm object by assignment
M2 = lm;
M2.n = 2;
M2.v = [2 1 1 2];
M2
```

4. Consider Example 5.9:

```
% Initialize
k = 2;
MN = lmn(k); % negation
MI = lmi(k); % implicaiton
MC = lmc(k); % conjunction
MD = lmd(k); % disjunction
ME = lme(k); % equivalence
MR = lmr(k); % power-reducing matrix
MU = lmu(k); % dummy matrix
 options = [];

% Dynamics of Boolean network
    % A(t+1) = MN*MD*C(t)*F(t)
    % B(t+1) = A(t)
    % C(t+1) = B(t)
    % D(t+1) = MC*MC*MN*I(t)*MN*C*MN*F(t)
    % E(t+1) = D(t)
```

```
       % F(t+1) = E(t)
       % G(t+1) = MN*MD*F(t)*I(t)
       % H(t+1) = G(t)
       % I(t+1) = H(t)
   % Set X(t)=A(t)B(t)C(t)D(t)E(t)F(t)G(t)H(t)I(t), then

   eqn = {'MN MD C F',
          'A',
          'B',
          'MC MC MN I MN C MN F',
          'D',
          'E',
          'MN MD F I',
          'G',
          'H'};

   % Set the variables' order, otherwise they will be sorted in
       the dictionary order
   options = lmset('vars',{'A','B','C','D','E','F','G','H','I'});

   % Convert the logical equations to their canonical form
   [expr,vars] = stdform(strjoin(eqn),options,k);

   % Calculate the network transition matrix
   L = eval(expr)

   % Analyze the dynamics of the Boolean network
   [n,l,c,r0,T] = bn(L,k);

   fprintf('Number of attractors: %d\n\n',n);
   fprintf('Lengths of attractors:\n');
   disp(l);
   fprintf('\nAll attractors are displayed as follows:\n\n');
   for i=1:length(c)
       fprintf('No. %d (length %d)\n\n',i,l(i));
       disp(c{i});
   end
   fprintf('Transient time: [T_t, T] = [%d %d]\n\n',r0,T);
```

5. Consider Example 14.1:

```
% Initialize
k = 3;
MN = lmn(k); % negation
MI = lmi(k); % implicaiton
MC = lmc(k); % conjunction
MD = lmd(k); % disjunction
ME = lme(k); % equivalence
MR = lmr(k); % power−reducing matrix
MU = lmu(k); % dummy matrix
options = [];

% Dynamics of Boolean network
```

```
    % A(t+1) = A(t)
    % B(t+1) = MI*A(t)*C(t)
    % C(t+1) = MD*B(t)*D(t)
    % D(t+1) = MN*B(t)
    % E(t+1) = MN*C(t)
% Set X(t)=A(t)B(t)C(t)D(t)E(t), then

eqn = {'MU E A',
       'MI A C',
       'MD B D',
       'MN B',
       'MN C'};

% Set the variables' order, otherwise they will be sorted in
    the dictionary order
options = lmset('vars',{'A','B','C','D','E'});

% Convert the logical equations to their canonical form
[expr,vars] = stdform(strjoin(eqn),options,k);

% Calculate the network transition matrix
L = eval(expr)

% Analyze the dynamics of the Boolean network
[n,l,c,r0,T] = bn(L,k);

fprintf('Number of attractors: %d\n\n',n);
fprintf('Lengths of attractors:\n');
disp(l);
fprintf('\nAll attractors are displayed as follows:\n\n');
for i=1:length(c)
    fprintf('No. %d (length %d)\n\n',i,l(i));
    disp(c{i});
end
fprintf('Transient time: [T_t, T] = [%d %d]\n\n',r0,T);
```

Appendix B
Proofs of Some Theorems Concerning the Semi-tensor Product

The proves in this appendix are cited from [1] with the permission from Science Press.

(1) *Proof of Theorem* 2.1

Proof The first part (distributive law) can be proven by a straightforward computation, so we prove only the second part (associative law).

First, we show that if F, G, and H have feasible dimensions for $(F \ltimes G) \ltimes H$, then the dimensions are also feasible for $F \ltimes (G \ltimes H)$.

Case 1. $F \succ G$ and $G \succ H$. The dimensions of F, G, and H can be assumed to be $m \times np$, $p \times qr$, and $r \times s$, respectively.

Now, the dimension of $F \ltimes G$ is $m \times nqr$, which works for $(F \ltimes G) \ltimes H$. On the other hand the dimension of $G \ltimes H$ is $p \times qs$, which works for $F \ltimes (G \ltimes H)$.

Case 2. $F \prec G$ and $G \prec H$. The dimensions of F, G, and H can be assumed to be $m \times n$, $np \times q$, and $rq \times s$, respectively.

Now, the dimension of $F \ltimes G$ is $mp \times q$, which works for $(F \ltimes G) \ltimes H$. On the other hand the dimension of $G \ltimes H$ is $npr \times s$, which works for $F \ltimes (G \ltimes H)$.

Case 3. $F \prec G$ and $G \succ H$. The dimensions of F, G, and H can be assumed to be $m \times n$, $np \times qr$, and $r \times s$, respectively.

Now, the dimension of $F \ltimes G$ is $mp \times qr$, which works for $(F \ltimes G) \ltimes H$. On the other hand the dimension of $G \ltimes H$ is $np \times qs$, which works for $F \ltimes (G \ltimes H)$.

Case 4. $F \succ G$ and $G \prec H$. The dimensions of F, G, and H can be assumed to be $m \times np$, $p \times q$, and $rq \times s$, respectively.

Now, the dimension of $F \ltimes G$ is $m \times nq$. To make this feasible for $(F \ltimes G) \ltimes H$, we need:

Case 4.1. $(F \ltimes G) \succ H$, that is, $n = n'r$. This works for $F \ltimes (G \ltimes H)$.
Case 4.2. $(F \ltimes G) \prec H$, that is, $r = nr'$. This works for $F \ltimes (G \ltimes H)$.

The dimension of $G \ltimes H$ is $pr \times s$. To make this feasible for $(F \ltimes G) \ltimes H$, we need:

Case 4.3. $F \succ (G \ltimes H)$, that is, $n = n'r$. This is good for $(F \ltimes G) \ltimes H$.
Case 4.4. $F \prec (G \ltimes H)$, that is, $r = nr'$. This is good for $(F \ltimes G) \ltimes H$.

D. Cheng et al., *Analysis and Control of Boolean Networks*,
Communications and Control Engineering,
DOI 10.1007/978-0-85729-097-7, © Springer-Verlag London Limited 2011

Next, we prove associativity. We will do this case by case. Since Cases 1–3 are similar, we prove only Case 1, that is, $F \succ G$ and $G \succ H$.

Let $F_{m \times np}$, $G_{p \times qr}$, and $H_{r \times s}$ be given. Based on the definition we can, without loss of generality, assume that $m = 1$ and $s = 1$. Then,

$$F \ltimes G = (F_1, \ldots, F_p) \ltimes \begin{pmatrix} g_{11}^1 & \cdots & g_{1q}^1 & \cdots & g_{r1}^1 & \cdots & g_{rq}^1 \\ \vdots & & & & & & \\ g_{11}^p & \cdots & g_{1q}^p & \cdots & g_{r1}^p & \cdots & g_{rq}^p \end{pmatrix}$$

$$= \left(\sum_{i=1}^{p} F_i g_{11}^i, \ldots, \sum_{i=1}^{p} F_i g_{1q}^i, \ldots, \sum_{i=1}^{p} F_i g_{r1}^i, \ldots, \sum_{i=1}^{p} F_i g_{rq}^i \right).$$

We then have

$$(F \ltimes G) \ltimes H = (F \ltimes G) \ltimes \begin{pmatrix} h_1 \\ \vdots \\ h_r \end{pmatrix}$$

$$= \left(\sum_{j=1}^{r} \sum_{i=1}^{p} F_i g_{j1}^i h_j, \ldots, \sum_{j=1}^{r} \sum_{i=1}^{p} F_i g_{jq}^i h_j \right). \qquad \text{(B.1)}$$

On the other hand,

$$\begin{pmatrix} g_{11}^1 & \cdots & g_{1q}^1 & \cdots & g_{r1}^1 & \cdots & g_{rq}^1 \\ \vdots & & & & & & \\ g_{11}^p & \cdots & g_{1q}^p & \cdots & g_{r1}^p & \cdots & g_{rq}^p \end{pmatrix} \ltimes \begin{pmatrix} h_1 \\ \vdots \\ h_r \end{pmatrix}$$

$$= \begin{pmatrix} \sum_{j=1}^{r} g_{j1}^1 h_j & \cdots & \sum_{j=1}^{r} g_{jq}^1 h_j \\ \vdots & & \\ \sum_{j=1}^{r} g_{j1}^p h_j & \cdots & \sum_{j=1}^{r} g_{jq}^p h_j \end{pmatrix}.$$

Then,

$$F \ltimes (G \ltimes H) = (F_1, \ldots, F_p) \ltimes (G \ltimes H)$$

$$= \left(\sum_{j=1}^{r} \sum_{i=1}^{p} F_i g_{j1}^i h_j, \ldots, \sum_{j=1}^{r} \sum_{i=1}^{p} F_i g_{jq}^i h_j \right),$$

which is the same as (B.1).

Since Cases 4.1–4.4 are similar, we prove Case 4.1 only. Let $F_{m \times npr}$, $G_{p \times q}$, and $H_{rq \times s}$ be given. We also assume that $m = 1$ and $s = 1$. Then,

$$F = (F_{11}, \ldots, F_{1r}, \ldots, F_{p1}, \ldots, F_{pr}),$$

where each F_{ij} is a $1 \times n$ block.

$$G = \begin{pmatrix} g_{11} & \cdots & g_{1q} \\ \vdots & & \\ g_{p1} & \cdots & g_{pq} \end{pmatrix}, \qquad H = (h_{11}, \ldots, h_{1r}, \ldots, h_{q1}, \ldots, h_{qr})^{\mathrm{T}}.$$

A careful computation shows that

$$(F \ltimes G) \ltimes H = F \ltimes (G \ltimes H) = \sum_{i=1}^{p} \sum_{j=1}^{r} \sum_{k=1}^{q} F_{ij} g_{ik} h_{kj}. \qquad \square$$

(2) *Proof of Proposition 2.5*

Proof Note that the elements a_{ij} of $V_{\mathrm{r}}(A)$ are arranged by the ordered multi-index $Id(i, j; m, n)$, and in $V_{\mathrm{c}}(A)$ they are arranged by $Id(j, i; n, m)$. Now, since the columns of $W_{[m,n]}$ are indexed by $Id(i, j; m, n)$ and its rows indexed by $Id(j, i; n, m)$, by the construction of $W_{[m,n]}$, it moves the (i, j)-element in the order of $Id(i, j; m, n)$ to (i, j)-position in the order of $Id(j, i; n, m)$, which is (j, i)-position in $Id(i, j; m, n)$. That is,

$$W_{[m,n]} V_{\mathrm{r}}(A) = V_{\mathrm{r}}(A^{\mathrm{T}}).$$

The first equality then follows from (2.11). Multiplying both sides of the first equality by $W_{[n,m]}$ yields the second equality. $\qquad \square$

(3) *Proof of Proposition 2.9*

Proof A simple computation shows that for a row vector X and a column vector Y with proper dimensions, we have

$$\langle X, Y \rangle_L = \left(\langle Y^{\mathrm{T}}, X^{\mathrm{T}} \rangle_L \right)^{\mathrm{T}}. \tag{B.2}$$

Consider $A \ltimes B$. Denote the rows of A by A^i and columns of B by B_j. It is then clear that the (i, j)-block of $A \ltimes B$ is

$$\langle A^i, B_j \rangle_L,$$

while the (j, i)-block of $B^{\mathrm{T}} \ltimes A^{\mathrm{T}}$ is

$$\langle B_j^{\mathrm{T}}, (A^i)^{\mathrm{T}} \rangle_L.$$

Using the definition, we see that the transpose of the (i, j)-block of $A \ltimes B$ is exactly the (j, i)-block of $B \ltimes A$. The conclusion then follows. $\qquad \square$

(4) *Proof of Proposition* 2.10

Proof We prove the first case. The proof of the second case is similar.

Denote by b_i the ith column of B, that is,

$$B = [b_1, b_2, \ldots, b_q].$$

Note that

$$B \otimes I_n = [b_1 \otimes I_n, b_2 \otimes I_n, \ldots, b_q \otimes I_n].$$

Using the block product law, we can then assume that $m = 1$ and $q = 1$. We then have

$$\begin{bmatrix} a_1 & a_2 & \cdots & a_{np} \end{bmatrix} \left[\begin{pmatrix} b_1 \\ b_2 \\ \vdots \\ b_p \end{pmatrix} \otimes I_n \right].$$

A straightforward computation shows that this equals $A \ltimes B$.   □

References

1. Cheng, D., Qi, H.: Semi-tensor Product of Matrices—Theory and Applications. Science Press, Beijing (2007) (in Chinese)

Index

D. Cheng et al., *Analysis and Control of Boolean Networks*,
Communications and Control Engineering,
DOI 10.1007/978-0-85729-097-7, © Springer-Verlag London Limited 2011